# ECOLOGY OF WOODLANDS AND FORESTS

## Description, Dynamics and Diversity

Taking a functional rather than an ecosystem or a utilitarian approach, Thomas and Packham provide a concise account of the structure of woodlands and forests. Using examples from around the world – from polar treelines to savannas to tropical rain forests – the authors explain the structure of the soil and the hidden world of the roots; how the main groups of organisms that live within them interact both positively and negatively. There is particular emphasis on woodland and forest processes, especially those involving the flow and cycling of nutrients, as well as the dynamics of wooded areas, considering how and why they have changed through geological time and continue to do so. This clear, non-technical text will be of interest to undergraduates, foresters, ecologists and land managers.

PETER A. THOMAS is senior lecturer in environmental science at Keele University, UK, where his teaching encompasses a wide range of tree and woodland related topics including tree design and biomechanics, tree and woodland ecology and woodland management. His research interests focus on tree ecology, dendrochronology and forest fires. He is the author of *Trees: Their Natural History* published by Cambridge University Press.

JOHN R. PACKHAM is Emeritus Professor of Ecology at the University of Wolverhampton, where he headed the Woodland Research Group for many years. He has special interests in forestry, was a founder member of the Continuous Cover Forestry Group (CCFG) and has worked extensively in English and Scandinavian forests. His research is particularly concerned with virgin forests, the ecology of the woodland field layer, and the establishment of attractive and diverse communities in new woodlands. Executive editor of *The Ecological Flora of the Shropshire Region* (1985), he was the first author of two major books on woodland and forest ecology and one on coastal ecology, and an organizing editor of *Ecology and Geomorphology of Coastal Shingle* (2001).

Forest type African elephant *Loxodonta cyclotis* feeding on acacia canopy. (Formerly regarded as a subspecies of the African elephant *Loxodonta africana*, the forest elephant is now considered a separate species.)

# ECOLOGY OF WOODLANDS AND FORESTS

## Description, Dynamics and Diversity

PETER A. THOMAS
and
JOHN R. PACKHAM

CAMBRIDGE UNIVERSITY PRESS
Cambridge, New York, Melbourne, Madrid, Cape Town, Singapore, São Paulo

Cambridge University Press
The Edinburgh Building, Cambridge CB2 8RU, UK

Published in the United States of America by Cambridge University Press, New York

www.cambridge.org
Information on this title: www.cambridge.org/9780521834520



First published 2007

Printed in the United Kingdom at the University Press, Cambridge

*A catalogue record for this publication is available from the British Library*

ISBN 978-0-521-83452-0 hardback
ISBN 978-0-521-54231-9 paperback

To our wives, Jody and Mary, who have supported our work so thoroughly. Also to those planting the forests of the future and conserving those of the present.

# Epigraph

For ye shall go out with joy, and be led forth with peace: the mountains and the hills shall break forth before you into singing, and all the trees of the field shall clap their hands.

*(Isaiah 55, v.12)*

# Contents

# Preface

As its subtitle implies, the aim of this book is to provide within a relatively small compass an account of the structure of the woodlands and forests of the world, the relationships between the main groupings of organisms which live within them, and a discussion of the significance of plant and animal diversity at both the community and regional level. There is particular emphasis on woodland processes, especially those involving the flow of energy and cycling of nutrients. An attempt has also been made to show how communities dominated by trees, together with their constituent animals and plants, have gradually evolved during geological time.

Foresters and conservationists have of necessity to be far-sighted, and are usually both cheerful and philosophical. While Isaiah 55, v.12 presents a somewhat unusual view of tree behaviour, it does convey a very positive approach, one well suited to the major forest tasks which have to be dealt with in this new century. One function of this book is to provide a background against which foresters, ecologists, land managers and others can view the past and plan for the future. This book, while drawing on previous work, is wherever possible based upon the most recent research, in the hope that those familiar with our other books will find something more of value here. It uses the ecosystem approach and endeavours to show how various organisms, often diverse in space and time, have employed basically similar strategies, sometimes resulting in the repeated evolution of special features that enable them to exploit particular environmental niches. It is intended to provide undergraduates, teachers, and all those interested in vegetation dominated by trees, with a concise account of woodlands and how they operate. The more society at large gets to know about these systems, which never cease to fascinate the authors, the greater is the chance that rare species and habitats – and in particular old woodlands – will be effectively protected.

A great deal of interest attaches itself to the study of particular ecosystems. Amongst the ultimate aims of a plant ecologist, however, must be the ability to predict the ways in which vegetation will change, and the achievement of an understanding of the general rules which govern plant and vegetation behaviour. Many eminent scientists have devoted much of their research to studies along these lines, notably Grime whose *Plant Strategies, Vegetation Processes, and Ecosystem Properties* (2001) is a seminal reference. In his preface, Grime quotes MacArthur (1972) 'To do science is to search for repeated patterns not accumulate facts'. We hope that the examples described in this book are illustrative of the general patterns that are the basis of woodland and forest ecology.

**Bold type** is used to emphasize key ideas and concept words when first explained, while entries involving definitions are printed in bold in the index. Much of the book is written directly from our own experience. Where the work of others is quoted, the names of the authors are given together with dates of publication, so that the article can be looked up in the references at the end of the book.

# Acknowledgements

We are indebted to the many colleagues with whom we have discussed topics of interest. These include Frans Vera, colleagues at Forest Research, Alice Holt, Farnham and many others including Håkan Hytteborn and Roland Moberg who introduced JRP to Scandinavian forests. We are grateful also to the late Arthur Willis, who read the whole draft, and to many who helped with discussion and the provision of information and diagrams including Ena Adam, Martyn Ainsworth, Ian Baillie, Posy Busby, John Campbell, Eleanor Cohn, Bill Currie, Ed Faison, Geoff Hilton, Jonathan Humphrey, Jim Karagatzides, Keith Kirby, Andy Lawrence, Pat Morris, Brooks Mathewson, Robert McDonald, Tony Polwart, Jack Putz, Glenn Motzkin, Tim Sparks, Brian Stocks, Robin Stuttard, Sarah Taylor, Jill Thompson, Ian Trueman and Ruth Yanai. Alan Crowden at Cambridge University Press is thanked for his constant encouragement as is David Harding for contributions to Chapters 2, 5 and 7, as well as his long continued friendship and co-operation from long before the publication of Packham and Harding (1982). Peter Hobson produced elegant drawings, comments on the draft and information regarding African forests; Nick Musgrove provided long continued assistance with data acquisition and computing; Malcolm Inman and Richard Homfrey also gave expert help. PAT gratefully acknowledges that much of his contribution to this book was written while a Bullard Fellow at Harvard Forest, Massachusetts; David Foster and his staff are thanked for their help and encouragement. We are both particularly grateful to Peter Alma for his initial suggestions regarding the original plan of the book and for his comments on the final draft.

# Metric equivalents

Metre = 39.37 inches = 3.28 feet
Kilometre = 0.6214 statute mile
Hectare = 2.4711 acres
Kilogram = 2.2046 lb
Tonne = 0.985 ton
1 mile = 1.6093 km

# 1

# Introduction: Forest basics

## 1.1 Characteristics of woodlands and forests

### *1.1.1 Wooded environments*

Forests often appear monumental and unchanging. This is, however, mostly an illusion caused by our short human perspective. The earliest green plants possessing both roots and tissues specially adapted for the transmission of water belonged to the Psilopsida, which gave rise to the ferns and fern allies. It is from the ancestors of this group, which arose in the Silurian (*c.* 440 million years ago), that all trees – both ancient and modern – are ultimately derived (see Fig. 1.1). Amongst the many evolutionary trends found within this group were tendencies towards the production of (a) tall trunks and (b) seeds from which young plants, including trees, could develop relatively rapidly. Tree ferns, cycads, maidenhair trees, conifers, palms and the very large number of broadleaved genera remain in our woodlands and forests to this day (further detail on past forests can be found in Chapter 9). The amount and composition of the world's wooded areas have changed continuously over geological time, sometimes more rapidly than at others, and continue to do so, helped especially now by human activities. This book is mainly concerned with understanding today's forests in that light.

Wooded land currently covers between 30–35% of the world's land surface (depending on what is counted as forest) or around 39–45 million $km^2$. The Food and Agriculture Organisation of the United Nations (FAO) figures used in Table 1.1 give 30.3% or 39.5 M $km^2$ of the world's land area as forested, with 2.8% of that being under plantation (i.e. purposefully planted). Forests are obviously not equally spread around the globe, their distribution being very dependent upon climate (this is expanded upon in Section 1.6 below). This can be seen in Table 1.1 which breaks down forestry cover into world regions. Some of the least forested countries have primarily desert environments (Gulf

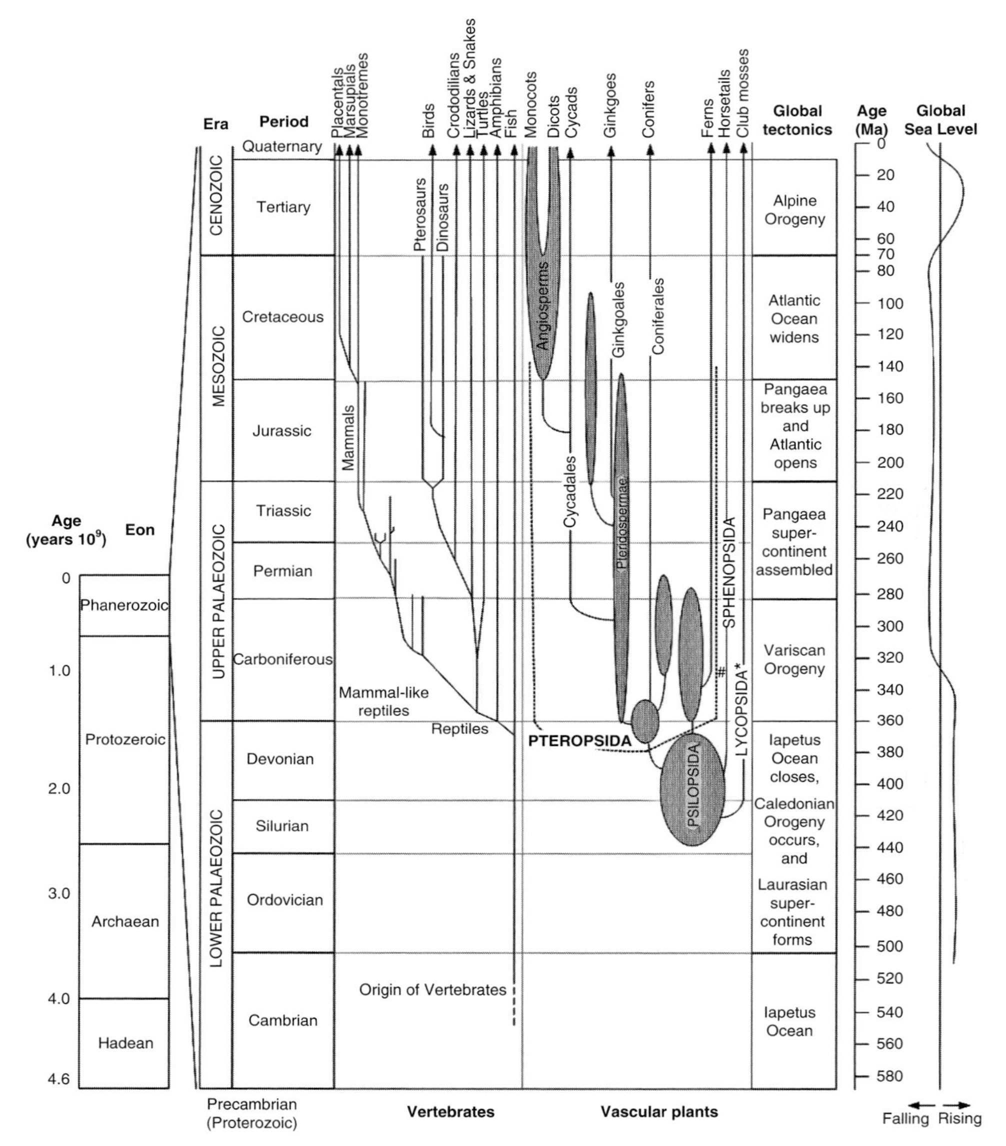


Figure 1.1 The stratigraphic column. Geological eons, eras, periods and time scale, indications of the origin and duration of the major vertebrate and vascular plant groups, together with estimated variations in sea level. Earth became solid *c.* 3.9 billion years ago at the end of the Hadean, and the beginnings of life were present within another 50 million years. Stromatolites and blue-green algae (known as cyanophytes or cyanobacteria) were present early in the Archaean era. The latter were the first photosynthesizing organisms on Earth. At first the oxygen they produced combined with iron-forming ferric oxides which sank to the bottom of the primitive seas. It then transformed the initially very adverse atmosphere and provided the oxygen required by animal

Table 1.1. *Forest cover in world regions, defined as including all natural forests and plantations. Taken from FAO (2005)*

| World region | Forest area (M $km^2$) | Per cent of land area forested | Per cent of forest area that is under plantation |
|---|---|---|---|
| Africa | 6.4 | 21.4 | 2.5 |
| Asia | 5.7 | 18.5 | 7.8 |
| Europe | 10.0 | 44.3 | 2.2 |
| N and Central America[a] | 7.1 | 32.9 | 2.5 |
| Oceania[b] | 2.1 | 24.3 | 1.9 |
| South America | 8.3 | 47.7 | 1.4 |
| **World** | **39.5** | **30.3** | **2.8** |

*Notes:*
[a] Including Greenland.
[b] Including Australasia and surrounding islands.

countries such as Kuwait and Egypt are all below 0.3% cover, and according to FAO figures, Oman and Qatar have no forest cover), or cold inhospitable climates (Iceland also has 0.5% forest cover). At the other end of the scale, the highest forest cover is found in northern boreal climates (Finland 74%) and on moist islands in the equitable Pacific such as the Cook Islands (67%) and the much larger Solomon Islands (78%). The five most forest-rich countries (the Russian Federation, Brazil, Canada, the United States and China) account for more than half of the world's total forest area (21 M $km^2$ or 53% – FAO, 2005). Deforestation and afforestation play an overriding part in how much forest is left in many areas of the world. After the last ice age the UK would

---

Caption for Figure 1.1 (cont.)
life. The Cambrian explosion, which began the Lower Palaeozoic, involved the sudden and abrupt production of myriads of life forms, including the complex and very varied trilobites, from simple precursors. Note that the Quaternary Period consists of the Pleistocene Ice Age, in which there have been a number of interglacials including the ongoing Holocene, which began 10 000 years ago and is part of the present Flandrian temperate stage with its global warming and ice melt. Our views on tracheophyte (vascular plant) relationships are constantly being modified as more and more fossil evidence accumulates: Bell and Hemsley (2000, p. 141) should be consulted for a more complex view. The Lycopsida (shown by the asterisk) included the clubmoss trees *Lepidodendron* and *Sigillaria*, and the Sphenopsida included *Calamites* (#). (Mielke, 1989; Benton, 1991; Briggs *et al.*, 1997; and After Bryson, 2004.)

have been fairly extensively forested but by World War I was down to around 6% cover, a figure which has now increased to 11.8%. From Table 1.1 it can be seen that the degree of planting of artificial plantations (as opposed to regenerating 'natural' forest) varies tremendously around the world. Forest area under plantation is greatest in Asia (7.8%); elsewhere it is less than 2.5%, giving a figure of 2.8% for the world forest as a whole. This adds further complexity to the way in which humans have influenced forest cover.

Nevertheless, it is estimated that about half of the forest that has grown under modern climatic conditions since the end of the Pleistocene, around 8000 years BC, has been lost, largely due to human activities. The spread of agriculture and domesticated animals, increasing population and cutting of forests for timber and fuel have all taken their toll. Some 13 M ha (0.13 M km$^2$) of forest are being lost globally each year (FAO, 2005). When new forests are taken into account the net loss of forest between 2000 and 2005 was still 7.3 M ha per year, an area the size of Sierra Leone or Panama. The only silver lining is that we are not losing forests as quickly as we did between 1990 and 2000 when the net loss was 8.9 M ha per year. Global net loss of forest has been estimated as 0.18% per year between 2000–2005 (FAO, 2005). It is perhaps not surprising that the UNEP World Conservation Monitoring Centre has identified over 8000 tree species that are threatened with extinction at a global level and are concerned for the estimated 90% of all terrestrial species that inhabit the world's forests. See Section 11.1.2 for rain forest losses.

### *1.1.2 Differences between woodlands and forests*

The terms forest and woodland are commonly used almost interchangeably, and if there is any differentiation, then most people see a forest as a remote, large, dark forbidding place while a woodland is smaller, more open and part of an agricultural landscape. These views are very close to the normally accepted definitions of the two terms. A **woodland** is a small area of trees with an open canopy (often defined as having 40% canopy closure or less, i.e. 60% or more of the sky is visible) such that plenty of light reaches the ground, encouraging other vegetation beneath the trees. Since the trees are well spaced they tend to be short-trunked with spreading canopies. The term **forest**, by contrast, is usually reserved for a relatively large area of trees forming for the most part a closed, dense canopy (although canopy closure as low as 20% is accepted in some definitions). A forest does not have to be uniform over large areas, and indeed is often made up of a series of **stands**, groups of trees varying in such features as age, species or structure, interspersed with open places such as meadows and lakes and areas where grazing animals are limiting tree development. **Since these terms overlap, throughout this book we will use the term forest as the collective term for wooded areas, including woodland, unless otherwise specified.**

These definitions are obviously based on the trees as the dominant organisms. This is a convenient way of setting wooded areas apart but it should be borne in mind that a complete forest or woodland is the sum of the tens of thousands of other plants, animals and microbes. More recently, definitions of forests as complete ecosystems have tried to take this holistic message to heart (see Helms, 2002). However, even the simple definitions given above are not without their problems. The figures given in Table 1.1 from the FAO are based on defining a forest as having just 10% tree cover or more with a minimum size of 0.5 ha and so include some very open areas. Amongst European countries the minimum requirements to be called a forest vary widely: a cover of 5–30%, area of 0.05–2 ha and a width of 9–50 m (Köhl *et al.*, 2000). The addition of estimates from individual countries gives western Europe 1 256 000 km$^2$ of forest, but using the extremes of the definitions above results in a variation of 113 000 km$^2$ (9%) around this figure. Such vagueness in definitions makes international comparisons very difficult and hampers conservation efforts. Lund (2002) suggests that there are at least 624 definitions of 'forest' used around the world!

In Britain, care is needed to distinguish the above from **Forest** (with a capital letter). From early Norman times (1070), a Forest was an area reserved for hunting usually by the monarch and administered under Forest Law. This definition says nothing about trees and indeed many Forests, such as the Royal Forest of Dartmoor in south-west England, were, and still are, almost treeless. As Oliver Rackham (1990, p. 165) neatly put it '. . . a Forest was a place of deer, not necessarily a place of trees'.

## 1.2 The value of woodlands and forests

As noted above, woodlands and forests cover between 30–35% of the world's land surface. Agriculture covers another 40% but since wooded areas are structurally bigger (i.e. taller and more complex), it is the wooded land that holds most living material of all the land vegetation types. Global wooded land holds in excess of 422 billion ($10^9$) tonnes of biomass just in the wood. Because they are so large and extensive, with many niches, it is inevitable that the world's forests are among the most important repositories of terrestrial biodiversity.

Forests also provide a wide array of goods and services. Forest products play a central role in the life of many rural communities: timber and fuel (in the

1990s, 3.5 billion $m^3$ of wood were consumed globally each year, with more than half used as fuelwood), food, animal fodder and medicines. Forests also play an important cultural role, in many ways defining some cultures such as in indigenous peoples of rain forests; without the forest their culture is diminished. Forests are also important in reducing soil erosion and in water conservation (see Chapters 2 and 6).

Urban dwellers benefit tremendously from forests. Global trade in primary forest products such as logs, sawn wood, panels, pulp and paper reached nearly $273 billion in 1997. It is not just timber; a large number of fruits and spices we use come from trees and woodland plants. Wild forests are still a valuable source of some of these. For example, almost all the Brazil nuts (*Bertholletia excelsa*) we eat (around 40 000 tonnes a year – Mori and Prance, 1990) are still collected in the wild (see Section 6.3.4 also). Forest plantations can also be a rich source of edible fungi; Chilean radiata pine plantations are already exploited in this way. At least 46 types of mushroom and nine types of truffle grow in forests and are potentially a most valuable food source. Wooded areas also have a large part to play in global carbon storage and sequestration (see Chapter 11).

On an individual level, trees and urban woodlands are beneficial to people. To name a few examples, they:

- Produce oxygen (a mature beech *Fagus sylvatica* produces sufficient oxygen over a year for ten people).
- Release many compounds into the atmosphere including monoterpenes which seem likely to have positive health effects (see Maloof, 2005).
- Absorb noise, dust, pollution and carbon dioxide.
- Reduce skin cancer (by blocking out sunlight), ironically in the mid-twentieth century the medical profession in the western world advocated extensive sunbathing to increase vitamin D levels.
- Reduce mental health problems by improving our moods and outlook.
- Improve post-surgery recovery rates in hospital wards which overlook wooded settings.

## 1.3 Tree biology and how it influences woodland ecology

### *1.3.1 Fitness of various species for particular uses*

All tree species possess a unique combination of morphological, physiological and reproductive traits which fit them for particular niches in the ecosystems they occupy. In the case of exotic plantations the silviculturally and economically most suitable tree species may have originated in a far distant country. Two quite different examples illustrate this.

The **Monterey or radiata pine** *Pinus radiata* is an impoverished and stunted tree in its natural range on a handful of sites on the coast of California, mostly notably, the Monterey Peninsula. It has been left marooned in less than favourable growing conditions as its range has been reduced by climate changes since the last ice age. Yet elsewhere it is capable of magnificent growth and has been extensively planted in many parts of the world including New Zealand and South America. In New Zealand it now accounts for some 90% of the exotic trees grown and develops so rapidly that it can be harvested at an age of 25 years, whereas even Douglas fir (whose timber may sell for roughly double the price) would normally have to grow for 45–60 years under the same conditions. It also has excellent form, wounds incurred when lower side branches are pruned heal rapidly, and it does not coppice so any unwanted trees die when felled. Its seeds are easy to collect and store and have a high germination rate, while bare-root seedlings and cuttings can be grown rapidly without shading and withstand weeding with herbicides. Planting stocks have a survival rate >95% on a wide range of sites; the tree grows well and predictably even on infertile soils, its vigorous early growth often outstripping gorse and other weeds while in some situations its roots penetrate to a depth of 5 metres. The tree also has a degree of tolerance to frost, snow, salt winds and severe drought. Its genetics have been widely studied, clones being developed to suit particular conditions. This shade-intolerant species is most easily grown under clear-felling regimes and tends to shade out other species when planted in mixtures: the multi-species commercial forests that thrive in Europe are not found in New Zealand.

The genus *Eucalyptus*, which consists of around 500 species of trees and shrubs, has a native distribution largely confined to Australia, but extending into New Guinea, eastern Indonesia and Mindanao (Hora, 1981). **Eucalypts** show a most remarkable range of size and habitat and various species of this vigorous and adaptable tree, which evolved in isolation even from New Zealand, are now widely planted in many parts of the world, especially California which has the largest range of eucalypt species in the USA. The smallest is less than a metre in height, whereas mountain ash *E. regnans* can live for 300 years and is the tallest hardwood tree in the world, growing to more than 100 m on deeper well-watered soils in the foothills of Victoria, South Australia. The river red gum *E. camaldulensis*, a robust tree up to 35 m high, is found in most of Australia and can live for 500 years; older trees shelter parrots in their cavities. Tasmanian blue gum *E. globulus*, which reaches 35–45 m in height, is the species most widely planted in the Mediterranean area and California. Conditions in tropical North Australia vary from the normally extremely hot and dry to the suddenly deeply flooded when tropical rainstorms cause the rivers to overflow. This is an area well suited to extremely territorial frilled lizards which feed mainly on insects and make rapid two-legged dashes from tree to tree to avoid attack by predatory birds. Ghost gums *E. papuana* grow in New Guinea and arid parts of northern Australia, while in the southern state of Victoria snow gums *E. pauciflora* grow high on the Australian Alps, tolerating winter temperatures as low as −20 °C and providing food and shelter to parrot populations which feed on their fruits. It is on the lower slopes of these hills that the mountain ash flourishes on good soils with adequate water.

Almost all eucalypts are evergreen, having leaves that are hard and rich in nutrients. Apart from the koala (see Section 5.7.2), few animals can digest them. The bark of ghost gums is shed to the ground leaving a strikingly white surface, which reflects sunlight, and the leaves tend to hang down, thus staying cooler. Fire affects almost the whole of Australia, whose trees are well adapted to it, many of them having developed the ability to coppice or sucker in response to millennia of natural fires. Eucalypt fires develop rapidly and burn intensely. Many trees can survive all but the most severe fires and some species need fire to release their seeds. Buds buried beneath the bark produce new leaves and branches and life often resumes within a few weeks. Jarrah *E. marginata* is one of many species with a lignotuber, like a huge wooden radish, which enables it to coppice.

### *1.3.2 Tree morphology*

Trees have arisen independently in a large number of plant families as a strategy which outgrows other plants in competition for light, and in so doing have evolved a large **perennial woody skeleton** to display, in a large tree, over half a million leaves. Such a tall structure is also a good platform for displaying flowers to wind or animal pollinators, and the height helps falling seeds to disperse further. These evolutionary trends have resulted in trees being the tallest and largest of all living things. The **tallest trees** in the world are currently the **coastal redwoods** of California (*Sequoia sempervirens*) at 115.5 m (358 ft) although the tallest tree ever, a mountain ash *Eucalyptus regnans* in Australia, may have been over 150 m (500 ft) – see Section 1.3.1 above. California also boasts the **largest tree, a giant sequoia** *Sequoiadendron giganteum* growing in the Sierra Nevada Mountains, called General Sherman. The General is 83.8 m tall with a width at the base of 11 m, giving an estimated mass of 2030 tonnes (by comparison, the blue whale, the largest animal, weighs only around 100 tonnes).

In order to grow so large, most trees (palms and other monocotyledonous trees being an exception – see Thomas (2000) for details of these) have a similar

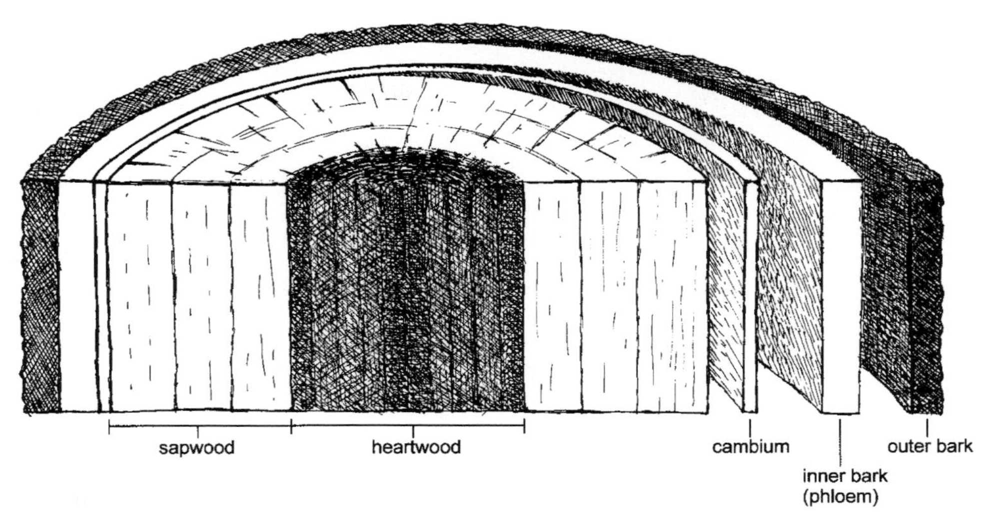


Figure 1.2 Tree cross-section. (From Thomas, 2000. *Trees: Their Natural History*, Cambridge University Press. Reprinted with permission.)

and distinctive structure (Fig. 1.2). The **outer bark** on the outside serves a dual function as a waterproof layer (to keep water in) and a defence against attacking organisms. The cork cambium in the bark is responsible for the production of corky cells whose walls are suberized by fatty compounds which render them impervious to gases and liquids. The bark is punctuated by lenticels, however, which allow vital gas exchange. In the cork oak *Quercus suber*, which is extensively grown in Portugal and Spain, this outer layer grows so rapidly that a thick layer of cork, renowned for its use for wine bottle corks, can be stripped off every 8–12 years.

The **inner bark or phloem** is made up of living cells that transport sugary sap usually from the **sources** (the leaves) to the **sinks** (the growing points and food stores). This nutrient-rich layer is utilized by a wide range of insects and pathogens which can have important repercussions to a woodland, such as the transmission of Dutch elm disease by the elm bark beetles (*Scolytus* spp.), discussed further in Section 5.4. Inside the phloem is the **cambium**, responsible for growing new phloem on the outside and new xylem on the inside. The phloem tends not to accumulate because it is stretched and crushed by the expanding tree, but the xylem on the inside accumulates each year to form the wood of the trunk. In seasonal climates, annual rings are created as the wood grows and the age of a tree can thus be determined by counting the rings. **Dendrochronologists** also measure the width and density of these rings and have linked these to various climatic factors, allowing them to calculate past yearly weather patterns. The influence of pathogens on tree-ring development

Figure 1.3 An old bristlecone pine growing in the White Mountains, California, one of the oldest living things in the world. (Photograph by Peter A. Thomas.)

is described in Chapter 5; masting is also likely to have an influence. The **oldest recorded living trees** are the **bristlecone pines** *Pinus aristata* var. *longaeva*[1] (Fig. 1.3) growing at 3000–3500 m in the White Mountains of California. The oldest known individual, Methuselah, has been dated at over 4700 years old by counting the rings, yet in 1974 it is reported to have given rise to 48 living seedlings. The oldest recorded specimen, inadvertently felled, lived to

[1] A difference of opinion exists over the correct scientific name. Americans like to call the bristlecone pine, *Pinus longaeva*. *The World Checklist of Conifers* (Welch and Haddow, 1993), however, calls it *Pinus aristata* var. *longaeva*.

4862 years at 3275 m on Mount Wheeler, Nevada. Since dead trees at this altitude decay so slowly, dendrochronologists have been able to cross-match the ring patterns of dead trees to the living trees and so create a chronology of ring widths extending back almost 9000 years.

The **xylem** or wood is divided into the outer sapwood and the central heartwood. Water is pulled up through the water-conducting tubes of the sapwood while the heartwood is usually considered to be the main component holding the tree up. As sapwood is converted to heartwood, the proportion of heartwood accumulates with age, and is used by a number of organisms – animals, bacteria and fungi – as a food source and habitat. While this appears detrimental to the tree, wood is remarkably resistant to decay and contains barriers severely limiting the spread of rot (see Thomas, 2000). Moreover, heartwood rot may in reality be beneficial since a number of trees are able to grow roots down through the rotting wood to absorb nutrients released by decomposers, and so in effect recycle themselves to gain a competitive advantage over solid neighbours. Hollow trees are also somewhat better at resisting strong winds as was demonstrated in southern Britain in October 1987 when 15 million trees blew down in gusts of up to 122 mph (196 kph); a large proportion of the oaks left standing were hollow. If the tree becomes too hollow, however, the trunk may physically collapse, losing the tree its place in the canopy even if it regrows from the remaining portions. So despite the advantages of being hollow, energy is invested in controlling the rot, reaching a compromise.

The woody skeleton is a liability to the tree in another way. The tree grows by adding a new layer of wood underneath the bark every year, which, like putting Russian dolls together, inevitably must be bigger than the previous year's layer. Thus, the expense of growing the new layer, and its upkeep in terms of respiratory burden, grow larger each year. At the same time, once the tree reaches mature size, its photosynthetic capacity becomes fixed since only a set number of leaves can be held on its woody frame: while the tree's income becomes fixed its outgoings escalate year by year. Inevitably, portions of the skeleton die. These dead branches, normally at the top of the tree, added to those branches lower down that die from being increasingly shaded as the tree and its neighbours grow, create a reservoir of dead wood utilized by a number of invertebrates and vertebrates. As will be seen later, normally a quarter of the biomass in a natural forest is dead wood.

The root systems of most trees are shallow (usually no more than 2 m down) and wide spreading, up to 2–3 times the width of the canopy as a norm. Conifers in northern forests have the shallowest rooting depths (80–90% of roots in the top 30 cm); most temperate and tropical trees have 26% of roots in the top 10 cm and 60% in the top 30 cm (Jackson *et al.*, 1996). Deeper roots grown by some trees become important in forest subject to periodic drought. In a woodland, it is inevitable that roots criss-cross past each other, and will graft where they touch, especially if they are roots from the same species. Counter-intuitively, this is often mutualistic rather than competitive and allows trees to share resources. It is not uncommon that cut stumps with no foliage of their own will continue to grow in diameter, fed through root grafts with surrounding siblings (see also Section 2.3.1). Conversely, above ground, tree canopies do not usually intermingle but remain separate, referred to as **canopy shyness**. The primary reason is that as canopies sway in the wind, ends of branches clash and rub, breaking off buds and inhibiting growth in that direction.

Trees create their own microclimate under the canopy; it is often darker, buffered against extreme temperature change and with a higher humidity. Different trees allow different amounts of light through the canopy due to variations in leaf density and degree of canopy shyness. This is also affected by the spacing of trees, influenced by soil conditions, water competition or external agents such as fire and wind. The percentage of sunlight that reaches the ground in summer can be as high as 20–50% in a birch wood, down to 2–5% under dense hardwoods such as beech, and even as little as 0.2% under a tropical rain forest. As a general rule of thumb, plants need 20% of full sunlight to grow maximally and at least 2% of sunlight for photosynthesis to match respiration needs (the compensation point). From this, it can be seen that forest interiors in summer are dark places for plants. We tend not to appreciate this because our eyes have irises that open to let more light in. The best way to see a forest from a plant's view is to rush headlong into a forest on a sunny day and for a fleeting moment before our eyes adjust, and as we trip over the first log we come to, we will appreciate just how dark it is. Or take photographs on the same exposure setting inside and outside the forest.

The amount of light reaching the ground seasonally is also affected by **evergreen and deciduous strategies** (see Section 3.6). Evergreen trees, which keep their leaves/needles for several years, cast a deep shade all year round. Conversely, deciduous trees often provide a window of high-light conditions at ground level in the early spring and late autumn which is utilized by understorey plants, discussed in Section 1.4.2 below. Evergreen species have to invest heavily in producing robust leaves that will survive harsh environmental conditions and the prolonged attacks of herbivores and pathogens. When their leaves eventually fall, they are usually hard, acidic and difficult to decompose, leading to a deeper litter and humus layer beneath.

## 1.4 Spatial structure

### 1.4.1 Vertical structure above and below ground

Though some may be missing, in most forests it is possible to recognize four distinct layers of vegetation starting with the **tree canopy** at normally 5+ m (Fig. 1.4). Under this are the underlying **shrubs**, including climbers (< 5 m), the **field** or **herb layer** of herbaceous plants (< 1 m made up of herbs – plants without woody stems – and short woody plants such as brambles), and a **ground** or **moss layer** of mosses and liverworts (bryophytes), lichens and algae. Forests can be more complicated than this and other layers have been

Figure 1.4 Vertical stratification of shoot systems in the Wyre Forest, Shropshire, UK, a good example of a temperate deciduous forest (Packham, 1998). The oak trunks have the basal curve characteristic of individuals which have regrown from coppice stools, while the mosses and liverworts of the bryophyte layer are too small to depict. The most acid area is on the left, while that near the centre is an aspect society (see page 17) in which bluebell, creeping soft-grass and bracken follow each other in a seasonal sequence. VM, bilberry; CV, heather; Q, oak (*Quercus petraea* or *Q. robur*); OA, wood-sorrel; CA, hazel; RF, bramble; HN, bluebell; HM, creeping soft-grass; PA, bracken; LP, honeysuckle growing over weak oak coppice regeneration; TS, wood sage; EA, wood spurge; B, silver birch. (Drawn by Peter R. Hobson. From Packham *et al.*, 1992. *Functional Ecology of Woodlands and Forests*. Chapman and Hall, Fig. 1.1. With kind permission of Springer Science and Business Media.)

recognized by various researchers. Tropical forests provide the biggest challenge in this respect and there have been various attempts at imposing a structural identity, ranging from the identification of a large number of horizontal and vertical layers to the more simple division of the canopy into two along the **morphological inversion surface** (Fig. 1.5). This separates the lower layer of densely packed tree crowns, which are often interwoven with lianas, from the more open layer above where the broad crowns are widely spaced. Solé *et al.* (2005) have developed this into the critical height $h_c$ below which gaps are scarce and form a fragmented landscape, and above which gaps are plentiful and form a 'fully connected' landscape. However, none of these schemes have proved inclusive enough to be useful over a range of forests and, as Whitmore (1984) notes, 'few concepts are likely so quickly to divide a roomfull of [tropical] ecologists into two vociferously opposed groups as this one'.

As noted above, one or more of the layers may be missing in a forest, mainly because the layering of vegetation is determined primarily by light availability. Thus, dense plantations may have no layers beneath the canopy except perhaps for a thin ground layer of shade-tolerant bryophytes. Similarly, a dense shrub layer, such as the introduced rhododendron in many acidic British woodlands, may preclude any layers below. Each layer provides niches for suitably adapted animals. For example, a study in an English woodland (Fig. 1.6) clearly showed that each layer in the woodland had its own species of foraging birds. Many other examples could equally well have been used.

Dense field and ground layers, although aesthetically pleasing and providing numerous animal niches, can cause problems for tree regeneration, swamping small seedlings, even though most trees have large seeds which aid early growth (see Chapter 3). This is one reason why, in temperate rainforests, seedlings are often most common on **nurse logs** that are continuously damp enough to provide moisture and lift the seedlings above the dense field layer (Fig. 1.7).

Vertical zonation continues below ground. In the Wyre Forest example used above (Fig. 1.4) the roots of the creeping soft-grass *Holcus mollis* would be in the top 2 cm of the soil, bracken rhizomes *Pteridium aquilinum* around 5 cm down, and bluebell bulbs *Hyacinthoides non-scripta* down to 10–12 cm. Such vertical stratification of roots of herbaceous plants may help in reducing competition between them.

The situation is made more complex by the **hydraulic lifting** shown by a number of trees and a few grasses. Here, water is raised at night from moist areas lower in the soil (flowing along a hydraulic gradient) to roots nearer the surface. (The process can happen the other way round if the soil surface is very wet, compared with deeper layers and so should really be referred to as hydraulic redistribution.) Hydraulic lifting is commonest in savannas and

Figure 1.5 Profile through a mixed rainforest, Rivière Sinnamary, French Guiana. The dotted line shows the morphological inversion surface (see text). Crowns with thick outlines are mature trees, stippled trees are immature. The trees with the dotted outline are rooted outside the plot. (From Richards, 1996. *The Tropical Rain Forest* (2nd edn). Cambridge University Press.)

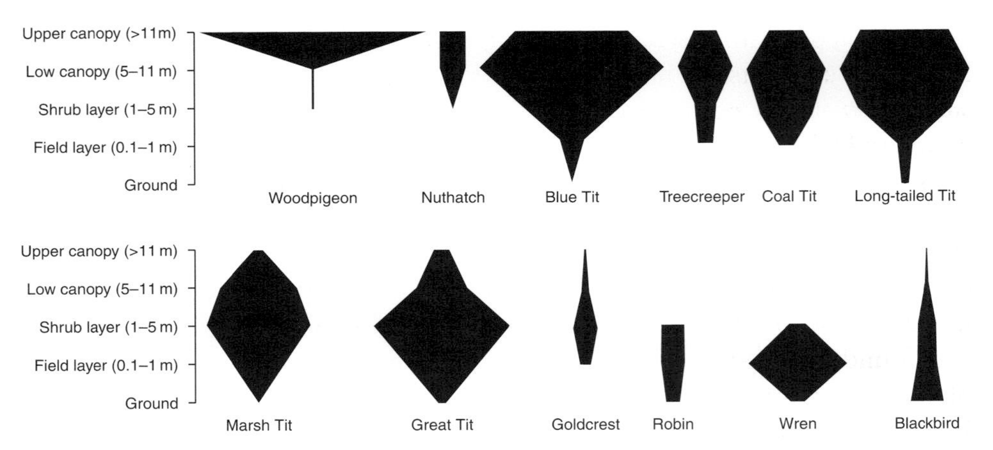


Figure 1.6 Vertical distribution of selected bird species during the winter in Bagley Wood, Oxfordshire, UK. The width of each bar indicates the relative abundance in different layers. The layers are as described in the text, though here the canopy layer has been divided into an upper and a lower portion. (Redrawn from Colquhoun and Morley 1943. From Fuller, 1995. *Bird Life in Woodland and Forest*. Cambridge University Press.)

Figure 1.7 A curiously straight line of mature western hemlock *Tsuga heterophylla* in the temperate rain forest of the Olympic Peninsula, Washington State, USA. These started as seedlings growing on a nurse log which has since rotted away. (Photograph by Peter A. Thomas.)

other xeric (dry) woodlands, especially amongst older trees (Domec *et al.*, 2004). The amounts can be significant; in a 19-m high mature sugar maple *Acer saccharum* it has been measured at around 100 litres per tree each night compared with a water use (transpiration) of 400–475 litres the following day (Emerman and Dawson, 1996). Of interest here is that other plants are known to benefit. In Mediterranean Aleppo pine forests *Pinus halepensis*, the shrub *Pistacia lentiscus* had improved water status when growing beneath large pines (Filella and Penuelas, 2003–4). Water movement in black pine *Pinus nigra*, in north-east Spain, has been tracked using deuterium marking; enriched water was found in the stem water of the pines themselves and also neighbouring plants of holm oak *Quercus ilex* and the top 15 cm of the soil (Penuelas and Filella, 2003).

As well as a vertical spatial pattern, above and below ground, there is also a temporal pattern in which active growth commences at different times in particular species, leading to the formation of **aspect societies** in woodland glades where the plant community is dominated by one or several species at a particular season and whose appearance changes markedly over just a few weeks in spring. In many English woods, including those of the Wyre Forest, bluebells appear first in spring, followed by creeping soft-grass, with bracken considerably later.

### *1.4.2 Ecological problems of understorey plants*

**Understorey plants** are by definition those living below the tree canopy. Unless the canopy is very open, as in birch woodlands, light will be in short supply for most of the growing season (see Section 1.3.2 above), hence it is referred to as the **dark phase** of the year. Deciduous canopies will go through the **light phase** in the non-growing season when the leaves have fallen but, of course, most understorey plants will not then be growing. Evergreen trees have a year-round dark phase.

In the English Midlands the light phase usually lasts at least 5 months and is referred to by Rackham (1975) as the **bare half-year**, the dark phase corresponding to the **leafy half-year**. It is no accident that temperate deciduous woods are floristically at their best early in the year; many 'pre-vernal' plants gain extra light by growing in early spring before the trees unfurl their leaves – see Box 3.1 and Fig. 3.9 which show the spring-time phenology of other species.

Lack of light is not the only problem for understorey plants. Water can also be in short supply; beech forests are often bare below as much from intense water competition as from the dense shade of the canopy. Pollination can also be problematic. Wind speeds are drastically reduced below the woodland canopy, especially when in leaf. In British woodlands the most prominent wind-pollinated herb is the summergreen dog's mercury *Mercurialis perennis* which flowers early in February and March when wind speeds are relatively high. Most woodland herbs are insect pollinated and face the problem of being found by insects in a dark woodland. The solutions are to flower before the canopy closes (the commonest option), produce pale or white flowers to show up in the gloom (such as wood sanicle *Sanicula europaea*) or fragrant flowers (such as wood sage *Teucrium scorodonia*).

The bryophytes of the ground layer are remarkably tolerant of dry, dark conditions and grow as and when conditions are suitable. Their main problem is being swamped by falling leaves in the autumn which, under trees with hard acidic foliage such as beech, sycamore and other maples, may effectively put them in the dark for much of the year. In such a woodland, the bryophytes are largely restricted to rocks, raised roots and dead wood from which the leaf litter readily slips or is blown.

**Herbivory** can be a distinct problem for wintergreen and evergreen plants since in the winter they are a scarce source of food within reach of hungry animals. In such situations, investment in chemical and physical defences is advantageous, in line with Grubb's **scarcity-accessibility hypothesis** (Grubb, 1992). This states that such defences will be grown where the plant is either scarce or readily accessible to herbivores and the cost of investment in defences is less than the cost of regrowth or if death is likely. This may explain why the European holly, a typical evergreen understorey tree, invests in leaf prickles to deter deer in winter (most hollies around the world are deciduous and prickleless), and why the slow-growing evergreen yew invests in potent chemical defences. It also helps explain why holly loses its prickles above around 3 m, above the height to which native herbivores can reach.

Having outlined these numerous problems, we are left with the question of just why so many plants *do* grow under trees. The simple answer is that they often grow in woodlands because they cannot grow elsewhere. Woodland plants can cope with strong shade and the other limitations imposed by a tree canopy, but would not survive the intense competition which they would meet in the denser field layers of other vegetation types.

### *1.4.3 Horizontal structure*

**Vegetation mosaics** of different scales exist in all forests and woodlands. These develop as a consequence of gap formation and the revegetation of former gaps, whose size varies from that created by the death of a single tree up to landscape scale openings created by agents such as hurricanes and forest fires. There is also horizontal variation in light and water caused by the mosaic of tree canopies. Thus oaks can crowd together with small gaps between them while ponderosa pines in dry areas of western North America are widely spaced due to root competition for water. Variations in soil nutrients, pH, moisture and aeration, together with herbivore grazing patterns (from insects killing particular trees or shrubs to deer creating large grazed lawns) also have an effect. Even apparently uniform plantations contain small-scale variations in conditions leading to patches of understorey plants and insects. It should not be surprising, therefore, that on the scale that we experience when walking through woodlands, some are seemingly uniform while others are variable over very short distances.

## 1.5 The woodland ecosystem: food chains, food webs and the plant, animal and decomposition subsystems

Trees dominate the woodland and forest communities in which they grow but hosts of other organisms – including fungi and bacteria – which evolved in parallel with them, live beside, beneath and in them in an interacting whole. The types of animals involved in such communities are illustrated in Fig. 1.8; in many other parts of the world the major differences from western European forests involve the presence of much larger herbivores and carnivores and of primates such as monkeys and gorillas.

The very many thousands of different organisms found in woodlands and forests would be quite unable to exist apart from their non-living (abiotic) surroundings, and it was in recognition of this that Tansley (1935) first used the term **ecosystem** in print. This is a level of organization that integrates the living and non-living components of communities and their environments into a functional whole. The term ecosystem was in fact suggested to Tansley by Clapham, then a young man in the Department of Botany at Oxford, in the early 1930s, and was used by Tansley without acknowledgement. This is made clear by Willis (1997), who treats the ecosystem as an evolving concept viewed historically. Ecosystems require an almost constant flow of energy if their living communities are to survive; nutrients are simultaneously cycled round them as Fig. 1.9 demonstrates.

The simplified food web shown in Fig. 1.10 illustrates the nature of the relationships involved in terms of what eats what. Food webs are by their nature complex and what an organism feeds upon can vary with season, weather, cycles of abundance and many other factors. To avoid the complexity obscuring the main principles, it is possible to look at energy and matter

| | | Feeding habits |
|---|---|---|
| **I Vertebrates** | | |
| **Mammals**[a] | | |
| 1 Fox | *Vulpes vulpes* | C |
| 2 Stoat | *Mustela erminea* | C |
| Polecat | *Putorius putorius* | C |
| 3 Feral cat | *Felis domesticus* | C |
| 4 Wild pig | *Sus scrofa* | O |
| 5 Red deer | *Cervus elephas* | H |
| Fallow deer | *Dama dama* | H |
| Roe deer | *Capreolus capreolus* | H |
| 6 Badger | *Meles meles* | O |
| 7 Hedgehog | *Erinaceus europaeus* | C |
| 8 Pipistrelle bat | *Pipistrellus spp.* | C |
| Noctule bat | *Nyctalus noctula* | C |
| 9 Squirrel | *Sciurus vulgaris* | H |
| 10 Hare | *Lepus capensis (europaeus)* | H |
| 11 Mole | *Talpa europaea* | C |
| 12 Shrews | *Sorex* spp. | C |
| 13 Voles, mice | *Cleithrionomys* spp. *Arvicola* spp. | H(O) |
| | *Sylvaemus* spp. | H(O) |
| Reptiles | | |
| 20 Adder | *Vipera berus* | C |
| Amphibia[b] | | |
| Toad | *Bufo bufo* | C |
| Newt | *Triturus* spp. | C |

| | | Feeding habits |
|---|---|---|
| **Birds** | | |
| 14 Buzzard | *Buteo buteo* | C |
| Sparrow hawk | *Accipiter nisus* | C |
| 15 Tawny owl | *Strix aluco* | C |
| 16 Pheasant | *Phasianus colchicus* | H(O) |
| 17 Wood pigeon | *Columba palumbus* | H(O) |
| 18 Green woodpecker | *Picus viridus* | C |
| Cuckoo | *Cuculus canorus* | C |
| Jay | *Garrulus glandiarius* | O |
| Magpie | *Pica pica* | O |
| Blackbird | *Turdus merula* | O |
| Songthrush | *T. ericetorum* | O |
| 19 Chaffinch | *Fringilla coelebs* | O |
| Bullfinch | *Pyrrhula pyrrhula* | O |
| Robin | *Erithacus rubecula* | C |
| Great tit | *Parus major* | O |
| Wood warbler | *Phylloscopus sibiliatrix* | C |
| Dunnock | *Prunella modularis* | C |
| etc. | | |

| | | Feeding habits |
|---|---|---|
| **II Invertebrates**[c] | | |
| **Mollusca** | Snails, slugs | H |
| **Arachnida** | Spiders, mites | C,H,D,X,Y |
| | Harvestmen | C |
| **Insecta** | | |
| Collembola | Springtails | H,D |
| Orthoptera | Cockroaches, locusts | H |
| Psocoptera | Book lice | D |
| Anopleura | Sucking lice | C,X |
| Thysanoptera | Thrips | H,D |
| Hemiptera | Cicadas, plant bugs, aphids, mealy bugs | H,Y |
| Lepidoptera | Moths and butterflies | H |
| Diptera | Flies | H,C,D,X,Y |
| Coleoptera | Beetles | H,C,D |

| | | Feeding habits |
|---|---|---|
| **Insecta** | | |
| Hymenoptera | Bees, wasps, ants | H,C,X,Y |
| Siphonaptera | Fleas | X |
| Myriapoda | Millipedes, centipedes | C,D,H |
| Crustacea | Slaters, hoppers | D,H |
| Annelida | Earthworms | D |
| Rotifera | Rotifers° | D |
| Nematoda | Roundworms° | D,C,H,X,Y |
| Platyhelminthes | Flatworms | C,D,X |
| Protozoa | Amoeba etc.° | D,X |

[a] The larger mammals are present only in extensive areas of forest, or in remote localities.
[b] Only in localities where there is standing water.
[c] Only the main groups are listed. Very large numbers of species and of individuals from certain groups are present. They are not shown on the diagram but, broadly, spiders and certain insect groups are arboreal, while most of the rest occur on the ground or in the soil. In particular, many species of insects are present, filling many different roles. Often their larvae feed in very different ways from the adults.

o, microscopic; C, mainly carnivorous; O, omnivorous; H, herbivorous; D, detritivorous; X, parasites on animals; Y, parasites on plants.

Figure 1.8 The main groups of vertebrates from western European oak–beech forest. The much smaller invertebrate animals (not depicted here) also play important roles in the ecosystem. (From Burrows, 1990. *Processes of Vegetation Change*. Unwin Hyman.)

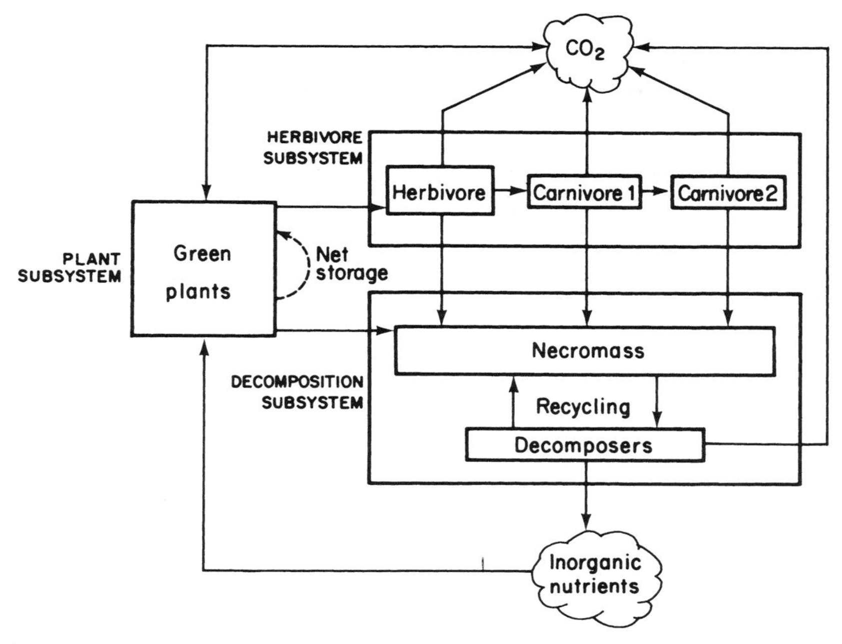


Figure 1.9 Generalized model of a woodland ecosystem showing the three subsystems. Arrows indicate major transfers of matter between organic matter pools (rectangles) and to and from inorganic pools ('clouds'). (Modified from Swift, Heal and Anderson, 1979. From Packham *et al.*, 1992. *Functional Ecology of Woodlands and Forests*. Chapman and Hall, Fig. 1.4. With kind permission of Springer Science and Business Media.)

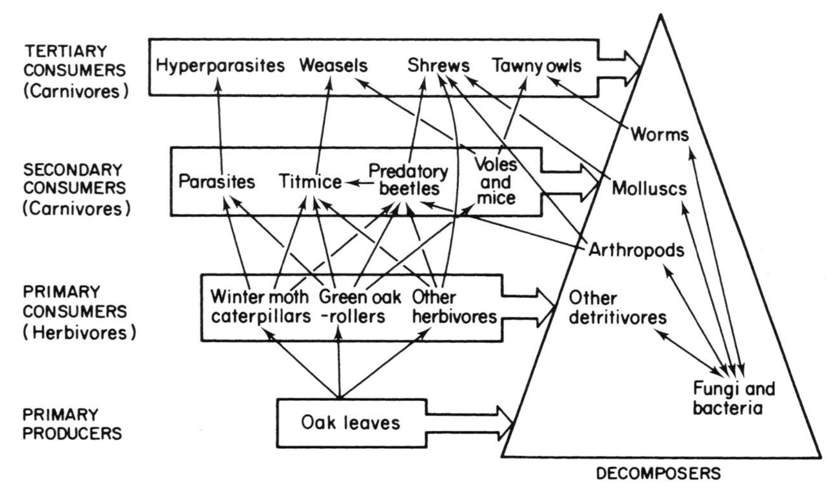


Figure 1.10 Simplified food web based on oak at Wytham Wood, near Oxford, UK. Note the two-way interactions among the decomposers. (Data from Varley, 1970. From Packham and Harding, 1982. *Ecology of Woodland Processes*. Edward Arnold.)

transfers through the woodland ecosystem using a systems approach concentrating on larger subsystems.

The **plant subsystem** is made up of all the green plants (the primary producers or photoautotrophs). They use photosynthesis to convert carbon dioxide and water into sugars and starch, which are used to grow new leaves and wood, and fund reproduction with some stored for the future. The total amount of material they create is referred to as the **gross primary production**. Not all of that is available to herbivores, however, because a significant proportion (typically 90%) is used in respiration (the running costs of keeping the plants alive). The overall amount of carbon fixed after respiratory costs is referred to as **net primary production** (see Chapter 3). In wooded areas, the amount varies from around 5 t $ha^{-1}$ $y^{-1}$ in the northern coniferous forests to 25 t $ha^{-1}$ $y^{-1}$ in tropical rain forest. Plant net production is potentially available to the **heterotrophs** of the herbivore and decomposition subsystems, which are unable to make their own food by photosynthesis or chemosynthesis as **autotrophs** do.

The **herbivore/carnivore subsystem** consists of all consumers feeding on the plants or each other. This encompasses herbivores (primary consumers) feeding directly on the plant subsystem (leaves, flowers, wood, etc.), including bacteria and fungi, and the carnivores (secondary and tertiary consumers) that feed on the herbivores and each other (see Fig. 1.8). As before, up to 90% of the energy eaten is used up in respiration (usually more in herbivores than carnivores because their food is harder to digest and their activity is more constant – Section 8.2), leaving just 10% at most to be passed on between subsystems. The number of levels that can be sustained is obviously at least partly determined by the amount of energy available to be passed from level to level. And by necessity each successive level of consumers must be of a smaller overall mass. This explains why a forest contains very few big cats compared with the number of herbivorous deer. It should also be noted that different organisms within the subsystem can fit into different levels of consumer; for example, the omnivorous black bear of North American forests is a herbivore feeding on berries and roots but is also carnivorous, as foresters are well aware!

The **decomposition (or detritus) subsystem**, described in Chapter 7, is often ignored but is as crucial as the two noted above: it is after all where most primary production eventually goes in natural ecosystems! Huge amounts of decomposition are possible by abiotic processes (such as fire) but in most ecosystems, the breaking down of dead material is carried out by living organisms. The decomposers are traditionally categorized into three size classes (see Fig. 1.11): the **microbiota** (less than 0.2 mm – fungi, bacteria, protozoa, slime moulds); the **mesobiota** (0.2–10 mm – nematodes, rotifers and small arthropods such as mites and insects like spring tails

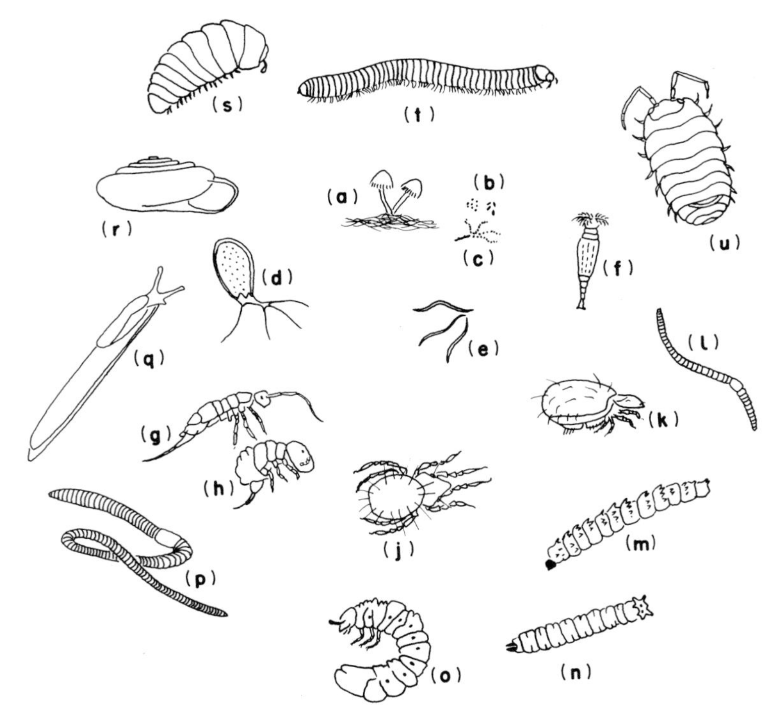


Figure 1.11 Representatives of the major groups of decomposers in litter and soil (not drawn to the same scale). **Microbiota**: (a) fungi; (b) bacteria; (c) actinomycetes. **Mesobiota**: (d) shelled amoeba; (e) nematodes; (f) rotifers; and including the **microarthropods**: (g, h) *Collembola* (springtails); (j, k) Cryptostigmata (oribatid mites). **Macrobiota**: (l) Enchytraeid (potworm); (m, n) Diptera larvae (bibionid and tipulid); (o) Scarabaeid beetle larva (white-grub); (p) lumbricid earthworm; (q, r) molluscs; (s, t) millipedes; (u) woodlouse. (Drawn by David Harding. From Packham and Harding, 1982. *Ecology of Woodland Processes*. Edward Arnold.)

*Collembola* and ants); and the **macrobiota** (>10 mm – millipedes, centipedes, snails, earthworms and large insects such as beetles and termites). They utilize all the dead components of the woodland: dead plant material and wood, dead bodies and all the faeces produced by the living. Dead material may be cycled several times through this subsystem before it is gone. Beetles, for example, chew on dead wood extracting some energy and nutrients, and their faeces in turn are reworked by other sets of organisms, extracting more energy and nutrition, and so on. Generally the larger organisms physically break down or **comminute** material suitable for the smaller organisms. Eventually all that is left are some of the more inert remains which form the black humus of the soil and which may persist for millennia (referred to by Ashman and Puri (2002) as something like a chemical 'junk yard'!). As decomposition proceeds, all the contained energy is lost in respiration but the nutrients are released in an inorganic (mineralized) state, available to the plant subsystem, ready to start the cycle again.

It is interesting to observe the activities and distribution of particular groups of **detritivores** (a term used here in the wide sense to describe decomposer animals that exist primarily on fungi or bacteria, as well as those which ingest the dead material known as **necromass** – see Section 7.5.2) such as the mites. Wallwork (1983) gives an account of oribatid mites in forest ecosystems, while Anderson (1971, 1975) describes both their vertical distribution in two woodland soils and their role in decomposing leaf litter. His examination of the gut contents of many individuals of these **microarthropods** enabled him to place various species of oribatid around the margins of a triangle whose corners represented 100% fungal material, 100% amorphous material and 100% higher plant material. Packham *et al.* (1992), who describe the decomposition of wood, leaves and other substrates, as well as the experimental use of cellophane, also give an account of succession amongst detritivores.

Though both are many-legged arthropods, the carnivorous centipedes (Class Chilopoda) have one pair of legs per segment, while the essentially vegetarian millipedes (Class Diplopoda), of which the larger species may also eat dead animal matter at times, have two. A good modern account of millipedes, which are divided into five major groups, is provided by Hopkin (2004). These animals are common and widespread, being found in leaf litter, rotting wood, and under bark in woodlands. Their common diet of dead plant material and fungi assists decomposition and the return of essential nutrients to the soil, though they may occasionally damage garden crops or plants. Millipedes in the soils of temperate countries are of moderate size, the pill millipede (*Glomeris marginata*) rolling into a ball the size of a pea. This species is very widely distributed throughout the British Isles and is a significant contributor to decomposition in deciduous woodlands, consuming 10% of its body weight in dead plant material per day and greatly enhancing the bacterial content of the remaining faeces. All millipedes secrete repellent chemicals; some tropical forms reach 30 cm and should not be handled, as the chemicals they exude can cause human skin to shed leaving an open wound. Fossil remains of these animals have not been discovered but those of their tracks have, the most famous being that of the Devonian *Arthropleura*, whose tracks suggest it reached a length of 1.8 m.

Waterlogged and anaerobic conditions prevent decomposition and lead to peat formation and ultimately coal. Acidic forest soils tend to have a build-up of organic matter on the surface (**mor humus**) caused by the absence of earthworms below pH 4.8 and reduction in fungal and bacterial activity.

Earthworms are instrumental in processing large quantities of dead material and making it available to microbes (see Sections 2.2.2 and 7.2.1).

Although forest ecosystems repeatedly cycle the vital nutrients, they are not entirely self-sufficient. Nutrients leak in and out of the forest when litter is blown around, and when forest fires cause nutrients in smoke and ash to be deposited elsewhere. Stream water often carries nutrients out of woodlands, while rain brings in new supplies. Ecosystems are not hermetically sealed.

## 1.6 Forest types and classification

### *1.6.1 Distribution in relation to climate: biomes*

Living organisms have very distinct distributions; there are at least 400 000 species of flowering plants but not one occurs everywhere in the world. There are, however, communities of plants and animals found over often extensive regions of the world that have a similar and characteristic appearance (or **physiognomy**) as defined by their life form (see Section 3.1.1) and principal plant species. These major ecological communities are known as **biomes** (Fig. 1.12) and their distribution depends very greatly on regional climate, particularly temperature and rainfall. Similar regional climates produce similar morphological adaptations in plants. The term biome was first used by Clements and Shelford in 1939; an interesting team as Clements was a botanist and Shelford a zoologist. Whittaker (1975) lists six major terrestrial physiognomic types – forest, woodland (see Section 1.1.2 for definitions of these), shrubland, semi-desert scrub, desert and grassland. Trees dominate the first two of these types, and occur so widely around the world that more than one biome type is defined within the wooded areas of the world on the basis of climate. Species in separate biomes sometimes belong to the same genera (e.g. *Nothofagus* in Chilean and New Zealand rain forest) suggesting a former proximity of the continents.

**Tropical rain forests**, which are of especial interest because of their extreme species richness (including the greatest animal diversity of any terrestrial biome) and the fact that all other forest biomes are thought to have evolved from them, are described in Section 2.5. It is estimated that 50% of all Earth's species are found in the 7% of land covered by tropical forests, and predominantly in rain forests (see Woodward, 2003 for a more detailed description of terrestrial biomes). They are at their best in the warm, moist circumtropical lowlands where there is little seasonal variation in climate. These are the **lowland or equatorial evergreen forests**, sometimes referred to as **aseasonal or perhumid** rain forests (care is needed with perhumid since it is sometimes used

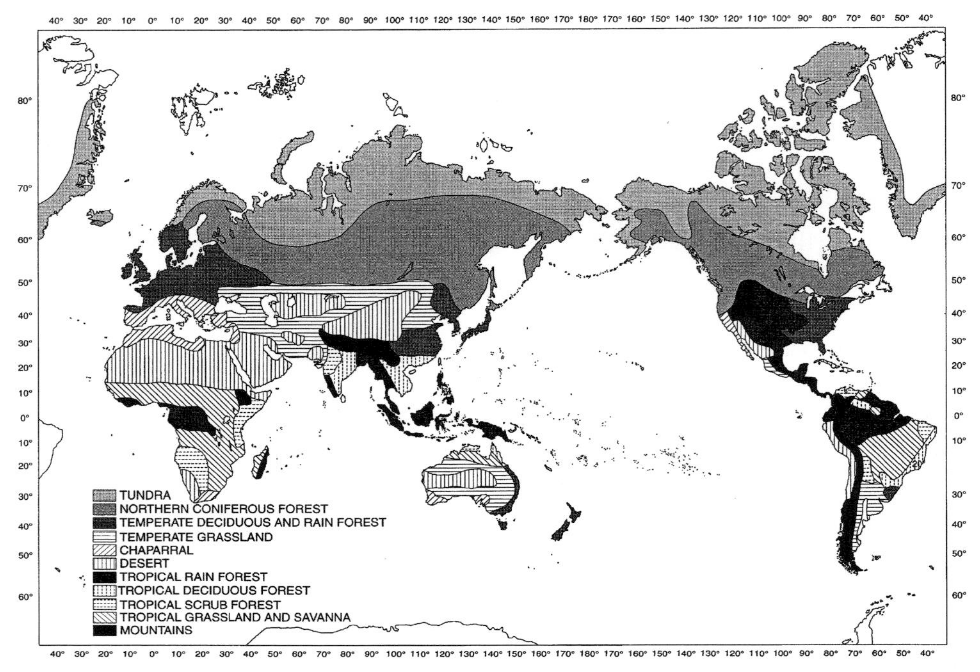


Figure 1.12 Terrestrial biomes of the world. (From Lincoln *et al.*, 1998. *A Dictionary of Ecology, Evolution and Systematics* (2nd edn). Cambridge University Press.)

to mean very wet and sometimes aseasonal). The vegetation is dominated by a rich variety of multi-layered broad-leaved evergreen trees, with up to 100 or more tree species per hectare (2.47 acres), though some forests are dominated by one family of trees, such as the Dipterocarpaceae in Indo-Malayasian rain forests.

In areas where rainfall is more seasonal (sometimes referred to as the **moist or humid tropics**), two types of forest are found. Above 1000 m the forest changes to a more seasonal **montane or cloud forest** where much of the precipitation comes as mist or fog and consequently epiphytes (see Section 5.5.1) are abundant. In the lowlands, the evergreen trees are progressively replaced by deciduous species that drop their leaves during the dry (or drier) season, resulting in **semi-evergreen rain forest** and, where drier still, **tropical deciduous forest**, also called seasonal tropical rain forest or monsoon forest. Here, the forest structure becomes simpler with fewer epiphytes and lianas. As moisture declines further and becomes even more seasonal, especially where there is summer drought and cool moist winters, evergreens again come to dominate. Fire also becomes increasingly common and gives rise to **Mediterranean**

**ecosystems** dominated by **sclerophyllous** trees and shrubs (evergreens with hard, thick, small, leathery leaves, resistant to hot, dry conditions). The biome is named for the largest area of this evergreen forest centred around the Mediterranean Sea where much of the former sclerophyllous chaparral woodland of holm oak (*Quercus ilex*) has been replaced by dense scrub (**maquis**) or more open heathy **garrigue** with many short aromatic shrubs. Similar Mediterranean-type vegetation is also found in the **chaparral** of California, the **matorral** of Chile, the **fynbos** of the Cape Province of South Africa (as distinct from the shrubby **veld**) and finally the bush vegetation of southern and south-western Australia. All these areas have a similar appearance, dominated by hard-foliaged and twiggy shrubs and small trees. Yet the species they contain are quite distinct (that is, there is a large number of endemics, species found only in that area and nowhere else); for example, of the 113 woody genera (with 169 species) in Chile, only 13 genera are amongst the 109 genera (with 272 species) in California. In fact, the Mediterranean ecosystems contain about 20% of the world's plant species.

Where the climate remains warm but becomes drier still with several months of severe drought, the tree cover thins out into the **tropical savannas**. These cover 20% of the world's land surface and are the home of many large grazing and browsing animals, especially in East Africa with elephants, zebras, antelopes and hippos. The largest savannas are indeed in Africa, where they cover 65% of the continent (see Box 1.1), but are also found in South and Central America and northern Australia. Fire is one of the main forces that controls the distribution of savanna, so much so that in areas of recurrent fire, savanna may push into areas that would otherwise be rain forest. The difference in vegetation is quite distinct and abrupt as demonstrated by the Olokomeji Forest in Nigeria. This rain forest has a life-form spectrum in which over 90% of the species present are phanerophytes (growing points – perennating buds – more than 25 cm above ground – see Section 3.1.1). The trees and shrubs have very thin bark so that the cambium is very easily damaged by fire. In grassy savanna woodland only 5 km away, less than one-third of the species are phanerophytes and most plants either avoid fire by being annual (one-quarter are annual plants or therophytes) or hide the growing points at ground level (hemicryptophytes – one-fifth of the flora) or below it (geophytes). The percentage of geophytes is four times that of the forest, where it is already higher than average because the climate is dry. Bark of tree and shrub savanna species is usually more than a centimetre thick, giving protection from low-intensity fires, and many woody savanna species have enormous root systems which sucker readily when the shoots have been burnt to ground level. All these features of savanna plants show great adaptation to fire not found in the rain-forest plants.

### Box 1.1 The African savannas

The **Miombo savanna**, which lies to the south of the East African plains, the nyika thornbush, and the Congo forest is one of the most extensive African biomes. Measuring 2575 km (1600 miles) from east to west, and over 1287 km (800 miles) from north to south, it covers southern Tanzania, most of the southern Congo, Angola, Zambia, Zimbabwe and Malawi. It is largely found on the elevated inland plateaus of southern central Africa on deep, semi-podzolic sandy soils of little use to cultivation. Typically it rains for one half of the year while the other is dry and dusty. The miombo is sparsely populated, and is in a part of Africa where sleeping sickness, caused by the protozoan *Trypanosoma*, continues to spread despite all attempts to eliminate its vector, the tsetse fly. This savanna is effectively a woodland with a field layer dominated by grass, though in many places the trees are tall and may form contiguous forest. At the onset of the dry season tree leaves wither and fall, the grass dies back and fire often sweeps through the area, baring the ground and charring the lower trunks of the trees. Just before the rains come, the trees burst into leaf, those of *Brachystegia* and *Julbernadia*, the two most dominant genera, being a brilliant red. *Julbernadia globiflora* often grows with *Brachystegia spiciformis*, which is very common and often dominant. *Ziziphus rhodesica* and *Gardenia spatulifolia* commonly occur in *Brachystegia* forest; their saplings often commence growth on termite mounds.

Scattered through the miombo landscape are prominent bosses of rock known as inselbergs or **kopjes**. These dark, strongly weathered rocky outcrops have floral and faunal communities distinct from those of the surrounding savanna matrix. Two fig trees, *Ficus burkei* and *F. capensis*, occur here; the former is also found in woodland and the latter in riverine vegetation. Whilst the miombo savanna supports almost twice as many species of animals as that of Guinea, it comes a poor second to the East African plains, in part because of the poor quality of its grasses, most of which are unpalatable 'sourveld' in the dry season. However, on the fertile flood plains of the rivers the biomass of buffaloes, hippos and elephants must rival that of the river basins of East Africa.

Dry grass-poor **mopane** *Colophospermum mopane* **woodland**, a special type of miombo found on heavy clay soils, borders the flood plains of the Zambezi, Kafue, Luangwa and other rivers of this region. This forms a favourite feeding ground for elephants, herds of which work their way through it, pushing over big trees, and snapping boughs and branches off smaller ones. This large-scale disturbance pattern plays a major role in the ecology of the grass-impoverished mopane, many of whose tree species readily produce substantial epicormic growth following stress or damage. Bushy growth on trees damaged by fire or elephants provides good protein-rich feed for animals that cannot reach the top of the big mopane trees. Impala and other herbivores rare in *Brachystegia* forests are frequently found here. Browsing of the black rhino, an animal particularly fond of the large fruits of the sausage tree *Kigelia africana*, is largely dependent on the activities of elephants. The very wide-lipped white rhinoceros is, in contrast, a grazer rather than a browser. Neither rhino is black or white: white was an imperfect translation of the Boer word for wide.

Despite its apparent dry and harsh nature, mopane supports a number of animals that are rare or absent in typical miombo. Impala and giraffe are the most obvious; others include greater and lesser kudu, bushbuck and duiker. The presence of these smaller herbivores is due to both the drier and more open character of the woodland and to the way the ecosystem has been changed, almost 'engineered', by elephants.

The dominant trees can be very different in different parts of the world. In parts of Mexico the savanna is dominated by a palm *Paurotis wrightii*, in Belize it is the Caribbean pine *Pinus caribea*, and in Africa acacia species.

At higher latitudes where the climate becomes cooler, especially in winter, and more seasonal but still moist, **temperate broadleaf forests** are found. In the northern hemisphere this seasonality has produced the **mixed deciduous forests** of eastern North America, western and central Europe (described in Section 2.4.1) and eastern Asia. These three areas were once part of a continuous forest belt and so share many closely related plants, dominated by a range of trees including beeches, oaks, limes, ashes, maples and a good number more. East Asian forests are the most diverse (suggesting they were the least affected by Quaternary cold periods), and those of Europe are the least rich (e.g. 729 species in 77 genera in temperate east Asia, 253 species in 90 genera in eastern North America, 68 species in 37 genera in western North America and 124 species in 43 genera in Europe – Rees *et al.*, 2001). In the southern hemisphere and parts of Asia (Japan, South Korea, southern China) the same conditions have favoured **broadleaved evergreen forests**, typified by the eucalypt forests of eastern Australia and the evergreen *Nothofagus* forests of South America and New Zealand. However, pockets of deciduous forest are found in Chile and Argentina dominated by other southern beeches such as roble *N. oblique* and rauli *N. procera*.

At higher altitudes still in the northern hemisphere, where the climate is too unfavourable for deciduous trees, the **boreal forest** takes over across huge swathes of North America (Shugart *et al.*, 1992) and Eurasia (Hytteborn *et al.*, 2005), making up almost a third of the world's forest area. The southern boundary is defined as where the summer becomes too short (less than 120 days with daily average temperatures of $>10\,^{\circ}\mathrm{C}$) and winters too long ($>6$ months) and harsh for deciduous forest. The transition is fairly broad, giving rise to the transitional mixed or **hemiboreal forests** of places such as Prince Edward Island in Canada (Section 4.1.2), and the pre-settlement southern forest of north-east Minnesota, USA.

Boreal forests show a number of transitions at their borders; that with wetland leads to mire and bogs in the northern regions. Such transitions tend to be in a state of flux, with forests becoming mires at one point in time while bogs and mires become afforested at another. Stratigraphically mires are important; the pollen trapped within the peat has told us much regarding Holocene changes.

The circumpolar boreal forest is dominated by evergreen cone-bearing, needle-leaved trees. Four genera of **conifers** cover the bulk of both regions: evergreen fir *Abies*, spruce *Picea* and pine *Pinus* with some deciduous larch *Larix*. Some broadleaved deciduous species are of secondary importance, including alder *Alnus* and especially birch *Betula* and poplar *Populus*. The value of aspen *P. tremula* stands in helping to conserve the diversity of boreal members of a number of animal and plant groups in Scotland is discussed in Box 6.1. Siberian dwarf pine *Pinus pumila* covers large areas near the eastern arctic and alpine belts; its North American equivalent is whitebark pine *P. albicaulis*. Although reasonably rich in tree species, the boreal forest is dominated by relatively few of them: 14 in Fennoscandinavia and the former USSR, 15 in North America. Even across the disjunct continents formed by the breakup of Pangaea in the Jurassic Period, the same genera form repeated elements.

Site conditions within these forests range from the extremely cold, dry continental regimes of Siberia and interior Alaska, to the warmer, moist oceanic climates of eastern Canada and Fennoscandinavia. The trees found are closely adapted to the conditions of their particular sites; in terms of soil moisture content the range in North America runs from the dry-tolerant jack pine to the moisture-loving lowland black spruce and tamarack. **Taiga** (a Russian term) is sometimes taken as being equivalent to the whole of the boreal forest, but is best restricted to its northern edge. The border between taiga and tundra to the north (i.e. the tree line) is a wide zone (see Section 3.5.3), in which the trees gradually become less prominent (Sirois, 1992).

There is no southern hemisphere equivalent of the boreal forest owing to the northerly position of the southern continents and the limited geographical range of southern conifers, exemplified by the monkey puzzle (*Araucaria araucana*) native to comparatively small areas of southern Chile and western Argentina (see below).

### *1.6.2 Coastally restricted forests*

There is a special interest in coastally restricted forests in which a few tree species are found growing in relatively narrow ribbons bordering an oceanic edge, usually within 250 km of a coast (Laderman, 1998). Many of these tree species have unusual morphological, physiological or ecological characteristics, which enable them to survive conditions – including toxic soils – in which others either do not flourish or fail altogether. What do these species have in common, and why are they not naturally found inland? The soils they grow in are often so damp as to cause problems with root aeration (see Section 2.5), but are never salt-laden like those in which **mangroves** establish along the coasts and estuaries subject to daily tidal inundation (Section 4.2.3).

**Laurel forests** form a transition between tropical and temperate regions where the climate is very humid with rainfall year round but with a distinct cool season. These specialized evergreen rain forests exist mainly on the eastern side of continents in southern China and southern Japan, South America, southern Australia and south-east Africa where they are influenced by trade and monsoon winds. In Japan, these forests consist of evergreen oaks and chinquapins *Castanopsis* in the beech family and, in the laurel family, *Machilus* spp. These ecosystems resemble those of Macaronesia (see the description of the **Madeiran cloud forests** in Box 1.2; Fig. 1.13) in being humid forests with typically laurel-like evergreen leaves, but are very different in other ways, usually possessing far more and different species of both animals and plants.

The southern tip of Florida and further west can also be included in this consideration of laurel forests. The **southern forested wetlands** of the USA, which extend through 12 states, have been more intensively studied than any others; descriptions of their economic importance, ecosystem properties, flora, fauna and possible future transition are given in Messina and Conner (1998). As with the Amazonian tropical rain forests, many of them have been converted for agriculture. Problems of root aeration, a particular difficulty here, are considered in Section 2.3.3. These areas are defined solely by the nature of the forest cover into three groups. Plots dominated by elm *Ulmus* spp., ash *Fraxinus* spp. and eastern cottonwood *Populus deltoides*, either singly or in combination, are designated as elm–ash–cottonwood forest. With oak–gum–cypress forests the trees present are tupelo *Nyassa* spp., sweetgum *Liquidambar styraciflua*, wetland oaks *Quercus* spp., and swamp cypresses *Taxodium distichum* and *T. distichum* var. *nutans*. Pond pine forests are dominated by pond pine *Pinus serotina*, either singly or in combination with other species.

The bald cypress–water tupelo *Taxodium distichum–Nyassa aquatica* association is important in the freshwater wetlands of southern USA including Mississippi, where it is associated with a number of other tree species. As in other cases the natural balance of individual communities is influenced by the degree of inundation and the type of soil present. There is a continuum from

## Box 1.2 Ancient Laurel Forests of Madeira: a continuity with the Tertiary

The forests of the Island of Madeira were not subject to the ice ages which ravaged north-west Europe in the Quaternary: this has led to the survival of a laurel forest or **Laurissylva** that is almost identical to that existing in the Tertiary period (see Fig. 1.1). A considerable portion of this forest survives virtually intact, though its integrity is threatened by the introduction and spread of acacias, eucalypts, Douglas fir and other forest trees (Section 5.6). The present state of knowledge regarding the origin, history and dynamics of this remarkable rain forest is outlined in Packham (2004). The laurel forests form a mosaic, heavily influenced by relative humidity, which differ in their structure and floristic composition, as Costa Neves *et al.* (1996) demonstrated. Lianas are uncommon while ferns are abundant; hare's foot fern *Davallia canariensis* and Macaronesian polypody *Polypodium macaronesicum* are amongst those being epiphytic on trees. The rich damp soils support a wide variety of native herbs, including four of the five orchids native to Madeira. There is, however, a tendency for non-native species, such as cape sorrel *Oxalis pes-caprae* and greater periwinkle *Vinca major* to advance into the laurel forest.

Before fires started in 1420, when the island was first colonized, the laurel forests covered most of the higher parts of the island, but they are now mainly in the northern half where they grow on steeply sloping mountainsides. These humid forests flourish in the mists of the cloud zone and although only four of their species belong to the laurel family, almost all their trees are evergreen, frequently multi-shooted and with glossy coriaceous (leathery) leaves. Many produce new shoots directly from the roots and coppice well. Huge foetid laurels *Ocotea foetens* still occur in southern parts of the island, as do Canary Island laurels, many being remnants of the former southern forest. The forest vegetation, which still covers rather more than 20% of the island, is dominated by Canary Island laurel *Laurus azorica* that grows to a height of 20 m and often bears the fruiting bodies of the fungus *Laurobasidium lauri* which grows only on this species of tree. Madeira mahogany *Persea indica*, which has been exploited for its valuable reddish timber, and babusano *Apollonia barbujana* can both reach a height of 25 m, but mature foetid laurels are massive trees reaching 40 m which tower over their relatives.

The laurel forests which developed on Madeira after its formation in Late Tertiary times differed markedly from those which covered much of Europe and are well known from fossil deposits, particularly those in Bohemia and Austria. All four members of the laurel family mentioned above as being present now in the **Laurissylva** of Madeira, along with other of its components, have been found as fossils aged about 5.3 million years in the south of France, but many heavy-fruited tree species never reached the islands of Macaronesia. Madeira itself did not erupt above the surface of the sea as a volcanic island until 5.2 million years ago and the whole of its biota built up subsequently as a result of immigration by air or sea.

Oaks, sweet gums, maples, magnolias, walnuts and palms have reached Madeira only in the last 600 years. The natural fauna is also very restricted; rodents, deer and elephants did not exert the significant influences they did in Europe while the bird population was also more limited.

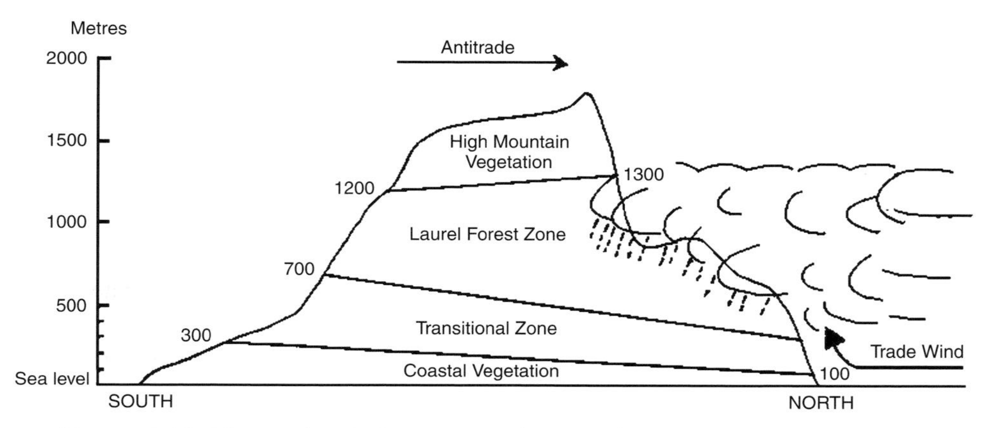


Figure 1.13 The trade-wind system and vegetation zones of Madeira; the southern end of the transect terminates in the city of Funchal. (Adapted from Szeimer, 2000. From Packham 2004. *Arboricultural Journal*, 28).

pure bald cypress to pure water tupelo. When the cypress is clearcut it is often succeeded by pure stands of tupelo.

Western seaboards in temporal zones have similar moist conifer forests, often termed **temperate rain forests**. In the southern hemisphere, southern Chile bears forests containing a number of conifers (Patagonian cypress *Fitzroya cupressoides*, Chilean incense cedar *Austrocedrus chilensis*, monkey puzzle *Araucaria araucana*) and many species of southern beech (*Nothofagus*). The *Eucalyptus–Nothofagus* forests of eastern Australia and Tasmania, and the kauri pine *Agathis australis* forest of New Zealand are also in this category (although the kauri forests are verging on subtropical).

The forests of the western seaboard of North America have been particularly well studied (e.g. Laderman, 1998). A high proportion of the species involved are evergreen needle-leaved conifers that have poor competitive ability in their early years. In view of the high economic and ecological value of these forests it is clearly essential to understand the successional processes which operate within them and to treat them accordingly. The seven species of false cypress *Chamaecyparis* are prominent in the coastally restricted forests of the Pacific Northwest and south-eastern USA, Japan and Taiwan. Yellow cedar (also called nootka cypress) *C. nootkatensis* is a most valuable timber tree in south-east Alaska, where it is also found as a low-growing prostrate form with upright stems a metre high which may be 100 years old. Such plants reproduce asexually by layering, while seedlings develop on better-drained sites. In Alaska, yellow cedar is currently suffering a forest decline which began around a century ago; this has been intensively investigated but its causes – biotic or abiotic – are still unclear (Hennon *et al.*, 1998). Further south, yellow cedar grows under moist, humid conditions on neutral to acidic soils (including ferro-humic podzols) from sea level to altitudes of 2000 m in British Columbia. The poor reproductive success of this tree is a problem in the maintenance and expansion of its forests (many of which are at high altitude). The tree is tolerant of shade, frost and high soil-moisture and in nature its longevity and allocation of resources to defensive compounds also promotes its long-term survival. It is not, however, a good competitor in its younger years and when mixed yellow cedar–hemlock *Tsuga heterophylla* forests are felled they often regenerate to pure hemlock.

Port Orford cedar (or Lawson's cypress) *Chamaecyparis lawsoniana* is a highly adaptable species, which thrives on diverse soil types adjacent to the Pacific coast and whose wood commands premium prices; its susceptibility to root disease is considered in Section 5.4.4. Atlantic white cedar (or white cypress) *C. thyoides* is an obligate wetland species with a natural range that extended along the Atlantic and Gulf coasts from southern Maine to northern Florida and through Mississippi. The failure of natural regeneration following intensive logging in swampy areas, which were considered worthless after felling, has led to a loss of as much as 90% of the area which this species occupied in Northern Carolina two centuries ago. Efforts are now being made to re-establish this large and splendid tree using both seedlings and rooted cuttings (Phillips *et al.*, 1998). Two other false cypresses are prominent in the coastally restricted forests of Japan. These are the hinoki cypress *C. obtusa* and the sawara cypress *C. pisifera*, both of which are prominent in Shikoku Island (Yamamoto, 1998).

Although one of the most distinctive and tallest trees in the world (up to 115 m), coastal redwood *Sequoia sempervirens* grows only along a narrow portion of the Pacific coast of California and Oregon. The tree lives for several thousand years, grows very rapidly and can form pure stands in which no other tree species can compete successfully. Its tapered trunk is protected by spongy bark up to 30 cm thick and the tree, while lacking a tap root, forms an extensive fan of horizontal roots just below the soil surface. This species rarely occurs along the edge of the coast, being intolerant of strong, salt-laden maritime winds, and requires deep moist soils.

## 1.7 Regional classifications of forests and woodlands

Dividing woodlands and forests into biomes is useful at the global scale, but within a biome there is a good deal of local variation and thus a need for a much more detailed and local regional classification. It is useful to be able to distinguish a beech from an oak woodland even though they are in the same temperate deciduous woodland biome. Some regional classifications are, like the biomes, based on physiognomy but most are based on what is living on the ground which really means the vegetation rather than other components such as the animal or microbe life, since the vegetation is most easily observed. There are many examples of regional classifications that could be chosen; two are given below.

### *1.7.1 The British National Vegetation Classification*

The need for an overall system of description for the British mainland and the many associated islands which were to be ecologically assessed led to the development of the **National Vegetation Classification (NVC)**, which is the basis of the five volume description entitled 'British Plant Communities'. The first of these (Rodwell, 1991) deals with woodlands and scrub and has 25 main communities (W1–W25). These are in turn divided into subcommunities. Maps show the national distribution of the communities, which were delineated on the basis of Domin cover-abundance records for all the vascular and non-vascular plant species present, i.e. all plants, not just the dominant ones. Keys and detailed descriptions enable fresh sites to be allocated to the appropriate community and subcommunity. The distribution of particular vegetation types, one of which is described below in Box 1.3, is influenced by several factors amongst which climate and soil type are pre-eminent (see Section 2.4.1).

### *1.7.2 New England*

Vegetation classification on the eastern seaboard of the USA was, as elsewhere on this continent, spearheaded by foresters. Many ecologists in New England consequently still use the vegetation classification put forward by Westveld and colleagues in 1956 in which New England is divided into six forested zones (Fig. 1.14). Unlike the NVC described above which uses the current vegetation, Westveld's scheme is based on natural forest vegetation zones: that is, the vegetation that would naturally be found across the area if human interference is ignored (a difficult concept to apply in the UK). The scheme is useful because it contains identifiable and usable categories, easily identified on the

**Box 1.3 An example of the UK National Vegetation Classification (NVC): Wistman's Wood, Dartmoor, Devon**

Wistman's Wood, owned by the Duchy of Cornwall, has been managed since 1961 under a nature reserve agreement with what is now Natural England, the Government conservation body. This long narrow 3.5 ha wood is divided into the North, Middle and South Woods, all situated in the valley of the West Dart river. Its vegetation in general conforms to W17, an upland Oak–Birch–*Dicranum* moss woodland (*Quercus petraea–Betula pubescens–Dicranum majus* woodland, subcommunity *Isothecium myosuroides–Diplophyllum albicans*), a grouping in which non-vascular plants are particularly important. The wood is dominated by pedunculate oak *Quercus robur*, with occasional rowan, holly, hawthorn, hazel and willow (*Salix aurita, S. cinerea*). Note that the pedunculate oak is not the oak from which the community takes its name – the sessile oak *Q. petraea*. This shows the robustness of taking into account *all* the vegetation; a community which otherwise looks like any other W17 woodland is not misclassified by one main species being 'wrong'.

Cattle and sheep have free access to this high altitude wood in an area of high rainfall; species sensitive to grazing such as bilberry, ivy and polypody sometimes occur with the lichens and mosses growing on tree branches. The granite boulders on which the wood has arisen are largely covered with lichens, especially *Parmelia saxatilis, P. laevigata, Sphaerophorus globosus* and various species of *Usnea*, and patches of moss in which *Dicranum scoparium, Hypnum cupressiforme* (*s.l.*), *Isothecium myosuroides, Plagiothecium undulatum* and *Rhytidiadelphus loreus* are frequent. Acid grassland with wavy hair-grass, common bent, sheep's fescue, creeping soft-grass, heath-bedstraw, tormentil and sorrel, develops where soil has accumulated. Woodland clearance in this area of Devon was extensive by the Iron Age and largely complete by the thirteenth century.

Wistman's Wood is surrounded by Bronze and Iron Age hut circles and has been influenced by humans and their grazing animals for thousands of years. An excellent summary of this wood is provided by Mountford *et al.* (2001); Wistman's Wood also provides an example of forest change in Section 9.4.1. Fragments of similar high altitude Dartmoor oakwood occur at Black Tor Copse and Higher Piles Copse.

ground (or from the map if human disturbance is great) and for foresters conveys information on what trees will grow in a particular area. However, given the size of New England (almost exactly half the area of the British Isles), the classification is rather coarse with the consequence that each zone contains much variation. For conservation (where the complete range of habitats is of interest) and for resource management (where small differences in vegetation may be important), something with a finer scale and which preferably includes all the other habitats, not just forests, is needed.

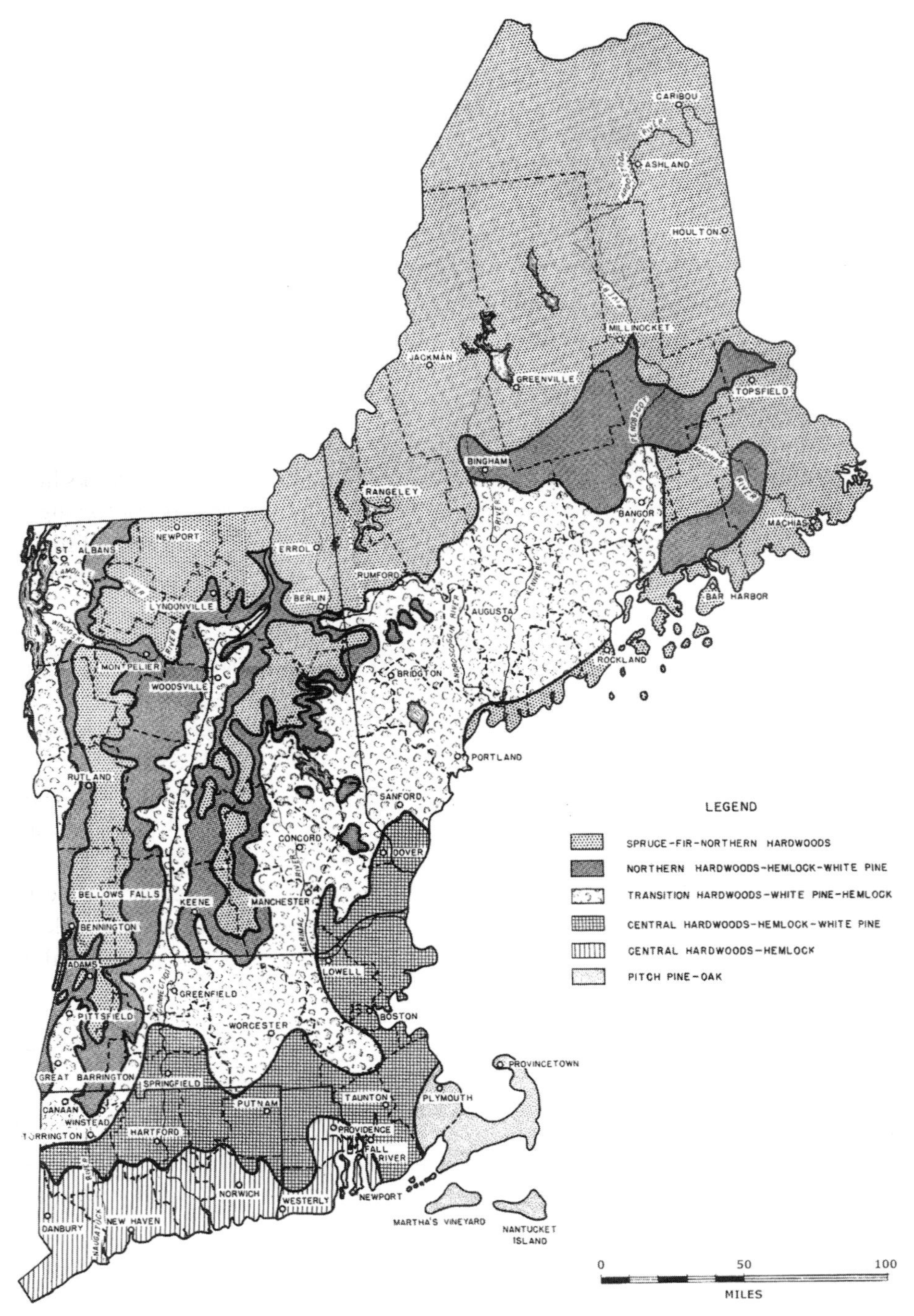


Figure 1.14 Natural forest zones of New England as compiled by the Society of American Foresters in 1955. (From Westveld *et al.*, 1956. *Journal of Forestry*, 54.)

At the state level, a number of finer resolution schemes have been produced, based on a variety of criteria but predominantly the current vegetation, both woody and herbaceous. So, for example, in Massachusetts 105 vegetation types, not including aquatic habitats, have been identified by Swain and Kearsley (2000) compared with the four zones of Westveld in the same area. Although attempts are made to cross-reference (crosswalk) classifications between nearby states, each state typically uses its own names for communities, compounded by geographical differences in species distributions, making it difficult to determine if communities in two states are really equivalent. It is clear from this that some sort of national classification would assist effective communication, but given the size of the USA and the huge number of habitats, it is a tremendous undertaking (and much like the current attempt to create an integrated classification across Europe). Country-wide classification schemes for forest types have been in place for decades but something with a wider inclusion of habitats is needed. A number of schemes are in progress aiming to produce hierarchical classifications so that forests and woodlands at the continent scale can be subdivided repeatedly down to individual habitat types so everyone will be happy! These schemes include:

- The US National Vegetation Classification (USNVC) which is part of the International Vegetation Classification (IVC), a continent-wide scheme (see: www.esa.org/vegweb). These schemes are based primarily on current vegetation (physiognomic at coarser levels – what the vegetation looks like – and floristic at finer levels – what species are there).
- The Ecological Systems Classification (ESC) put forward by NatureServe which takes the finest levels of the IVC (plant communities) and builds them into progressively bigger systems based on similarity of ecological processes, substrates and/or environmental gradients (see: www.natureserve.org).

# 2

# Forest soils, climate and zonation

## 2.1 Soils and trees

Soils are often given superficial treatment, and yet without them forests would quickly cease to function. As well as physically supporting plants, soils act as refuse collectors, processing organic waste and thereby recycling nutrients, a major influence on the productivity of forests. Without functioning soils, forests would rapidly be choked with dead wood and other material, and the bulk of nutrients needed by plants and animals would be locked up and unavailable.

Moreover, soil is not simply a loose collection of 'dirt', it is a complex mix of living and non-living components, consisting of air (soil gases: typically 25% by volume), water (25%), mineral particles (45%) and organic matter (5%); the last can be subdivided by weight into around 10% organisms, 10% roots and 80% humus. As described further in Chapters 1 and 7, various soil animals, such as earthworms and arthropods and the micro-organisms, including fungi and bacteria, decompose dead material to release nutrients and form the left-over, rather inert black humus of the soil.

Soil takes a long time to form, usually thousands of years, and its quality is one of the most important conditions governing the growth of trees, smaller plants and associated organisms in any site. As Fig. 2.1 demonstrates, soils have distinct morphologies each with a characteristic profile (a sequence of horizontal layers or **horizons** from ground surface to unaltered bedrock or sediment). Soils are the product of the integrated effects of **five soil-forming factors**: climate, parent rock, vegetation and associated organisms, relief of the land and time.

These set the conditions under which **physical, chemical and biological processes** operate to produce the distinct layers or horizons found in soil profiles. The first three main groups of the processes involved, **weathering, translocation** and the **organic cycle** are constructive and tend to build up the horizons, while the last two – **erosion** and **deposition** – tend to blur profile morphologies

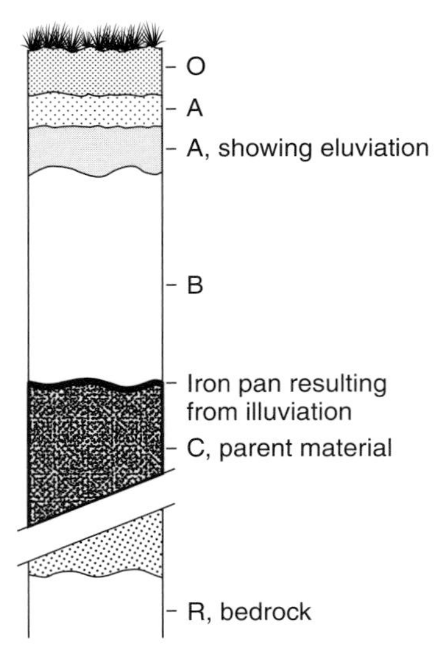


Figure 2.1 Idealized soil profile showing the main horizons. The **O horizon** is dominated by humus and plant litter at various degrees of decomposition. The **A horizons** are characterized by humus mixed with the mineral particles and include zones of **eluviation**, the movement by water of fine particles (such as clays and humus) and dissolved substances down further into the soil. The A horizons are commonly composed of an upper horizon and a lower, darker horizon showing eluviation (**$A_1$**). The **B horizon** is where these particles and possibly the dissolved chemicals accumulate in a process called **illuviation**. Here iron and aluminium oxides may build up enough to form a water-impermeable hard iron pan. Below this the **C horizon** is made up of weathered parent material not affected or moved by the pedogenic processes, and the **R horizon** is composed of the bedrock.

particularly on hills and in valleys. Weathering is the process whereby the mineral components of the soil are broken down either physically into smaller particles or chemically, releasing nutrients or other chemicals into the soil. Translocation is the movement by water of fine particles and dissolved substances down into the soil (**eluviation**) and their deposition lower down (**illuviation**). Some of these components, especially in the case of dissolved chemicals (solutes), may be taken away by the flow of groundwater, a process termed **leaching**. The organic cycle is the decomposition and redistribution of organic matter to form distinctive horizons in different soils. Erosion of material from one place by wind or water will obviously leave behind a truncated profile with one or more horizons missing (particularly where wind is involved) or a profile divided by channels and valleys (as with water erosion). Deposition gives rise to a range of new fairly mixed (azonal) soils, named according to the agent of movement: alluvium (water), loess (wind) and colluvium (gravity).

The weathering of fresh rocks releases reasonable amounts of the nutrients required by plants (see Section 2.2.1) apart from nitrogen. This element is typically needed in greatest quantities, its supply is often the main factor limiting forest growth, and it is usually most lacking in young soils. Despite making up 78% of the atmosphere, gaseous nitrogen cannot be used by plants until it has been 'fixed', either as nitrate ($NO_3$) produced by lightning, or more importantly as ammonia by specialized microorganisms (or by the Haber process for artificial fertilizers). **Nitrogen fixation** is therefore a crucial process (see Chapter 8). Inorganic nitrate produced by lightning and received in rain will gradually build up in the developing soil, but it is significant that plants which establish first, and so commence the **organic cycle** on which other organisms depend, often have their own nitrogen-fixing mechanisms (see Chapter 8). Such plants release soil nitrogen for their successors when they drop litter, die and decompose, and the forest then depends primarily on organic nitrogen being recycled. The increasing amounts of ammonia and oxides of nitrogen in the atmosphere released from agriculture and industry have resulted in many forests becoming saturated with nitrogen, rather than limited by it.

Organic materials also give rise to **humus**, which provides a reserve supply of nitrogen, forms part of a base-exchange complex (see Section 2.2.1), and helps retain soil moisture. The speed at which shed tree leaves are disintegrated by biological and chemical processes varies enormously with climate; in tropical rain forests such as the Congo, 50% of such a cohort may be completely broken down – largely by fungi – within twelve weeks, a process likely to take years in Lapland. The nutrient quality of the litter is also important, especially the C/N ratio: litter rich in N breaks down faster than litter poor in N. Inevitably there are feedback links, between the type of litter, the climate, decomposition and availability of nutrients in the soil; this is discussed further in Chapter 7.

Much of the organic matter contributed to the soil by trees is in the form of shed leaves and dead roots which accumulate at or near the surface. Tree roots transport mineral nutrients from within the soil profile. These are deposited on the soil surface through leaf fall or near the surface in roots and the organic cycle makes them available to forest herbs.

## 2.2 Features of forest soils

### *2.2.1 Soil profiles and properties*

Soil profiles of the kind shown in Fig. 2.2 occur in many of the world's temperate zones (see also Box 2.1). **Soil reaction or pH** is an important property influencing the availability of important mineral elements in different

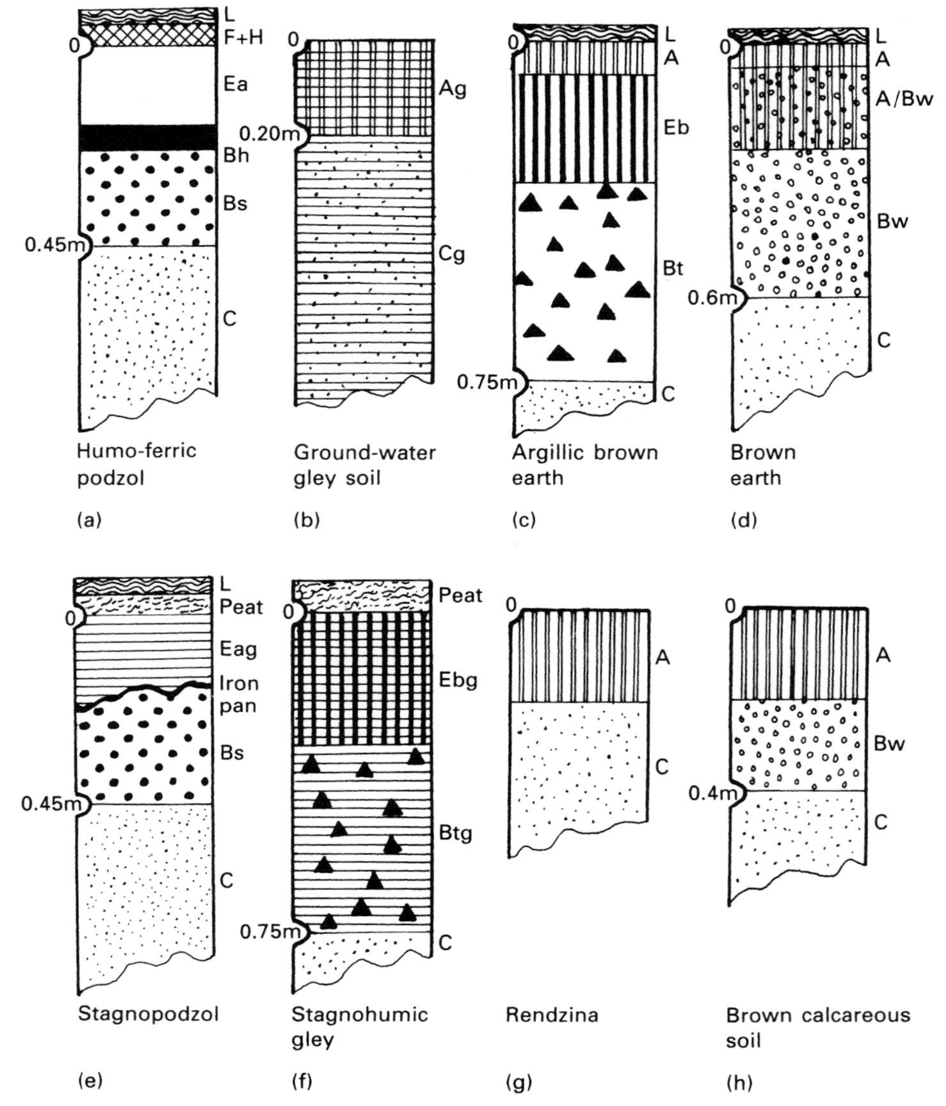


Figure 2.2 Profiles of common British soil types. The organic O horizon can be divided into three distinct horizons as shown depending upon the degree of decomposition of the litter and its integration into the soil below. The A and $A_1$ horizons shown in Fig. 2.1 are here further subdivided into those without eluviation (A and Ap – the latter not shown in the figure) and with eluviation (Ea and Eb). Similarly, the B horizon is differentiated into four discrete horizon types. The three humus types forming the O/A horizons (mull, moder and mor) are discussed in this section. **Rankers**, like rendzinas, are lithomorphic soils (that is, having a shallow AC profile) but in this case the soils are non-calcareous and over hard rock. (Redrawn from Burnham and Mackney, 1964. Modified from Packham and Harding, 1982. *Ecology of Woodland Processes*. Edward Arnold.)

**Soil horizon notation**

*Organic surface horizons*

| | |
|---|---|
| L | Plant litter, only slightly comminuted |
| F | Comminuted litter |
| H | Well decomposed humus with little mineral matter |

*Organo-mineral surface horizons*

| | |
|---|---|
| A | Dark brown, mainly mineral layer with humus admixture |
| Ap | Ploughed layer |

*Eluvial horizons that have lost clay and/or iron and aluminium*

| | |
|---|---|
| Ea | Bleached or pale horizon which has lost iron and/or aluminium |
| Eb | Relatively pale brown friable horizon which has lost some clay |

*Illuvial horizons enriched in clay or humus or iron and aluminium*

| | |
|---|---|
| Bt | Horizon enriched in clay |
| Bh | Dark brown or black horizon, enriched in humus |
| Bs | Orange or red-brown horizon, enriched in iron and/or aluminium |
| Bf | Thin iron pan |

*Other subsoil horizons*

| | |
|---|---|
| Bw | Weathered subsoil material, not appreciably enriched in clay, humus or iron, distinguished from overlying and underlying horizons by colour or structure or both |
| C | Little-altered parent material |

*Notes*

| | |
|---|---|
| g | The addition of 'g' denotes mottling or greying thought to be caused by water-logging |
| A/Bw | Indicates transitional horizon |

***Mull*** is a characteristic A horizon, which may be covered by a thin L horizon, but F and H horizons are scanty or absent
***Moder*** characteristically has an H horizon thicker than the L and F combined
***Mor*** (raw humus) has thick L and F layers

Figure 2.2 (cont.)

soils. It is measured on the **pH scale**, pH 7 being chemically neutral. Since soils in humid zones are inherently acidic (due to carbon dioxide dissolving to form carbonic acid, organic acids produced by roots, fungi and decomposition, and the ready uptake by plants of nutrients, most of which are positive ions), an ecologically neutral soil has been regarded by various authors as having an average pH of between 5 and 6.5. Soil pH values normally vary between 4 and 10 with most forest plants growing best between values of 5.5 and 8.5. Strongly acidic soils below pH 4 often have toxic levels of soluble aluminium, iron and manganese and fewer nutrients since these are very soluble and readily leached. In strongly alkaline soils with a high pH, many of the essential nutrients are

**Box 2.1 International equivalent names for soils. Those used in the text are British soil classes; this table will allow conversion to systems more familiar to some readers. (Courtesy of Ian Baillie.)**

| | FAO/ISRIC World Reference Base for Soil Resources | USDA Soil Taxonomy | | | | |
|---|---|---|---|---|---|---|
| British soil class | 1st level (2nd level where necessary) | Order | Connotation of order name | Suborder & Great Group | Connotations of lower level names | Notes |
| Humo-ferric podzol | Podzol | Spodosol | **POD**zol | Orthod | Orthodox podzol | 'Pod' comes from Russian for ash, because of grey horizon |
| Ground-water gley | Gleysol | Inceptisol | **INCEPT**ion stage of soil development | Endoaquept | 'Aqu-' = wet 'Endo-' = at depth | USDA recognizes wetness only at 2nd level |
| | | Entisol | Rec**ENT**ly deposited parent material | Endoaquept | | |
| Argillic brown earth | Luvisol | Alfisol | Has some free **AL** and **Fe** | Udalf | 'Ud-' = humid climate | Especially calcareous brown earths |
| | Acrisol | Ultisol | **ULT**imate stages of leaching | Udult | 'Ud-' = humid climate | For more leached and acid soils |
| Brown earth | Cambisol | Inceptisol | **INCEPT**ion stage of soil development | Eutrudept or Dystrudept | 'Eutr-' = high base status 'Dystr-' = acid | Most British brown earths are in less leached 'Eutr-' groups |
| Rendzina | Leptosol (Rendzic) | Mollisol | Soft, easily worked ('**MOLL**ify') | Rendoll | | 'Rendzina' from Serbo-Croat |

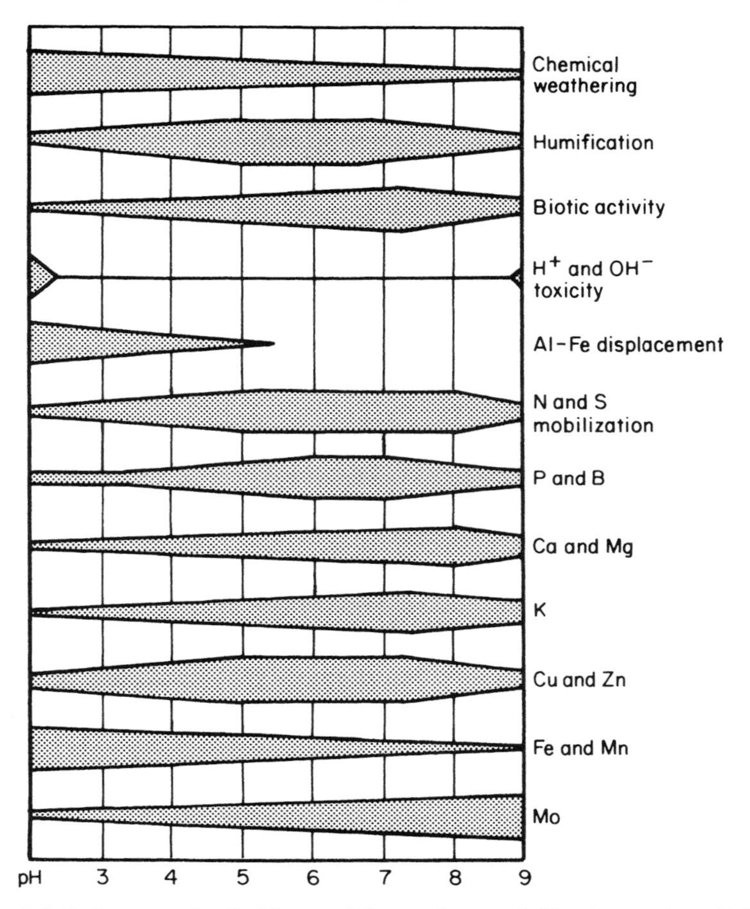


Figure 2.3 Influence of soil pH on soil formation, mobilization and availability of mineral nutrients, and the conditions for life in the soil. The width of the bands indicates the intensity of the process or the availability of the nutrients. Plant needs are divided into macronutrients and micronutrients, a somewhat arbitrary division since all are important but macronutrients are needed in larger amounts. **Macronutrients**: Nitrogen (N), Potassium (K), Calcium (Ca), Magnesium (Mg), Phosphorus (P) and Sulphur (S). **Micronutrients**: Chlorine (Cl), Iron (Fe), Boron (B), Manganese (Mn), Zinc (Zn), Copper (Cu), Molybdenum (Mo) and Nickel (Ni). (After Schroeder 1969. Redrawn from Larcher, 1975. *Physiological Plant Ecology*. Springer.)

insoluble and so unavailable to plants (Fig. 2.3). Neutral to acidic soils, such as the brown earths, often possess **mull humus** (where the humus is mixed with the mineral soil, primarily by soil animals – see Section 2.2.2) and may be quite rich in calcium ions adsorbed by colloids of the clay–humus complex. Calcium carbonate is especially abundant in **rendzinas**, alkaline or nearly neutral soils whose organic A horizon rests directly on calcareous parent rock.

Soil texture is also important in differentiating how different soils behave. The three important size ranges of particles are clay (< 0.002 mm), silt (0.002–0.02 mm) and sand (0.02–2 mm). Different combinations of these three soil components and their interaction with organic material important in their aggregation create the range of soil textures from fine clays to coarse sands. Sand is important for drainage but in terms of effect on plant nutrition, the clay component is the most crucial. Clays are normally 'secondary minerals': that is, formed from minerals dissolved in the soil water. Along with some of the humus, the clays form colloids, suspensions of particles too small to settle out. Both these inorganic clay colloids and the organic humic colloids have a very large surface area and also have a net negative charge on their surface. The negative charge means that cations (positive ions that include many of the essential plant nutrients such as calcium, magnesium and potassium) are absorbed from solution and, as conditions change, can be exchanged for others. The total amount of exchangeable cations that a soil can hold is measured as **cation exchange capacity (CEC)** or **base exchange capacity**. Broadleaved trees normally require soils richer in cations than do conifers. **Base saturation** is a measure of the extent to which soil exchange sites are filled by cations or bases rather than aluminium or hydrogen ions. **Basic soils** are rich in alkaline minerals, especially the cation calcium, and are typically rich in other nutrients as well.

The dominant exchangeable cation in temperate soils is calcium, which causes clay particles to aggregate into stable lumps: Ca clays are flocculated. In contrast sodium (Na) clays, which form when the sea penetrates coastal barriers and in depressions in arid zones, are highly dispersed into very small particles and percolate down the profile in suspension (illuviation). Such a 'clay shift' can also occur under other conditions, especially in neutral or slightly acid soils. However, the presence of free bicarbonate ions ($HCO_3^-$) inhibits such movement in more alkaline soils, while aluminium and dissociating ferric hydroxide ($Fe(OH)_3$) do so in acidic soils.

Soil colour is greatly affected by the presence of **sesquioxides**, the free oxides and hydroxides of iron, aluminium and manganese, which tend to coat sand particles. Partial or occasional waterlogging can dissolve these oxides, however, leading to an iron pan formation lower in the soil profile (see below and Fig. 2.1). Soils of the humid tropics are generally red or yellow because of the oxidation of iron or aluminium, while the combination of organic matter and iron oxides gives the brown colour typical of many temperate soils.

The amount of water available also makes a big impact on forest soils, especially when they are acidic. **Podzols** (meaning ash-like) are base-deficient, strongly acid and often support conifers. They develop relatively rapidly on acidic sandstones, gravels and other permeable parent materials when substantial amounts of water are available for leaching. Soils developing over acidic rocks have few earthworms (mostly absent in temperate soils below pH 5.0) and so develop a thick layer of relatively undecomposed mor humus on top of the soil (see Section 2.2.2). This is reinforced by the nutrient-poor, decomposition-resistant litter of the northern conifers and other evergreens typically associated with this type of soil in a circumpolar belt. Note though that if climatic conditions are wet enough with sufficient downward leaching, a peaty layer of mor humus will even form on boulder clay and other substrata. Water passing through the surface humus becomes so acidic with dissolved humic and carbonic acids that it can strip the iron and aluminium oxides from the underlying mineral particles, leaving a layer of quartz sand grains bleached to an ashy grey colour (an eluvial Ea layer). Quartz grains remain, often accompanied by some secondary silica. As the influence of mor humus increases and leaching becomes stronger the Si/(Al + Fe) ratio rises. As conditions change further down in the soil, the mobilized oxides and illuviated humus are deposited to create an orange or red-brown tiger-stripe (Bs) horizon or even a hard humus or iron pan (Bh) typical of the **humus-iron podzols** ((a) in Fig. 2.2). These pans can be impermeable enough to hold a **perched water table**, leading to waterlogging of the surface.

Long periods of waterlogging cause soils to become anaerobic and lose the bright red and ochre colours caused by the presence of ferric iron. Iron and manganese are reduced to ferrous and manganous forms, which are soluble and more mobile. If such **gleyed soils**, which are very common in low-lying areas and valleys, undergo seasonal drying, re-oxidation causes black manganese concretions and the yellow and rusty spots associated with ferric iron. Gley (g) horizons are often neutral or mildly acid, besides being normally deficient in phosphate. **Stagnopodzols** (peaty-gleyed podzols, (e) in Fig. 2.2), which possess a thin dense iron pan and a layer of peat, are very common in upland Britain and similar wet environments.

In addition to water, climate as a whole makes a big difference to soils. The British Isles exemplify this nicely. There is a strong correlation between the soil regions of Britain and the five climatic regimes defined by Burnham (1970) on the basis of mean annual precipitation (over or under 1000 mm), and mean annual temperature. Very cold was taken as under 4 °C, cold as 4.0–8.3 °C, and warm as above 8.3 °C. Clay eluviation under the warm, dry regime results in **argillic (leached) brown earths** ((c) in Fig. 2.2) that have previously been used for agriculture rather than forestry. Attitudes are changing and we may well in future establish plantations on the better soils. This eluviation process is initiated by the summer dryness so common in south and east England. Early winter rains bring clay particles near the surface of the soil profile into suspension. These are washed into the subsoil, in a mechanical eluviation process called **argilluviation** (arg = clay, luviation = washing), and the water is drawn into the dry lumps of soil (**peds**) leaving clay films on their external surfaces in a clay-enriched horizon.

Under a cold, dry regime (found, for example, in parts of north and east Scotland, the North York Moors, and areas on the eastern slopes of the Pennines) rainfall and potential evaporation are both low so what water is available tends to leach down through the soil. Parent materials in these areas are of low base content, so most soils are acidic, have mor humus and podzolization is widespread. On lower ground where the organic matter is more decomposed and integrated into the A horizon (mull humus or a less integrated transitional moder) soils tend to be less acidic and podzolization is not as strong. These sites, which originally would have supported oakwood, show intergrades (**semi-podzols = brown podzolic soils**) between brown earths and podzols. Heathlands now occupy most of the podzols on higher ground but, before the combined influence of climatic change and human interference, pinewoods were frequent here.

Under the warm, wet regime of south-west England, most of Wales and coastal areas in north-west England and south-west Scotland, there is a high degree of leaching but the warmth leads to more biological activity. Over acidic rocks weak podzolization can occur in places. Mostly, however, the litter dropped by plants is nutrient rich and easily decomposed, and the active soil fauna, especially earthworms, readily mix the humus into the mineral soil to create a neutral to moderately acidic (pH 4.5–6.5) **brown earth** ((d) in Fig. 2.2). Until cleared for cultivation almost all the area under this climatic regime was covered by deciduous woodland. Trees, especially of Sitka spruce *Picea sitchensis*, are planted quite extensively on some of the better **acid brown earths** and forest yields are high, although some of the more podzolized areas often need deep drainage ditches. The acidification of brown earths is helped by leaching by rain leading to a lower pH and calcium content at the surface than further down the profile. This **surface acidification** is particularly obvious where parent materials are uniform as in Hayley Wood, Cambridgeshire, whose soils are derived from chalky boulder chalk with a small admixture with loess (Rackham, 1975).

Areas with a cold, wet regime are covered by **stagnopodzols**. Here a deep layer of undecomposed litter builds up leading to intense podzolization and often waterlogging. The humus layer, aided by wet conditions, may become thick enough to be called peat. Such soils coated with peat on all but the steep slopes are classified as **blanket peats** characteristic of a **cold, wet regime**. Most of these areas are too exposed for tree growth and forestry. More favourably sheltered areas often require ploughing to mix in the surface peat layers and improve drainage before trees can be planted. Leaching is strong and weathering weak; and so the recycling of nutrients is slow. Fertilizer additions, especially of phosphates, considerably improve tree growth. Sitka spruce, sometimes planted with lodgepole pine *Pinus contorta*, grows well under these damp conditions, and is very successful in less exposed parts of this region. In contrast, the alpine humus soils developed under **very cold, wet conditions** are near or above the natural treeline in Britain and have never been forested.

The British Isles are relatively small and so do not show the even greater diversity of climate associated with such areas as North America or Eurasia. Here variation in rainfall, potential evaporation, irradiation, temperature and length of growing season lead to major differences in the soil types and vegetation present. The most striking of these different climates are those of the wet tropics.

Around one-third of tropical rain forests are in the **seasonal moist or humid tropics** and have some period when soil processes and plant growth are affected by moisture shortage (see Section 1.6.1). By contrast, soil processes in the **aseasonal or perhumid tropics**, which are characterized by high even temperatures and abundant rainfall year round, proceed rapidly and without prolonged and regular interruptions. These processes include the leaching of solutes, the weathering of primary rock minerals, the production and weathering of secondary clay minerals, and many rapid biological processes such as the comminution (breaking up) and chemical decomposition of organic matter. They produce what are commonly called **oxisols**. In the last few decades the sheer heterogeneity of these soils has gradually been appreciated. Unfortunately a generally agreed scheme of classification has not been produced but an idea of their diversity is given in Box 2.2; Baillie (1996) should be consulted for more detail on this complex and developing subject. Large areas of the tropics have geologically and tectonically stable land surfaces bearing old and intensively weathered soils. Typically these ancient soils are visually striking: deep and rather uniform and with a topsoil that has little organic darkening, but a bright reddish or yellowish subsoil. Few original minerals remain near the surface, while the well-weathered clay minerals include substantial amounts of sesquioxides, giving the soils their characteristic bright colours. The main group of clay minerals are the kaolinites. These are chemically and physically less active than other clay minerals, having low electrical charges on their surfaces and so a low cation exchange capacity. This is why these soils are generally very nutrient poor and have been referred to as 'wet or green deserts' or even a 'counterfeit paradise', a phrase coined by

**Box 2.2 Soils of the perhumid tropics**

**Kaolisols**; this group includes most intensively weathered and leached zonal humid tropical soils and roughly corresponds to the 'tropical red earths' of earlier descriptions.

*Modal kaolisols, clay kaolisols, basic oxidic clays, ultrabasic oxidic clays, limestone oxidic clays, alluvial kaolisols, ferricrete kaolisols.*

**Non-kaolisol mature terra firma soils**; all have low or very low nutrient contents.

*Podzols* develop from very sandy and quartzose raw materials including highly siliceous old volcanic deposits.

*White sands* have a thick mor layer of surface litter with deep bleached sandy subsoils, but lack an underlying humus pan.

*Acid planosols* are locally important in humid savanna; their subsoils are very compact and impermeable with little penetration by roots or water. Such soils have a sharp discontinuity between the pale, loose sandy upper horizons and the brightly mottled subsoils.

**Immature terra firma soils**; which have arisen on recent parent materials or on older deposits which have been recently truncated by erosion.

*Andosols, recent colluvial soils, recent alluvial soils, skeletal soils, shallow calcareous clays.*

**Poorly drained soils**; where water tables are high for at least part of the year and impede drainage. This results in restricted leaching, poor aeration, and the reduction of free iron from the trivalent to the divalent state. The reduced rate of decomposition of organic matter in these soils reaches its extreme with the formation of peat.

*Freshwater gleys, ferruginous semi-gleys, saline gleys, acid sulphate soils, peats.*

**Montane soils** (not subdivided)

(This simplified listing is based on table 10.1, in Baillie (1996). This also gives the equivalents in international soil classifications of the terms used above.)

Betty J. Meggers, the Smithsonian archaeologist (Mann, 2002), implying that the lush forests grow on impoverished soils – but see the *terra preta* soils discussed in Section 2.5.1. Aluminium is often the dominant exchangeable cation in these acid soils. They have open porous structures; aeration is therefore generally good, and water infiltrates the surface rapidly and readily percolates downwards.

There are, however, many important tropical soils that are not acid, although highly weathered, deep, reddish, and well drained, as Box 2.2 indicates. Amongst these are the young soils which develop on **alluvium** (unconsolidated granular sediments deposited by rivers), and the **colluvial** soils that originate at the surface of recent hillwash deposits on lower slopes.

### *2.2.2 The nature of forest soils and their influence on the ground flora*

Traditionally, soils of high quality have been used for agriculture, so those beneath long-established woodlands were usually not only of lower original quality, but have not been limed, fertilized and drained. Such human manipulations have sometimes influenced the soils of modern plantations to such a degree that this causes difficulties when attempts are made to introduce characteristic woodland floras (Section 11.6.2). Conversely, in many parts of the world abandoned agricultural land has reverted to forest and after a century or so it can superficially look like original untouched forest. However, the effects of agricultural practices such as ploughing on the soil can be long-lasting. For example, a study at Harvard Forest in central Massachusetts, USA (Motzkin *et al.*, 1999) found that farm land abandoned a century or more before still had clear evidence of a ploughed surface (Ap) horizon. Moreover, composition of the forest (from trees to bryophytes) was significantly affected by this historical land use, despite the length of time since it was last ploughed and intervening disturbances such as extensive damage by a hurricane in 1938, fire, the ravages of chestnut blight (Section 5.4.6) and widespread tree felling and planting. Former ploughed land is associated with such plants as white pine *Pinus strobus*, red oak *Quercus rubra* and red maple *Acer rubrum*, and undisturbed sites with eastern hemlock *Tsuga canadensis* and American witch hazel *Hamamelis virginiana*. Although agriculture intensely affects the surface of the soil, woodland soils have been affected to a far greater depth by plant growth; though most tree roots are less than 0.3 m deep, they can penetrate to 3 m (see Sections 1.3.2 and 2.3).

The impression given so far in this chapter may be that forest soils are fairly uniform over large areas, but this is not always so. In eastern England, **drift deposits** of glacial till are a mosaic of boulder clay, sands and gravels, from 1–70 m thick, laid down over a range of bedrocks. The drift is very variable; boulder clay is usually heavy clay with lumps of chalk, yet woodland soils on it include some of the most acid soils in Britain and are often sandy or silty. This is because much of the till or solid geology is overlain by wind-blown deposits, up to a metre thick, of silt and fine sand, which are almost certainly loess deposits blown in from eastern and central Europe during great dust storms in late glacial times. Again, ploughing has played a role with extensive mixing of such material into agricultural soils, so that its influence is diluted. Besides this homogenization, a combination of ploughing and sheet erosion tends to leave a low cliff at the downhill edge of ancient woodland in the UK.

Soils of old forests are more layered (have clearer horizons) than those modified by agriculture, but they may be disturbed when old trees blow over, pulling their root plates out of the ground often with large amounts of soil attached. In other cases, trunks fall after the major roots have decayed and the soil is not greatly disturbed, or unfelled trees merely rot while still standing, each simply leaving a stump and fragmented debris. But even these have their effect on shaping the soil: the vertical and horizontal channels created by rotting roots act as easy conduits for animals and water. Rooting animals, particularly wild pigs, often cause superficial disturbance, but when grazing is heavy or many walkers use an area (see Section 11.5) the soil becomes compacted. This reduces pore space in the topsoil and increases surface water runoff. The opposite effect is caused by earthworms, moles and other animals which live in the soil and help integrate organic matter and to mix the soil horizons.

**Humus** type is a cause of diversity both below and above ground (Ponge, 2003), besides being most important in determining plant growth. As mentioned above, acidic soils (such as podzols supporting conifers, or even acid brown earths in temperate forests beneath oak and beech) have few earthworms or other large animals and so litter accumulates as a thick organic deposit, in which L (litter), F (comminuted i.e. fragmented litter) and H (well decomposed amorphous humus with little mineral matter) layers can often be distinguished (Fig. 2.2a), sitting discretely on top of the mineral soil. The L layer may be a few centimetres thick but it is the F layer rich in fine roots and fungal growth and the H layer that make up the bulk of the organic deposit. As discussed in Chapter 7, the hard nutrient-poor litter typical of plants growing on these soils also contributes to organic build-up. The **raw humus** or **mor** of such systems is of low fertility and possesses little soil nitrogen; its C/N ratio is often above 20 and even 30–40. Endomycorrhizas are common (see Section 5.4.1), probably because the fungi involved are known for their efficient use of organic nitrogen. **Mull humus**, in contrast, has passed at least once through the gut of one of the larger soil animals, usually an earthworm's, to form stable mineral–organic soil crumbs with virtually no litter (L) layer. It develops under deciduous or mixed forests on moderately well-drained soils with adequate calcium and a high pH, and also in forests of cedars and of those spruces whose litter has a high calcium content. The C/N ratio of mull is usually about 10. **Moder humus** is intermediate or transitional between mor and mull, characteristic of fairly acidic soils often under trees with hard litter such as beech or oak. Here, earthworms are usually absent and the litter is mainly decomposed by fungi and arthropods (ants, mites, springtails, millipedes, woodlice/sowbugs, etc.), leaving a mix of plant fragments and mineral particles held together by arthropod faecal material. This gives a loose, crumbly texture to the humus.

Trees themselves have a direct influence on the nature of soil humus. In some places in the UK the soils of birch plantations have mull humus, while soils beneath adjacent Scots pine on the same parent materials undergo podzolization and mor humus is formed. Even trees in the same genus vary in their influence on soil pH. Topsoil beneath eastern red cedar *Juniperus virginiana* has a raised pH, while that beneath common juniper *J. communis* in the same Connecticut old fields is reduced by the low base status and acid nature of its leaf litter (Spurr and Barnes, 1980). Soil pH of the root zone of the eastern red cedar is lowered, but considerable amounts of calcium and other bases absorbed by its roots are returned to the soil in the leaf litter. A similar effect of leaf litter has been seen in woodlands over the acidic Coal Measures near Sheffield, UK. Mean pH of the surface soil (0–3 cm) in 80 samples taken was higher than that at 9–12 cm (Packham and Willis, 1976). Bases contributed by leaf litter are again likely to have caused this effect.

Specific examples of how the underlying geology and soil type influences forest development and the ground flora present can be found around the world. In some cases the link is very visible. For example, in Swaziland, southern Africa, sandstones with sandy topsoils and tough clay pans are widespread, but areas of igneous rock occur which carry a brown, clay-rich soil. The dominant clay pan soils support a savanna of thorny *Acacia* spp. which are largely absent from the clay soils, being replaced by a woodland savanna containing the marula plum *Sclerocarya birrea*, which is noted for the intoxicating effect of its fermenting fruits on the local elephants!

Usually, however, the picture is more complex. The Ercall is a prominent hill in Shropshire, UK made up of a variety of rocks which, with variations in slope and water regime, result in a complex set of woodlands within an area of about 50 ha. The woodlands are largely of oaks but with differences across the site (Fig. 2.4). The north-west side of the hill is composed of acidic igneous rocks with a podzol supporting oak woodland (pedunculate oak *Quercus robur* and sessile oak *Q. petraea*) with an acidic field layer including heather, bilberry and wavy hair grass (A in Fig. 2.4). On the south-east slope, sedimentary quartzite again supports a podzol but better drainage and a warmer (and hence drier) slope results in an open woodland dominated by birch with a dense covering of bracken beneath ($B_1$ and $B_2$). The eastern flatter slopes are on siltstones and shales and are damper and richer in nutrients giving a good brown earth soil (C). These support a rich oak woodland with a variety of other trees and a diverse ground flora. In some places boulder clay creates an argillic (clay-enriched) brown earth with a more neutral pH and greater cation exchange capacity (see above) holding an abundance of plant nutrients (D). Not surprisingly, here is an even richer mixed woodland, with diverse

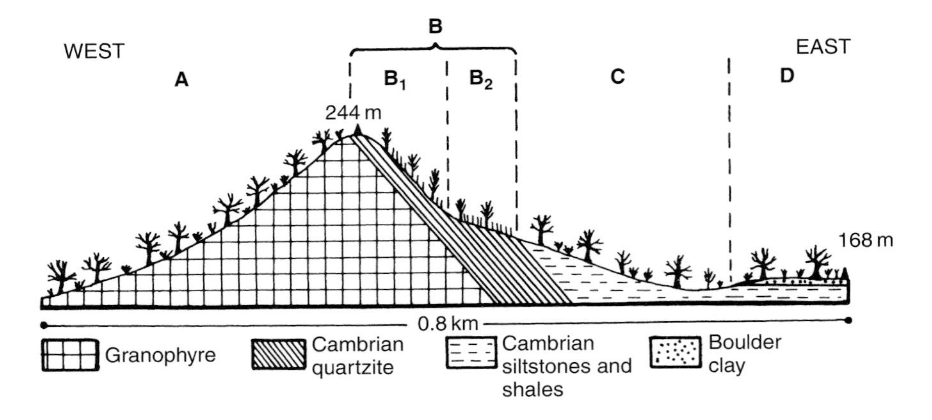


| | Soils | Vegetation | pH | |
|---|---|---|---|---|
| | | | Topsoil | Subsoil |
| A | Podzols and brown podzolic soils | Oak-birch coppice; heather, bilberry, bracken or wavy hair grass locally dominant. The mosses *Leucobryum glaucum* and *Plagiothecium undulatum* are conspicuous | 3.9 | 4.5 |
| $B_1$ | Humo-ferric podzols and podzolic rankers | Bracken, some birch scrub and heather | 3.7 | 4.4 |
| $B_2$ | Gley-podzols | Bracken, some birch and bluebell | 3.8 | 4.6 |
| C | Typical brown earths | Oak-birch coppice, with rowan and holly, very variable field layer including male fern, ivy, honeysuckle, bluebell, bramble, creeping soft-grass, wood sorrel, yellow archangel | 4.1 | 4.3 |
| D | Gleyic argillic brown earths | Mixed deciduous woodland with ash, birch, wych elm and alder, with hazel, hawthorn, oak and ash regenerating; field layer includes dog's mercury, enchanter's nightshade, male fern, tufted hair grass, wood sanicle | 5.2 | 5.7 |

Figure 2.4 Relationships between underlying rock, soils and vegetation on The Ercall, Shropshire, UK. Leaching has decreased the pH of the topsoils. (Redrawn from Burnham and Mackney, 1964. From Packham and Harding, 1982. *Ecology of Woodland Processes*. Edward Arnold.)

tree, field and ground layers, including the calcium-loving wood sanicle *Sanicula europaea*. Where a stream passes through the site, the extra moisture and alluvial deposits result in a wet woodland with alder, birch and hazel; a $5 \times 5$ m area looked at contained 27 species of vascular plants and 9 species of mosses and liverworts (bryophytes) with many more species within the immediate area (Burnham and Mackney, 1964).

Sometimes the effects of soils on vegetation may not be visually obvious but are nevertheless no less important. The mixed dipterocarp forests in Sarawak, Malaysian Borneo, are among the most species-rich tropical forests. A long-standing problem is to try to understand how so many different species can co-exist without one or two eventually coming to dominate. Part of the answer, it seems, is to do with soils. Potts *et al.* (2002) looked at 105 different 0.6 ha plots in an area approximately 500 × 150 km and found 1762 different tree species. Sixty species were judged common as they were found amongst the ten most abundant species in at least five of the plots. Of these, 23 species were confined to sandstone soils with a surface layer of humus and another 20 were restricted to clay-rich soils over shale with no humus layer. The other 17 were not exacting as to which soil they grew on. These patterns are not easy to see on the ground because of the bewildering variety of trees but differences in soils do seem to partly explain tree distribution.

### 2.2.3 Influence of trees on the degradation of forest soils

**Soil degradation** can be a particular problem with plantation forestry and is often suspected if the growth of a subsequent similar crop is inferior to that of the original crop. The physical, biological and chemical properties of the soil, including its structure, level of organic matter, presence of toxins, and the distribution and availability of nutrients may all be involved. In forestry, gross problems are overcome by using artificial fertilizers and although in small quantities compared with agriculture and mainly involving potassium, they can still be a matter for concern, which is discussed in Section 8.6.2. However, there is evidence that not all plantations are detrimental to soils. In New Zealand, some areas have now had three cycles of Monterey (radiata) pine *Pinus radiata*. Maclaren (1996) concludes that there has been no decrease in productivity in subsequent rotations of planted forests and, indeed, Woollons (2000) found enhanced growth in the second rotation, though primarily because of better establishment standards and more favourable climatic conditions rather than any soil amelioration. Moreover, a global study by Evans (1999) produced similar conclusions, noting that where lower yields were found it appeared that poor silvicultural practices were responsible.

Certainly, trees can improve soils in some cases. A matter of some interest is the changed availability of nutrients when pasture is afforested. In the case of pine forests it seems that availability of nitrogen, phosphorus and sulphur in the top 10 cm of soil is usually enhanced, perhaps due to the release of enzymes by fungi associated with tree roots. In contrast, surface levels of available calcium, magnesium and potassium may decline, especially in the period before canopy closure. The implication is that there should be an improved growth of pasture species if pine forest is converted back to pasture. This was supported by a trial in which the yield of grass and clover grown on pasture soil was far less than that on soil from beneath adjacent pines. Ovington (1958) found that hardwood and coniferous trees both produce a similar reduction in soil calcium levels. Certain species, including birches, hornbeam *Carpinus betulus*, sweet chestnut *Castanea sativa* and larch *Larix* spp. have reputations as 'soil improvers', and earthworm numbers increase considerably as birch crops age.

Soil acidity increases if land is planted with pines, but this is not necessarily a direct cause for concern as most forest trees are more successful on acid soil. Work done on reclamation of china clay waste, development of vegetation on glacial moraines, marine and fresh water sand dunes, etc. has shown that decline in pH is a consistent phenomenon. The main problem is that of podzolization. For example, in New Zealand, earthworms (which are all introduced) drop in number to a very low level when pastures are densely planted with Monterey pine. This is probably linked to increased acidity and decreased calcium levels. Inevitably this leads to a build-up of mor humus and a locking-up of nutrients. Though improved harvesting methods and use of new machinery have lessened their impact, **soil compaction**, which commonly occurs when stands are harvested, reduces productivity as does erosion.

In the hot, wet climates of the tropics, intense leaching and weathering change iron, manganese and aluminium minerals to insoluble compounds, often near the surface, while silica and other minerals are carried downwards (Baillie, 1996). In consequence the effects on drainage and soil nutrient status caused by the removal of teak and other tropical forests are often very adverse.

### *2.2.4 Trees and erosion*

There is a mutual interaction between soils and the trees which root in them. Soils provide physical stability to the trees and enable them to absorb water and nutrients. Trees reduce soil erosion, and the decomposition of their dead parts and litter leads to forest renewal, as will be seen in Chapter 7. Soil erosion is particularly severe in areas with steep slopes, as in the Mediterranean island of Crete, where little of the original cypress and native pine forests now remain. It is also a major problem in New Zealand where it is again largely associated with land clearance, as it is in many other areas with hilly and mountainous countryside. A considerable proportion of New Zealand's pasture land is unsustainable in its present use; even when vegetated the roots of pasture species do not bind the soil strongly enough to prevent loss by surface wash and mass movement from water running over the surface, and fluvial erosion (i.e. by rivers). Erosion in exposed areas of tropical rain forests can be even more severe, sometimes leading to gullies more than 10 m deep along the lines of logging tracks.

Trees do much to reduce erosion; their roots dry out the soil due to the high water demands of the canopy and bind it far more effectively than pasture species. Hubbard Brook data suggest a 41% increase in water running from a catchment after clear felling, due to loss of the evapotranspiration component (Hornbeck *et al.*, 1997). Soil beneath a dense tree cover will usually be drier than one where it is less dense, so the nature of the canopy can be critical where a major object of planting is to reduce erosion. There is little point in planting deciduous poplars if canopy interception of precipitation during the winter is critically important. If root strength is an important factor, a choice of Douglas fir, poplar or manuka *Leptospermum scoparium* (a 'tree' up to 4 m high native to New Zealand) will give roots twice as strong as those of Monterey pine. Roots can reinforce slopes in three main ways. Situations often occur in which vertical 'sinker' roots pin unstable soil horizons to the underlying substrate, but in some cases the 'failure plane' is deeper than the deepest roots. Even then afforestation can reduce soil flow rates by a factor of 10–30. This is explained by the 'rafting hypothesis': a shallow but semi-rigid raft of laterally connected root-plates (tree roots and associated soil) literally floats on the unstable underlying substrate and prevents it moving. In the third mechanism the soil mantle upslope from the tree is buttressed by the stump and larger roots. This will be effective only if sufficient trees are present and their sinker roots penetrate the stable bedrock (Maclaren, 1996).

Interaction of trees with the **water cycle** (see Section 3.3.1) also reduces erosion and the incidence of flooding. The canopies of trees and shrubs normally intercept 20–40% of rain which evaporates without reaching the ground and so reduces the volume of water available for runoff and erosion. Tree canopies can also be moderately effective in protecting the soil surface from **splash erosion**, as they intercept incoming rainfall and absorb much of its kinetic energy before it hits the ground. Similarly, a normally small proportion of the rainfall will reach the ground as stemflow, further reducing splash erosion – see Section 3.3 for more details. However, this softening effect is sometimes exaggerated and the reality is that throughfall drops are larger than raindrops and are potentially more erosive. Fortunately, overriding all this in importance is the surface litter which protects the soil surface from drop impact whatever its origin. A good layer of litter does much to reduce splash erosion.

## 2.3 Roots, foraging and competition

### *2.3.1 Variation in root systems*

The radicle (embryonic root) of a seedling often continues to grow downwards as a primary or **tap root**, but in most trees the main root system consists of secondary **lateral roots** which grow sideways away from the stem, usually no more than 1–2 m below the soil surface but with some deeper roots. Different species have different rooting patterns and in this respect three common European conifers differ greatly in their ability to adapt to varied soil conditions. Young silver fir *Abies alba* has the least branched roots, and the adult has a deep root system with a dominant tap that does not adapt well to shallow soils. In its first year Scots pine *Pinus sylvestris* has the longest primary and the largest number of spreading secondary and tertiary roots. Its many small roots enable it to flourish in dry barren soils, yet the great plasticity of the main root system, which can be reliant on long laterals but whose tap root can penetrate very deeply in suitable soils, allows this tree to adapt to diverse habitats from dry sand to wet peat. The third species, Norway spruce *Picea abies* has a mature root system composed of shallow laterals and its primary root stops growing after 5 years; consequently this tree is commonly uprooted by wind (Fig. 2.5) but with its widespread root system it does well on a wide range of soils. The rooting patterns of different species will vary depending upon what other trees are growing near: for example a number of studies in Germany have shown that when Norway spruce or ash *Fraxinus excelsior* are growing with beech *Fagus sylvatica*, the spruce and ash tend to produce their fine roots more shallowly than when grown alone, and the beech produce fine roots more deeply.

In most forests, the canopy of each tree is discrete from surrounding trees. This 'canopy shyness', first mentioned in Chapter 1, is caused primarily by physical interference of approaching branches as they knock into each other when blowing in the wind. In the relative stillness below ground, however, roots are not contained in the same way and continue to grow past each other. In temperate trees the spread of roots away from the trunk is around 2–3 times the width of the canopy, and up to four times on dry sandy soil. This wide spreading of the roots (with a radius of 30 m or more in the case of an oak) is what allows the tree to absorb enough water and nutrients to keep its huge structure functioning. However, the strategy for optimizing uptake of these essential components varies considerably between different trees. The ash has very long branched lateral roots that exploit a large volume of soil. Trees like this use their long coarse roots to extend great distances into large volumes of soil, and have **extensive root**

Figure 2.5 Root-plate of Norway spruce *Picea abies* blown out of the peat at Sweat Mere, Shropshire, UK. (Photograph by John R. Packham.)

**systems**. These root patterns work well in winter-rain regions where tree roots, unlike those of grasses, can draw water from great depths in dry summers; they are also very effective in stony soils whose water is not uniformly distributed. Trees with **intensive root systems** rely on the very effective use of the water in a much smaller volume of soil; as exemplified by the beech *Fagus sylvatica* with its shorter laterals with numerous short and extremely fine terminals. Though both these systems function well under normal conditions, beech suffers more than most trees in drought and in southern England its stands were badly affected by the very dry summer of 1976. This suggests that having absorbed all the water near it, the tree does not extend its laterals toward remaining damp soil quickly enough. Beech does have a couple of insurance policies. On easily penetrated rocks, such as fissured chalk, roots can grow many metres down and extract water from nearer the water table. Also beech can grow **internal roots** within damp, decayed and hollow trunks allowing extra water absorption. A very large internal root system was present in a tree that blew down in the Wyre Forest in central England some years ago. These are examples of **adventitious roots**, roots which arise directly from pre-existing stems. Internal roots are also found in a variety of other temperate trees such as elms, yews and large trees of yellow birch *Betula lutea* in humid North

America but especially in tropical trees. Undoubtedly these roots give the trees a competitive advantage by allowing extra water uptake and recycling of nutrients from the otherwise not very useful body of dead heartwood.

Adventitious roots above ground have several other uses in forest trees. A number of rain-forest trees, both tropical and temperate, produce **canopy roots** from the trunk and branches which grow down the outside of the tree and exploit humus pockets collected by the numerous epiphytic plants. These roots can also reabsorb nutrients that have been leached out of foliage higher up and washed down the tree. More widespread around the world, some trees and shrubs develop adventitious roots on stems touching the soil and this can result in vegetative reproduction. This **layering** often gives rise to plants that eventually become functionally – and often physically – independent of the parent. An extreme case of this is seen in the layered beech at Arley Arboreum, Worcestershire, England where an erect tree planted at least two centuries ago has now completely disappeared, but its layered remains form a contorted series of vigorous trunks and branches that now cover around a quarter of an acre (0.1 ha). Roots above soil level are, of course, used by a variety of epiphytes (see Section 5.5.1). Aerial roots of epiphytic orchids possess an outer covering (**velamen**) of dead cells that absorbs water from the atmosphere; these roots can also carry out photosynthesis. Northern rata *Metrosideros robusta* trees in North Island, New Zealand, start their life as epiphytes, grow to the ground, and eventually outlive and replace the host trees.

A consequence of tree roots spreading so far is that in a forest they inevitably intermingle. For example, in the mixed hardwoods of Harvard Forest, Massachusetts, Lyford and Wilson (1964) found the roots of 4–7 trees below the same square metre of ground and mostly close together. Where these roots touch, they are likely to fuse together. These **root grafts** commonly weld the lateral roots of an individual tree into a rigid root plate which helps hold the tree up. But equally, such grafts can join together the below-ground systems of adjacent trees of the same species and, more rarely, different species. These are sometimes responsible for the spread of disease (such as Dutch elm disease – see Section 5.4.5), while the remaining connected tree will sometimes take over the root system of a neighbour whose trunk has been cut down. A thinned conifer plantation south of Exeter in south-west England came to contain a healed stump devoid of leaves, which contined to live and to increase slightly in girth. Such stumps are clearly receiving photosynthate from neighbours to which their roots are joined. Cases like this also occur in Douglar fir plantations in New Zealand. Such stumps have been kept alive via root grafts for several decades. It is interesting here that rather than competition allowing the stronger individual to take from the weakened stump, the exchange is from the dominant individuals to the underdog, but probably not enough to disrupt the overall dominance pattern. Cut-down hardwoods would of course recover by coppice growth but may be equally helped by surrounding uncut trees.

Adding all these factors together can result in impressive interactions between trees and soil. Lateral pressures exerted by hurricanes and other strong winds on tall trees are very great and so their roots have to be correspondingly strong. Although the coastal redwood *Sequoia sempervirens* can reach a height of 115 m and the giant sequoia (big tree) *Sequoiadendron giganteum* 95 m, these closely related species native to California have shallow root systems that rarely go deeper than 2 m from the surface. Despite this, available soil resources are effectively exploited by roots which often spread tens of metres from the trunk. An undisturbed layer of thick damp mulch on the forest floor is beneficial to the health of both these trees that take their mineral nutrients and 85–90% of their water requirement from the soil, the remaining water being absorbed directly from the atmosphere by the leaves.

### *2.3.2 Root competition and specialist adaptations*

The roots of individual plants compete with those of the same or other species when **foraging** for water and **essential nutrients**. The stratification shown by the aerial parts of forest plants is paralleled by the **vertical zonation** of their roots as described in Chapter 1, those of the largest plants usually penetrating to the greatest depths. Vascular plants use their roots to forage for water and mineral nutrients and also to anchor them in the ground. Trees tend to split this function: the shallower feeder roots absorb most of their mineral nutrients, often assisted by mycorrhizal associations (see Chapter 5), while the deeper roots take up water and anchor the trunks. This vertical layering is made use of in agroforestry where trees and crops are grown together with the assumption that the crops are using water near the soil surface and the trees are using deeper water, maybe 20–30 cm down. However, this assumption has never been fully proved and inevitably there is still some competition between the crops and trees; crop yield is normally lower near the trees. Although **root layering** may diminish competition between herbs and trees it certainly does not eliminate it, so competition can be very important in the success of establishing young trees. Bare areas in many young beechwoods result from the permeation of the ground by their roots as well as the heavy shading of an almost unbroken canopy: seedlings face stiff competition for resources both above and below ground. The **mosaic** of plants on the forest floor is strongly influenced by the availability of light in space and time, but variations in humus and nutrient contents, soil pH, aeration and moisture also influence the distribution of shrubs, herbs and bryophytes, which can often be used as **indicators** of environmental conditions.

Modern studies have placed an increasing emphasis on the importance of below-ground competition. Working with seven herbaceous understorey species from an unproductive old field in Michigan, Rajaniemi *et al.* (2003) found that increased productivity resulting from fertilizer application was associated with a loss in plant diversity. These changes were caused almost exclusively by root competition; above-ground competition from more vigorous growth had small effects on the community but did not contribute to changes in diversity. This has important implications for forests; indeed in Britain NPK fertilizer drift from fields caused by the wind has had an adverse effect on many woodland margins, whose soils now support larger populations of stinging nettle *Urtica dioica*, a strongly competitive species, and have rather lower biodiversity, than they did a few decades ago.

An important aspect of root competition is the ability of a root to determine whether another root is from the plant to which itself belongs: **self/non-self root discrimination** as Falik *et al.* (2003) put it. They used experimental pea *Pisum sativum* plants that had been pruned to leave just two roots. These were planted in small pots such that a pot contained two roots either of the same plant or of two different plants. When such a plant was put in a pot by itself, root development was significantly less than when the two roots within a pot came from two physiologically different plants. These results point to an ability to respond to competition from roots of another plant (by growing more vigorously towards the offending roots in preferentially exploiting the soil) while avoiding it between its own roots; and, since the roots respond to each other without actually touching, to the production of a signal of some kind that allows this discrimination. Evidence suggests that this also happens in a range of other plants and it will be interesting to see just how widespread this self recognition is in forest plants.

Gersani *et al.* (2001) developed a game-theoretic model that considered the effects of intra- and inter-plant competition (i.e. within one plant and between plants) on root proliferation and reproductive yield. They predicted that if soil space and resources per individual were held constant, then plants would produce more roots and give less reproductive yield (fewer seeds) per individual as the number of plants sharing the combined space increased. This theory turned out to be compatible with the results obtained when tested using containers of soybean *Glycine max* plants grown both in isolation and together under glasshouse conditions. Even though plants growing together each had the same volume of soil as when growing alone, sharing resulted in more roots being grown. Sharing plants grew 85% more root mass (weight) but since overall plant mass did not change this was at the expense of above-ground growth. Sharing plants also produced 30% less mass of seeds. The implications of different types of root behaviour are important; these authors suggest that the plants appear to invest their resources effectively and make greater root growth when there is competition; different roots and parts of a plant assess and respond to opportunities in a way that maximizes the good of the whole plant. If a plant is operating as a co-ordinated whole it should first produce roots in unoccupied soils, then in soil occupied by a competitor (as found by Falik *et al.*, 2003), and lastly in soil where its own roots are already present.

Narrowly endemic species (i.e. those found only in a particular geographical region) are often restricted to distinctive **edaphic** (soil) environments, as in the case of two species of hakea (*Hakea oldfieldii* and *H. tuberculata*) which occur in winter-wet shrublands growing on skeletal soils 0–20 cm deep overlying massive ironstone rock in Mediterranean south-west Australia. This area is the major centre of diversity for this Proteacean genus, which consists of woody perennials ranging in size from shrubs to small trees. Poot and Lambers (2003) studied these together with five other species of *Hakea* found on more common soils (including two from eucalypt woodland) growing their seedlings in pots 40 cm deep and making harvests at 62, 125 and 188 days. Initially the ironstone endemics allocated a significantly greater proportion of dry mass to their roots, although this difference evened out later. At the last harvest the two rare ironstone endemics had an average of 64% of their roots in the bottom 10 cm of the pots, as opposed to a mean of 35% for the other five species. These two species had considerably greater root lengths for a given plant mass because their average root diameter was less; only in these two species did the main root axis continue to grow at the same rate after reaching the bottom of the pot. Specializations shown by the two ironstone endemics undoubtedly increase the chances of obtaining access to water before the onset of severe summer drought in their native habitats. In more common soils these same traits, which compromise competitive ability both above and below ground, are likely to reduce the chance of survival.

### *2.3.3 Soil and root aeration*

Aeration has an important influence on the chemical status and biota of soils; it also greatly affects the species of vascular plants that can grow on them (Grosse *et al.*, 1998). The diffusion rate of oxygen in water is low (some 10 000 times less than in air) and any oxygen present in flooded soils is rapidly exploited by the bacterial population. Consequently roots growing in frequently or permanently flooded sites often encounter saturated soils in which oxygen partial pressure is zero (anaerobic). Under such conditions root oxygen demand is not met from the **rhizosphere** (the region immediately surrounding the roots – see Section 5.5.1 – which normally provides a protective oxygenated sheath against the toxic effect of anaerobic conditions), and can be satisfied only by an internal supply from a part of the plant exposed to air, akin to using the plant as a snorkel. Adaptations which improve the oxygen supply of tree roots in wetlands include a variety of aerial roots (**pneumatophores**) including the prop or stilt roots of red mangroves *Rhizophora* spp. (see Section 4.2.3) that grow from the trunk down into the mud, the short pencil-shaped peg roots of the black mangroves *Avicennia* spp. which stick up like snorkels, the knees (a loop of root arching into the air) of a range of tropical hardwoods and the temperate bald or swamp cypress *Taxodium distichum*, as well as the plank and buttress roots of other trees and the formation of aerenchyma (air passages through the plant tissue) in submerged plant structures (Crawford, 1990; Thomas, 2000). Adventitious roots, produced near the water's surface, or at the top of saturated soil, are grown by a diverse range of trees from eucalyptus of the southern hemisphere to the tupelos, willows, elms and ashes of the north. These can be used to replace roots that rot when flooded in the growing season.

### *2.3.4 Vascular plants, soil pH, mineral nutrients and microorganisms*

Figure 2.3 emphasizes the important influence of soil pH; chemical weathering, for example, proceeding very much faster in the most acid soils. Humification (the breakdown of dead plant material to humus) and biotic activity in general increase towards the mid-range of pH and then drop away as soils become alkaline. Soil bacteria and fungi react differently to variation in soil pH, though it strongly influences both. The optimum pH for soil bacteria is slightly on the alkaline side of neutrality (Clark, 1967); though they commonly tolerate a range between pH 4 and 10, some species have a relatively narrow tolerance to soil reaction. Soil fungi are better adapted to acid soils than bacteria, and bacteria do poorly on distinctly acid soils even when the competitive rivalry from fungi is restricted experimentally by the use of fungicides. This intolerance of bacteria for acidic conditions causes the decay of organic matter in acid soils to proceed relatively slowly. Moreover, although anaerobic conditions are the main cause of mineralization by bacteria (the conversion of organic compounds to soluble inorganic chemicals – see Chapter 7), very acidic conditions, with abundant positive hydrogen ions, lead to the accumulation of ammonia as the ammonium ion ($NH_4^+$) rather than nitrate nitrogen ($NO_3^-$) simply because the ammonium carries a positive charge. Direct and indirect effects of soil pH on the soil fauna also influence decomposition patterns (see Section 7.2).

The influence of soil pH on soil nutrients varies greatly. As can be seen in Fig. 2.3, the availability of iron and manganese diminishes steadily as pH rises, sometimes to the stage where they limit the growth or performance of particular plants. In contrast, the availability of molybdenum is greater in alkaline soils as is that of most other nutrients until very alkaline conditions are met. Larcher (1975) found the protoplasm of most vascular plant roots to be severely damaged outside the range pH 3–9. Within this range, various plant species show a wide range of tolerance of soil pH, and the resulting changes in soil chemistry, in the same way that they do to drought or waterlogging.

Sir Arthur Tansley observed in the early 1900s that some plants were restricted to calcium-rich soils (**calcicoles** – calcium loving) while others were not (**calcifuges** – calcium fleeing); see Lee (1999) for a detailed review of the subject. The best examples are perhaps seen in grassland species but the principle applies equally to forests on acid and alkaline soils. The causes are very complex and have been traced most significantly to the effects of pH, primarily through the agencies of iron and aluminium rather than calcium itself. On acidic soils both aluminium and iron are highly soluble and potentially toxic even at low concentrations but calcifuge plants have effective **chelating mechanisms** (which combine a metal ion with substances in the root) that prevent harmful levels of absorption. Iron cannot be completely locked up since it is an important nutrient and this inevitably lets some aluminium in despite its lack of a role within the plant (it is interesting to note the paradox that aluminium is the commonest metal ion in soils and the one metal not used by plants). However, on alkaline, calcium-rich soils, similar mechanisms lock up what little iron is available and calcifuges look pale and sickly (chlorotic) due to iron deficiency. Conversely, calcicoles lack such efficient chelating mechanisms and so can scavenge the small amounts of iron available on alkaline, calcium-rich soils; but on acidic soils have no defence against toxic amounts of aluminium. Other factors may complicate the issue in some plants such as tolerance of boron poisoning, which develops in markedly alkaline soils, and excessive amounts of calcium uptake on alkaline soils. Acid secretions by roots into the rhizosphere (Section 5.5.1) may also play a vital role; calcicoles produce an abundance which, although readily broken down by microbes, may mobilize iron, phosphorus (P) and cations and so improve uptake; conversely, calcifuges which produce few acid exudates can be deficient in P on alkaline soils (Zohlen and Tyler, 2004).

The majority of vascular plants are **amphi-tolerant** (amphi – 'both sides'), competing successfully with others in the mid-range of soil pH. They do best, however, when they are growing at the pH that suits their physiology, so slight changes in pH can result in quite different plant communities or even a monoculture field layer. Species unable to compete at mid-range pH are in nature often confined to habitats where competition is less intense. Though creeping soft-grass *Holcus mollis* has a physiological optimum around pH 6, in central European woods it is frequently crowded out by other species (Ellenberg, 1988). In English oakwoods *H. mollis* is unable to compete effectively with wavy hair-grass *Deschampsia flexuosa* below pH 4.0, so it is restricted to a rather narrow soil pH band by competition from other species.

Wavy hair grass is described as **acidophilic-basitolerant** (acid loving and tolerant of alkaline soils), and outcompetes the **acidotolerant** creeping soft-grass. On soils which are not excessively acidic, wavy hair grass also outcompetes **acidophilic** heather *Calluna vulgaris* whose roots are restricted to acidic soil horizons. Colt's-foot *Tussilago farfara*, a ruderal (plant of disturbed areas) sometimes found along woodland rides, is in contrast **basophilic acidotolerant**. Interestingly, bearberry *Arctostaphylos uva-ursi* competes best on acid and basic soils, being uncommon on those of intermediate pH.

Fungal associations with tree roots are almost universal in long-established woodlands, in which **mycorrhizas** are of great importance (see Section 5.4.1). Fungi of ectotrophic mycorrhizas form a compact sheath of hyphae over the roots, which are stimulated to form the numerous stubby branches commonly seen in beech, oak, eucalypts and pine. Mineral nutrients (notably N and P) and water absorbed from the soil by the fungi are passed to the trees, from which the fungi receive simple sugars. Tree growth is often limited by the level of available phosphate; association with mycorrhizal fungi greatly improves the situation. In difficult situations, such as the establishment and maintenance of agro-forestry in semi-arid regions, inoculation of young trees with mycorrhizal fungi can be a very wise investment. The height of seedlings in their first year from new forest nurseries in New Zealand was always very variable until the soil had become inoculated with suitable mycorrhizal fungi.

## 2.4 Forest zonation and site quality

### *2.4.1 Influence of climate and soil type*

This section is concerned with zonation within forested regions. The ecology of forest margins and the features which mark both the beginning and the end of tree communities along climatological and altitudinal gradients are dealt with in Section 3.5. However, these **timberlines**, the biological boundaries marking the limit of a forest either high up on mountain sides, in frigid polar regions or adjacent to grasslands or deserts, are the subject of intense interest, particularly as there is evidence that their location is changing in response to global warming.

Soils are complex entities, shaped in response to underlying rock, climate and the influence of vegetation. In return, plants respond to the wide variations in soil that can occur in composition, structure, moisture, oxygen, pH and nutrition. All these factors together are summed up as **site quality**. Standing back and looking at the landscape, it is possible to see an overall **zonation**, a change within the vegetation as site quality alters. Sometimes, a change in the forest is abrupt, as with a striking change in the underlying rock type, but normally transitions in forest composition and structure are gradual, producing almost endless variation. Being human, however, we like to impose a pattern on this continuum, to divide up the endless variation in forests into discrete types that we can name and understand.

The woodland and scrub communities of Britain provide a large-scale example of how forest zonation and site quality are influenced by climate and soil type. Figure 2.6 shows 13 main NVC communities (National Vegetation Classification communities – see Section 1.7.1) grouped in relations to these two factors, with the six mixed deciduous and oak–birch woodlands lying centrally. It is important to realize that the name used for each group applies to the normal dominant assembly of species. A particular W17 woodland, for example, may not contain birch *Betula*, while oak *Quercus* is a common tree in many W8 communities named for the dominant ash *Fraxinus*, field maple *Acer* and dog's mercury *Mercurialis*. Within this central group, the soils diminish in fertility from left to right corresponding, in the warm, dry lowlands of the south-east, to a W8 Ash–field maple–dog's mercury – W10 Oak–bracken–bramble – W16 Oak–birch–wavy hair-grass woodland sequence. As Fig. 2.6 shows, the woodland sequence in the cool, wet north-western submontane zone runs W9 Ash–rowan–dog's mercury – W11 Oak–birch–wood-sorrel – W17 Oak–birch–*Dicranum* moss woodland. The latter includes, despite its south-western position, the vegetation on the bank of large granite boulders (clitter) of Wistman's Wood which is described in Box 1.3 and Section 9.4.1. Here pockets of acid, free-draining brown earth soils have accumulated amongst the clitter; these contrast with the wet peats and gleys of the adjacent floodplain and grass-moor plateau.

At the bottom of the figure are the W13 yew woodlands whose origins on the thin dry calcareous chalk soils of south-east England are discussed in Section 9.4.4. Above them are placed the three communities characteristic of the zone of natural beech dominance (W12, W14 and W15), again arranged in order of decreasing fertility from left to right, with dog's mercury *Mercurialis perennis* being

| Rendzinas and brown calcareous earths | Brown earths of low base status | Rankers, brown podzolic soils and podsols | |
|---|---|---|---|
| W20<br>*Salix–Luzula* scrub | W19<br>*Juniperus–Oxalis* woodland | W18<br>*Pinus–Hylocomium* woodland | COLD NORTHERN UPLANDS AND SUBALPINE ZONE |
| W9<br>*Fraxinus–Sorbus–Mercurialis* woodland | W11<br>*Quercus–Betula–Oxalis* woodland | W17<br>*Quercus–Betula–Dicranum* woodland | COOL AND WET NORTH-WESTERN SUBMONTANE ZONE |
| W8<br>*Fraxinus–Acer–Mercurialis* woodland | W10<br>*Quercus–Pteridium–Rubus* woodland | W16<br>*Quercus–Betula–Deschampsia* woodland | WARM AND DRY SOUTH-EASTERN LOWLAND ZONE |
| W12<br>*Fagus–Mercurialis* woodland | W14<br>*Fagus–Rubus* woodland | W15<br>*Fagus–Deschampsia* woodland | ZONE OF NATURAL BEECH DOMINANCE |
| W13<br>*Taxus* woodland | | | LOCALLY IN SOUTHERN BRITAIN |

Figure 2.6 The influence of climate and soil type on the distribution of three north British communities (W18–W20), the six mixed-deciduous and oak-birch woodlands (W8–W11 and W16–W17), the woodlands in the zone of natural beech dominance (W12, W14 and W15) and the southern yew woods (W13) of the National Vegetation Classification (Data from Rodwell, 1991. *Woodlands and Scrub. British Plant Communities*. Vol. 1. Cambridge University Press.)

associated with the most fertile, and wavy hair-grass *Deschampsia flexuosa* with the least. In contrast to these are W20 (willow–woodrush scrub), W19 (juniper–wood-sorrel woodland) and W18 (pine–*Hylocomium* moss woodland), which form the sequence of decreasing fertility shown for the cold northern uplands and subalpine zone of the north of Britain.

The 12 NVC woodland communities not shown in Fig. 2.6 are woodlands W1–7, which are characteristic of damp places and involve alder *Alnus*, willow *Salix*, downy birch *Betula pubescens* and ash *Fraxinus excelsior*, on the one hand, and the scrub and underscrub communities of W21–25 on the other.

Dog's mercury and hart's tongue fern *Phyllitis scolopendrium* grow well on chalk soils. Both were present in the sycamore wood on the chalk escarpment at Arundel, Sussex. Stinging nettle *Urtica dioica*, also found here, will flourish only if the soil has an adequate phosphorus content. While most natural soils have been subjected to extended leaching, **excessive fertility** is often a problem when former agricultural land is converted to forest. This is particularly the case when the intention is to create a high-quality herb layer beneath the developing tree canopy. A modern solution, widely and successfully used in Denmark and the Netherlands, is **top-soil inversion**. A huge plough is used to turn over a spit a metre deep, thus burying the nutrient-rich surface soil and bringing poor subsoil to the surface. This method has recently been employed at Wheeldon Copse, near Chester in Cheshire, England. This 7-ha former arable site is being developed by the Woodland Trust and Landlife, an environmental charity. Here cornfield annuals – corncockle *Agrostemma githago*, corn marigold *Chrysanthemum segetum*, cornflower *Centaurea cyanus* and a sprinkling of poppies *Papaver rhoeas* – were sown before the young trees were planted, with the intention of diminishing the invasion of rank weeds and grasses from the field edges while the trees established.

### *2.4.2 Fiby urskog: soils, topography and zonation of a Swedish primitive boreal forest*

The majority of northern hemisphere forests have now been heavily influenced by humans, so those such as Fiby urskog (Hytteborn and Packham, 1985; Fig. 2.7) which have been relatively little affected in this way, are especially interesting in showing local natural zonation and variation of forest vegetation. Though the highest parts of the reserve are now 60–65 m above mean sea level, they were exposed to wave action when the land first emerged from the sea, so drift and other sediments were washed down from the present hilly areas. Coarser down-washed sediment is now present amongst the huge granite boulders, while silt and clay occur lower down. Peat formed later. The hill

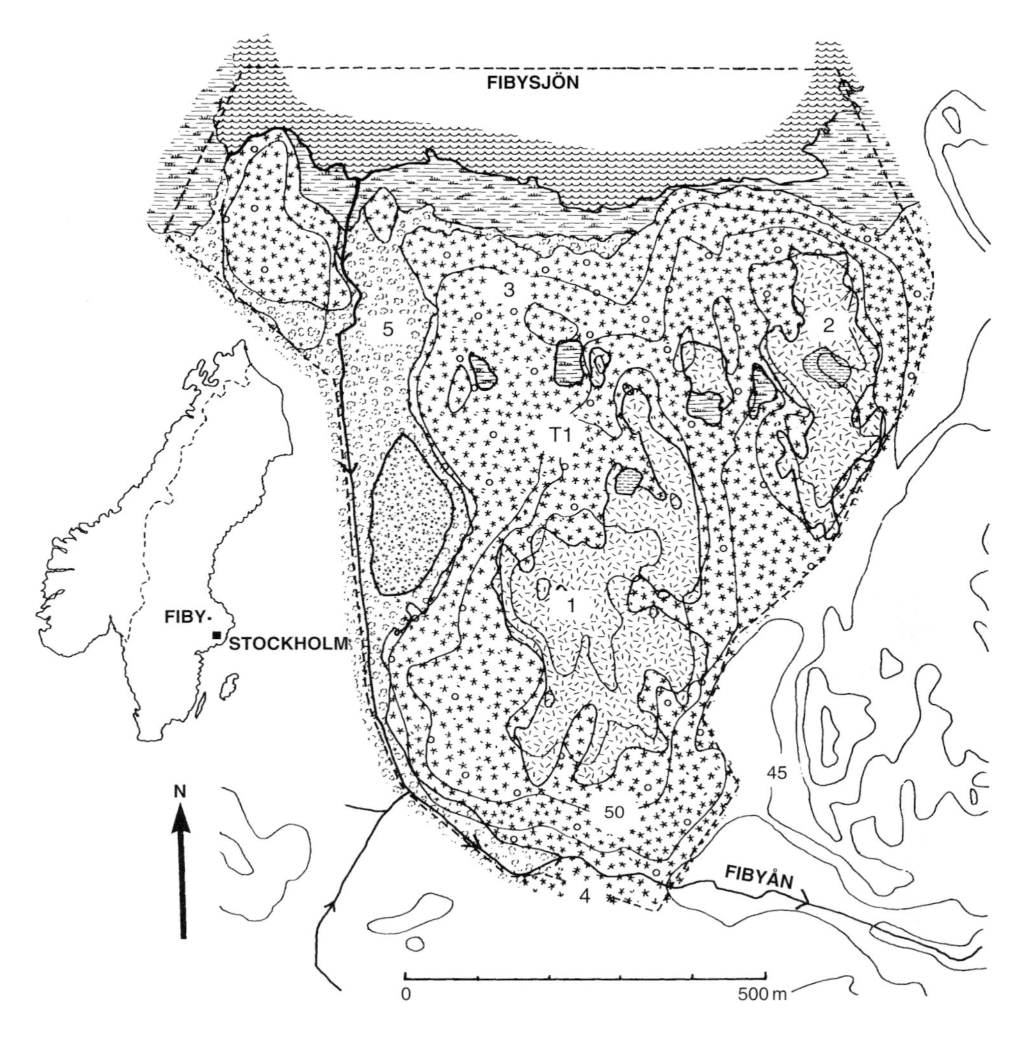


- Forest dominated by Norway spruce, but with some Scots pine and a few scattered aspen and birch, both silver *Betula pendula* and downy *B. pubescens.*
- *Hällmarkstallskog*: granite ridge system dominated by Scots pine but with downy birch and small fragments of mire vegetation also.
- Downy birch forest on peat with some alder. Norway spruce is invading while willow species are dominant along the lake margin.
- Scots pine dominant on peat.
- Minerotrophic mire. Rich fens surround Fibysjön. The small areas in the forest are of poor fen, sometimes with scattered spruce and birch.
- Scots pine–Labrador tea *Ledum palustre* bog (ombrotrophic mire).

Figure 2.7 The primitive forest of Fiby urskog, which lies 16 km west of Uppsala, central Sweden, in an area underlain mainly by the Precambrian Baltic Shield of granite and gneissic crystalline rocks. The reserve consists of 65 ha of forest and 13 ha of the lake to the north of it, whose surface is normally some 40 m above mean sea level. T1 is the midpoint of the N–S transect shown in Figs 2.9 and 2.10; 1–5, positions of the forest relevés described in Fig. 3.3. Sernander's Plot 1 (see Fig. 9.10) was in the southwest of the area dominated by spruce. (Data of Hytteborn and Packham.)

slopes and lower areas are dominated by Norway spruce *Picea abies* growing on boulder moraine, glacial till and later sedimentary deposits, while Scots pine *Pinus sylvestris* is the most important tree in the relatively open forest on drier granitic sites of the central area. Shallow soils of low nutrient quality in depressions in the granite have mire vegetation with hare's-tail cottongrass *Eriophorum vaginatum*, Labrador-tea *Ledum palustre*, cowberry *Vaccinium vitis-idaea* and bog moss *Sphagnum* spp. Outside the depressions, and with the exception of well-worn paths, the granite under the trees is usually covered by a cryptogamic mat of lichens and mosses (see Fig. 2.8). **Cryptogams** are

Figure 2.8 A young Scots pine established in the cryptogamic mat on Lichen Hill, Fiby urskog. This figure shows part of a 0.25 m$^2$ quadrat dominated by the lichen *Cladina arbuscula*; other important species were *Cladonia uncialis*, *C. gracilis*, *C. squamosa*, *Cladina stellaris* and the acrocarpous moss *Polytrichum juniperinum*. In total ten lichen and two moss species were recorded in this quadrat. (Photograph by John R. Packham.)

plants reproducing by spores rather than seeds; they include the ferns and fern allies. The area of mature forest occupied by broadleaved trees and shrubs is low, though several species are present, including the normally more southern wych elm *Ulmus glabra* and small-leaved lime *Tilia cordata*. The uneven size and age structure, rotting logs and standing dead trees are all typical of primeval (= old growth) forest.

The lake to the north of Fiby urskog is humic and **dystrophic** (very acidic and lacking in important mineral nutrients); it is fringed by wetland vegetation including willows such as the grey *Salix cinerea*, bay *S. pentandra*, and creeping *S. repens*. Plants with the hydrophytic and helophytic life forms described in Section 3.1.1 grow here. At the western margin of the reserve is a small stream with a tract of seral birchwood growing on peat along its eastern edge. This developed after the dam at the southern end of the reserve fell into disrepair in the 1930s, allowing the water level of the stream to fall. Understorey vegetation shows an equally clear zonation around the site. Dwarf shrubs such as heather and bilberry are common in drier areas while mosses (*Hylocomium splendens*, *Plagiothecium curvifolium*, *Pleurozium schreberi*, *Ptilium crista-castrensis* and various *Dicranum* species) are prominent in the ground cover of the spruce forest. Wavy hair-grass often has its tillers well scattered in this moss layer but also does well in drier reasonably open areas. There is a good deal of variation on a small-scale creating a **mosaic structure** (see Section 6.4.3); within the cryptogamic mat adjacent 0.01 $m^2$ quadrats often show appreciable differences.

It is important to bear in mind that this zonation is not static. Long-term weathering and soil development, together with climate change (natural or otherwise), will lead to changes. In the shorter-term, successional changes are also important: the mature spruce forest is prone to tree-loss by storms, creating gaps which are recolonized (described in Section 9.3), and spruce is becoming increasingly important in the streamside birchwood.

The above description may tend to imply that forest types are uniform with sharp boundaries. In practice even a particular community such as spruce forest is far from uniform across its area; all the plants within it are responding individually to their immediate environments. This is illustrated by the transect in Fig. 2.9, a 1-m wide strip, recorded on a slope in the spruce forest; see also Fig. 2.10 which shows the main features of the canopy, positions of tree trunks and fallen logs. It can be seen that topography and soil conditions strongly influence distribution patterns of herbs and dwarf shrubs along the transect. Thirty-one species were recorded along the transect (only some of which are shown in the figure) with species density being highest near the bottom of the slope. Some species (such as wood horsetail) are very restricted

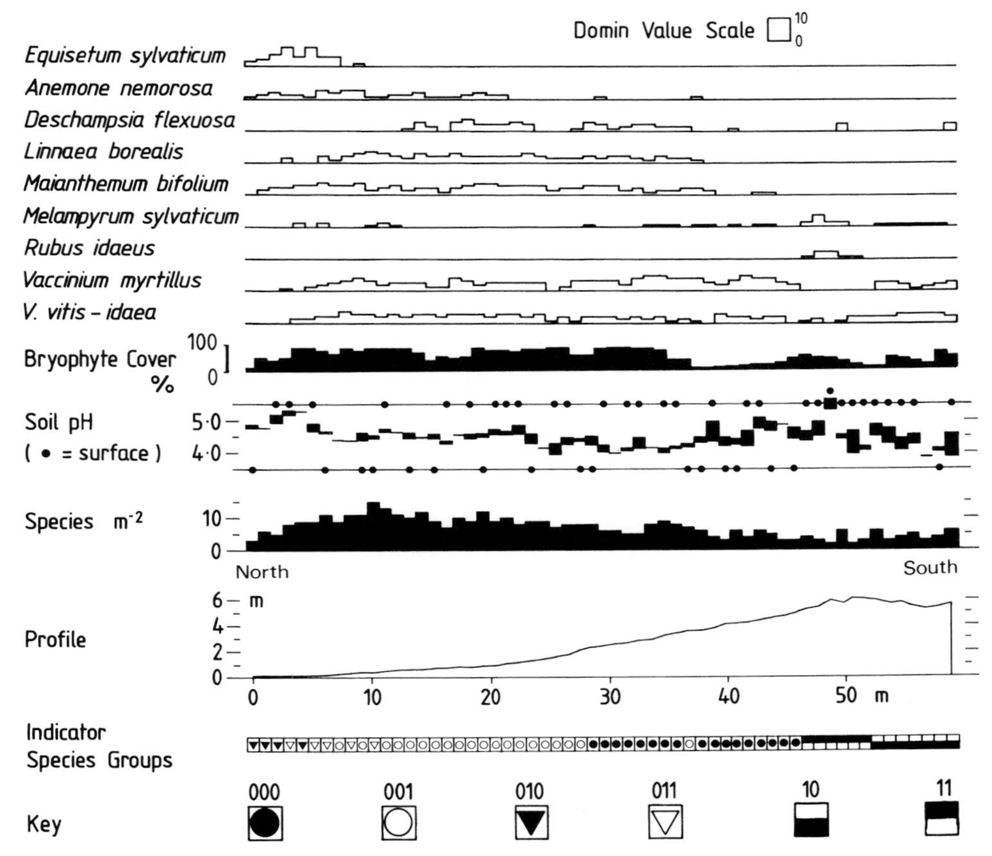


Figure 2.9 Transect 1 at Fiby urskog along a sloping area of old-growth spruce forest. The diagram shows: (1) abundance values (using the Domin scale) for the major field layer species: wood horsetail *Equisetum sylvaticum*, wood anemone *Anemone nemorosa*, wavy hair grass *Deschampsia flexuosa*, twinflower *Linnaea borealis*, May lily *Maianthemum bifolium*, small cow-wheat *Melampyrum sylvaticum*, raspberry *Rubus idaeus*, bilberry *Vaccinium myrtillus* and cowberry *V. vitis-idaea*; (2) the percentage of ground covered by bryophytes within each $m^2$ quadrat; (3) the difference between the soil pH at 0–3 and 10 cm depth (the dot for each quadrat is placed at the value for the surface soil so it is easy to see which of the two values for a particular quadrat was the greater); (4) species density per $m^2$; (5) topographic profile; (6) the indicator species groups to which the stands belong (see text). (Data of Packham, Hytteborn and Moberg. From Packham *et al.*, 1992. *Functional Ecology of Woodlands and Forests*. Chapman and Hall, Fig. 4.10. With kind permission of Springer Science and Business Media.)

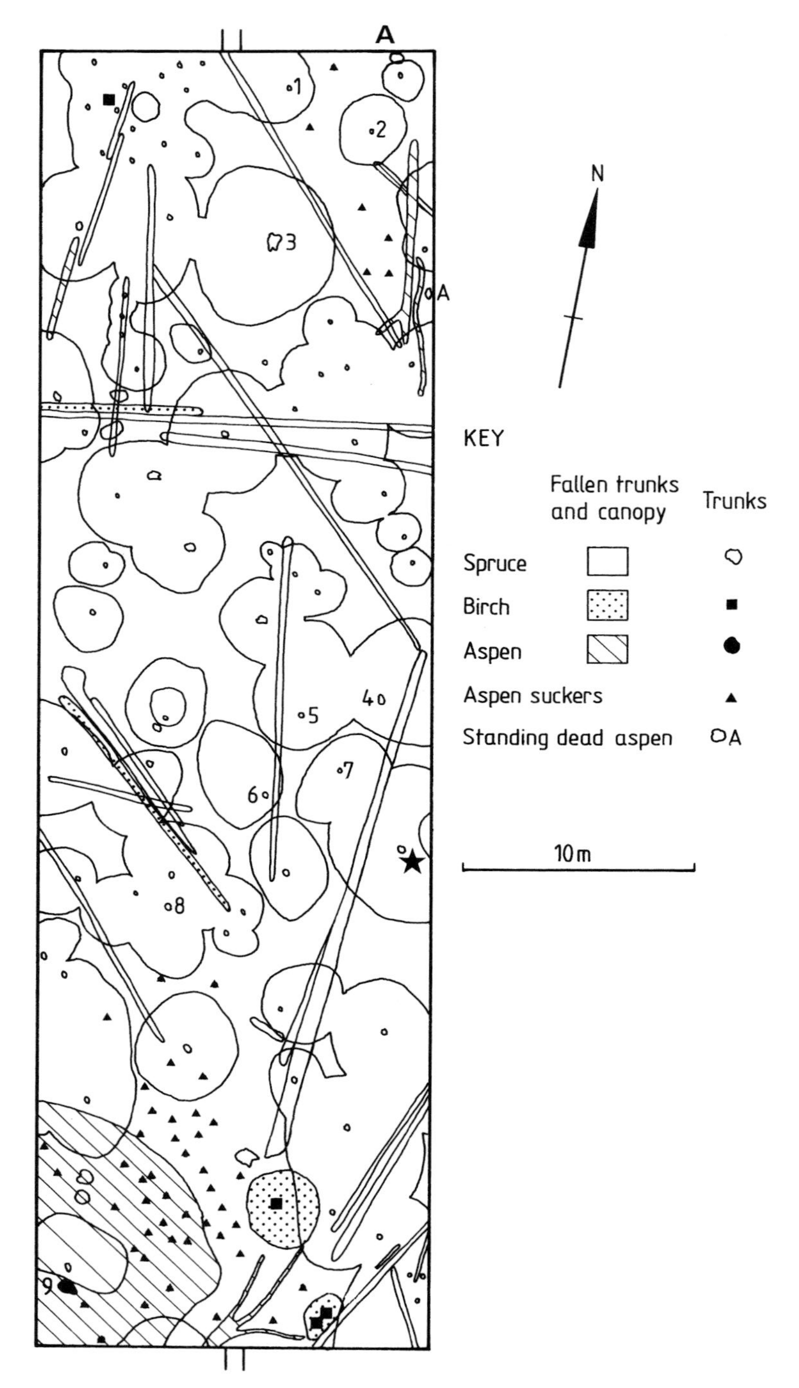


Figure 2.10 Fallen trunks, tree and shrub canopy in the forest area surrounding Transect 1, Fiby urskog. Paired marks at the top and bottom of the map show

in their distribution, while others such as bilberry and May lily are less exacting in their requirements and consequently more widespread along the slope. Variation in the distributions of the 31 species present has been used to divide the quadrats observed in the field layer vegetation into successively smaller groups using the statistical technique of indicator species analysis (Hill *et al.*, 1975). This creates 'polythetic' groups, so called because all the species data available are taken into account.

The six major polythetic groups identified along the transect in Fig. 2.9 run smoothly from the bottom of the hill in a sequence given the numbers 010, 011, 001, 000, to 11 and 10 towards the top. This is a hierarchical grouping: the first digit shows that the first four groups (Group 0) are quite different from the last two (Group 1). Within Group 0, the second digit shows that the first two groups (010 and 011) are different from the second two (001 and 000), and each of these groups is finally subdivided by the last digit. Indicator species for Group 0 are wood anemone, oak fern *Gymnocarpium dryopteris*, twinflower, hairy wood-rush *Luzula pilosa*, chickweed-wintergreen *Trientalis europaea* and bilberry. Wood anemone is found only in Group 0, and is therefore an excellent indicator. Though not such a good indicator, bilberry is considerably more likely to be found in Group 0 than Group 1. Common cow-wheat *Melampyrum pratense*, small cow-wheat and raspberry are indicators for Group 1, which dominates the upper southern portion of the transect where the field and bryophyte layers are rather open. Soil conditions vary considerably here; the lowest quadrats (1–9 m) have a shallow peat layer resting on clay with the clay continuing until 14 m. The lower slopes are the most moist and this favours higher species density, though this is reduced by the very marked shade exerted by small spruces at the bottom of the hill. Low boulders are prominent from 24–59 m.

A number of species grow in very distinct habitats, for example downy willow *Salix lapponum* is found on a quaking mire, while the liverwort *Ptilidium*

---

Caption for Figure 2.10 (cont.)
the position of the metre-wide strip recorded for herbs and dwarf shrubs. The star marks the mature spruce (no. 7), released by the great storm of 1795, whose increment diagram is shown in Fig. 9.11. Numbers 1–8 refer to spruce whose diameters at 1.3 m (= breast height) in cm, heights (m), and estimated age in years are as follows **1**: 4.6, 6, 37; **2**: 8, 12, 110; **3**: 50, 30, 132+; **4**: 42.6, 28, 167; **5**, 46.6, 30, 181; **6**, 35.5, 28, 170+; **7**: 35, 26, 166+; **8**: 35, 30, 144+; **9**: 64, 30, 160 (aspen). (Data of Packham, Hytteborn, Claessen and Leemans. From Packham *et al.*, 1992. *Functional Ecology of Woodlands and Forests*. Chapman and Hall, Fig. 4.11. With kind permission of Springer Science and Business Media.)

*pulcherrimum* occasionally occurs on rock and is frequent on fallen trunks and wood. The larger *P. ciliare* is fairly common in the bryophyte mat.

## 2.5 Rain forests: climate, soils and variation

### *2.5.1 Tropical rain forests: the changing archetype*

Corner (1964) considered tropical rain forest to be the cradle of flowering plant (angiosperm) evolution and that all other forest types were derived from it – it is the archetype, the earliest common ancestor. Early trees such as those of the Coal Measure forests (see Section 9.1 and Fig. 1.1), as well as the first seed plants and the first flowering plants, do indeed appear to have evolved under conditions of high temperature and constant humidity. Aseasonal tropical rain forests (see Section 1.6.1 for definitions) occur naturally in equatorial regions receiving ample rainfall, averages ranging from 1700 mm to over 10 000 mm annually. Temperatures range between 21 and 34 °C in tropical rain forests, and the daily (diurnal) temperature range is greater than the annual range, in marked contrast to boreal forests. Shelter, freedom from fire, suitable soils and appropriate seed sources are also essential if rain forests are to establish. As with all vegetation everywhere, tropical rain forest has evolved, moved and established in relation to major climatic changes. Both its species composition and its very location on the surface of the Earth have changed: it is far from being a stable archaic relic.

It is impossible to describe the general structure of tropical rain forests without mentioning the work of P. W. Richards (see Willis, 1996), who developed the technique of taking a strip of trees and recording their position, height, diameter and outline drawn to scale to give a profile diagram. Such forests often have three or more tree canopy layers that provide shade, shelter and physical support for smaller plants, as Fig. 2.11 illustrates. This interlocking canopy reduces the rate of airflow, tending to maintain high and constant atmospheric humidity inside the forest. The lack of annual herbs in this 'closed' forest is due to the low amount of light reaching the forest floor; nevertheless the presence of openings facilitates the existence of both light-demanding and strongly shade-tolerant tree seedlings (see Section 3.2.1). These diverse ecosystems with their very many plant and animal species possess a number of specialized features, including those related to reproduction. Parrots and other rain-forest birds are attracted to red, black and blue fruits, whose contained seeds may be transported several kilometres before being deposited in the faeces of the birds. The **dust seeds and spores** of epiphytic orchids and ferns, which commonly grow high in the tree canopy, are carried long distances by wind and air currents within the forest. **Bats** assist with both

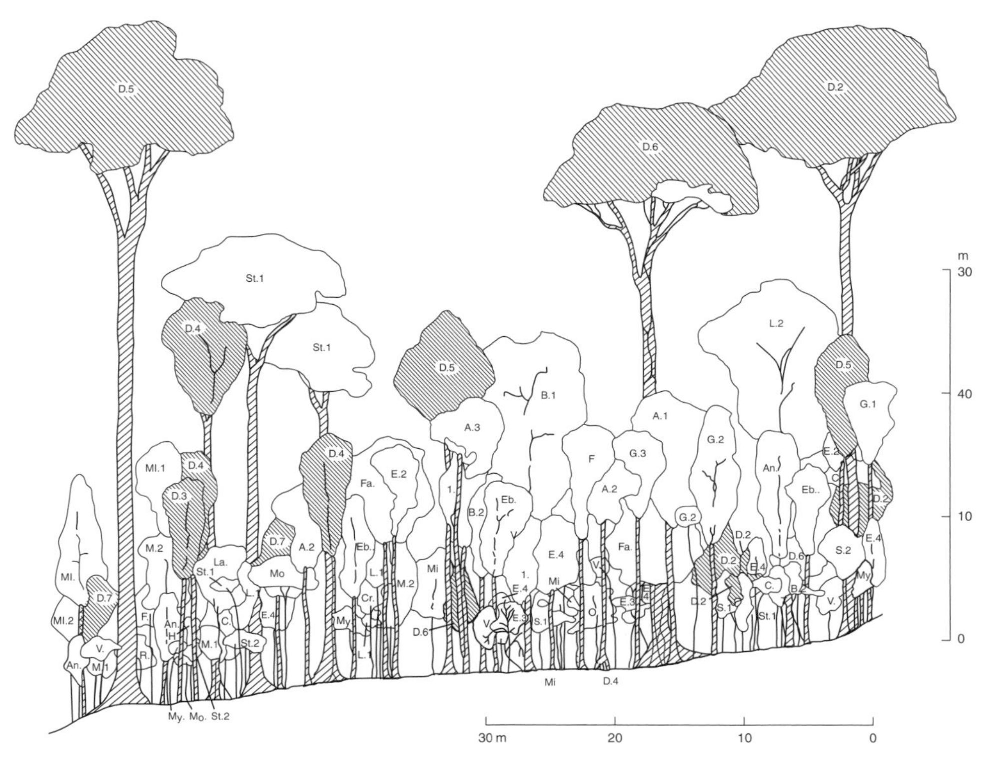


Figure 2.11 Profile of lowland evergreen dipterocarp tropical rain forest at Belalong, Brunei, so called because it is dominated by dipterocarp trees (a large family of tropical Asian trees). The two ends of the plot, which measured $60 \times 7.5$ m, were occupied by mature forest and had younger growth (i.e. building phase) invading a gap between them. All trees over 4.5 m are shown, but there is no attempt to show the seedlings and young saplings which allow gap regeneration to proceed. The canopy top is formed by three giant mature dipterocarps: *Shorea laevis* (D5), *S. parviflora* (D6) and *Hopea bracteata* (D2) up to 45 m high. Other dipterocarp trees below are hatched. Each symbol on a tree outline refers to a particular species. From this, the incredible species richness of the forest can be judged. (From Ashton, 1964. *Oxford Forestry Memoirs*, 25 in Whitmore, 1998.)

pollination and seed dispersal; seeds of figs and other rain-forest fruits may be transported as much as 20 km from the parent trees. Winged seeds, such as the shuttlecocks of dipterocarps, glide away from the parent tree for variable distances depending upon the species and the weight of the seed. Butterflies and other insects also act as pollinators and the flowers they visit possess attractive devices of one sort or another. **Ants** have remarkable symbioses with **myrmecophilous plants**, which receive protection from predators and epiphytic growths including fungi, and in return provide food (nectar) and galleries to house the insects. Weakened patches of cork provide access to trunk galleries in ant trees, while hollowed leaves, stems, branches and galls of epiphytic ant plants can all provide shelter.

Ants also appear to be responsible for the legendary single-species 'devils' gardens' of the tree *Duroia hirsuta* in the Amazonian forests of Peru. These trees live for hundreds of years and grow together in quite large groups in a forest famous for its otherwise remarkable diversity of trees, vines, shrubs and forbs. Local legend has it that these areas are maintained by an evil forest spirit. In reality, an ant *Myrmelachista schumanni* nests in the *Duroia* trees and poisons all other plants with formic acid, maintaining the monoculture of *Duroia*. The ants benefit from this pruning because the young *Duroia* saplings which grow in the cleared sites provide new nesting sites for the expanding ant colony. A colony can have as many as 3 million workers and 15 000 queens and the tree/ant colonies can be over 800 years old (Frederickson *et al.*, 2005).

Many of the trees possess **buttresses** at the base (Fig. 2.11), which undoubtedly give them mechanical stability and help to stabilize those on unstable substrates. They also enable the tree to absorb adequate amounts of oxygen: roots in the frequently sodden soil have to compete with decomposer organisms for the often limited amount of soil oxygen. Flowers commonly develop directly on the trunk (**cauliflory**) or on branches (**ramiflory**). This makes them readily available to pollinators including beetles, and also facilitates the support of large fruits, e.g. jackfruit. Water drains more rapidly from leaves with the **drip tips** common in rain-forest trees.

Much of the **nutrient capital** of tropical rain forests is thought to have accumulated when plant roots were in contact with fragmented parent rock, perhaps as far back as the Tertiary (1.6–65 Ma) (Walter, 1973). The above- and below-ground distribution of mineral nutrients in five tropical rainforests is shown in Fig. 2.12, which shows the popular belief that most of the nutrients in a tropical rain forest are in the biomass is seldom true except for certain minerals in certain forests. Moreover, decomposition rates in tropical forests are not necessarily higher than in temperate forests. Tropical soils are, however, often nutrient-poor and acidic (pH 4.5–5.5), while – as we have seen in Section 2.2.1 – the leaching of basic ions and silica, frequently results in the accumulation of iron, manganese and aluminium sesquioxides. Moreover, once the vegetation including roots is removed, leaching of nutrients is rapid; hence the expression mentioned in Section 2.2.1 that rain forests are really 'green deserts' – a rich forest grown over poor soils. Termites and fungi, especially decomposer basidiomycetes, destroy dead wood, while nutrient recycling is accomplished by mineralization and root absorption often assisted by mycorrhizas.

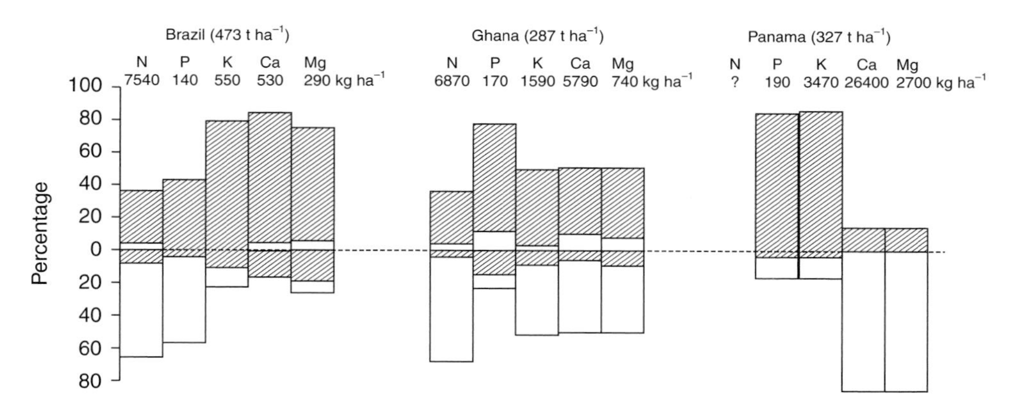


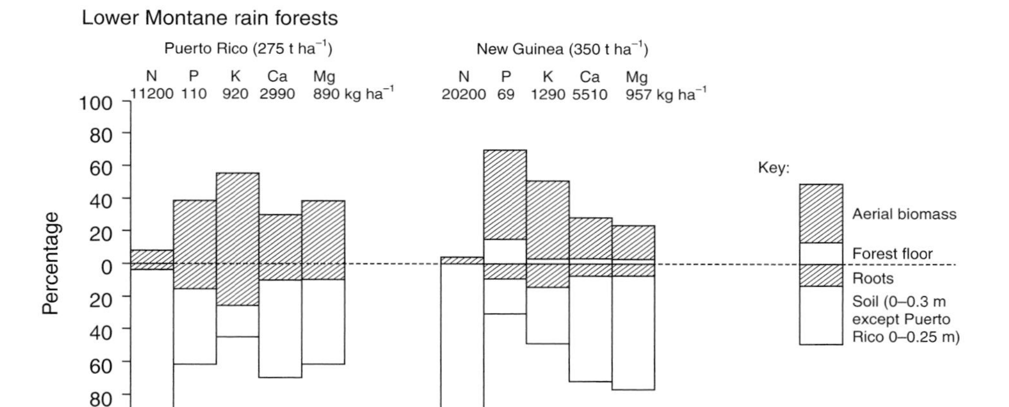


Figure 2.12 Distribution of inorganic nutrients above and below ground in five tropical rain forests. Biomass figures are in parentheses. This figure demonstrates that the popular belief that most of the nutrients of a tropical rain forest are in the above-ground biomass, although true for the soluble cations potassium (K), calcium (Ca) and magnesium (Mg) in the Brazilian forest shown here, is usually incorrect. N, nitrogen; P, phosphorus. (Redrawn from Whitmore, 1998. *An Introduction to Tropical Rain Forests* (2nd edn). Oxford University Press.)

In fact humans have influenced these forests for far longer than is generally supposed. Some soils in the Amazon basin are rich, black and remain nutrient-rich when cleared of vegetation. These *terra preta dos Indios* (Indian dark earth) soils, which cover perhaps up to 10% of Amazonia (an area the size of France), are thought to be a relict of Amerindians from pre-European contact, dating back several thousand years and may point to early populations being much larger than once thought (Mann, 2002). The soils appear to have been formed by partial burning of the forest ('slash and char' rather than modern 'slash and burn') and integration of the resulting charcoal, which has a high cation exchange capacity (Section 2.2.1), into the top 40–60 cm of the soil. The incorporation of forest organic matter into the soils as charcoal rather than as litter or ash may help to explain the durability of these soils, in comparison with the short-lived increases in fertility of slash and burn. Nutrients were added in the form of faeces and other animal wastes particularly fish, perhaps with an inoculation of microorganisms from an existing *terra preta* soil which seem important in aiding soil formation. Whether these nutrient-rich **anthrosols** (anthropogenic soils) were deliberately created or developed inadvertently is open to debate, but current evidence points to the former (Erickson, 2003). The resulting dark soils are higher in phosphorus, calcium, sulphur and nitrogen than surrounding tropical red soils, they retain moisture and nutrients better, and are not as rapidly exhausted by agriculture when well managed. Extensive research is currently looking at how these soils were formed and can be created anew.

Rain forests have in recent centuries suffered severely at the hands of humans, and today even the vast Amazonian rain forests are rapidly losing ground. They form the most complex terrestrial ecosystems that have ever existed and are of potentially immeasurable economic, pharmaceutical and scientific value, besides posing many intriguing questions regarding their biodiversity, variation, stability and former distribution.

### 2.5.2 Rain-forest history: the Australian story

Millions of years ago much of Australia was covered with closed moist forests similar to the rain forests which now cover a mere two million hectares, equal to about a third of the area of Tasmania. Though their present extent is so small, the rain forests of Australia are of exceptional interest and the history of how they have waxed and waned, as well as the communities to which they have given rise, is fascinating. The answers to many of the questions concerning the nature and origin of tropical rain forests lie in the geological past. Some 280 million years ago (Ma) the fossil tree *Glossopteris* existed in what are now Antarctica, Africa, South America and Australia, suggesting a similar climate. Moreover the southern beech genus *Nothofagus* is found as a fossil in Antarctica and various species grow in Australia, New Guinea, New Zealand, New Caledonia and South America. Why did the climate of Australia change to that of the most arid of the unfrozen continents, and what accounts for these tree distributions?

The answer to these questions lies in the theory of **continental drift**, which was first put forward by Alfred Wegener in 1912. This drift, caused by **plate tectonics**, is driven by convection currents in the mantle below the tectonic plates of the Earth's crust, of which the continents form part (see Section 9.1.1). When flowering plants first appeared some 135 Ma, in a period when the world's vegetation was dominated by cycads and conifers, Australia formed part of the massive southern supercontinent **Gondwanaland**. This former huge southern hemisphere landmass was at that time made up of what are now Africa, South America, India, Madagascar, Arabia, Australia and Antarctica. Around 45 Ma, Australia finally broke free from Antarctica and began its separate drift 27° northwards, finally slowing when it hit the Asian continental plate some 15 Ma. Australia then became very much drier, largely because of a worldwide change in atmospheric and oceanic circulation patterns. The more recent history of the Atherton Tableland, Queensland, has been worked out from the proportions of fossil pollen grains of various species found in soil profiles going back 120 000 years. These enable us to discover, in a period interspersed by ice ages, the proportions of land dominated by rain forest as opposed to sclerophyll vegetation (see Section 1.6.1), and show that rain forests can advance into neighbouring eucalypt forests when conditions are suitable. Around 26 000 years ago rain forest almost disappeared on the Atherton Tableland, only to begin picking up again around 13 000 years ago.

Many ecologists had thought that the current flora and fauna of Australian rain forests were the result of migration from south-east Asia in the geologically recent past. Though this is true for a small percentage of Australian rain-forest species with Asian relatives, it is now clear that the southern beeches *Nothofagus* spp. and many other rain-forest plants originated in Australia when it was much cooler and wetter than it is now, before migrating northwards to what are now other continents. Australian vegetation is now dominated by hard-leaved (**sclerophyllous**) species such as gum trees *Eucalyptus*, waratahs *Telopea* and wattles *Acacia*. The earliest fossil records of eucalypts (see Sections 1.3.1 and 1.3.2) are no older than 38 Ma, but it is now considered that the first sclerophyllous species evolved from rain-forest plants soon after the separation from Antarctica 45 Ma. The junction between rain forest and woodland dominated by eucalypts is often remarkably sharp (although sometimes there is a transitional belt). A number of reasons have been put forward, including fire which burns readily through the eucalypt woodland but not into the rain forest (Bowman, 2000), but soil also seems to play a part. Rain forests are mostly restricted to relatively fertile soils since the rain-forest trees have a higher nutrient requirement, especially for phosphorus, than the eucalypts. But the precise boundaries are known to have moved in the past, either by rain forest invading the woodland, or cyclones opening up rain-forest areas to eucalypt invasion, so other factors apart from soil are obviously important.

Present-day Australian rain forests are arranged in a discontinuous string along the northern and eastern coasts of Australia from near Cooktown, Queensland southwards to New South Wales and Victoria. This discontinuity is as much to do with variations in rainfall caused by topography as it is to human clearance. These rain forests can be classified into many groups on the basis of their structural components, but it is simpler to place them into the broad climatic categories of Tropical, Subtropical, Warm-temperate, Cool-temperate and Dry. **Tropical rain forests** with an uneven, multi-species canopy of at least three layers (although these are not always discrete and apparent) occur in Northern Queensland. In the north they often receive 2400 mm annual rainfall, although 1600 mm is adequate for rain forests in cooler areas further south where evapo-transpiration rates are lower (Cameron-Smith, 1991). This rainfall is seasonal, so these forests are all part of the moist or humid tropical rain forests (Section 1.6.1). **Subtropical rain forests**, which have many species in the tree canopy, but are somewhat simpler than tropical rain forests, can survive on a yearly average as low as 900 mm. They are found from the cooler uplands of northern Queensland to the coastal lowlands of New South Wales. **Warm-temperate rain forests** grow in cooler climates, at higher altitudes, and on less fertile soils. Their composition is less diverse and more uniform with 3–15 tree species in the canopy; they also lack the buttress roots characteristic of tropical and subtropical rain forests.

**Cool-temperate forests** are simpler still, often with only one species – frequently of a southern beech *Nothofagus* – in the upper canopy and few species in the lower layer. These commonly high-altitude, cool forests with very high rainfall are often cloaked with mist, and their dominant trees usually have a thick coat of mosses, liverworts, ferns and lichens. Tree ferns and thorny bushes are often common. Tree leaves are small and simple, frequently with toothed margins. Palms and stranglers are absent, as are root buttresses, although trunk bases are sometimes massive. Such forests are found from the McPherson Range on the New South Wales–Queensland border to Tasmania. Small-leaved southern beech, notably antarctic beech *N. moorei*, abound in the northern cool-temperate rain forests. Pinkwood *Eucryphia moorei* then dominates rain forests down to the border with Victoria. Further south, myrtle beech *N. cunninghamii* extends into Tasmania where deciduous beech *N. gunnii* also occurs (Cameron-Smith, 1991).

Tiny remnants of the strangely named **dry rain forests** occur in regions of Australia with distinct wet and dry seasons. The number of tree species in the low to medium canopy layer varies, but may be large. Scattered larger trees known as **emergents** rise above this layer. The ability to shed leaves helps some species to survive temporary water shortages. Rain forests sheltered behind coastal dunes or headlands are unusual, but such **littoral rain forests**, often with a wind-sheared upper canopy, do occur on Australia's eastern coast. Related to subtropical rain forests, they grow on a wide range of parent materials including slate, basalt and deep beds of sand. Where shelter is insufficient, salt spray kills exposed leaves and branches, leaving dense thicket.

Rain-forest communities and their distributions have changed enormously over time. There is now considerable concern for them on a world basis and this is discussed in Section 11.1.2.

# 3

# Primary production and forest development

## 3.1 Plant life forms and biological spectra

### *3.1.1 Variation in vascular plant and bryophyte life form*

In addition to taxonomic classifications which endeavour to place closely related species in the same family, botanists have for centuries attempted to distinguish particular life forms, any one of which may be adopted by quite unrelated species. The simplest of these is the distinction between woody and herbaceous plants. Raunkiaer (1934) developed the most widely known scientific description of life forms, and then used it to initiate the use of **biological spectra** to compare different floras. The main feature of this ecologically valuable system is the position of the vegetative perennating buds or persistent stem apices during the cold winter or dry summer forming the unfavourable season of the year. The main life forms shown in Fig. 3.1 form a sequence showing successively greater protection from desiccation, indicating the position of the vegetative buds when the plant is dormant.

It was assumed that the flowering plants evolved when the climate was more uniformly hot and moist than it is now, and that the most primitive life form is represented by the **phanerophytes** which still dominate tropical vegetation. These large terrestrial plants can grow continually forming stems, often with naked buds, projecting high into the air. Those whose buds are protected from cold or desiccation by bud scales are considered to be more highly evolved. Tropical evergreens like the gum tree *Eucalyptus orientalis* lack the protective bud scales of evergreen phanerophytes of the temperate zone, such as holly *Ilex aquifolium* and Scots pine *Pinus sylvestris*. Ash and larch belong to a third group formed by deciduous phanerophytes with bud scales. These large plants are divided into four height classes: **nanophanerophytes**, woody plants with perennating buds between 0.25 and 2 m above the ground; **microphanerophytes**, between 2 and 8 m; **mesophanerophytes**, between 8 and 30 m; and

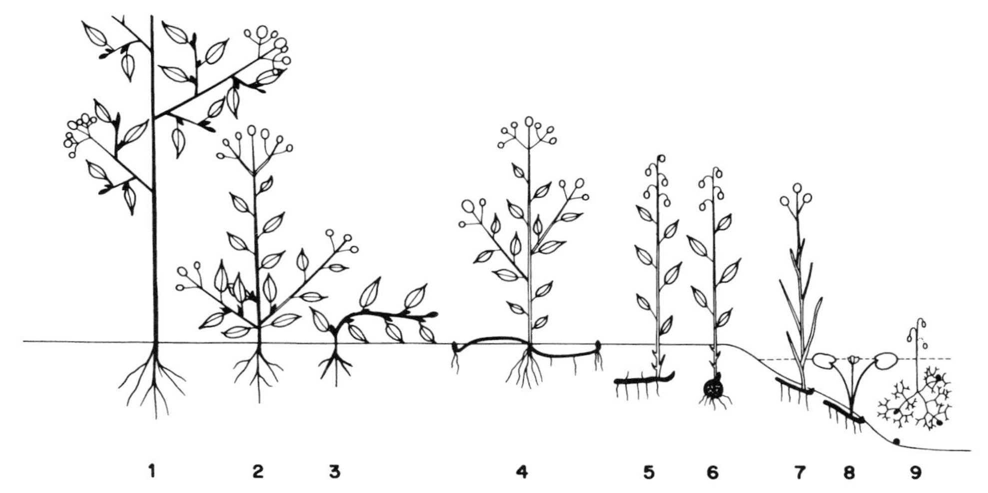


Figure 3.1 Diagram of the chief types of life form, apart from the therophytes which survive the unfavourable season as seeds. 1, phanerophytes; 2–3 chamaephytes; 4, hemicryptophytes; 5–6, geophytes; 7, helophytes; 8–9, hydrophytes. Parts of the plant which die in the unfavourable season are unshaded; while the persistent axes and perennating buds are in black. The sequence represents an increasing protection of the surviving buds, which are most exposed in the phanerophytes. (From Raunkiaer, 1934. *The Life forms of Plants and Statistical Plant Geography*. Clarendon Press.)

**megaphanerophytes** of over 30 m. Using the two criteria of height and bud protection, Raunkiaer originally divided the majority of phanerophytes into 12 groups, but he also recognized others, such as the epiphytic forms (including many aroids and orchids) which often grow on the trees of tropical and subtropical forests.

Woodward (1989) considered that temperature is the most important factor controlling leaf type and duration in the major type of tree present in any area with adequate rainfall. Broadleaved evergreens tend to dominate from the equator to the Mediterranean region. Of these the holm oak *Quercus ilex* can tolerate the lowest temperatures, surviving brief periods at −15 °C, but none are present in northern forests. Broadleaved trees, such as pedunculate oak *Quercus robur*, that can survive minimum temperatures below −15 °C are almost all deciduous; many can survive −40 to −50 °C. Forests where minimum temperatures fall below −50 °C are dominated by needle-leaved conifers – typically pines, firs and spruces – of which a few (larches – *Larix* spp.) are deciduous. Even here a few broadleaved deciduous species of birch and poplar manage to survive.

Woody climbers such as ivy *Hedera helix*, honeysuckle *Lonicera periclymenum*, Old Man's beard *Clematis vitalba* and the tropical lianas are specialized phanerophytes profiting from the stature of their neighbours. Mueller-Dombois and Ellenberg (1974, p. 449) give a more complex life form system using five main stem or trunk forms – **normal woody trees, tuft trees, bottle trees, succulent** and **herbaceous stem trees** – to subdivide the phanerophytes. Trees can also be classified as grading from **pachycaul** forms, including the tree ferns and palms, with thick, unbranched or little-branched main stems, bearing a terminal crown of large compound leaves, to 'twiggy', much-branched **leptocaul** species bearing smaller undivided leaves, such as the elms. Common ash *Fraxinus excelsior*, with its pinnate leaves and stubby twigs, has a tendency to the pachycaul habit, but no British tree has the massive frost-sensitive apical meristem of a true pachycaul. Perhaps the most well-known pachycaul is the baobab (*Adansonia digitata*) of the African savannas, with a trunk so massive that hollow specimens have been used as rooms and even a jail.

**Chamaephytes** are low-growing woody or herbaceous plants whose perennating buds are on aerial branches not more than 25 cm above the soil, and frequently much lower, where the wind is not so strong and the air is damper (Fig. 3.2). Perennating buds of **hemicryptophytes** are at the surface of the soil where they are even better protected, while those of **geophytes** are buried beneath the soil on rootstocks, rhizomes, corms, bulbs or tubers. **Therophytes** survive unfavourable periods as seeds, being abundant in deserts and open habitats. Common in the early stages of reversion of bare land to scrub, they become rarer as it progresses to mature woodland.

Life form is essentially an adjustment of the vegetative plant body and plant life history to the habitat; it is primarily determined by heredity and selection. Under some circumstances the environment directly influences life form, as when severe winter or other conditions kill the upper buds so that individual plants fall into the life form below that normal to the species, resulting in the dwarfing of trees at high altitudes or in almost constant wind (see Section 3.5). Conversely, stinging nettle *Urtica dioica*, normally a hemicryptophyte, may overwinter as a herbaceous chamaephyte in mild winters.

Fine divisions in life form can also be categorized using leaf size. **Leptophylls**, with an area of up to 25 mm$^2$ per leaf, were the smallest of the six sizes of leaf used by Raunkiaer (1934). The upper area limits of the next four members of the series – **nanophyll, microphyll, mesophyll** and **macrophyll** – increase by a factor of nine in each instance. The largest leaves (**megaphylls**) all exceed a nominal value of 164 025 mm$^2$ in area, equivalent to a leaf about 40 $\times$ 40 cm. The tropical aroid genus *Monstera* (climbing shrubs often maturing as epiphytes with aerial roots reaching the soil, and sold as houseplants under the name 'Swiss cheese plants') has megaphylls which have gaps or rounded holes when mature. Leaves tend to be large in the hot wet tropical rain forests, medium-sized in temperate

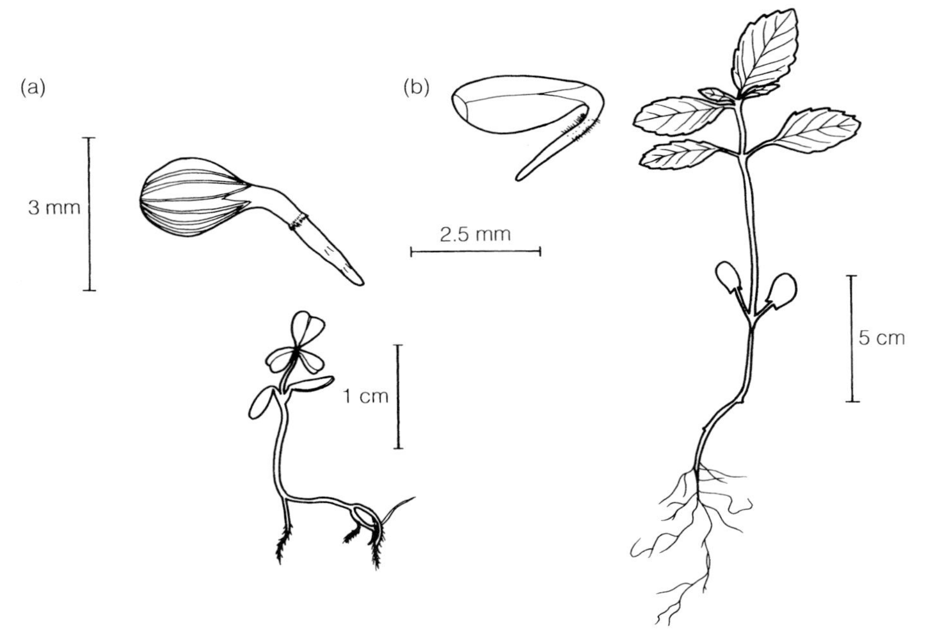


Figure 3.2 Seedlings of two woodland chamaephytes (a) wood-sorrel *Oxalis acetosella* and (b) yellow archangel *Lamiastrum galeobdolon*. The seeds of these species appear to require a period of chilling to break dormancy and are difficult to germinate experimentally. In nature considerable numbers of seedlings often appear together in early spring. Very occasionally *Lamiastrum* seedlings have three cotyledons. Wood-sorrel can also exist as a rhizome geophyte or a rosette hemicryptophyte. Bluebell, an example of a bulbous geophyte, is shown in Fig. 10.9. (Drawn by John R. Packham. From Packham *et al.*, 1992. *Functional Ecology of Woodlands and Forests*. Chapman and Hall, Fig. 2.2. With kind permission of Springer Science and Business Media.)

woodlands, and small in the cold or dry conditions of tundra and heaths. In tropical rain forests at least 80% of the species of trees present have leaves of the mesophyll class (2025–18 225 $mm^2$): most are also unlobed **sclerophylls** ('hard leaves') with pronounced drip tips. Microphylls predominate in montane rain forests where the climate is cooler. This gives some justification for using a **leaf size spectrum**, based on percentages of the different leaf sizes present, to characterize different vegetation types. However, light intensity and soil conditions, particularly available nitrogen and phosphorus, also have an important influence on leaf size even within the same genotype.

Life form systems can also be of use with lower plants, indeed Gimingham and Birse (1957) developed a very useful **life form classification for bryophytes**. This involved five main types: **cushions** (dome-shaped) e.g. *Leucobryum*

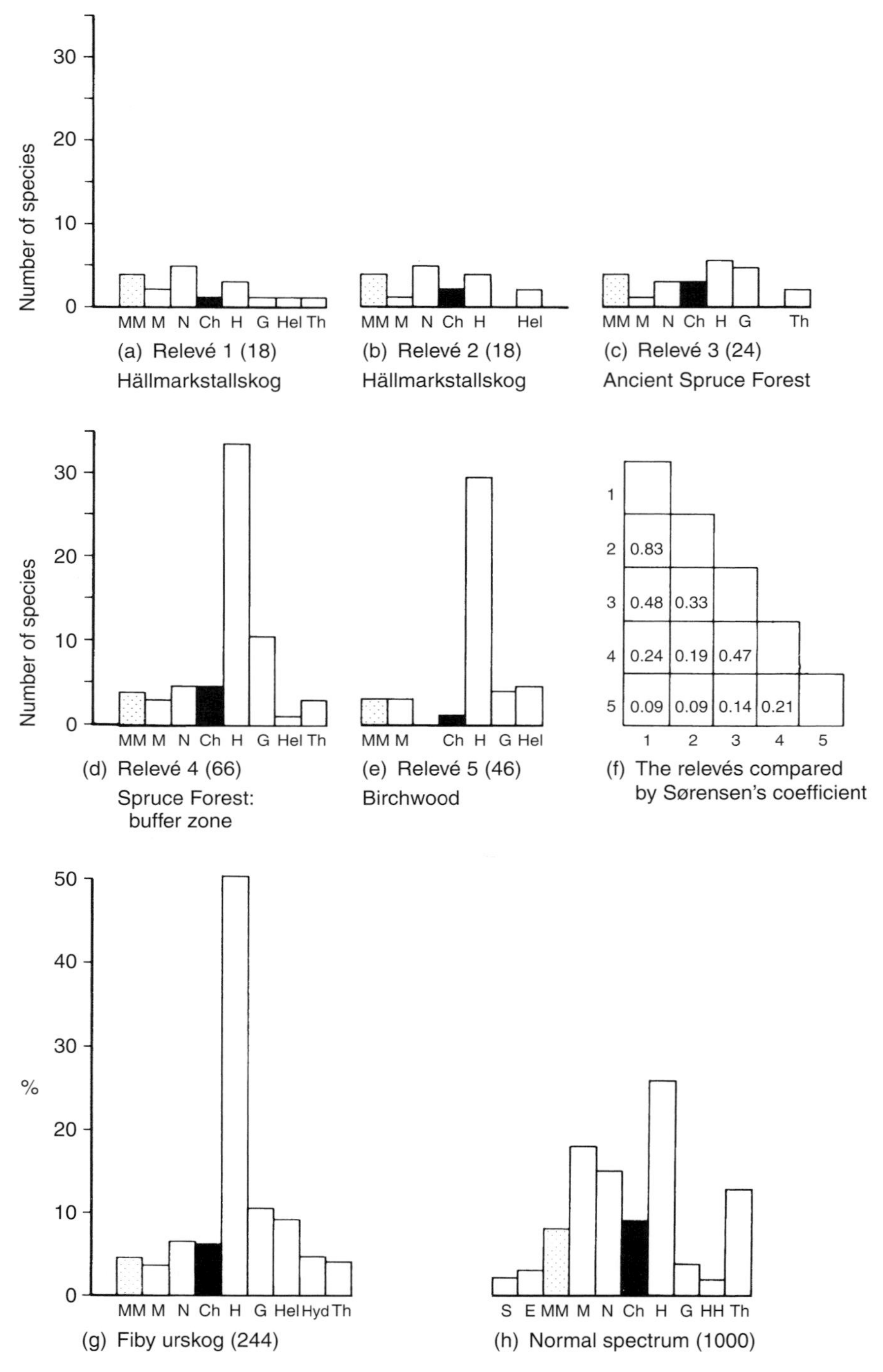


Figure 3.3 (a–e) Plant life form analyses for the 5 relevés whose positions in the primeval forest of Fiby urskog are shown in Fig. 2.7. Relevés 2 and 3 each had an area of 0.4 ha, those of relevés 1, 3 and 5 were 0.3 ha. (f) A comparison of the 5 relevés by means of the Sørensen coefficient of community (see text for details).

*glaucum*, **turfs** (like the pile of a carpet) e.g. *Dicranum majus* and *Polytricum formosum*, **canopy formers** (producing a raised leafy canopy or **dendroid form** such as *Climacium dendroides* and *Thamnium alopecurum*), **mats** and **wefts** (dense and loose interwoven shoots, respectively). Rough mats, smooth mats and dendroid forms are the commonest types on rocks, but a wider variety grow on soil. The influence of light levels and soil conditions on bryophyte distributions is discussed in Section 6.5.1.

**Biological** (or **life form**) **spectra** for particular areas are constructed by expressing the numbers of species in each life form class as percentages of all the species present. These can be compared with the **normal spectrum** that Raunkiaer derived from 1000 plants taken as a representative sample of the world flora (Fig. 3.3h). There is a strong correlation between the climate of an area and the life forms of the plants present; a **phytoclimate** is characterized by the life form which most greatly exceeds the percentage for its class in the normal spectrum. Like that of most of the cool temperate zone including Britain, the phytoclimate of Raunkiaer's native Denmark is **hemicryptophytic**. This is because although the vegetation may be visually dominated by trees, in the phytoclimate classification each species carries equal weight in this system, regardless of abundance or importance. Thus, although the relatively few species of trees native in Britain since the last glaciation were dominant almost everywhere until the advent of large-scale agriculture, these phanerophytes appear quite insignificant when compared with those of the normal spectrum.

The regions of the world in which the four major world climates occur, with their subdivisions, can be delimited by lines along which biological spectra are similar. Where rainfall is not deficient the phytoclimate of the tropics is **phanerophytic**, and within this zone a higher proportion of the larger forms are present in the wetter areas. Subtropical desert areas are **therophytic** and burst into life after very occasional periods of heavy rain. Geophytes are best represented in regions with a Mediterranean climate where the unfavourable season is the hot

---

Caption for Figure 3.3 (cont.)

(g) Biological spectrum of the entire reserve compared with (h) the normal spectrum taken by Raunkiaer (1934) from the world list. Figures in parentheses are numbers of species. S, stem succulent phanerophytes; E, epiphytic phanerophytes; MM, meso- and megaphanerophytes (perennating buds 8–30 m and > 30 m above ground, respectively); M, microphanerophytes (short woody plants with buds 2–8 m above ground); N, nanophanerophytes (smaller shrubs with buds 0.25–2 m above ground); Ch, chamaephytes; H, hemicryptophytes; G, geophytes; Hel, helophytes; Hyd, hydrophytes; HH, helophytes and hydrophytes; Th, therophytes. (Data of Packham, Hytteborn, Claessen and Leemans. From Packham *et al.*, 1992. *Functional Ecology of Woodlands and Forests*. Chapman and Hall, Fig. 2.3. With kind permission of Springer Science and Business Media.)

dry summer, but do not give their name to a phytoclimate. Plants with this life form are at a disadvantage where the soil warms slowly in spring and the growing season is short. Resting buds of **hemicryptophytes** in the cool temperate zone are often protected by snow in hard winters, but warmed by the sun as soon as it melts. The **chamaephytes** give their name to the cold zones near the poles where cushion forms in particular derive protection from snow, growing again as soon as it melts in spring. A survey of temperate broadleaved woodlands in Britain and North America showed that hemicryptophytes were abundant in all, but especially in northern areas, while cryptophytes with their buried buds were more important in southern and drier sites.

### *3.1.2 Local life form variation within a primitive forest*

Figure 3.3 illustrates life form variation within the primeval forest of Fiby urskog, Sweden. This shows the number of species for each life form in five areas of the reserve whose positions are shown in Fig. 2.7 (note that a **relevé** is a description of a visually uniform area). Relevé 4 is in species-rich spruce forest of the edge buffer zone. This has been subject to considerable interference in the past, has many more herb species than the other conifer-dominated relevés and the greatest number of additional species are hemicryptophytes. Other parts of the forest are equally species-rich, however. The next most important group is that of the geophytes. Many are rhizome geophytes such as oak fern *Gymnocarpium dryopteris*, baneberry *Actaea spicata*, wood anemone *Anemone nemorosa*, the liverleaf *Hepatica nobilis*, May lily *Maianthemum bifolium* and chickweed-wintergreen *Trientalis europaea*. Abundant in northern spruce forests, wood-sorrel *Oxalis acetosella* is found as an herbaceous chamaephyte, a rosette-hemicryptophyte or a rhizome geophyte according to the particular environment of the individuals concerned. **Helophytes** (marsh plants permanently rooted in mud) such as tufted sedge *Carex elata*, water horsetail *Equisetum fluviatile*, common marsh-bedstraw *Galium palustre*, yellow loosestrife *Lysimachia vulgaris* and marsh cinquefoil *Potentilla palustris* are common in relevé 5, a damp downy birch *Betula pubescens* woodland now being invaded by Norway spruce *Picea abies*. Therophytes are absent here.

The index of similarity, otherwise known as the Sørensen coefficient of community, has been used to illustrate the differences between these communities. It is calculated from the number of species found in one community (*a*), the number in the other (*b*), and the species common to both (*c*):

$$\textbf{Index of similarity} = \frac{2c}{a + b}$$

The floras of the two areas of ridge-ground pine forest (relevé 1 and 2) are very similar. Though both possess only 18 species of vascular plants, there are many bryophyte and lichen species, particularly where the cryptogamic mat is undamaged. Both are progressively less similar to the ancient spruce forest (relevé 3), the spruce forest of the buffer zone (4) and the seral birchwood (5). The normal spectrum taken from the world list has less than half the representation of hemicryptophytes and geophytes as that for the Fiby urskog reserve. The comparatively large representation of helophytes and hydrophytes in the reserve is, like that of so many Scandinavian forests, due to the vegetation of the lake and mires.

## 3.2 Light and shade

### *3.2.1 Influence of shade on tree development*

Owing to their height and complex structure, there can be less light near the ground in forests than in any other terrestrial vegetation type. Shade can thus be a very important factor in determining the forest dynamics. Very great variations occur in the light levels reaching the floors of, for example, Scandinavian forests in which periodic storms produce mosaic patterns of light and shade. Tree fall greatly increases the amount of light reaching many young saplings and dwarf trees of Norway spruce *Picea abies*, some of which develop on the rotting logs of trees that fell long ago, and are thus able to grow away rapidly after decades of relative inactivity. Careful inspection, however, often reveals skeletons of saplings for which the additional light came too late.

Light intensities below a forest (normally expressed as a percentage of full sunlight outside the forest) vary greatly. In Europe, an open birch forest may still have 20–50% of sunlight at ground level, dropping to 2–5% under beech *Fagus sylvatica*. As beech is deciduous, there is of course more light available during the leafless winter (see Section 1.4.2), but the trunks and branches still block some light such that light levels are still likely to be below 70–80% full sun. Evergreen forests tend to cast similar shade year-round (although the absolute amount of radiation reaching the ground is less in winter since the sun's apparent output is less and the lower sun angle leads to more interception by the atmosphere and a longer pathway through the forest). In Europe, summer light levels below natural Scots pine *Pinus sylvestris* forests are usually around 11–13% and below Norway spruce *Picea abies* they can be as little as 2–3%. In tropical rain forests, light levels at the forest floor may be even lower, reaching just 0.2–2.0%.

The shade cast by mature evergreen spruce forest, or by deciduous beech *Fagus sylvatica* forest in summer, is very intense and really active growth beneath it is virtually impossible. Far more light reaches the understorey of the relatively open eucalypt forests of Australia. The mature leathery (sclerophyllous) leaves of mountain ash *Eucalyptus regnans* are long and narrow, and hang downwards so preventing heating in the hot Australian sun. This keeps water loss within reasonable bounds and allows the growth of other species beneath it. Beech would overheat and suffer impossibly large water losses in such an environment.

Most plants need 20% of full sunlight for maximum photosynthesis (the **saturation point**) and require 2–3% sunlight to reach the **compensation point** where respiration costs are just balanced by photosynthesis, i.e. the plant is just breaking even and starts to grow actively. These are gross generalizations (variations are discussed below) but it demonstrates that the average light intensity at the forest floor is often below the compensation point of most woodland plants and seedlings, and yet they survive. Part of the solution to this paradox rests in **sunflecks**, patches of sunlight passing through gaps in the canopy, which can give up to 50% of full sunlight. Evans (1956) found that in a tropical rain forest in Nigeria 20–25% of ground area was illuminated by sunflecks at solar noon (when the sun is directly overhead) and that these made up 70–80% of the total solar energy reaching the ground. These flecks are thus seen to be important in a forest, especially to shade plants (see Section 3.2.2) that are capable of responding quickly to these brief flurries of light. Another consideration is that plants vary in their compensation point; Bates and Roeser (1928) found that while the pinyon pine *Pinus edulis* of the open forests of the arid American south-west has a compensation point of 6.3% full sunlight, the coastal redwood *Sequoia sempervirens* that grows in deep shade requires just 0.62% sunlight. Plants may also persist in light conditions below their compensation point (i.e. they are sustaining a net loss of energy) while their food reserves last. For some plants of the woodland flora, this may be a seasonal problem, enduring deep shade during the summer and growing in the spring and possibly the autumn (see Section 3.2.3). For seedlings it may be a waiting game, forming a persistent seedling bank (see Section 4.2.2) able to take advantage of an opening in the canopy if it should come before they die from lack of light.

Many investigations of **rain-forest tree seedlings** have emphasized the contrasts between those which are strongly **light-demanding** and those which are strongly **shade-tolerant**. A comparison of seedlings of 15 species belonging to the latter group, grown at three different light levels (10%, 0.8% and 0.2% full sunlight) for up to a year, affords an insight into the nature of competition between its members. Seedlings of the 15 species differed in their height and mass (weight) at the beginning of the experiment. Figure 3.4 shows final harvest values for seedling dry mass and height. Those of *Prunus turneriana* (labelled pt in the figure) were tallest at the start and remained so in both the 10% and 0.8% treatments. As seed size varied so much, the relative growth rate of biomass ($RGR_m$) was critical. At 0.8% that of the fastest-growing species, *Gillbeea adenopetala* (ga) was 30 times that of the slowest-growing species *Cryptocarya murrayi* (cm, a true laurel) even though the final mass of the latter is higher (Fig. 3.4). Intuitively, it might be expected that big seeds would result in fast-growing seedlings but the seed reserves of *G. adenopetala* were the smallest of all the species and 118 times smaller than those of *C. murrayi*. This suggests that in these forests rapid growth to outcompete other seedlings is more important than maintaining reserves of energy to last longer in deep shade, perhaps because gaps appear less frequently than in other forests. Differences in growth, especially height, between species at 10% full light were much less; seedling mortality in both these treatments over the full period of the experiment were negligible. At the lowest light level (0.2%) mortality increased greatly, and in a number of species reached 100% by the end of the experiment, reinforcing the idea of a desperate upward race for light.

Some species gained height much more rapidly than others, which they might well shade out under appropriate circumstances. However, the results also showed a change in height ranking (and thus in the competitive hierarchy) across these shade-tolerant species over time, especially under high-light regimes. Thus, the impact of a changing environment upon such competitors is likely to favour different species in different situations. At the lower end of the height-increase scale, species with greater survival rates under adverse conditions have an improved chance of outliving their neighbours and growing into adult trees when a forest gap forms. This study thus helps our understanding of how different combinations of morphological and physiological traits possessed by these 15 species of tree seedlings enables their **continued co-existence** in the same **shade-tolerant niche**.

Aerial photographs make it possible to investigate **long-term canopy dynamics** over considerable periods; indeed Fujita *et al.* (2003) have already done so over a period of 32 years in a 4-ha permanent plot in an old-growth evergreen broadleaved forest in the Tatera Forest Reserve, south-west Japan. This reserve is dominated by Japanese chinquapin *Castanopsis cuspidata* and the dawn isu tree *Distylium racemosum* at low altitudes and the Japanese evergreen oak *Quercus acuta* on the hills. In the low-altitude study plots of this truly primeval forest, canopy height varies between 20 and 30 m, while the diameter at breast height (dbh) of some trees exceeds a metre. Aerial photographs taken in 1966, 1983, 1993 and 1998 enabled the creation of digital elevation models on a 2.5 m grid of the canopy surface of a 10-ha area. This

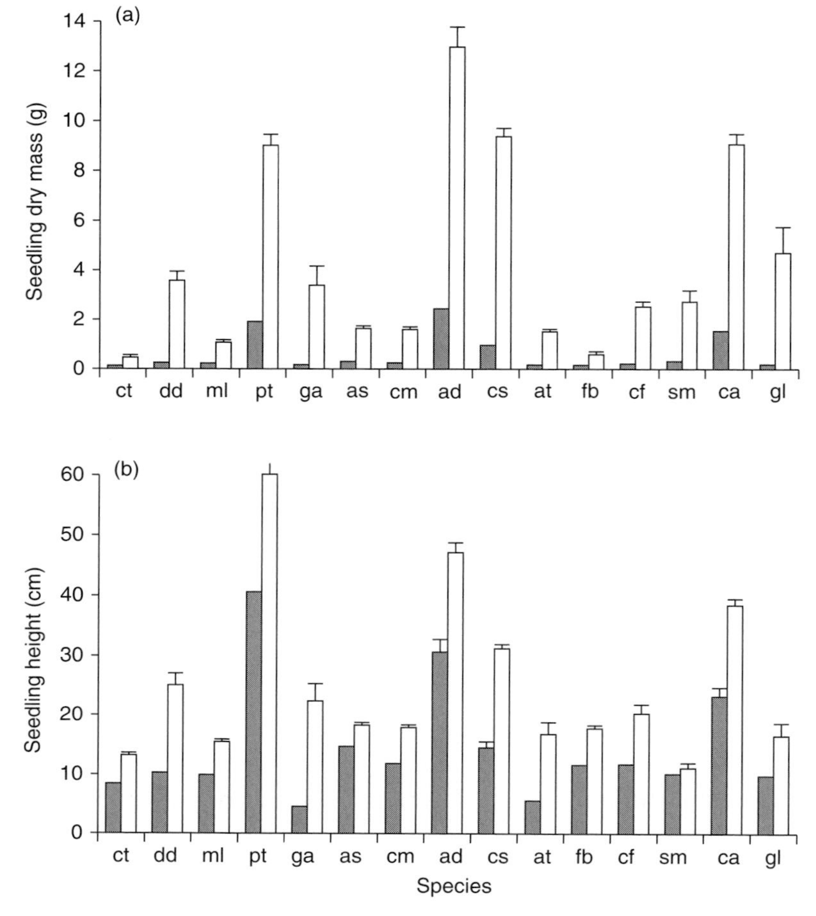


Figure 3.4 Final harvest values of (a) seedling dry mass and (b) seedling height for 15 shade-tolerant Australian tropical rain forest species grown from seed in either 10% (open bars) or 0.8% (dark bars) of full sunlight. Species are arranged according to length of their growing season, from 7 months on the left to 12 months on the right (From Bloor and Grubb, 2003. *Journal of Ecology* 91, Blackwell Publishing.)

| | | | |
|---|---|---|---|
| *Gillbeea adenopetala* | ga | *Darlingia darlingiana* | dd |
| *Cryptocarya murrayi* | cm | *Prunus turneriana* | pt |
| *C. aff. triplinervis* | ct | *Flindersia bourjotiana* | fb |
| *C. mackinnoniana* | ml | *Castanospora alphandii* | ca |
| *Argyrodendron trifoliolatum* | at | *Cupaniopsis flagelliformis* | cf |
| *Aglaia sapindina* | as | *Guioa lasioneura* | gl |
| *Athertonia diversifolia* | ad | *Synima macrophylla* | sm |
| *Cardwellia sublimis* | cs | | |

included the 4-ha plot which was surveyed in order to create a topographic map of the ground surface in 1990. The 3.4% of the 6400 2.5 × 2.5 m quadrats which remained as gaps throughout the 32-year period tended to be in the centres of large old gaps, whereas those experiencing re-disturbance (2.5%) were along the edges of old gaps; 74.5% of the area, in which gaps were constantly forming and closing, remained as closed canopy throughout. The establishment of deciduous and shade-intolerant pioneer species in predominantly evergreen broadleaved forests like this was seen to be dependent on the existence of long-term, large openings.

The important influence of shade upon the development of young trees has been thoroughly investigated, but complications resulting from **herbivory** have been less well recognized. Survival of shade-tolerant juvenile trees in the **understorey** depends on resource allocation strategies which divert resources to the recovery from damage inflicted by deer and other herbivores, but slow their upward growth. In contrast, resource allocation should enable rapid height growth in **canopy gaps** where competitor saplings grow rapidly. Using the shade-tolerant dipterocarp *Shorea quadrinervis*, a dominant canopy tree in Borneo, Blundell and Peart (2001) investigated the interaction between canopy gaps and simulated herbivory on juvenile plants (<1 cm dbh) in an area where caterpillars and grasshoppers cause the most damage to its foliage. Under natural conditions *S. quadrinervis* juveniles often spend many years under low-light conditions in the understorey, where on average they grow upwards at a rate of only 1 cm a year, and experience an average mortality rate of approximately 10% a year. Such young trees allocate a much higher proportion of their resources to the roots than do those growing in gaps and consequently have higher **root/shoot ratios**. While this slows their growth towards the light, this less-vulnerable below-ground tissue can be used as a source of energy for leaf replacement in the event of herbivory. These tendencies were clearly shown in the results of a shading experiment depicted in Fig. 3.5, which involved the use of 24 juveniles that were collected from the understorey around an adult tree in the forest used in the main experiment. Half were subsequently grown under shade conditions corresponding to the understorey and the other half placed in an artificial gap for 8 months.

In the same study, **simulated herbivory** experiments involved the removal of the apical meristem or 10, 50 or 90% of the foliage from plants in the forest itself. Survival of juvenile plants in the gaps was high over the 8 months of the experiment and not significantly affected even by removal of 90% of the foliage. Of the plants growing in the understorey beneath the forest canopy survival of those suffering up to 50% defoliation was again not significantly affected, but some 70% of plants suffering 90% foliage removal died within

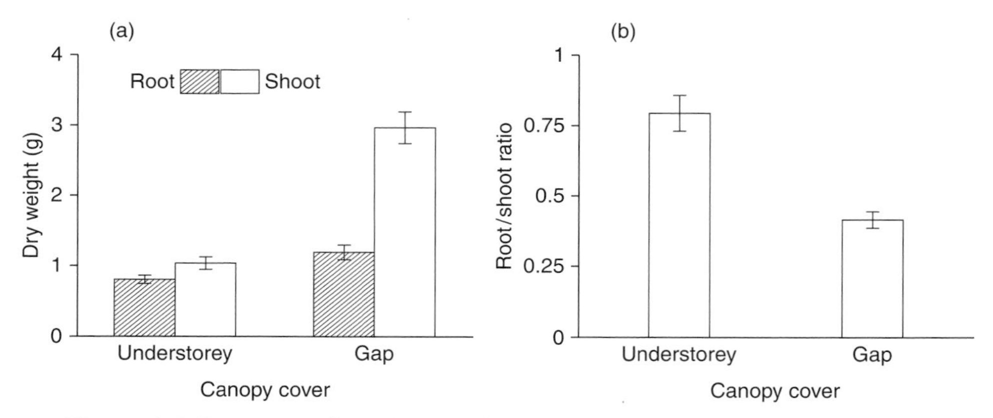


Figure 3.5 Resource allocations and root/shoot ratios in juveniles (<1 cm dbh) of the dipterocarp tree *Shorea quadrinervis* grown in gaps ($n = 12$) and the understorey ($n = 12$). Values shown are means $\pm 1$ SE. Experiments were run over 8 months at Gunung Palung National Park, Kalimantan Barat, Indonesian West Borneo. (From Blundell and Peart, 2001. *Journal of Ecology* 89, Blackwell Publishing.)

the 8 months. The authors suggest that most understorey individuals experiencing a single severe defoliation would probably die before they had been observed in an area survey.

In the understorey, growth was negligible even in controls (0% of leaf removal) and there was no effect of the removal treatments (Fig. 3.6). Height growth in the juveniles placed in the gaps was actually increased by herbivory, particularly removal of the apical meristem or 10% of leaf tissue and was reduced only after 90% removal. Conversely, under the shade of the canopy, height growth of all individuals, both control and experimental, was negligible (Fig. 3.6). *Shorea quadrinervis* can be seen to be adapted to rapid growth in open gaps. But the increased growth of gap plants after herbivory is unexpected considering that the reduced leaf area must lead to less photosynthesis. This indicates that the plant is adapted to invest as much energy as possible in new height growth in order to maintain its competitiveness for light against its neighbours. Higher damage levels increased subsequent net leaf loss. Although leaf production was much higher, leaf retention was much lower in the gaps than in the understorey.

Traditionally, trees and other plants have been classified as **shade tolerant** or **intolerant** (see Box 9.3) which gives useful information to foresters and others interested in regeneration in forests. Trees such as European beech *Fagus sylvatica* and the sugar maple *Acer saccharum* from North America are very tolerant of deep shade while birches and poplars grow best under high light intensities. However, it is now apparent that the ability to tolerate shade can

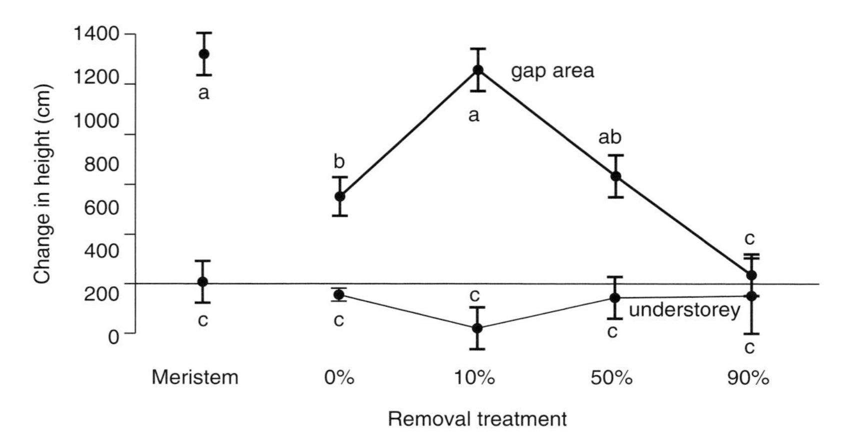


Figure 3.6 Effect of simulated herbivory on change in height over 8 months of the dipterocarp tree *Shorea quadrinervis* at Gunung Palung National Park. Plants in gaps are indicated by thick lines, those in the understorey by thin lines. Simulated herbivory involved removing just the apical meristem or removing various percentages of tissue from all leaves; 0% removal were the untouched controls. Change in height plotted as residuals (means ±1 SE) of change in height vs. initial height. Different letters indicate statistically significant differences. (From Blundell and Peart, 2001. *Journal of Ecology* 89, Blackwell Publishing.)

change through the life span of a tree. The ways in which the light requirements of a developing individual tree change as it progresses from a seed to a seedling to the juvenile (sapling) stage and finally to an adult are discussed by Poorter *et al.* (2005a). They measured crown exposure (i.e. degree of exposure to direct sunlight) for 7460 trees belonging to 47 different species in a Liberian lowland rain forest on the Atlantic coast of West Africa, following individual trees for periods varying from 2.8–9.8 years. Figure 3.7 illustrates the way in which the average irradiance of light at population level can be low, intermediate or high, and how this can change as the forest matures. A species with a low light requirement throughout its life could have a **height–light trajectory** (**HLT**, a fitted curve relating canopy exposure to tree height) running from bottom left to top left, while one with a high light requirement throughout its life would have a vertical trajectory on the right side of the diagram. In practice, as the diagonal arrows on the diagram indicate, light requirements of particular species often vary between different stages.

Poorter *et al.* (2005a) were particularly concerned with the transition between the juvenile and adult stages, and found that all nine of the height–light trajectories indicated on the diagram were represented in the 47 tree species they examined. This is important because it shows that later requirements of

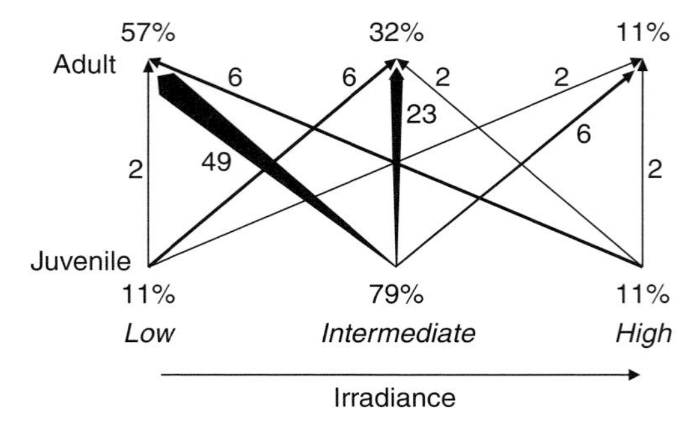


Figure 3.7 Relative commonness of different height-light trajectories for 47 rain-forest tree species. The relative light levels of juveniles (below) and adults (above) are shown. Light levels are classified as *relatively* low (lower than the average forest level), intermediate or *relatively* high. The number of species that follow a particular light trajectory is indicated by the thickness of the arrow and the corresponding percentages. (From Poorter *et al.*, 2005a. *Journal of Ecology* 93, Blackwell Publishing.)

species classified as pioneer or shade tolerant on the basis of seed and seedling behaviour may be very different. Two strategies were common. The majority (57%) of species started in intermediate light environments in the juvenile stage and passed to low light environments as adults; these are typically shorter trees that become progressively stuck under taller canopy trees. The other common strategy was to stay in intermediate light conditions from juvenile to adult. The whole-life shade-tolerant and the whole-life shade-intolerant niches were filled by just one species each (2% of the 47 species investigated). Large stature species tend to possess relatively slender stems and narrow crowns, grow rapidly and so achieve increased crown exposure earlier.

### *3.2.2 Light and shade plants: growth analysis*

It was shown above that the adult trees differ in their degree of tolerance to shade and that the degree of shade tolerance may change in an individual tree over its life. It should also be noted that trees can differ in their shade tolerance within an individual canopy. Trees that live in shaded environments may have just one layer of leaves on the outside of the canopy like the fabric on an umbrella – mono-layered trees. Most shade-intolerant trees, however, have several layers (multi-layered trees) one inside the other. In fact it may be difficult to distinguish discrete layers but nevertheless multi-layered trees have a thick canopy where some leaves are grown below others. Thus, the

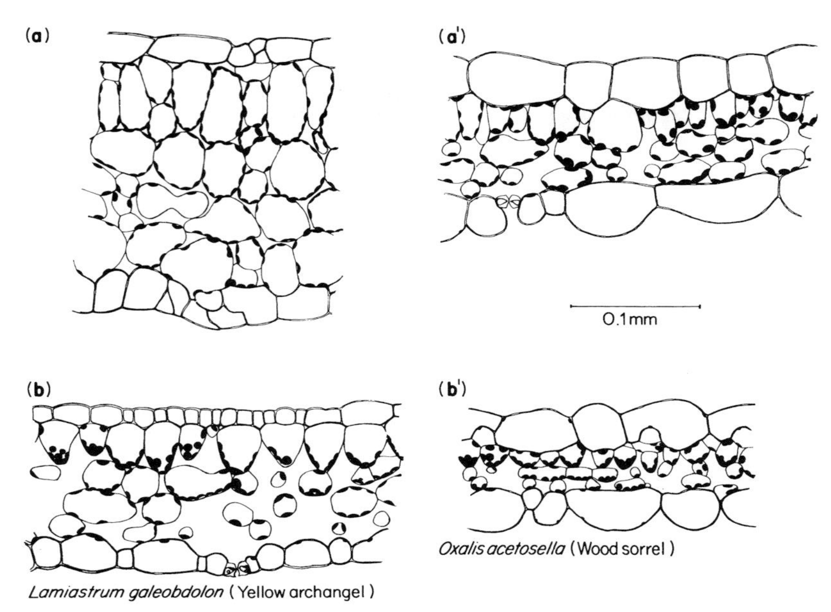


Figure 3.8 Transverse sections through leaves of yellow archangel *Lamiastrum galeobdolon* ssp. *montanum* (left) and of wood-sorrel *Oxalis acetosella* (right). Both species are **hypostomatous**, i.e. have their stomata restricted to the undersides of the leaves. (a), (a′) sun leaves; (b), (b′) shade leaves, whose palisade mesophyll consists of funnel cells (see text for an explanation). (From (right) Packham and Willis, 1977. *Journal of Ecology* 65; (left) Packham and Willis, 1982. *Journal of Ecology* 70, Blackwell Publishing.)

shade-intolerant silver maple *Acer saccharinum* is multilayered while the shade-tolerant sugar maple *A. saccharum* is a monolayered tree. Multiple layers of leaves works because, as noted above, most leaves need around just 20% of full sunlight to photosynthesize at their maximum rate (this does vary between species and the degree of environmental stress leaves are under; for example photosynthesis is less efficient in water-stressed leaves). This broad figure of 20% is a quite small requirement for leaves, so providing each layer of leaves is open enough to allow more than this past them, then lower layers of leaves can survive. This is also helped by the leaves inside the canopy being shade leaves capable of utilizing light more efficiently than those growing in full sun – the sun leaves (Fig. 3.8, and see below for a discussion of how these leaf types differ). Care is needed when considering light below leaves that only wavelengths of light usable in photosynthesis are considered. Yellow and blue light (peaking at 430 and 662 nm wavelength, respectively) is used by chlorophyll and these frequencies make up the bulk of the **photosynthetically active radiation (PAR)**. Given that the soft green light (around 600 nm) which we appreciate under a

TREE LAYER

*Carpinus betulus* hornbeam
*Alnus glutinosa* alder
+ *Tilia cordata* small-leaved lime
*Acer pseudoplatanus* sycamore
*Fagus sylvatica* beech
*Quercus robur* pedunculate oak
*Fraxinus excelsior* : ash

SHRUB LAYER

*Ribes uva-crispa* gooseberry
*Sambucus nigra* elder
*Lonicera xylosteum* fly honeysuckle
*Crataegus* sp. hawthorn
^ *Prunus padus* bird cherry
* *Prunus avium* wild cherry
+ *Acer campestre* field maple
*Coryllus avellana* hazel
*Humulus lupulus* hop
^ *Euonymus europaeus* spindle
*Rubus idaeus* raspberry
*Rubus caesius* dewberry
*Cornus sanguinea* dogwood
*Hedera helix* ivy
*Ilex aquifolium* holly

HERB LAYER: GREEN IN SPRING

*Leucojum vernum* spring snowflake
^ *Gagea lutea* yellow star-of-bethlehem
* *Anemone nemorosa* wood anemone
* *Adoxa moschatellina* moschatel
*Ranunculus ficaria* lesser celandine
*Corydalis cava* hollowroot
*Veronica hederifolia* ivy-leaved speedwell
*Arum maculatum* lords-and-ladies
* *Allium ursinum* ramsons

EARLY FLOWERING SUMMER GREEN

*Pulmonaria officinalis* lungwort
^ *Primula elatior* oxlip
*Mercurialis perennis* dog's mercury
* *Ranunculus auricomus* goldilocks buttercup
* *Paris quadrifolia* herb paris
*Ranunculus lanuginosus* downy buttercup
^ *Polygonatum multiflorum* solomon's-seal
*Geum urbanum* wood avens
*Stellaria nemorum* wood stitchwort
*Geranium robertianum* herb-robert
* *Vicia sepium* bush vetch
*Phyteuma spicatum* spiked rampion
* *Melica uniflora* wood melick
*Galium aparine* cleavers
*Silene dioica* red campion
* *Poa nemoralis* wood meadow-grass
*Aegopodium podagraria* ground-elder
* *Carex remota* : remote sedge
*Stachys sylvatica* hedge woundwort
*Dactylis glomerata* cock's-foot
*Chaerophyllum temulum* rough chervil

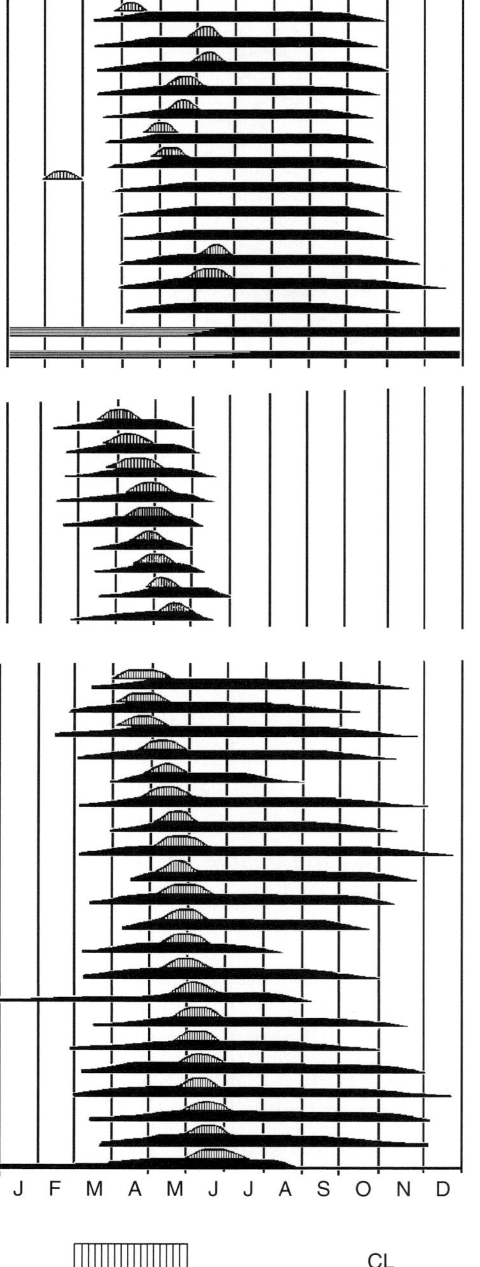


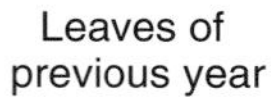
Leaves of previous year

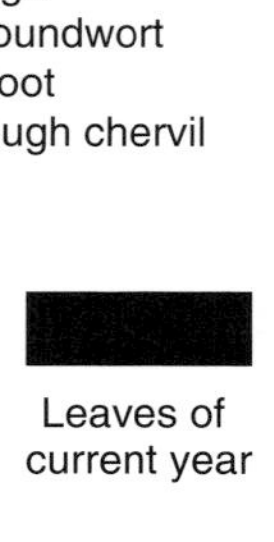
Leaves of current year

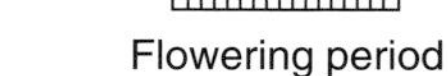
Flowering period

CL
Period when closed (cleistogamous) flowers produced

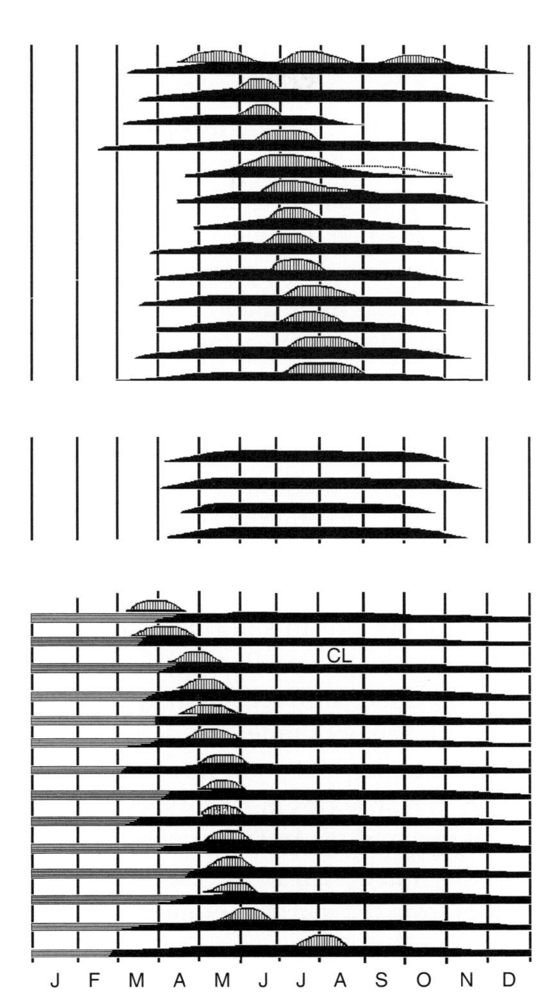


Figure 3.9 Phenological development of different species in the tree, shrub and herb layers of central European damp oak–hornbeam *Quercus–Carpinus* forests. * implies that the species concerned was one of the 23 shown in this diagram which were listed by Rose (1999) as Ancient Woodland Vascular Plant Indicators (AWVPs) in all four regions of southern Britain (see Section 6.4.1). ^ means that the plant is not listed in all four regions. + means these species should not be used as AWVPs in England unless they are well within a wood and do not appear to have been planted. See Sections 5.9 and 6.4.1 for more detail. Data are an average of 3 years. (Modified from Ellenberg, 1988. *Vegetation Ecology of Central Europe*. Cambridge University Press.)

dense canopy is of little use to plants, we may often fail to appreciate just how dark it can be for an understorey plant.

Light is the most generally limiting factor for herbs, shrubs and young trees in woodlands, where many of them show remarkable adaptations to different levels of PAR. In British woodlands, **shade evasion** is found in such plants as bluebell *Hyacinthoides non-scripta*, wood anemone *Anemone nemorosa* and lesser celandine *Ranunculus ficaria* that have their main period of growth before the tree canopy expands (Fig. 3.9). Truly **shade-tolerant** species such as enchanter's nightshade *Circaea lutetiana* frequently have their shade leaves so modified, by increasing the light-catching area and chlorophyll content, that for the same expenditure of weight they achieve much the same photosynthetic rates as sun leaves receiving considerably higher levels of radiation. Species such as yellow archangel *Lamiastrum galeobdolon*, which grows along hedgerows as well as in woodlands, possess great **phenotypic plasticity** (the capacity for marked variation in appearance caused by environmental conditions) and show differences between their sun and shade leaves considerably greater than those in wood-sorrel *Oxalis acetosella*, a species well adapted to shade but not to sun (Packham and Willis, 1977, 1982). These and many other primarily woodland species, show very marked variation when they develop under different light regimes, tending to become wider and thinner under the heavy shade of a dense woodland canopy. Their **sun leaves**, which develop under high light intensities, are smaller, thicker and often yellowish-green. **Shade leaves** in many species commonly have a spongy mesophyll (the main tissue of the leaf) with very large air spaces and palisade cells (the main photosynthetic tissue directly under the top epidermis) which taper downwards from the margin adjoining the upper epidermis and are known as **funnel cells** (Fig. 3.8). They also have larger epidermal cells, a lower vein density and a thinner cuticle than sun leaves. In addition to filtering out wavelengths usable in photosynthesis, leaves also change the red/far-red (R/FR) ratio of the light by preferentially absorbing red light (red light is approximately 660 nm in wavelength, far-red is 700–1000 nm); the ratio is typically 1.2 in daylight and is reduced to around 0.2 in the green light filtering through leaves. Mitchell and Woodward (1988) demonstrated that yellow archangel responds to these differences in the light quality. When receiving PAR filtered to give a low R/FR ratio, as under a deciduous tree canopy in summer, this species flowers less vigorously and its internodes and petioles are more elongated compared with plants receiving the same amount of unfiltered light. Foraging for PAR amongst ground vegetation is assisted by these responses. The R/FR ratio is also important in the germination of seeds. Short exposures to red light tend to break dormancy while far-red promotes dormancy. This helps ensure that

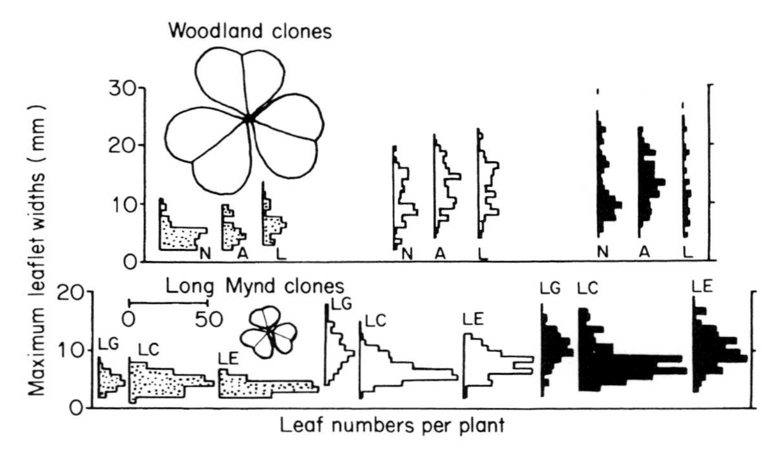


Figure 3.10 Variation in leaflet widths of clones of wood-sorrel *Oxalis acetosella* from two contrasting sites in Shropshire, UK. Clones N, A and L were from a woodland streamside near Telford; clones LG, LC and LE were from beneath bracken *Pteridium aquilinum* in a montane pasture on the Long Mynd. Clones were gathered from the wild in late March 1971 and grown in cold frames under three light regimes: under clear polythene only (stippled bars), under light further reduced by a white muslin shade (clear bars); and in light greatly reduced by a black muslin shade (filled bars). The plants were harvested in August 1972. (Redrawn from Packham and Willis, 1977. From Packham and Harding, 1982. *Ecology of Woodland Processes*. Edward Arnold.)

seeds germinate when they are receiving direct unfiltered sunlight when the seedlings will have most chance of survival.

Figure 3.10 shows variations in leaflet widths of clones of wood-sorrel from two contrasting habitats in Shropshire, mid-England when grown under three different light regimes. The results of this experiment demonstrate both the effects of phenotypic plasticity, with the leaflets of clones from both sites becoming larger as shading increased, and of genetic factors. Leaflet width/leaflet number relationships between the two sites differed; plants originating from the woodland streamside near Telford having fewer but larger leaves than those from the more exposed montane site, where wood-sorrel grew beneath a cover of bracken *Pteridium aquilinum*. Individual clones from both sites also differed considerably amongst themselves; note particularly the marked contrast between the Long Mynd clones LG and LC.

**Growth analysis** (Table 3.1) affords valuable techniques for studying the ways in which plants respond to changes in light level, humidity regime, temperature and other features of the environment that are altered by shading. Apart from cases where shading is so extreme as to result in the production of relatively small, etiolated leaves, **chlorophyll content** per unit dry weight is usually higher in shade

Table 3.1. *Definitions used in growth analysis*

1. **Unit leaf rate** (ULR) $= \frac{dW}{dt} \times \frac{1}{L_A}$

2. **Leaf weight ratio** (LWR) $= \frac{L_W}{W}$

3. **Specific leaf area** (SLA) $= \frac{L_A}{L_W}$

Where $L_A$ = total leaf area
$L_W$ = total leaf dry weight
$W$ = total plant dry weight
$\frac{dW}{dt}$ = rate of dry weight increase of the whole plant

**Relative growth rate** $RGR = ULR + LWR + SLA$

**Leaf area ratio (LAR)** is a morphological index of plant form, the leaf area per unit dry weight of the whole plant.

4. Leaf area ratio (LAR) $= \frac{L_A}{W} = LWR \times SLA$

In contrast, **unit leaf rate** (ULR), the rate of increase in dry weight of the whole plant per unit leaf area, is a physiological index closely connected with photosynthetic activity. A high LAR together with a low ULR is characteristic of heavily shaded woodland herbs in temperate forests.
**Relative leaf growth rate** (RLGR) is analogous to RGR and is the rate of increase in leaf area per unit leaf area.

5. **Leaf area index** (LAI) $= \frac{L_A}{\textit{Ground area occupied by plant}}$

6. **Stomatal index** $= \frac{\textit{Number of stomata per unit area}}{\left(\begin{array}{l}\textit{Number of stomata per unit area} + \\ \textit{Number of epidermal cells per unit area}\end{array}\right)}$

leaves than in sun leaves. Differences in chlorophyll contents between sun and shade leaves appear much greater when expressed on a dry-weight than a fresh-weight basis, while chlorophyll values *per unit area* may actually be lower for shade leaves than sun leaves. Shade plants do not always have thin leaves; many rain-forest species such as *Cordyline rubra* and *Lomandra longifolia* have thick leaves with unusually large chloroplasts concentrated in the upper palisade, giving them a high chlorophyll content and a low specific leaf area (SLA: see Table 3.1). Stomatal frequency and pore size vary amongst rain-forest plants, but differences between sun and shade species are not significant.

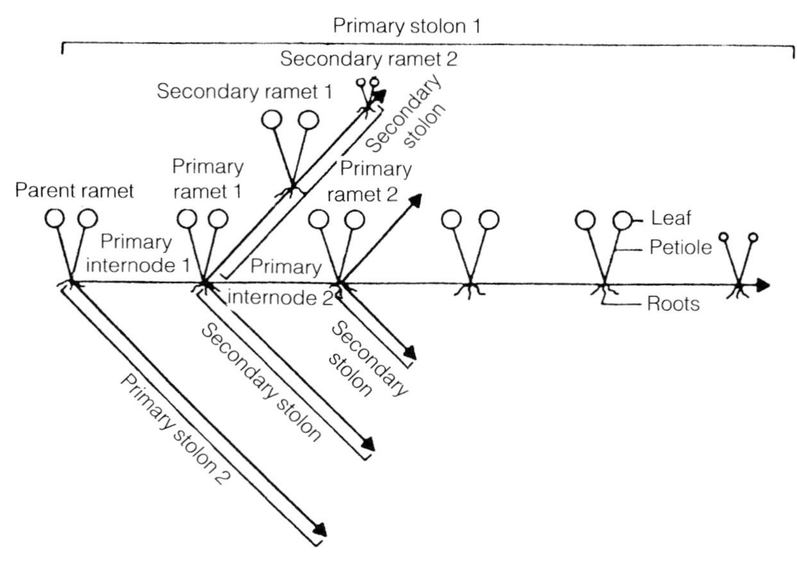


Figure 3.11 Schematic diagram of the basic vegetative morphology of ground ivy *Glechoma hederacea*. (From Hutchings and Slade, 1988. *Plants Today* 1, Blackwell Publishing.)

Plants and animals both carry out **foraging**, a process in which an organism searches or grows within its habitat so as to obtain essential resources. The main resource-gathering structures of plants are roots (see Section 2.3) and leaves, whose distribution in two- or three-dimensional space is partly determined by genetic 'growth rules' which influence many features of form. This is particularly well seen in the woodland labiates ground ivy *Glechoma hederacea* and yellow archangel, where the distance between adjacent nodes on growing stems, the probability of branching at nodes, and the angle between branches are important (Fig. 3.11). In such plants the basic plant units, known as **ramets**, consist of two horizontal leaf blades (laminas), each attached to a rooted node by an erect petiole. If planted, a parent ramet produces two primary stolons with new ramets at each node, and under favourable conditions a branched clonal system soon develops.

Clonal integration, plasticity and the effects of different nutrient and light levels on foraging in ground ivy are of particular interest, as this plant can be seen as a model for other woodland species of similar growth form and ecology. Slade and Hutchings (1987a,b,c) found that mean numbers of stolon branches, ramets per clone and mean weight per stolon length were all greatest in a high-light, high-nutrient treatment; they were least with low light and high nutrients. The latter plants had the highest percentage dry-weight allocation to petioles (leaf stalks) and the lowest to roots; the relatively low-mass stolons and the few side branches allowed them to forage for light extensively rather than intensively as did plants in the high-light, high-nutrient treatment whose behaviour would tend to keep them within any favourable area encountered naturally. Plants in the high-light, low-nutrient treatment possessed the highest root/shoot ratio of the three treatments; this again is an effective response, one that promotes effective foraging for nutrients.

Hughes (1959), using artificial shading under controlled conditions, showed that the meristematic activities of the sun and shade leaves of small balsam *Impatiens parviflora* are fundamentally similar. The shade leaves simply expand more; because of this greater expansion the cells are wider so the stomatal frequency is lower, though the **stomatal index** (Table 3.1) remains the same. Working with the same species, Young (1975) found that **leaf weight ratio** (LWR) was markedly affected by the rooting medium, but little altered by changes in total daily light. Specific leaf area (SLA) varies with temperature, rooting medium, day length, total daily light and the 'physiological age' of the leaf or plant. Because SLA is, over a substantial range of daily light level, inversely proportional to total daily radiation received while **unit leaf rate** is directly proportional to it, the net effect of these two relationships is that the **relative growth rate** (RGR) of small balsam remains approximately constant over the light range concerned.

The **root/shoot ratio** of woodland plants appears to be related to soil moisture content as well as intensity of radiation, but in shade plants the proportion of plant weight devoted to roots is commonly low. Temperate shade species growing under lower PAR tend to possess high fresh-weight/dry-weight ratios, and SLAs, while RGR is low, and the niches in which many such plants occur are often not particularly favourable physiologically; their presence in them is related to their ability to compete under the conditions concerned. In nature shade plants often occur at light levels well below the optimum for the species. Heavy shading has, however, eliminated competition from sun species of much higher potential RGR.

Unlike shade plants that cannot utilize the additional light, sun plants can continue to increase net photosynthesis until much higher light levels are reached. This higher capacity for carbon dioxide fixation per unit leaf area in sun plants is directly related to the greater amounts of carboxylating enzymes and the greater volume of the leaf per unit leaf area. The proportion of chlorophyll *a* to chlorophyll *b* increases in plants grown at high light intensities; this makes sense because chlorophyll *a* is responsible for capturing the energy from light, so more of the extra energy in high-light environments can be utilized. Even fully grown leaves may partially adapt to a change in average light level by modifying their enzyme systems.

Boardman (1977) reviewed the wide variety of morphological, physiological and biochemical mechanisms involved in the phenotypic plasticity that enables a given genotype to adapt to various light levels. His conclusion was that the light levels to which a genotype can adjust reflect a genetic adaptation to the conditions of its native habitat and that adaptation for great photosynthetic efficiency in strong sunlight precludes high efficiency in dense shade.

### *3.2.3 Seasonal changes and aspect societies*

**Phenology** is the study of the onset and duration of the activity phases of animals and plants throughout the year (Sections 4.6.1, 11.3.2). These are largely synchronized to the weather so the dates on which they occur differ from year to year. Though the order in which the various plant species unfold their buds, flower, fruit and senesce is much the same; different individuals of the same tree species frequently commence leafing out (**flushing**) earlier or later than their fellows. This influences, amongst other things, their liability to defoliation by insects (Section 5.3). However, as described in Section 1.4.1, this sequential activity of different species leads to the formation of **aspect societies** where the plant community is dominated by one or several species at a particular season and whose appearance changes markedly over just a few weeks in spring.

Although the total number of species is much greater, the broad **phenological groups** of the damp oak–hornbeam woodlands of central Europe (Ellenberg, 1988) correspond well with the periods of active vegetative growth and flowering of those in an oak–hornbeam wood in Hertfordshire, southern England by Salisbury (1916a,b) long ago. Salisbury noted that several species (see Box 3.1) had **pre-vernal flowering** that was almost completed before the leaves of hornbeam *Carpinus betulus*, which formed part of the deciduous tree canopy, were fully expanded in mid-May. Aerial parts of lesser celandine *Ranunculus ficaria* and wood anemone *Anemone nemorosa* die down completely by the end of June, but dog's mercury *Mercurialis perennis*, a **summer-green** plant, remains active throughout the summer and shoots are commonly present during the winter in sheltered British woodlands. Bluebell *Hyacinthoides non-scripta* has **vernal flowering**, continuing to flower until mid-June, but then dies down very abruptly although its seed capsules are held high above the ground until early autumn. Pignut *Conopodium majus* commences vegetative growth at the beginning of March, has summer (**aestival**) flowering starting in May, and continues active growth through the summer. The influence of climate on flowering time is illustrated by lesser celandine, which usually flowers in late February in Hertfordshire, early April in Germany and late April in the Ukraine, which has a continental climate. It is also interesting to see the same niche occupied by different species in different

**Box 3.1**

Variations on the theme of understorey plants growing before and after the deciduous tree canopy casts the deep shade of summer, using European examples. See also Fig. 3.9 which shows the timing of spring growth of understorey plants in an oak–beech wood in relationship to the leafing-out of the trees and shrubs.

**Pre-vernal** ('before spring') and **vernal** ('spring') plants produce their leaves sometimes as early as January and by the beginning of July have completely died back above ground.

Bluebell *Hyacinthoides non-scripta*
Lesser celandine *Ranunculus ficaria*
Spring snowflake *Leucojum vernum* (shown in Fig. 3.9)
Wood anemone *Anemone nemorosa* (shown in Fig. 3.9)

**Summergreen** plants start with the pre-vernal plants and do most of their growing before the tree leaves appear but keep their leaves through the summer using what little light is available.

Dog's mercury *Mercurialis perennis* (shown in Fig. 3.9)
Meadowsweet *Filipendula ulmaria* (shown in Fig. 3.9)
Honeysuckle *Lonicera periclymenum*

**Wintergreen** plants keep at least a few green leaves year round, usually as a new basal rosette formed before the winter, so that growth can start as soon as spring conditions allow and can continue into a warm late autumn after leaf fall.

Wood-sorrel *Oxalis acetosella* (shown in Fig. 3.9)
Primroses *Primula* spp.
Yellow archangel *Lamiastrum galeobdolon* (shown in Fig. 3.9)

**Evergreen** plants are the true evergreens that keep all their leaves year round.

Stinking iris *Iris foetidissima*
Ivy *Hedera helix* (shown in Fig. 3.9)
Holly *Ilex aquifolium* (shown in Fig. 3.9)

**Parasites and myco-heterotrophs** solve the problem by acquiring their carbon from living green plants or fungi, so doing away with the need for light and photosynthesis.

Toothwort *Lathraea squamaria* (parasitic)
Bird's-nest orchid *Neottia nidus-avis* (myco-heterotroph)
Coralroot orchid *Corallorhiza trifida* (myco-heterotroph)
Yellow bird's-nest *Monotropa hypopitys* (myco-heterotroph)

**Myco-heterotrophs** were, for over a century, incorrectly described as saprotrophs and were believed to live on soil organic matter. In fact they are parasitic upon fungal carbon, a reversal of the normally accepted relationship between vascular plants and fungi (Leake, 2005). Over 400 plant species belonging to 87 genera, which lack chlorophyll are parasitic upon fungi, which they exploit as their principal source of carbon. It is also estimated that over 30 000 species depend upon **myco-heterotrophy** for establishment from dust-like seeds or spores during critical early phases, but produce green shoots on emerging into light from the soil or growing as epiphytes on other plants.

Recent research has yielded much useful information regarding the early stages of the plant/fungal relationships and the very specific associations between members of the two groups. Yellow bird's-nest, for example, when growing on the dune slacks at Newborough Warren NNR, Anglesey, Wales, is associated with the *Salix*-specific fungus *Tricholoma cingulatum* when present under its autotrophic co-associate creeping willow *Salix repens*. Virtually all the nutrients taken up by the yellow bird's-nest are acquired from the fungal sheaths on its roots, so it is thus indirectly parasitic on the creeping willow. When this herb is growing under Scots pine in the same area, the roots of both vascular plants are linked by *T. terreum*, a closely related fungus.

places. Hollow corydalis *Corydalis cava*, for example, grows on good soil in southern Sweden and central Europe but is not native in Britain where its place is taken by bluebell. Hollow corydalis flowers at the same time as lesser celandine, but its foliage, though produced slightly later, dies back earlier.

The conspicuous **chasmogamous** flowers of violets *Viola* spp. and wood-sorrel *Oxalis acetosella* (see Section 4.3), which are open and can be visited by insects, develop in spring. Later in the year both form **cleistogamous** flowers that are closed, much reduced and self-pollinated. Those of wood-sorrel have hooked peduncles and are often buried in plant litter; it is probable that most seeds of this species are produced by such flowers. Cleistogamy is an efficient means of reproduction, but its adoption reduces gene flow.

Phenological studies investigating dry-matter allocation to the organs of species characteristic of different aspect societies have been made both under cultivation (Packham and Willis, 1977, 1982), and in the wild. The geophytes ramsons *Allium ursinum* (Ernst, 1979) and bluebell *Hyacinthoides non-scripta* have been studied in this way in their native woodlands (Fig. 3.12). In both species the bulb is renewed annually. At the beginning of October bluebell bulbs, which are hidden well below the soil, have sloughed off the roots active in the previous summer. The new roots beginning to emerge through the side of each bulb arise from the base of a newly initiated 'daughter' plant in the centre of the bulb. These make their way by enzyme action through the leaf and bud scales of the 'parent' bulb, and by late autumn the bulb scales, specialized tubular scale leaves, foliage leaves, flower stalk (scape) and flowers of the new

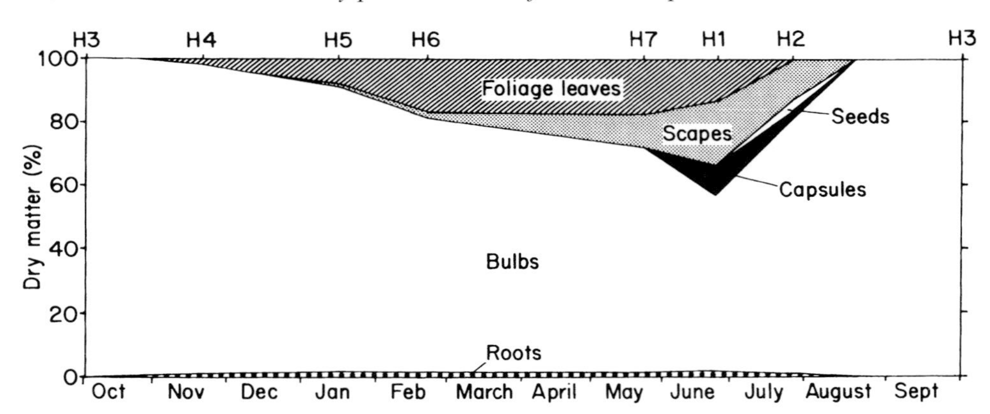


Figure 3.12 Percentage dry matter allocation to plant organs in bluebell *Hyacinthoides non-scripta* growing amongst bracken *Pteridium aquilinum* on an acidic sandy brown earth in an open region of Himley Wood, near Dudley, West Midlands, England. Note the high allocation to bulb weight. Each of the seven harvests (H1–H7) is based on the plants within a 0.1 $m^2$ area. Samples included plants of all the age classes present, but in this site there was a preponderance of mature bulbs, over 90% of bulb biomass occurring at a depth of 20 cm or more. (Reprinted from Grabham and Packham, 1983. *Biological Conservation* **26**, with permission from Elsevier.)

plant can be clearly seen. They increase in size as the old bulb withers, and by mid-February the shoot will often have risen above the soil and begun to exploit the relatively high light intensities found at the forest floor before the tree canopy expands. The plant rapidly accumulates the food reserves needed for flowering and a new, usually larger, bulb is formed. Flowering ends in June, capsules form and the now flaccid leaves come to rest on the soil surface, where they soon decompose. The cycle is complete by autumn and the seeds, which under favourable conditions take five years to develop into flower-bearing plants, have been discharged from the capsules.

In the European wild daffodil *Narcissus pseudonarcissus* establishment of new plants by daughter bulbs is high in comparison to that by seedlings, but in both bluebell and ramsons there is usually a high rate of seed output and seedling establishment per unit area of stand. Vegetative propagation is rare in both these species.

## 3.3 Water

### *3.3.1 Water yield and quality*

Water is of tremendous importance to all vegetation. On a global scale, small changes in precipitation produce major variations in the type of forest able to grow (see Section 1.6.1). Like the nutrient cycles described in Section 8.3, water moves in a global cycle, falling on the land, evaporating or running off back to the sea, before evaporating and falling back on the land again. This **water cycle** is important in the global picture of what precipitation falls where. On a smaller scale, foresters and forest ecologists are often more interested in what happens within a **catchment area (or watershed)**, a hydrologically discrete area that feeds all its water into a main river system (Fig. 3.13). With forests the interest often lies in water yield and quality. Within a watershed the water cycle can be summarized as:

$$Q = P - (ET + G + S)$$

Where $\boldsymbol{Q}$, the **water yield** reaching a stream, is determined by the amount of precipitation ($\boldsymbol{P}$) minus the water that is lost from the catchment by evaporation ($\boldsymbol{ET}$ – evapotranspiration – described below) and the water temporarily

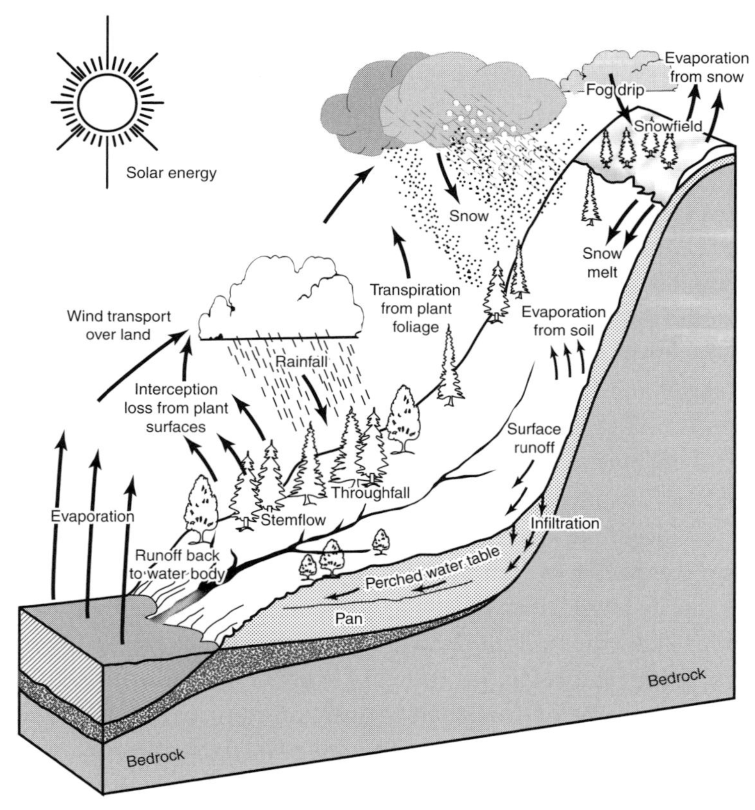


Figure 3.13 A summary of the major components of the water cycle. (Redrawn from Kimmins, 1997. *Forest Ecology*, Prentice Hall.)

retained in the catchment, either stored in the soil (***S***) or as groundwater (***G*** – i.e. in the water table).

The amount of precipitation falling in a catchment area is primarily determined by the prevailing climate and is largely unaffected by the vegetation. In the past many foresters held the view that planting forests increased rainfall in a watershed but in temperate areas at least it is likely to be insignificant and certainly $<5\%$ (Golding, 1970). On the continental scale, forests help to increase rainfall in the sense that they repeatedly recycle the amount of atmospheric moisture passing from the oceans to the land. Due to this, in the Amazon basin, for example, where much of the daily rainfall is immediately evaporated to generate clouds for rainfall downwind, it is highly likely that continual clearance of the forest will reduce rainfall elsewhere in the region since much of the water will enter rivers and be lost to the system. Moreover, it is now becoming known that the effects of such tropical deforestation have far wider repercussions in mid and high latitudes through **hydrometeorological teleconnections**, large-scale links in the water cycle and weather. Avissar and Werth (2005) have shown that deforestation of Amazonia and Central Africa severely reduces rainfall in the US Midwest in the spring and summer, and in the upper US Midwest during the growing season. Deforestation of south-east Asia similarly affects China and the Balkan Peninsula.

Some extra water can be captured as dew (the condensation of water vapour on cold surfaces). As this can reach around 1 mm of water per night it can be significant in keeping small seedlings alive. In humid coastal areas and at high elevations that intercept clouds, significant extra precipitation can arrive in the form of **fog-drip**. With their large surface area, trees are particularly good at sieving small particles of water from cloud and fog, and can increase precipitation by 30–50%.

The vegetation has most influence on what happens to precipitation once it has fallen. All vegetation intercepts rainfall which evaporates before it reaches the ground, but this process is much more important in forests than in pasture because of the larger surface area available to hold water, known as the **interception storage capacity**, which may be anywhere between 0.25 and 7.5 mm of rain in forests. The proportion of rain lost by this **interception** obviously depends upon the rainfall event. In light showers all the water may be evaporated from tree foliage so none reaches the ground and in general interception is in the order of 50–90% in showers of less than 20 mm of rain and 10–30% in heavier downpours. This is why we shelter under trees in light showers but find they leak heavily when the canopy storage capacity is full. Wind will also make a difference since it will shake drops from the canopy before they can evaporate. Over a year, the approximate interception loss is 20–25% from deciduous trees and 15–40% from conifers (the latter have a higher surface area and in evergreen species this is kept year round). Interception losses are lower from more open forests but some of the saving is lost by increased evaporation from the soil. In addition to interception losses, the vegetation also loses 'internal' water by evaporation from inside the leaves and to a lesser extent through the bark. This **transpiration** is under the control of the plants but in forests may still approach 15–20% of precipitation. When all forms of water loss by evaporation are combined, the overall losses expected from this **evapotranspiration** are something in the order of 30–60% of precipitation in deciduous forests, 50–60% in tropical evergreen forests and 60–70% in coniferous forests, compared with around 20% in grasslands. The remaining water either soaks into the soil or runs over the surface as **runoff**, either way eventually reaching streams. The important difference is that runoff is more likely to cause erosion and it speeds the delivery of water to the stream (the importance of which is discussed below).

Given the high rates of evapotranspiration from a forest, it is not surprising that discussion of water yield from forests relative to that of pastures is often of considerable importance, especially in New Zealand where the subject has been investigated on a national basis (Maclaren, 1996; Davie and Fahey, 2005). Forested areas do indeed have water yields (runoff or total streamflow) that are 25–80% lower than pastures; the highest differences being in dry South Island sites. Much work has also been carried out into the effect of forest removal on water yields. Most show that regardless of the forest type in question, removal of less than 20% of the trees has an insignificant effect on water yield, presumably because of increased soil evaporation replacing interception loss and transpiration (Brown *et al.*, 2005 provide a good summary). Bosch and Hewlett (1982) concluded that each extra removal of 10% of the forest led to an increase in stream yield of around 25 mm for deciduous hardwoods and around 40 mm for conifers and eucalypts – comparatively small amounts. Nevertheless, where water flow is low, afforestation is likely to reduce it even further, unless it assists aquifer recharge by superior **infiltration** (permeation of water into the soil rather than becoming runoff). Low flows can be increased by drainage of soils for forest planting.

As has been found at the cost of many human lives in Third World countries, forests are significant in **flood control**, which they help to reduce in small catchments and particularly in areas close to streams. In larger catchments they can help by reducing sedimentation. Unwise heavy logging of tropical rain forests growing over deep layers of unconsolidated sediments can lead to disastrous **mud slides** with heavy losses of human life. The main role of forests in flood control, however, is in quickly evaporating water back into the atmosphere, reducing the burden on streams after rain but particularly in aiding water infiltration at the expense of runoff. Since the pathways through the soil are small and tortuous, infiltration slows the passage of water to the stream (the **sponge effect**) thus dampening the sudden influx of water into streams from rain events. Infiltration in forests is aided by the porous surface of leaf litter and by the vegetation protecting the soil surface from the high-energy impact of rain drops which can break apart soil aggregates and wash fine material into soil pores, blocking them for further infiltration. Water dripping from the canopy (**through-fall** in Fig. 3.13) can be equally damaging because the drops are large but this is usually concentrated in a small area around the perimeter of the canopy. Least damaging of all is **stem flow**, water running along branches and then the trunk to reach the ground.

The **quality of water**, which can vary diurnally, seasonally or with the flow of water, must be considered in relation to its intended use so various water quality indices are employed for uses such as drinking, bathing, fish spawning and general use. In sparsely developed areas of many countries water is of very high quality, while in lowland reaches of agriculturally developed catchments nitrate levels in shallow groundwaters are sometimes too high. Many small lakes are **eutrophic** (over-enriched with dissolved inorganic nitrogen and phosphorus), and these often contain aquatic weeds including aggressive exotics such as *Egeria densa* and oxygen weed *Elodea canadensis,* as well as blooms of blue-green algae. Faecal contamination and the presence of disease-causing organisms affects water from areas of intensive dairying. In contrast, levels of nitrogen and phosphorus in waters draining from indigenous or exotic forests are almost always much lower, as is surface runoff and sedimentation. The latter is highest at times of road-making and logging. In agricultural areas **riparian buffer strips** or **forest zones**, narrow zones of trees bordering lakes and waterways, are of particular value in reducing erosion and sedimentation as well as maintaining the integrity of the aquatic habitats, and protecting water quality by moderating shade and water temperature. Forestry agencies in the USA usually recommend that riparian strips are 10–30 m in width; the lower end of this scale is useful for protecting the physical and chemical characteristics of the streams while wider strips will do more to maintain the complete ecological integrity of the aquatic habitats (Broadmeadow and Nisbet, 2004).

### *3.3.2 Swamp forests and peatlands*

Marshy forest known as **taiga** spreads across sub-arctic North America and Eurasia, with tundra to the north of it and steppe to the south (see

Section 1.6.1). Besides their own intrinsic interest, such peatlands preserve a temporal archive of community development. This makes it possible to reconstruct the history of the **swamp forests** through the analysis of long core samples of the sediments, and by developing detailed chronologies of successional change. Peatlands form when bodies of water fill with sediments and peat (**terrestrialization**), or areas of dry land are converted to peatland by flooding (**paludification**). Figure 3.14 is based on the study of three similar *c*. 10-ha peatlands, all with pronounced hummock–hollow topography and close to the border between New Hampshire and Massachusetts, eastern USA. The trees of Black Gum Swamp, Rindge Bog and Ellinwood Bog are dominated by red spruce *Picea rubens*, eastern hemlock *Tsuga canadensis* and red maple *Acer rubrum*. In all three sites peatlands initially formed as a result of terrestrialization and there are abundant **gyttja** (lake deposits) and fossils associated with them. Following this there was a sharp transition to shrub peat which represents an expansion of a mat of low woody vegetation over shallow water; it contains ericaceous wood fragments within a matrix of decomposed *Sphagnum* and fern rhizomes. The much more gradual change from shrub peat to wood peat, as trees colonized from the margins, is indicative of successional processes involving a range of shrub and forest communities.

It is, however, apparent that paludification also occurred later, as all sites have areas with up to a metre or more of wood peat deposited directly on mineral soil or till without a layer of intermediate gyttja. Moreover, radiocarbon dates indicate progressively younger basal samples away from the lake basins. Although successional sequences were similar in all three sites they were not synchronous. Rindge Bog, the largest and deepest basin, had floating vegetation 2000 years after Black Gum Swamp. In this region of New England, and possibly elsewhere, autogenic processes and local topography exert greater control over peatland development than **allogenic** (large-scale, externally imposed) factors such as climate change (Anderson *et al*., 2003).

## 3.4 Temperature and pollutant influences on tree growth

Trees in urban areas are exposed to higher temperatures, enrichment of carbon dioxide ($CO_2$) and nitrogen as well as greater levels of other pollutants than those in rural ecosystems. Each of these factors exerts a negative or positive influence on trees (see also Section 11.4.2); the important question is to ask what their net effect is in different parts of a large region. Studies by Gregg *et al*. (2003) used soil transplants, nutrient budgets, chamber

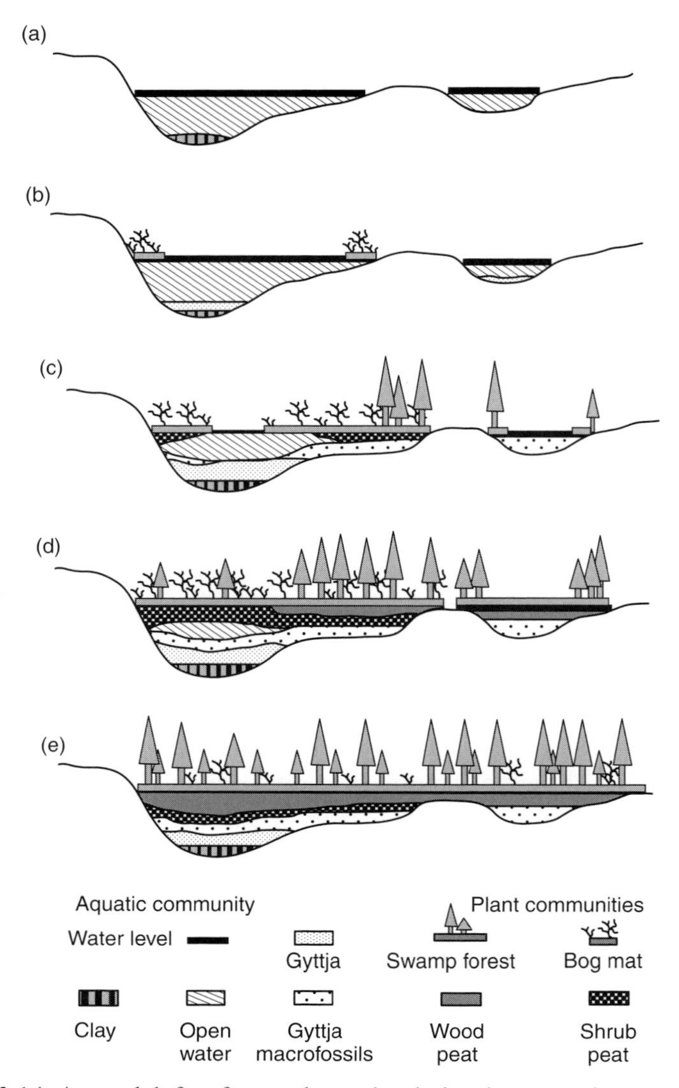


Figure 3.14 A model for forested peatland development in central New England. Gyttja (a Swedish word pronounced 'yut-tya') is a nutrient-rich organic deposit. Events in the two basins are not assumed to be synchronous. (From Anderson *et al.*, 2003. *Journal of Ecology* 91, Blackwell Publishing.)

experiments and multiple regression analyses to investigate this. They also used an inherently fast-growing clone of eastern cottonwood *Populus deltoides* as a '**phytometer**' to measure overall growth responses to multiple pollution sources both in urban New York City and in surrounding rural areas finding, surprisingly, that urban plant biomass accumulation so measured was double that of rural areas. Soils, $CO_2$ concentration, nutrient deposition, urban air pollutants and microclimatic variables could not account for increased growth in the city.

Their results demonstrate the overriding importance of **ozone** ($O_3$) as a gaseous pollutant, one which – contrary to expectation – was at a higher concentration and did more to reduce productivity in rural than in urban areas. These urban trees also benefited from higher temperatures and higher rates of nutrient and base-cation deposition. Though urban precursors fuel the reactions of $O_3$ formation, nitrous oxide ($NO_x$) scavenging reactions resulted in lower cumulative $O_3$ levels in urban sites than in agricultural and forested areas throughout north-eastern USA. This made a vital difference despite the considerable amounts of many gaseous, particulate and photochemical pollutants (including $NO_x$, nitric acid – $HNO_3$, sulphur dioxide – $SO_2$, sulphuric acid – $H_2SO_4$, $O_3$ and volatile organic compounds) in the city, where heavy metal pollution of the soil was also much greater.

## 3.5 Altitudinal zonation and timberlines

### *3.5.1 Alpine timberlines*

The polar margin of arctic forests and the upper margin of subalpine forests (i.e. forests on mountainsides just below the treeless, alpine zone) have **cold timberlines**. In reality these are facets of the same environmental conditions shown in Fig. 3.15a; the alpine timberline becomes lower in altitude away from the equator until it reaches sea level near the poles. The ultimate factor determining timberlines is decreasing temperature with altitude and latitude (in the northern hemisphere forest ceases to grow when the average temperature of the warmest month drops below 10 °C). The timberlines of subalpine forests, however, are also modified by higher wind speeds related to increasing altitude and snow accumulation (which offers shelter from the cold and wind). Since timberlines on mountains are thus ecologically more interesting, and are more likely to be seen by most people, these are first considered here (Fig. 3.15b). Every forest margin is bounded by an **ecotone**, a zone of transition between two ecosystems, in this instance alpine forest and open mountainside. This timberline ecotone above the forest limit is described by European ecologists as the **kampfzone** (struggle zone). In North America the term **timberline** or **treeline** is applied to the uppermost height at which trees reach at least 13 ft (3 m) at maturity, and is normally where the closed canopy forest ends. Natural treelines are usually sharp. Above this, isolated trees may grow until the **tree limit** is reached. These trees provide refuge for new seedlings and the

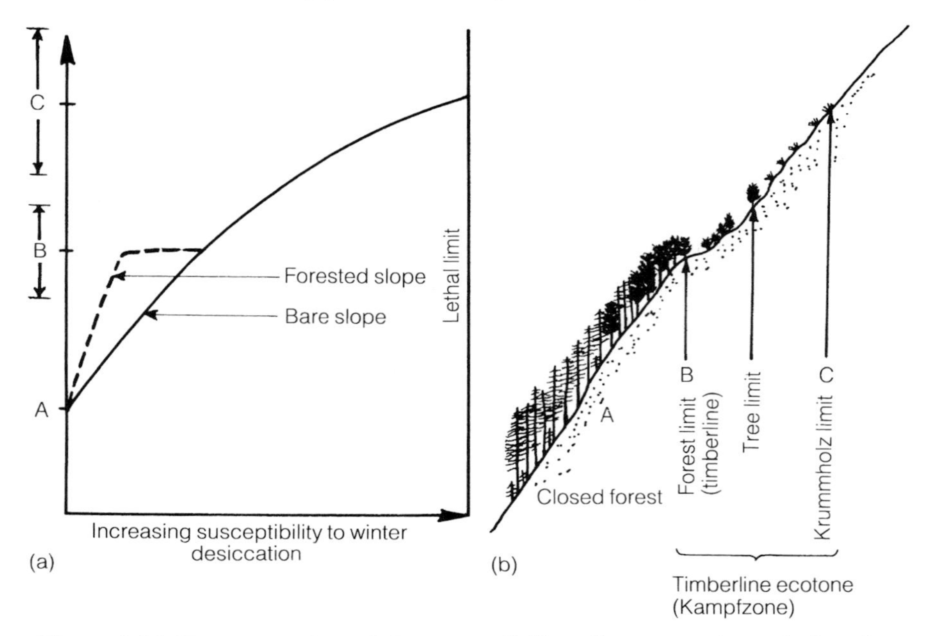


Figure 3.15 Representation of the susceptibility of trees to winter-desiccation on a mountain slope. In both (a) and (b) risk of damage begins at altitude A; above this height conditions become increasingly extreme and the growing season is shortened so there is less time for needles to form and develop cuticular protection adequate for drought resistance. Up to the **forest limit** a closed stand can gradually develop by natural regeneration. Within the forest winter desiccation is much less severe than on a deforested slope, but above the sharp **timberline** B, conditions rapidly deteriorate and trees of normal stature occur only on unusually favourable sites. C, the **krummholz** limit, is determined by the resistance to desiccation of the trees growing at the highest point; all trees near to this point are stunted and deformed. In diagram (a) the arrows above and below B and C indicate the height ranges within which these two limits are found after a series of warm (upward movement) or of cold (downward movement) seasons. The trees shown in diagram (b) of the European Alps are Norway spruce *Picea abies* and arolla pine *Pinus cembra*, of which the latter has its lethal limit for winter desiccation at the higher altitude. At Obergurgl, Austria, the timberline for arolla pine is at 2070 m. (Modified from Tranquillini, 1979. Redrawn version from Packham *et al.*, 1992. *Functional Ecology of Woodlands and Forests*. Chapman and Hall, Fig. 4.13. With kind permission of Springer Science and Business Media.)

rooting of branches sagging to the ground (layering) often resulting in oval patches extending away from the prevailing wind. As trees on the windward side are gradually killed by cold and desiccation of winter winds (causing branches that project above the winter snowline to have a strongly flagged

Figure 3.16 Subalpine fir *Abies lasiocarpa* forming krummholz in the Canadian Rocky Mountains. The upper part has flagged branches that grow only on the protected leeward side; branches on the windward side are killed by dry, cold desiccating winds which in winter carry ice spicules which wear away the needles' protective wax layer. The skirt of healthy branches around the base are covered by snow in winter and so escape wind and excessive cold. (Photograph by Peter A. Thomas.)

appearance), and new trees establish on the lee-side, these clumps gradually move across the landscape over centuries, literally blown by the wind. The trees at the **krummholz** ('bent or crooked wood' limit at the top of the kampfzone) are very contorted, often forming multi-stemmed cushions and so low in stature that in winter they are covered and protected by snow. Arno and Hammerly (1984) give a wide-ranging account of the alpine and arctic forest timberlines of North America and elsewhere, providing detailed illustrations of the ways in which tree form is influenced by extreme conditions, particularly cold and wind. Fig. 3.16 illustrates the extreme deformations suffered by trees living high on exposed mountains. Though environmental influences probably predominate, genetic factors may also be involved in this squat tree habit. Not all timberlines are marked by krummholz; **elfin woods** maintain their closed structure even as the individual trees are dwarfed by poor growing conditions. Good examples of these occur at 40 °S in the Argentine Andes where the southern beech *Nothofagus pumilio* becomes more and more stunted as it approaches the timberline, but even here forms dense stands with an almost closed front against the alpine level (Ellenberg, 1988).

Vegetational zones and timberlines are usually at higher elevations in large mountain masses than on isolated summits. It may be that wind velocities are lower or that snowfall is greater in such areas as the Rocky Mountains or the European Alps; either would result in more soil water being available for summer growth and less winter damage. As Fig. 3.15a indicates, winter desiccation is much less severe within the forest than on a deforested slope. Conditions within a forest that stretches well up a mountain from a much lower starting point vary greatly, and usually result in an **altitudinal zonation** in which different tree species dominate particular elevational zones as well as changing their form in response to environmental conditions. Forests of the north-eastern USA up to 750 m altitude are usually of the northern hardwood type, dominated by sugar maple *Acer saccharum*, American beech *Fagus grandifolia*, yellow birch *Betula lutea* and sometimes eastern hemlock *Tsuga canadensis*. Red spruce *Picea rubens* is increasingly important with increasing altitude and above 850 m the forest is mainly of this species and balsam fir *Abies balsamea*. Balsam fir becomes ever more important ascending the mountain on which the upper limit for forest growth is in the 1350–1700 m range, with the topmost individuals dwarfed to prostrate or shrubby krummholz forms. Small pockets of shrubs or shrubby trees occur occasionally in protected places in the alpine tundra above the kampfzone. Upper forests in this region frequently show wave regeneration in which mature trees subject to the influence of the prevailing wind die off at the front of the wave (see Section 9.3.2). Other types of disturbance, such as very heavy felling or severe spruce budworm *Choristoneura fumeriferana* outbreaks, which affect both spruce and fir, frequently give rise to regeneration forests in which the proportion of fir is higher than before.

The lower zones of the European Alps and other mountainous regions are frequently dominated by broadleaved trees with conifers occupying the upper forest zones. High altitude forests in parts of the Austrian Tyrol appear to have been felled in the Middle Ages to provide timber for mining (Tranquillini, 1979). Though the forest has gradually spread up the slopes the kampfzone is still unusually wide; in regions not so disturbed it remains very abrupt as in the Andean example quoted earlier.

In Europe the number and area of true old-growth montane forests are decidedly limited, so the investigations made by Piovesan *et al.* (2005) into the remnants of beech *Fagus sylvatica* forest growing near the tree line at 1600–1850 m in the central Apennines, Italy, are of especial interest. These authors used a combination of historical, structural and dendrochronological (i.e. tree ring) approaches in attempts to answer the following important questions. What are the structural attributes and the history of old-growth *Fagus* forest in Mediterranean montane environments? Which processes determine their structural organization? Are these forests stable in time and how does spatial scale affect our assessment of stability? How do these forests compare with other old-growth forests? The four patches of old-growth beechwood concerned, which escaped logging after World War II, were photographed from the air in 1945, 1954, 1985 and 1994, while the structure of the old-growth region of the forest was investigated using 18 circular and 2 rectangular plots. Living and dead components within this region were both within the range expected for old-growth forests of temperate biomes. Widths of annual rings (see Sections 1.3.2 and 10.1.2) observed in dendrochronological analyses using cores from 32 dominant or co-dominant trees revealed the roles of disturbance, competition and climate in structuring the forest. The identification of a persistent *Fagus* community in which gap-phase regeneration (see Section 9.2.2) has led to a single-species multi-aged stand at spatial scales of a few hectares, is of particular importance as far as the long-term conservation of such forests is concerned.

Timberlines will certainly move vertically as climate changes with time. The presence of 'fossil stands' of bristlecone pine *Pinus aristata* (the longest-lived of all organisms – Section 1.3.2) found well above present upper bristlecone timberlines in the White Mountains east of the Sierra Nevada of southern California, indicate that during a warmer climatic period from 2000–4000 years ago the upper margin of the forest was at least 300 m higher than it is now.

As a mountain is ascended conditions become more variable which in itself can be a distinct problem. In exposed parts of the northern Rocky Mountains seedlings of white spruce *Picea glauca* are killed by stem girdling caused by a few hours of high temperature at ground level, though at night temperatures drop sharply. Partial shading prevents most heat deaths in seedlings of this tree and drought is then the most likely cause of mortality, especially as growth at high altitudes is so slow that seedlings take several seasons to form tap roots long enough to enable them to survive a drought (see Box 3.2 for further discussion on survival at the timberline). Tree roots at the timberline show intensive development of **mycorrhizas** (Section 5.4.1). Almost all short roots of arolla pine *Pinus cembra* in central European timberlines are mycorrhizal, and mycorrhizal fungi have evolved high-altitude strains adapted to the high daytime temperatures encountered. Species as diverse as *Nothofagus solandri, Eucalyptus pauciflora, Picea engelmannii, Pinus contorta, P. flexilis* and *P. hartwegii* will establish near timberlines only if mycorrhizas are present.

Disturbance caused by fire can locally depress the level of upper timberlines or raise that of the lower timberline, often operating in conjunction with winds that remove snow needed to protect tree seedlings in a harsh continental

## Box 3.2 Influence of altitude on forest dynamics in northern Sweden

Figure 3.17 illustrates the influence of increasing altitude on the population dynamics of trees growing in the Vallibacken Forest, northern Sweden. This forest has storm gaps of a similar type but smaller size than those of the boreo-nemoral forest of Fiby urskog (see Section 2.4.2). Cumulative age distributions in two forest plots are shown for Norway spruce and downy birch as a semi-logarithmic diagram with 10-year age classes. Each **cumulative number curve** was created by cumulatively adding up the age of all the trees of the species concerned within the plot. The curve commences with the number of trees in the oldest age class, the numbers of trees in successively younger age classes were then added in turn until the youngest age class (0–10 years) was reached. Provided there is at least one good seed year with acceptable germinability during each 10-year period, and the mortality in each age class is constant, such curves can be accepted as

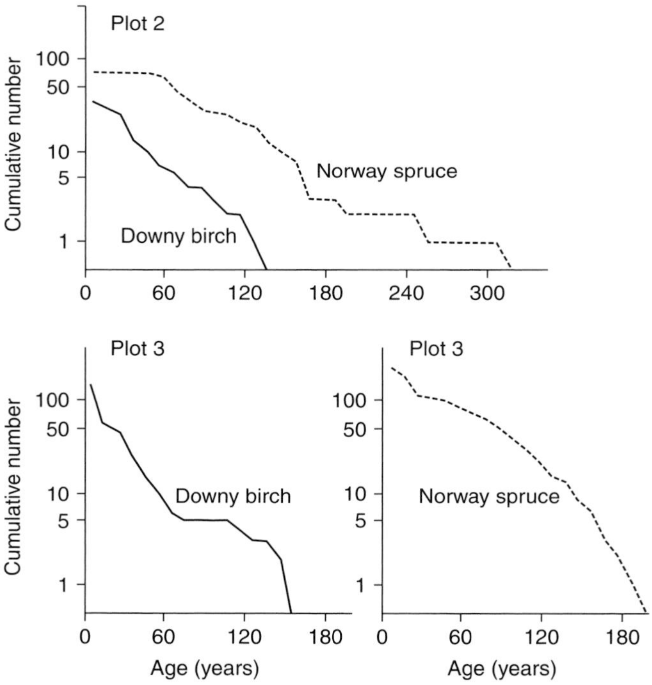


Figure 3.17 Cumulative age distributions of all individuals of downy birch *Betula pubescens* and Norway spruce *Picea abies* in two forest plots of similar area in the Vallibacken Forest, North Sweden with the vertical axis on a log scale. For clarity, the two species at plot 3 are shown separately. Plot 2 is at an altitude of 460 m and plot 3 at 335 m. The coniferous forest limit is at 580 m in this area. (From Hytteborn *et al.*, 1987. *Vegetatio* 72, Fig. 5. With kind permission of Springer Science and Business Media.)

static **survivorship curves** (Deevey, 1947) which convey an impression of how a typical generation of young saplings would fare over time. This is true of plot 3, of which 140 $m^2$ was a storm gap and the remaining 260 $m^2$ were covered by tree canopy. The mortality of both species in this lower plot is initially high, but after 70 years birch reaches a plateau that continues to 110 years. After initial high mortality in the very first age classes spruce died off more slowly. Its mortality rate increased later on, but not as fast as birch in the same higher age classes. The form of this curve reflects the ability of spruce to survive as **dwarf trees**, with poorly developed leading shoots and well-formed lateral branches. The survivorship curve for spruce in a plot from Fiby urskog containing storm gaps showed this effect even more clearly (Hytteborn and Packham, 1987).

There was no evidence of typical storm gaps in the considerably higher plot 2 or its surroundings, though the plot was open with a crown cover of only 15%. The cumulative age distribution of Norway spruce here cannot be interpreted as a survivorship curve because the number of cones produced at this altitude and latitude (67 °N) is both low and irregular, while mature seeds are produced in any quantity only in those rare years when the mean temperature between June and September exceeds 10 °C. Thus in plot 2 spruce growth and regeneration are dominated by environmental factors, while in plot 3 the biotic influence of the mature trees is overriding.

climate. Such a case developed over a century ago at 2290 m in the upper Larch Valley, Banff National Park, Canada. After fire levelled a stand of tall alpine larch *Larix lyalli* the area reverted to tundra; so far there has been only a slow re-invasion of krummholz conifers.

Forests can also have natural lower timberlines where they usually give way to dry grassland, sage-brush, tall shrubs or scrub oaks. This boundary is usually the result of inadequate moisture and termed a **drought-caused timberline**. The elevational distance between this and the alpine or **cold timberline** is only 600 m on the semi-arid Lost River Range in Idaho, USA. **Double timberlines** are found in western USA where some of the ranges are so dry that only a few species of drought-tolerant trees can survive, and then only within narrow elevational bands. When the elevational distributions of these trees do not coincide two separate forest belts occur; the lowest is typically of pinyon pine and juniper (*Pinus edulis* and most commonly the Utah juniper *Juniperus osteosperma*), commonly referred to as PJ. This is separated by a belt of treeless sage-brush above which is a narrow zone of bristlecone and limber pine *Pinus flexilis* extending from about 2740 to 3500 m.

In Iceland and the British Isles extensive clearing, heavy grazing and the burning of heath which took over forest land as a result of disturbance, have caused the loss of any semblance of a natural timberline. Remnants of subalpine Scots pine *Pinus sylvestris* forest still remain in Scotland, where the natural limit of this tree is between 610 and 700 m in the Cairngorms, though human activity has generally lowered it to around 500 m. Although British woodlands do not reach such high altitudes as their continental counterparts some, such as Wistman's Wood, which at 380–410 m on Dartmoor is amongst the highest oakwoods in Britain, are of remarkable interest. The Dartmoor oakwoods experience an oceanic or Atlantic climate which is both cool and wet; they are remarkable for their very extensive lower plant communities (Box 1.3). The sprawling and stunted form of the oaks present in Wistman's Wood at the beginning of the twentieth century, though undoubtedly influenced by grazing, was mainly caused by extreme exposure to wind and the very low amount of shading from older trees and other vegetation. This exposure was particularly severe during the 'Little Ice Age' of the seventeenth to mid-nineteenth centuries.

### *3.5.2 Temperature–moisture gradients below the timberline*

Altitude, and the changes in environmental conditions that go with it, plays just as important a role below the timberline as at the top. For example, there is a well-marked temperature–moisture gradient from the warm, dry lowlands to the cold, wet mountain peaks in the north-western region of the USA shown in Fig. 3.18. The lower limit for these trees usually shows a gradual transition, apparently set by soil moisture levels. Ponderosa pine *Pinus ponderosa*, Douglas fir *Pseudotsuga menziesii*, Engelmann spruce *Picea engelmannii* (not shown in Fig. 3.18) and alpine fir *Abies lasiocarpa* are found at successively greater altitudes and their roots will usually grow in increasingly moist soils. This equates with their drought resistance, which is highest in ponderosa pine and least in alpine fir. In this situation temperature does not seem to be a major factor with these species, all of which can be grown at low altitude if watered. Also, dry atmospheres produce little effect as long as there is sufficient soil moisture.

Figure 3.19 demonstrates a more complex situation from an area further south. Here as many as ten species of pine can be encountered up a landscape gradient from the foothills to the crest of the mountain range (Richardson and Rundel, 1998). The vertical axis simply represents the altitudinal position of the communities involved, but the horizontal one compounds the influence of topography and soil moisture. Even if it received the same rainfall a steeply

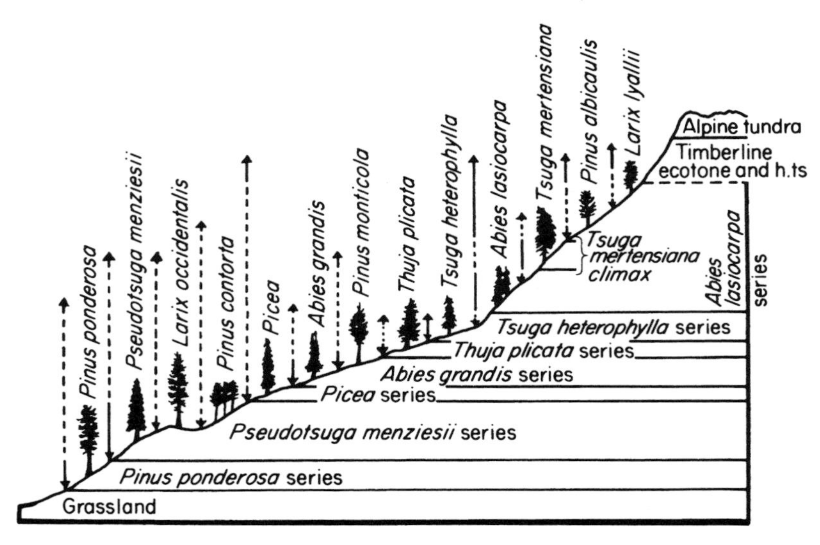


Figure 3.18 Usual order in which coniferous trees occur with increasing altitude in the Rocky Mountains of north-western Montana. The relative altitudinal range of each species is shown by arrows. The dashed portion of each arrow is where the species is seral (early successional: one of the first trees to grow on bare ground) and the solid portion of the arrow is where the tree is the potential climax dominant (late successional and there to stay in successive generations). The temperature–moisture climatic gradient runs from the lowlands where the conditions are warm and dry to the timberline where they are cold and wet. There are several timberline types (h. ts): all have a timberline ecotone with a krummholz above them (see Section 3.5.1). (After Pfister *et al.*, 1977. From Packham and Harding, 1982. *Ecology of Woodland Processes*. Edward Arnold.)

sloping site on a hill is likely, assuming the soils to be similar, to experience lower soil moisture level than one which is level, and so would be shown more towards the right (xeric or dry) side.

The blue oak *Quercus douglasii* woodlands of the foothills frequently include the gray or digger pine *Pinus sabiniana*, while ponderosa pine dominates the forest from 1500 m to as much as 2200 m on drier slopes and south-facing exposures. At these altitudes singleleaf pinyon pine *P. monophylla* exists as small relict populations on steep canyon slopes. In mesic (moist) sites coniferous forests at these middle elevations are dominated by a white fir *Abies concolor*/ mixed coniferous forest with sugar pine *P. lambertiana*, giant sequoia *Sequoiadendron giganteum* and incense-cedar *Calocedrus decurrens* as associates. Western white pine *P. monticola* is a common associate on the drier slopes and exposed areas are strongly dominated by Jeffrey pine *P. jeffreyi* from 2200–2800 m. Above 2500–2800 m, sierra lodgepole pine *P. contorta* ssp.

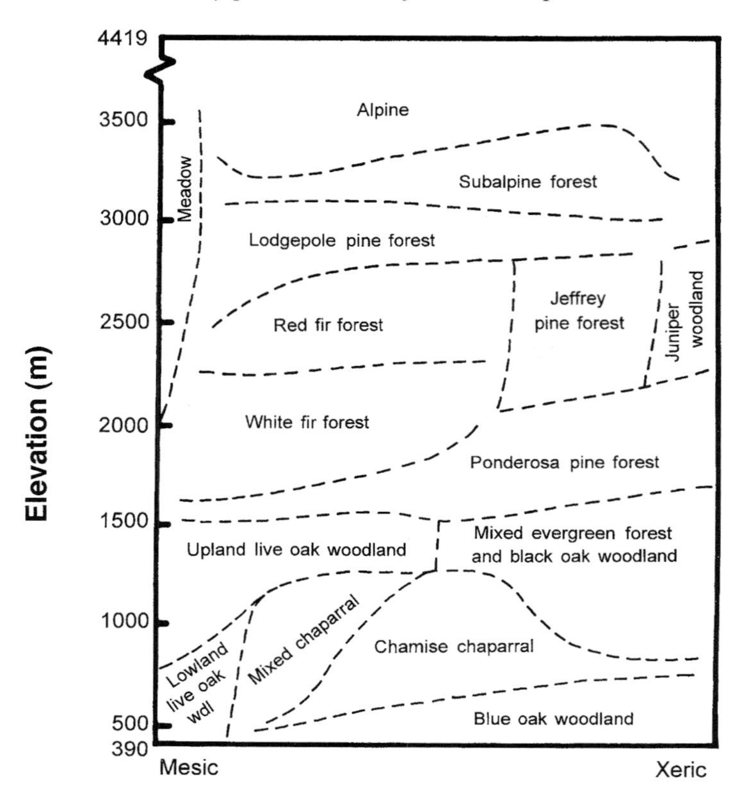


Figure 3.19 Diagrammatic representation of community positions along elevational and topographic–moisture gradients for the west slope of the southern Sierra Nevada: mesic = moist; xeric = dry. (From Richardson and Rundel, 1998. In *Ecology and Biogeography of* Pinus. Cambridge University Press.)

*murrayana* forms single-species forest. At the top, upper subalpine and timberline habitats above the lodgepole pine zone have foxtail pine *P. balfouriana* as the most important species, with smaller numbers of whitebark pine *P. albicaulis* and limber pine *P. flexilis*.

### 3.5.3 *Arctic timberline limits*

Arctic limits for tree growth vary from area to area; the most northerly is about 70 °N, though in western Alaska and Siberia it dips sharply south due to the cold Bering Sea and does the same around Hudson Bay in Canada. A broad transitional zone separates the lowland boreal forest from the arctic tundra (Sirois, 1992; see Section 1.6.1). The northernmost part of this zone is the **forest-tundra** with scattered patches of krummholz or stunted trees (Fig. 3.16), with larger trees along rivers and in sheltered sites, set in a matrix of tundra. This extends south to the **open boreal forest** or 'lichen woodland', whose open groves of erect trees are underlain by a rich lichen carpet (mainly *Cladonia* spp.). The proportion of tree cover to lichen mat increases southwards until reaching the **forest line**, where trees cover at least 50% of the landscape. The arctic timberline differs dramatically from its alpine equivalents in its great width across the lowlands (150 km or more), in contrast to a very sharp cutoff of up to 0.5 km or so on a mountain. Alpine timberlines are sometimes showered with tree seeds from stands below; in contrast regeneration from seed is rare in the arctic timberline. A number of important species of the undergrowth are abundant in the arctic timberline zone, including crowberry *Empetrum nigrum*, narrow-leaf Labrador tea *Ledum decumbens*, cloudberry *Rubus chamaemorus*, bog blueberry *Vaccinium uliginosum* and cranberry *V. vitis-idaea.*

Spruce are by far the most common trees of the arctic timberline of North America. Black spruce *Picea mariana* grows primarily on poorly drained boggy sites that are prevalent in the lowlands underlain by permafrost. This tree, which readily assumes a stunted or krummholz form on exposed sites, regenerates vegetatively by layering where the lower branches root into the abundant moss and become independent trees. In contrast white spruce, *P. glauca* grows on well-drained soils, especially along streams where the permafrost has been melted back by the flowing water, and usually forms a single erect trunk at the treeline. It mostly reproduces from seed.

The arctic and alpine timberlines, along with drought-timberlines (see Section 3.5.1), are of particular interest in marking the boundaries of the world's woodlands and forests, whose zonation is controlled by a number of factors. These include climate (especially the wind, water and temperature regimes), fire, soil type, altitude and biotic factors including the ever-extending influence of humans. Woodlands and forests include the most diverse communities known; the mechanisms which maintain this diversity are considered in Chapter 6.

## 3.6 Evergreen and deciduous strategies: aspects of competitive advantage

Almost all conifers – including the Wollemi pine *Wollemia nobilis* discovered in 1994 – are evergreen, the exceptions including the larches *Larix*, the Chinese swamp cypress *Glyptostrobus lineatus*, the three American swamp cypresses *Taxodium* (of which the Mexican swamp cypress *T. mucronatum* is evergreen in warm climates), and the dawn redwood *Metasequoia glyptostroboides*. This last species was described from fossil material in 1941 and then discovered as a living tree in 1945. As with most angiosperm (broadleaved) tree genera many oaks are deciduous; one of the numerous evergreen species is the European holm oak *Quercus ilex* that is much used for windbreaks. Nevertheless, this does not address the question of the advantages and disadvantages of these two strategies.

Net production is obviously a major criterion. If growing conditions are favourable all year round, as in tropical rain forests, then there is no advantage in being deciduous and so evergreen angiosperms dominate. In climates with a dry or cold season, it is cheaper to grow disposable leaves than to grow more robust leaves capable of surviving the off-season, so in most moist temperate areas deciduous trees dominate. However, if the environmental conditions get worse – such as a very short growing season – it may once again be more beneficial to grow evergreen leaves since none of the growing season is wasted. Evergreen leaves also occur on trees growing on nutrient-poor soils (since it is too expensive in nutrients to grow a new set every year), and in very dry Mediterranean climates where the thick, expensive leaves needed to survive the growing season are retained for several years to repay the energy invested in them. In areas with even worse growing seasons, such as near timberlines, deciduous leaves re-appear. Here, despite the costs, the winter is so severe that it is cheaper to build new leaves every year rather than attempt to keep leaves alive. Thus the northernmost trees in the Arctic are birches and there are often birches, willows and larches in the krummholz. Thomas (2000) gives a more detailed account of this subject.

A number of detailed studies have been made on the relative merits of these evergreen and deciduous strategies. The photosynthetic capacity of individual leaves of temperate deciduous trees such as beech is considerably higher than that of evergreen conifers, though whole trees of the latter frequently have high net production. This is partially due to the prolonged growth period; photosynthetic gains are made in the spring, winter and parts of the autumn when deciduous trees are bare of leaves. Leaf turnover rate is also important; when leaves are replaced annually the amount of photosynthate that can be put into the formation of wood and bark is greatly reduced. Photosynthetic capacity and respiratory activity are both high in deciduous trees. When young leaves are developing they respire intensely, have a small surface area, and are usually low in chlorophyll, making them a cause of overall carbon loss until their photosynthetic capacity is at its peak within a few days of full expansion. While the photosynthetic capacity of evergreen leaves slowly falls over the years, as has been shown for white spruce *Picea glauca* and balsam fir *Abies balsamea*, they continue to make a substantial contribution for a very long period.

A very thorough comparison of the growth of beech (Schulze, 1970) and Norway spruce (Schulze *et al.*, 1977a,b) growing within a kilometre of each other at 500 m on the Solling Plateau, North Germany was made as part of the

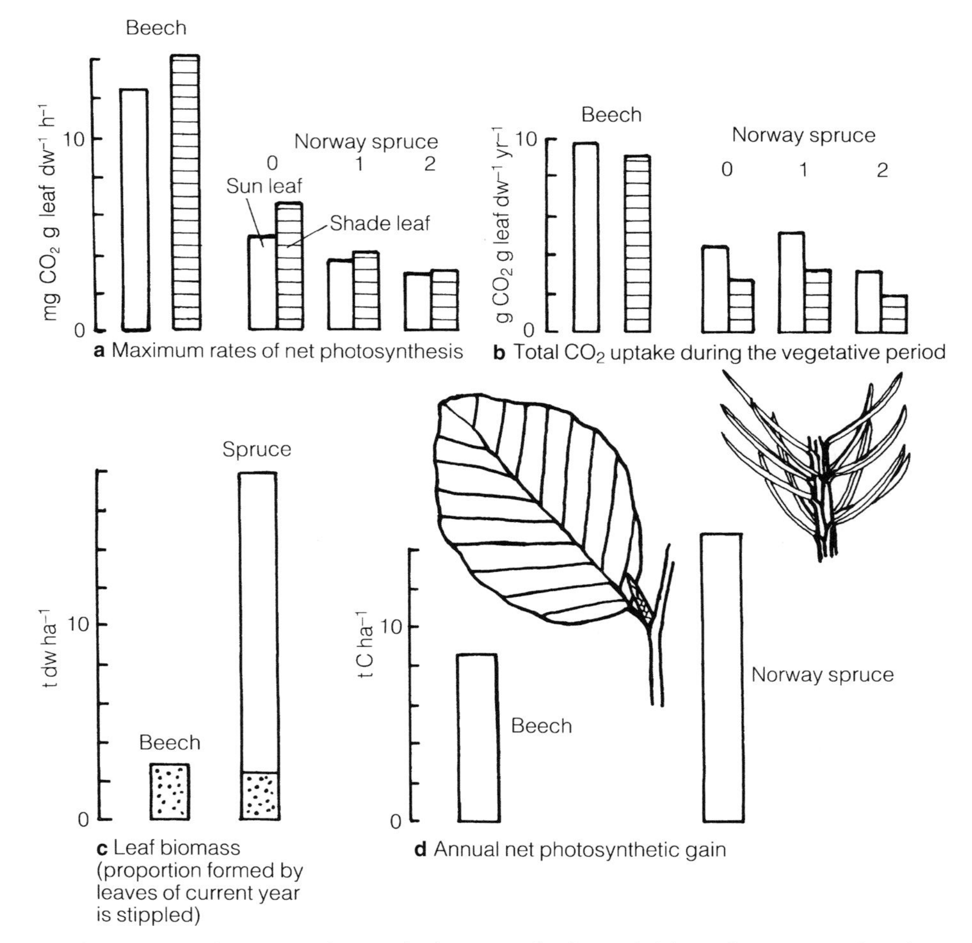


Figure 3.20 A comparison of photosynthetic activities of common beech *Fagus sylvatica* and Norway spruce *Picea abies* growing on the Solling Plateau, Germany. In (a) and (b), columns for sun leaves are shown white, those for shade leaves are hatched. 0, 1 and 2 correspond, respectively, to spruce leaves developed in the current year, and one and two years previously. Carbon fixation per unit of leaf dry weight (mass) is much greater in beech which, however, has a much smaller leaf biomass (c) and lower annual photosynthetic gain (d) than spruce. (Drawn from the data of Schulze *et al.*, 1977b. From Packham *et al.*, 1992. *Functional Ecology of Woodlands and Forests*. Chapman and Hall, Fig. 2.15. With kind permission of Springer Science and Business Media.)

International Biological Programme. Beech dominates many natural forests in central Europe, old-growth forests being particularly prominent in Poland, while spruce is the conifer most frequently planted in German re-afforestation programmes. The production values shown in Fig. 3.20 are thus of great interest. They are based on measurements of photosynthetic capacity of a typical tree of each species. The beech was 100 years old and 27 m high when examined in 1968, while the spruce investigated in 1972 was 89 years old and 25.6 m high; they differed in four major respects with regard to production. Both the sun and the shade leaves of beech had a much higher photosynthetic capacity per unit dry weight (mass) than even one-year-old needles of spruce. Beech had a short growing season, having a positive $CO_2$ uptake on 176 days in the year as opposed to 260 for spruce. Though beech had a higher annual production of leaves than spruce, the latter had a much greater photosynthesizing biomass, some of its needles surviving for 12 years. Care has to be taken if such results are applied to a consideration of evergreen and deciduous forests elsewhere; increased levels of atmospheric pollution have reduced the useful life of conifer needles in many places.

In the Solling study the deciduous beech invested more dry matter in its leaves every year, but the long-term return for the investment made by the evergreen spruce was greater because its leaves continued to fix carbon so much longer, though at a slower rate. Thus in these terms the spruce has, despite its ancient lineage, a performance superior to that of its angiosperm rival, its primary production being 14.9 t C $ha^{-1}$ $y^{-1}$, while that of beech was 8.6 t C $ha^{-1}$ $y^{-1}$. In terms of the value of the two strategies it would be very interesting to make similar detailed comparisons on an evergreen and a deciduous species within the same genus such as oak *Quercus*.

Despite the high productivity of evergreen Norway spruce in the region, common beech *Fagus sylvatica* is frequently the dominant of many natural forests in central Europe. It is much more shade-tolerant, its seeds being able to germinate and grow where those of spruce cannot. Moreover, the deciduous habit entails a reduction of the surfaces on which ice can accumulate, so it is less affected by winter storms and by snow and ice breakage. The surface rooting of Norway spruce means that gales can topple it relatively easily, and its life is often shorter than that of beech, which competes powerfully with other trees and when growing actively suffers markedly lower mortality from fungal diseases than spruce. Clearly, relative growth rate, though important, is only a single aspect of competitive advantage, whose balance is often swayed by climatic factors.

Norway spruce flourishes in northern Scandinavia and Russia, where common beech would not ripen seed and is unknown as a natural forest tree. However, the deciduous habit and the production of strongly constructed, even xeromorphic, needles are both adapted to the water stresses of northern winters, as the presence of aspen *Populus tremula* and Norway spruce in arctic Norway demonstrates.

Stem photosynthesis must play a part in maintaining carbohydrate levels in sessile oak *Quercus petraea*, one of the deciduous trees in the stems of which rates of photosynthesis are still appreciable at low light intensities. Nevertheless, evergreen conifers dominate most boreal forests, clearly gaining from the ability of their needles to photosynthesize for several years. When they yield to larches, their deciduous relatives, it is frequently on well-drained and often gravelly soils. The leaves of evergreen trees in the very different climate of tropical rain forests have a low rate of $CO_2$ uptake but, because the active vegetative period is so long, can attain a high annual biomass gain.

## 3.7 Contrasts between three widespread tree genera: the pines, beeches and oaks

These three genera include some of the most economically and silviculturally important species to be found in the world's woodlands and forests. All three provide valuable timber; examples of their foliage and reproductive structures are illustrated in Fig. 3.21. Though the pines and oaks have adapted very successfully to a wide range of habitats, the former belong to a more ancient group and the form of their mature trees is rather similar, though the number of needles per dwarf shoot varies from one or two in singleleaf pinyon *Pinus monophylla* to as many as eight in Cooper pine *P. cooperi*. Many, but not all, have winged seeds. While both beeches and oaks belong to the Family Fagaceae, there are more than 800 species of oak but only a dozen or so of beech, a widespread and notably shade-tolerant genus with considerably lighter seeds than oak.

### *3.7.1 Persistence, variation and adaptation within the genus* **Pinus**

*Pinus* has been accurately described as the 'most ecologically and economically significant tree genus in the world' (Richardson and Rundel, 1998, p. 3). It has 111 species, mostly restricted to the northern hemisphere, and this ancient genus, of which the earliest representatives date from the Lower Cretaceous around 120 Ma, is still remarkably successful. Indeed several species, often through the intervention of humans, are currently increasing their range. By the end of the Mesozoic 65 Ma *Pinus* had diverged into two groups: (1) the hard or white pines which normally have 2 or 3 needles in a bundle (officially called the diploxylon pines with two veins or fibrovascular bundles per needle); and (2) the soft or yellow pines with 5 needles per bundle (the haploxylon pines with one vein and little resin in the wood). The names hard and soft refer to the density of the wood.

The geographical range of pines varies tremendously. The Scots pine *P. sylvestris*, an example of the first group, has the widest distribution of any

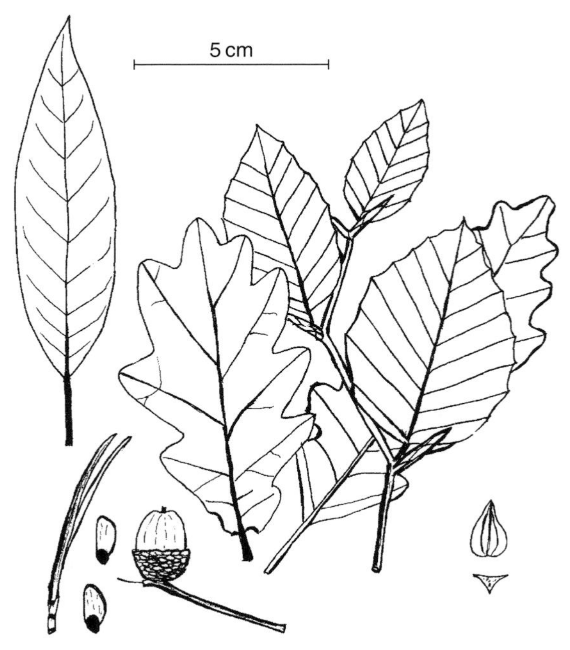


Figure 3.21 Tree foliage and reproductive structures, all to the same scale. The leaf on the left is that of the Japanese evergreen oak *Quercus acuta*. Beneath it is a 2-needled shoot and two diminutive winged seeds of mountain pine *Pinus mugo*. Next to it is an acorn and the underside of the leaf of pedunculate oak *Quercus robur*, which has short leaf stalks and auricles – shown in heavy line – at the base of the leaf blade. The adjacent sessile oak *Q. petraea* (under the beech leaf) has much longer leaf stalks but lacks auricles and its acorns are not borne on long stalks like those of pedunculate oak. Foliage of the common beech *Fagus sylvatica* is shown on the right along with its nuts viewed from the outside and in section. (Drawn by John R. Packham.)

pine, being found from the Scottish Highlands along the Atlantic seaboard to the Pacific coast of Eastern Siberia, with relict populations throughout the northern Mediterranean Basin. Conversely, the Canary pine *P. canariensis* (a hard pine) is found naturally on just five of the Canary Islands off the African coast with a natural range of probably less than 400 km$^2$. Size also varies greatly. The sugar pine *P. lambertiana* of the western USA (a soft pine) has the greatest height and girth of any pine, reaching more than 75 and 5 m, respectively. In contrast the dwarf mountain pine *P. mugo* of Europe (a hard pine) is no more than a shrub. Pines can also be long-lived: bristlecone pines *P. aristata* (a soft pine) are the oldest living organisms on Earth, growing on the White Mountains of California where some individuals have reached ages of nearly 5000 years (Section 1.3.2).

Pines, like many other conifers, typically grow in harsh environments that are cold and/or dry, from near the arctic timberline to acidic nutrient-poor soils in the tropics, perhaps driven to these environments by competition with the later-evolving and very successful angiosperm trees (see Section 3.7.4). The Siberian dwarf stone pine *Pinus pumila*, is extremely tolerant of cold and so is an important species covering large areas of the arctic and alpine belts in eastern Asia. It reaches a height of only 0.5–5 m. The main cold-adaptive feature of this species, which lacks a central trunk and forms clumps of connected branches, is its ability to lie prostrate beneath snow cover in winter. In order to flourish, this species also needs full sun in summer, good drainage and adequate aeration for its roots. When well-lit, dwarf stone pine produces very adequate crops of seed, but when shaded – as it is when growing in the lower canopy of larch forests – it develops vegetatively, aided by its adventitious roots on prostrate branches (Khomentovsky, 1998). With its strong reproductive potential, the Siberian dwarf stone pine is a most valuable pioneer species on a wide variety of substrates including fresh volcanic ash, burnt tundra and coastal dunes of climatically adverse north-eastern Eurasia. The North American equivalent of this species is the whitebark pine *P. albicaulis*, a closely related haploxylon pine. Further south in North America very dry soils of near deserts are the home of such species as ponderosa pine *P. ponderosa*, and the various pinyon pines such as *P. edulis* and *P. monophylla*. This continues onto the dry tropical soils of the savannas of Central America and the Caribbean where 47 species of pine form the greatest diversity of pines of any area in the world. Pines resemble other conifers in their tolerance of low-fertility stress, but are marked out by their ability to act as aggressive post-disturbance invaders.

Fire was (and is) a major feature of the surroundings in which the pines evolved and closely influenced the development of their life histories. Some pines meet repeated fires over very long periods; there are records of a ponderosa pine trunk showing 21 dated fire scars between 1659 and 1915. Fire regimes of low-, medium- and high-severity have led to the evolution of different physical and ecological characteristics in many species. Several species reproduce by basal resprouting after fire; the cut stumps of the canary pine *P. canariensis* produce numerous resprouts on the volcanic soils of Tenerife. **Serotinous cones** which remain closed for sometimes decades after the seeds mature, but release seeds once high temperatures have melted the resin sealing them shut, are important in the survival strategies of several species such as Monterey radiata pine *P. radiata* and knobcone pine *P. attenuata* (see also Section 4.6.2). The lodgepole pine *P. contorta* var. *latifolia* and a number of other trees possess both serotinous and non-serotinous populations.

Figure 3.22 Variation in cone size amongst pines. Sugar pine *P. lambertiana* (a) native to California is a five needle pine with a typical cylinder-shaped cone. Cones of three-needle pines such as the coulter or big-cone pine *P. coulteri* (b), gray or digger pine *P. sabiniana* (c) and Monterey pine *P. radiata* (d), all from California, tend to be large and globular while those of two needle pines, including dwarf mountain pine *P. mugo* (e) from Europe and jack pine *P. banksiana* (f) from Canada, are usually much smaller. Note that Monterey pine and jack pine are both serotinous and remain closed until heated. There is also enormous variation in wing size, seed weight and seed size within the genus. Sugar pine, Coulter pine, and gray pine have some of the largest winged seeds while jack pine has one of the smallest. (Photograph by Peter A. Thomas.)

Pines are amongst the most genetically diverse of organisms, showing great variations in form and stature as well as a more than 300-fold variation in seed weight. Cone size is also immensely variable and although there is often little correlation between seed size and cone size, ecological patterns can be picked out (Fig. 3.22). The soft pines (with 5 needles per bundle) tend to have long, soft cylinder-shaped cones, reaching up to 25–50 cm and even 60 cm in the sugar pine. The hard pines typically grow in the driest habitats (or bogs where water is physiologically hard to get due to restricted root growth) and produce hard, more globular cones. Two needle pines tend to produce small cones (often golf ball-sized such as in Scots pine) while the 3-needle pines produce the heaviest cones, typically bigger than a tennis ball and in the case of the coulter or big-cone pine *P. coulteri*, 25–30 cm in length and weighing up to 2.3 kg each. Species with the biggest cones tend to live in arid environments where food for herbivores is in short supply; it could well be that intense herbivore pressure on the large seeds needed for successful pine establishment in dry areas has made it cost-effective in evolutionary terms to invest in large expensive cones.

Pines are diploid (2n = 24), respond rapidly to selection, and have a mating system which promotes outcrossing (cross-pollination) and the consequent development of diversity while allowing self-pollination, enabling isolated individuals to reproduce and colonize. Seed dispersal is diverse. Initial dispersal by normal winds leaves a high concentration of seeds close to the seed source. If taken, moved, stored but left uneaten by rodents, these seeds can germinate later at more distant sites. On relatively rare occasions winged seeds may be carried long distances by turbulent winds, sometimes leading to new populations in fresh areas. In pines with wingless seeds, birds may disperse and store seeds over both long and short distances (see Section 4.2.2).

Although pines vary greatly in stature according to species and environment, they mostly have a rather similar basic narrow, pyramidal structure whether they are found on an arctic timberline or in the tropics. The pyramid shape helps intercept light from a low angle in high-latitude summers, and in southern dry areas reduces the interception of sun at midday which lowers water stress. Variations are, of course, numerous, such as the broad spreading canopy of the stone pine *Pinus pinea* of the Mediterranean, sometimes aptly called the umbrella pine. Here the shape (similar in many savanna trees) acts like an aerofoil, diverting winds around the canopy rather than through it, again reducing water loss but allowing vertical convective heat loss. Environmental differences have, however, led to considerable eco-physiological differences both between and within species, while individuals also adapt to changes in conditions. Pine seedlings growing in low-light conditions show a dramatic reduction in root growth, while older pines in shaded environments are more likely to suffer injury or death due to drought. Although the immediate cause of death is desiccation, the ultimate cause is the reduction in photosynthesis caused by low light. Other conifers such as Norway spruce, silver fir and western hemlock are much more shade-tolerant. There is considerable variation within the genus regarding cold-tolerance and water relations, while many recent studies have been concerned with air pollution (in the Jeffrey pine *P. jeffreyi*, elevated levels of ozone cause chlorosis) and increased concentrations of atmospheric $CO_2$.

### 3.7.2 Variation in the oaks and beeches

While both are in the beech family (Fagaceae) and occupy large areas of the world's warm temperate forests, the beech *Fagus* has only 8–13 species depending on the authorities consulted, whereas oak *Quercus* has more than 250 in the western hemisphere alone. Indeed, Mitchell (1974) states that over 800 oaks have been described, including many hybrids. The beeches are a closely related group of trees with rounded spreading canopies that reach heights of 30–45 m. Their trunks are strikingly cylindrical in form, the annual rings of the upper parts of the trunk being wider than those near its base. The bark is grey, relatively thin and very smooth, so the appearance of these trees is in great contrast to that of the oaks with their typically furrowed bark and more tapering trunks. Fresh cohorts of beech *Fagus sylvatica* (Fig. 3.23) usually arise after mast years in which fruiting has been exceptionally heavy (Section 4.4.1). European and

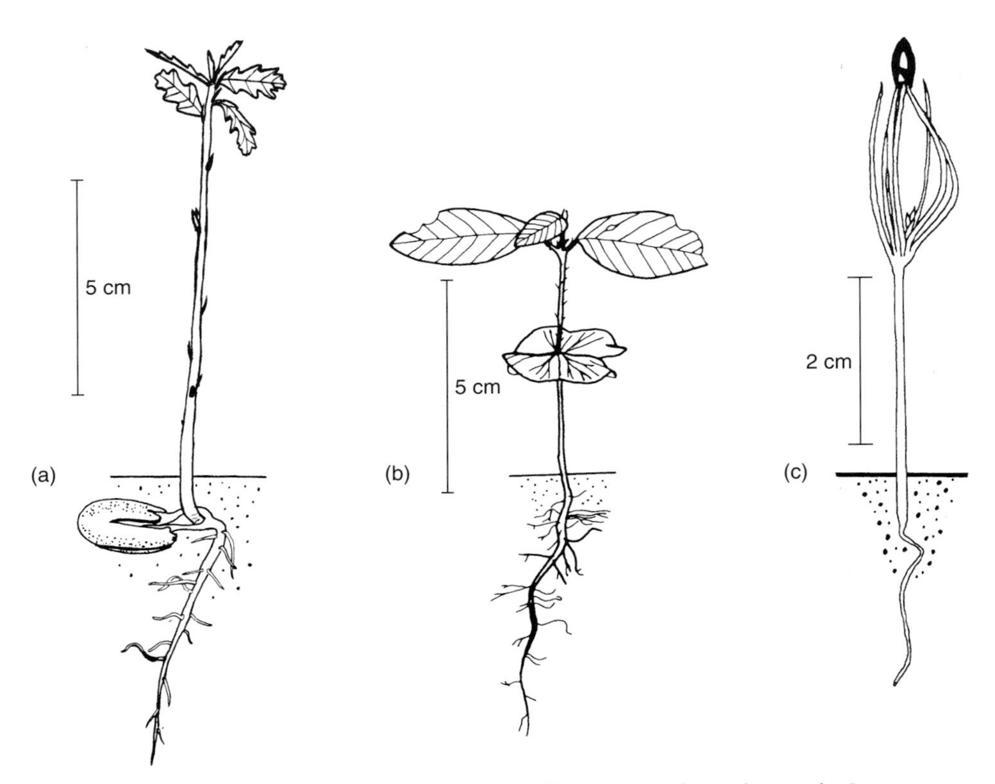


Figure 3.23 Two dicotyledonous tree seedlings (a) pedunculate oak *Quercus robur*, whose germination is **hypogeal** with its cotyledons remaining below ground (and largely concealed within the remains of the acorn) and (b) beech *Fagus sylvatica*, contrasted with that of a conifer (c) Scots pine *Pinus sylvestris* in which the hypocotyl has elongated and the tips of several cotyledons are still trapped within the testa. Three single needles borne on the developing plumule can just be seen. Like the Scots pine, beech has **epigeal** germination, the two semi-circular cotyledons are raised above ground and remain photosynthetically active for a considerable period. (Drawn by John R. Packham. From Packham and Harding, 1982. *Ecology of Woodland Processes*. Edward Arnold.)

Japanese beeches both coppice when cut down. In contrast the American beech *F. grandifolia* develops from suckers arising from its widespread roots (see Fig. 10.8b) so it is ecologically very different, its reproduction being much less influenced by seed predators. Beeches occur in forests spreading from central and eastern USA, through the UK and across to western Asia with a gap before the final occurrences in China and Japan. Though their forests are very extensive, beeches show much less variation than either the pines or oaks so it is not surprising to find that the range of conditions they can endure and the area they cover is considerably less than those of the other two groups.

The oaks, of which roughly half are evergreen, belong to three groups: the red and white oaks (both of which have some evergreen species), and an intermediate group all of whose species are evergreen. All three groups possess both tree and shrub species; the Mediterranean Kermes oak *Q. coccifera* (normally <2m tall) being an example of the latter. Some oaks are **semi-evergreen**, dropping all their leaves in spring and replacing them within a week from rapidly developing buds. This is seen in the long-established Turner's oak *Q. turneri*, a hybrid between the pedunculate *Q. robur* and evergreen *Q. ilex* oaks, which has long grown in Gibraltar. Sexual reproduction in the oaks involves the production of acorns (Fig. 3.23), which occur in only one other genus (*Lithocarpus densiflorus*, the closely related tanbark oak), and possess a cupule of bracts. Until late in the nineteenth century North American Indians collected acorns, removed the shell, ground them into a flour, leached this of tannins with running water and then used it to make a variety of foods. Beech nuts are also edible; their oil has been used for cooking.

Unlike the beeches, oak trees vary greatly in form, the trunks of some being very tall and upright while others are repeatedly forked from quite a low level. Oak leaves are usually cut or lobed, whereas those of beeches tend to be ovate. Oaks are found in a great variety of situations, from coastal plains to altitudes as great as 4000 m in the Himalayas. In regions where many species of oaks and other trees are present together, there is considerable interest in the precise attributes which lead to a species dominating a particular ecological niche while being absent from another. Contrasts between annual variations in temperature and moisture regimes between the eastern, central, western, northern and southern regions of the USA are very great and so is the range under which particular species are found. The **ecological amplitude** of the humid temperate Oregon white oak *Q. garryana* is very different from that of the semi-desert Gambel oak *Q. gambelii* of the south, for example. The same is true of the various species, including both oaks and pines, shown in Fig. 3.19.

### 3.7.3 Influence of soil conditions on pines, oaks and beeches

Soils influence the establishment and growth of trees but are in turn modified by activities of the trees themselves, notably by the leaf litter they produce. Beeches are lime-tolerant, growing well on chalk of the South Downs of southern England for example. They flourish on light to medium soils, but are tolerant of a wide range of substrates provided they have reasonable drainage. Most oaks are favoured by a reasonably rich soil, one that is not too sandy or dry; their leaf litter, like that of beeches, gradually improves the soil beneath them. The influence of soil conditions on the biodiversity and distribution of oaks in southeastern USA is described in Section 6.5.1.

Pines, on the other hand, are well adapted to acidic and nutrient-poor soils. They tend to acidify better soils (especially under the influence of acid rain – Section 11.4.2), the resulting nutrient depletion giving them an advantage over many angiosperms until soils become too acid even for them. Mycorrhizas play a major role in facilitating the passage of mineral nutrients, including phosphorus, from the soil to pine roots. It is thus important to inoculate the soil with appropriate fungi when establishing exotic pine plantations. Provided the climatic and soil requirements of a particular species are well understood, pines are frequently of great value in planting programmes, often on short rotation periods, a notable example being the introduction of Monterey radiata pine *P. radiata*, originally a narrow range endemic in North America, to Australia, New Zealand, Chile and many other parts of the world.

### 3.7.4 Competition between gymnosperm and angiosperm trees

The earliest fossil angiosperms (plants with enclosed seeds) date from the early Cretaceous, *c.* 120 Ma. They evolved rapidly and the world population is now believed to be around 400 000 species, many of them trees or shrubs. The beech family (Fagaceae) seem to have arisen in the montane tropics; its members migrated and diverged into the current living genera by the late Cretaceous some 60 Ma. Rapid speciation of the oaks began in the Eocene (40–60 Ma) in response to the expansion of drier and colder climates, and has continued in response to various other changes ever since. Though the gymnosperms (plants with naked seeds – the conifers and their relatives) arose as long ago as the Middle Devonian (365 Ma) there have never been more than a few thousand species. The expansion of the angiosperms and the concurrent decline of the gymnosperms has indeed been referred to as one of the most important phytogeographic processes in the history of the Earth (Richardson and Rundel,

1998). In the early Cretaceous 113 Ma, the angiosperms formed just 1% of the pollen records of the Atlantic coast of North America but within around 15 million years this had risen to 40%. Since the early angiosperms were almost all insect-pollinated (and so comparatively little of the pollen is carried by the wind to land on the ground) these figures do not show the true impact of angiosperms. When pollen reached 40% angiosperm, the vast majority of the leaf fossils preserved were angiosperm, showing how rapidly and completely they took over from gymnosperms (Ingrouille, 1992 gives a fuller description). This competition has resulted in the restriction of most gymnosperms, including pines and other conifers, to the harsher environments of the globe. Nevertheless, certain gymnosperms, especially the pines, are still notably successful and the tallest existing trees are all gymnosperms (though eucalypts appear to have a greater potential).

Competition between gymnosperm and angiosperm trees, which occurs in very many of the world's forests, is discussed by Bond (1989) who points out that by accumulating several cohorts of leaves, trees of gymnosperms often achieve higher productivity than those of angiosperms, as explained in Section 3.6. In contrast, solute flow (the mixture of water and nutrients moving up through the wood or xylem) is more efficient in angiosperms, and their seedlings are markedly more efficient than those of gymnosperms, which at this stage are dependent on a single cohort of relatively stereotyped and not very efficient leaves. Angiosperms tend to dominate favourable habitats (warm, high light, high nutrients), where their greater plasticity and initial rapid growth, frequently accompanied by use of insect pollination and seed dispersal by animals, confers marked advantages.

## 3.8 Ecology and significance of ageing trees

This topic is considered in detail by Fay (2002), who also considers the management of veteran trees and outlines the valuable work done in recent years by the Ancient Tree Forum (ATF) of the UK. Trees undergo complex changes during their very long lives, during which they play a series of different roles in the ecosystems they largely dominate. Though many foresters are mainly concerned with timber production and therefore in comparatively young trees, dead wood, whose decomposition is considered in Chapter 7, is essential at some stage in the life histories of many associated organisms and it is important to retain this **saproxylic habitat**. Humphrey (2005) suggests that the many spruce plantations in the British uplands (a mixture of Norway and Sitka spruce) develop features of old growth, including appreciable dead wood, after 80–100 years, not much longer than the normal stand rotation for wood production. At Windsor Great Park, southern England, an important site for insects of dead wood, felled trunks have been re-erected to increase the volume of habitat available (thus replacing the standing dead hulks left in truly natural forests where safety is not a consideration). This is important since wood in contact with the ground is often wetter and cooler than standing hulks, and so a habitat for different organisms. Wildlife here has also been enhanced by creating simple cavities for birds and bats. Additional nesting sites have also been produced by tying cut branches with suitable cavities into crown limbs. The leading figure behind this work considers that ancient trees could well be designated as Sites of Special Scientific Interest (Green, 2001). Some conservationists go further and suggest that healthy trees be deliberately injured by drilling or cutting to encourage pockets of dead wood and collection of water to create habitats for invertebrates and fungi (see Section 7.7.2).

Apical dominance is a major feature in the life of a tree as its branch and twig structure ramifies and becomes more complex in developing from a seedling to early maturity. The root system also develops rapidly at this stage, exploiting the available moisture and nutrients in the soil. The crown of the tree loses some peripheral vitality and becomes increasingly rounded during full to late maturity. A degree of branch loss is associated with dysfunction of conductive tissue. In the inner crown, however, there is often reiterative growth in which new branches grow vigorously towards the sky and are served by compartmented vascular channels linked to the root system. **Basifugal mortality**, in which the outer crown starts to die back, is accompanied by incipient root death with increased fungal colonization leading to internal trunk decay from below at full maturity.

It is the final stages, during which the living crown becomes denser and lower, that are of particular interest when considering the physiology and ecology of veteran trees and their influence on the ecosystems they inhabit. The outer crown continues to degenerate so that the outer and uppermost branches become leafless and skeletal; often falling during storms. New shoots develop from dormant or adventitious buds on the stem and main branch system, so primary production remains adequate, although the bark circumference becomes discontinuous as areas of vascular dysfunction in the main trunk develop further. In this strange last stage much of the ancient tree is dead or dying, while other regions grow vigorously. The continued life of the tree is now largely dependent on the mechanical stability of the crown and root anchorage; the two being united by the vascular columns of the trunk that serve the recently formed reiterative shoots that may effectively act as mini-trees borne on the remains of the ancient trunk. Box 3.3 gives more detail on how veteran trees can hang onto life.

## Box 3.3 Phoenix regeneration: new trees from old

Figure 3.24 Regeneration from the trunk of an English oak blown down in the storm that struck south-east England in October 1987. When photographed in late May 4 years later there were vigorous shoots along the trunk and the production of a number of new trees from the old trunk was about to commence by this means. (Photograph by Rodney Helliwell.)

Veteran trees frequently reproduce by means of seeds; under certain circumstances the cambial net may also enable them to give rise to one or more 'new' individuals vegetatively. The several different types of **'phoenix' regeneration** involve tree survival strategies in which active new trees develop from the remains of damaged trees or disintegrating ancient individuals, often as a result of the effects of increased light following frequently natural clearance. Under such circumstances dormant meristems may be activated with the result that reiterative growth occurs and a new crown is produced. Extensive **trunk layering** can occur as, for example, when beech blew over in the Forest of Fontainebleau, France; existing lateral branches developed into main trunks that acquired effective root systems of their own. Occasional examples of **trunk regeneration**, in which adventitious roots develop inside a disintegrating ancient stem, finally giving rise to a new succession trunk, occur in most broadleaved tree species. Sometimes almost the entire trunk of an ancient tree will die and decompose, but small active columns remain and are sufficient to form a small diameter trunk supporting a rejuvenated crown. Ash,

**Box 3.3 (cont.)**

beech, hawthorn, hornbeam and oak can all undergo **trunk remnant rejuvenation** of this type. Sometimes two trees can form from a single discontinuous parent trunk. **Lateral layering** arises from limbs which subside to ground level, give rise to root systems and form new trunks. **Basal regeneration** involves the formation of a new tree from the base of an old stem following partial or complete trunk collapse. Perhaps most remarkable is **phoenix crown regeneration**, which has been observed in yew, sweet chestnut, lime and oak. Here the collapsed crown roots and forms new radiating trunks. **Natural crown restoration** occurs when a tree's original canopy fragments, only to be replaced by a rejuvenated trunk and crown growth.

Now that the value of veteran trees has been more fully recognized, methods of enhancing their continued existence are being actively explored, notably by the ATF. Restoration pruning has been effective in promoting recovery in many species, especially where crown or root stability has been impaired. The aim is to increase the amount of light reaching particular regions of the crown by means of targeted pruning, with careful monitoring and further adjustment at intervals of 3–10 years. Inactive outer limbs are often trimmed to reduce stress on weak trunks using jagged 'coronet' cuts which simulate natural fracture and which seem to produce better stimulation of new growth than a clean saw cut (see Section 7.7.2). Natural fracture techniques are also used to shorten branches and thus maintain live stubs in suitable parts of the canopy. Attempts have also been made to promote adventitious bud formation by scoring branches with a timber scribe; it is too early for the value of this technique to have been proven. Many ancient trees in Britain have previously been pollarded (Section 5.7) and attempts to re-pollard after many years of neglect have been made to reduce the weight of the canopy and stimulate new growth. This is not always easy since complete re-cutting can kill the physiologically weak trees. Normally the canopy has to be cut in stages over several years.

In Sweden, which has the largest forest area in Europe and where it is the main exporter of forest products, a woodsman will test the internal condition of a tree by striking it with a hammer. By this process of 'sounding' it is possible to detect the presence of decay, a very necessary test since it is estimated that in southern Sweden 15% of the tree population is lost from fungal decay. It does not, however, provide information regarding the degree or stage of the decay. This information is vitally important in the case of ancient trees in public places, where a limited degree of decay may be permissible but safety is essential. **Detection of trunk decay** is a very complex subject and a wide range of methods and equipment have been used during the

150 years that this topic has been mentioned in forest literature. Ouis (2003) reviews the range of non-destructive techniques which can be used to detect decay in standing trees. Some employ vibro-acoustical techniques using either vibrations at frequencies within the acoustical bandwidth or sound waves at acoustical or ultrasonic frequencies. The second class of techniques uses methods based on electromagnetic radiation such as radar. Modern destructive testing methods in which only a small sample is extracted from the tree trunk are also of value.

# 4

# Reproductive strategies of forest plants

## 4.1 Plant strategies

### *4.1.1 Strategic response to competition, disturbance and stress*

The concept that living organisms display ecological 'strategies' has advanced rapidly in recent years. The **r–K continuum** of MacArthur and Wilson (1967) made an important early contribution in contrasting the opportunistic **r-species** (with rapid rates of population growth), which exploit temporary habitats, with the equilibrium **K-species** of stable habitats in which competitive ability and survival of the individual is more important than population growth. In the same way that differing species of organism have gradually evolved over long periods of time (Darwin, 1859), such strategies arose through the exertion of competitive natural selection upon varied populations in which differences from the previous norm constantly arose. The **CSR model** of Grime (1974, 1979) – standing for Competitor–Stress tolerator–Ruderal – made an important further advance in adding the **stress-tolerators**, organisms capable of exploiting continuously unproductive environments or niches. The competitors are equivalent to the K-species and the ruderals (living on disturbed sites) approximate the r-species. This theory also recognizes that in plants there is a separation of the established (adult) and regenerative (juvenile) strategies and they may respond differently to their environment. This theory has been very thoroughly applied over a long period of time in a number of ecosystems, and with the publication of *Plant Strategies, Vegetation Processes and Ecosystem Properties* (Grime, 2001), is becoming a major tool in the manipulation of vegetation and ecological prediction.

Stress and disturbance are the two main external factors limiting the amount of living and dead plant material in a habitat. **Stress** in this context consists of the external constraints limiting the rate of growth (productivity, i.e. dry-matter production) of all or part of the vegetation, including shortages of light, water and mineral nutrients and the influence of suboptimal temperature.

**Disturbance** limits plant biomass by causing its partial or total destruction through trampling, mowing, ploughing, the felling of trees, and the activities of pathogens and other herbivores. It also results from wind damage, frost, drought, soil erosion and fire.

Though no plants can long survive both high stress and high disturbance, there are three combinations of these factors under which plants continue to exist. These have led to the evolution of the **three primary ecological strategies** encapsulated in the name CSR. Grime (2001) lists 18 ways in which plants possessing these strategies differ from each other. These involve morphology, physiology, life history, palatability to unspecialized herbivores (Section 5.7) and litter production. If both stress and disturbance are absent then the composition of a plant community is determined by competition between species. i.e. it is made up of **competitors (C)**. Most trees are competitors, exploiting conditions of low stress and low disturbance. **Stress-tolerators (S)** endure high stress and low disturbance such as dog violet *Viola riviniana* in the UK, growing in acidic, damp, shady woodlands, while **ruderals (R)** such as annual meadow-grass *Poa annua* and *Funaria hygrometrica* (a moss common in woods, often on recently burnt land) grow under conditions of high stress and low disturbance. Competitors include herbs, shrubs and trees, whereas ruderals are herbs or mosses. Stress-tolerators include various trees, shrubs, herbs and particularly lichens (see Fig. 4.1).

The definition of **competition** adopted by Grime (1979) in the first edition of his well-known book, is that it is the tendency of neighbouring plants to utilize the same quantum of light, ion of mineral nutrient, molecule of water or volume of space. Competition between plants occurs both above and below ground, competition for light becoming particularly strong as the canopy begins to close so that shoots of one plant are shaded by those of another (Section 1.4.2). **Competitive ability** is a function of the area, the activity, and the distribution in space and time of the surfaces through which resources are absorbed and as such it depends upon a combination of plant characteristics. Very many plants possessing all the following four features are extremely competitive:

(1) tall stature;
(2) a growth form which allows extensive and intensive exploitation of the environment above and below ground, such as a densely branched rhizome as in stinging nettle *Urtica dioica* and creeping soft-grass *Holcus mollis*, or an expanded tussock structure as in tufted hair-grass *Deschampsia cespitosa* (of which a form is frequent in the heavy soils in the oak woods of southern England);
(3) a high maximum potential relative growth rate (RGR); and
(4) a tendency to deposit a dense layer of litter on the ground surface.

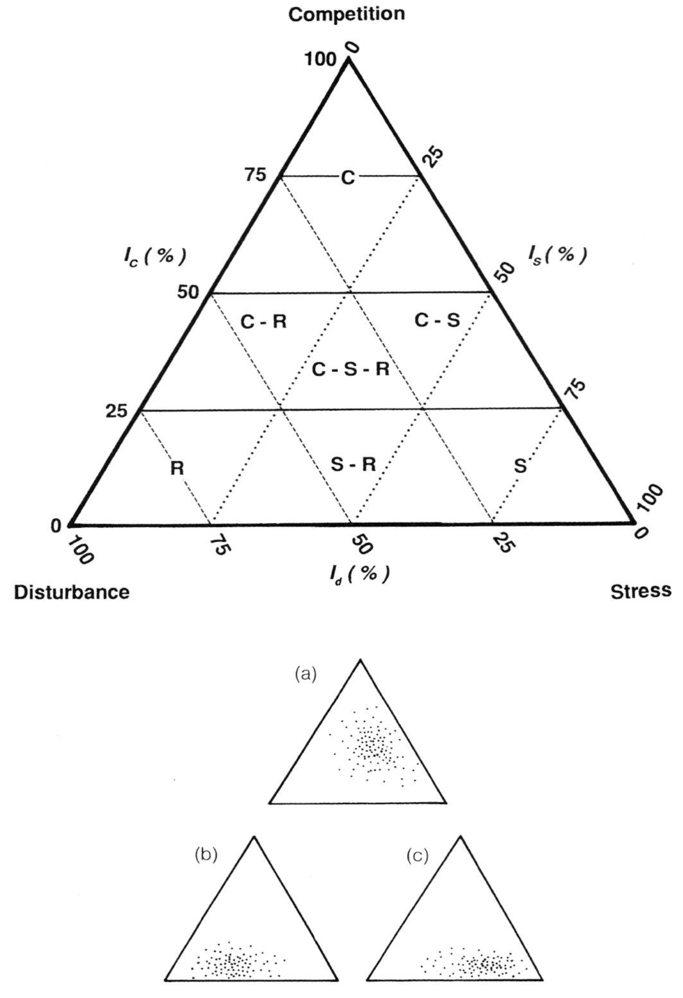


Figure 4.1 Model describing the various equilibria between competition, stress and disturbance in vegetation and the location of primary and secondary strategies. C, Competitor; S, stress-tolerator; R, ruderal; C-R, competitive-ruderal; S-R, stress-tolerant ruderal; C-S, stress-tolerant competitor; C-S-R, C-S-R strategist. $I_c$, relative importance of competition (—–); $I_s$, relative importance of stress (·········); $I_d$, relative importance of disturbance (- - - - -).The strategic range of three life forms is also shown (a) trees and shrubs (b) bryophytes and (c) lichens. (Redrawn from Grime *et al.*, 1988. *Comparative Plant Ecology*. Unwin Hyman.)

To represent these strategies, Grime *et al.* (1988) have developed a triangular diagram whose points are the maxima of competition, stress and disturbance (Fig. 4.1). It is here that the **three primary strategies** are situated. The position of a particular species is determined in relation to the indices of competition ($I_c$), stress

($I_s$) and disturbance ($I_d$). The four main types of **secondary strategy (C–R, C–S, S–R and the central C–S–R)** occur in intermediate positions. Grime *et al.* (1988) derive the position of a particular species by using a dichotomous key involving various morphological, behavioural and reproductive characteristics. Each strategy (also called a functional type) can be represented within the triangular diagram by a set of C, S and R co-ordinates. These co-ordinates relate to a whole set of attributes that contribute to a species' ability to survive under given conditions of productivity and disturbance.

Figure 4.2 illustrates the strategies employed by some common species of the UK woodland field layer. Woodlands are usually stable; indeed the only ruderal present in the main figure is common chickweed *Stellaria media* which is occasionally encountered at disturbed woodland margins. Like other vegetation dominants, the stinging nettle *Urtica dioica* can generate high **shoot thrust**, pushing aside foliage of other species and resisting physical displacement. The leaves of such perennial competitors, capable of rapid growth, also have high nitrogen contents; this coincides with high concentrations of the enzyme 1,5-biphosphate carboxylase/oxygenase (Rubisco) which appear to facilitate rapid rates of photosynthesis in shady conditions. This latter feature is also aided by foliar phosphorus concentrations that are much higher than in slow-growing species. Shoots of stinging nettle die back completely in winter so associates such as the annual competitive-ruderal goosegrass *Galium aparine*, low-growing rough meadow-grass *Poa trivialis* (Fig. 4.2) and the moss *Brachythecium rutabulum*, which climbs stems of other plants, are able to exploit relatively high light intensities in spring and autumn.

Though the main focus of attention in earlier studies of plant strategies, and indeed of Grime (1979), was that of life-history traits (i.e. the strategies of life-history events such as germination, early survival, seed production and longevity), it is now apparent that the primary strategies outlined above also concern other attributes (resource capture and utilization, tissue chemistry and life-span, anti-herbivore defence, rates of decomposition). As Grime (2001) remarks these 'have obvious and direct connections to the functioning of ecosystems'; much of his second edition is designed to establish the nature and usefulness of these connections.

### *4.1.2 Influence of forest clearance in Prince Edward Island, Canada*

The extensive survey made by Sobey (1995a, b) of the vegetation of Prince Edward Island in eastern Canada is of particular interest in its demonstration of the very long period required for a succession to culminate in climax hardwood forest here. Some reproductive strategies are exceedingly long term! The

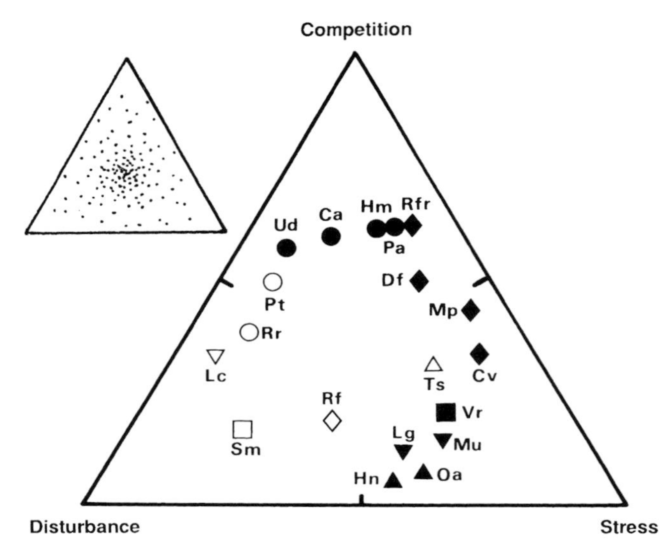


Figure 4.2 C-S-R ordination diagram of woodland field layer species in the UK. This describes the various equilibria between competition, stress and disturbance in vegetation. The point shown for each species indicates the position of its maximum percentage occurrence in a matrix of vegetation types classified according to the strategies of the component species in Grime *et al.* (1988). Where points are not provided the data for the species concerned are insufficiently clear. The small diagram shows the strategic range of herbs. (From Packham and Cohn, 1990. *Arboricultural Journal* 14.)

| | | | | |
|---|---|---|---|---|
| S–C | ♦ | **Cv** | *Calluna vulgaris* | Heather |
| C | ● | **Ca** | *Chamaenerion angustifolium* | Rosebay willowherb |
| (C–R) | | | *Circaea lutetiana* | Enchanter's nightshade |
| (C–R to C–S–R) | | | *Digitalis purpurea* | Foxglove |
| S–C | ♦ | **Df** | *Dryopteris filix-mas* | Male fern |
| C | ● | **Hm** | *Holcus mollis* | Creeping soft-grass |
| S to C–S–R | ▲ | **Hn** | *Hyacinthoides non-scripta* | Bluebell |
| S to S–C | ▼ | **Lg** | *Lamiastrum galeobdolon* | Yellow archangel |
| R to C–R | ▽ | **Lc** | *Lapsana communis* | Nipplewort |
| S to S–C | ▼ | **Mu** | *Melica uniflora* | Wood melick |
| S–C | ♦ | **Mp** | *Mercurialis perennis* | Dog's mercury |
| S to C–S–R | ▲ | **Oa** | *Oxalis acetosella* | Wood-sorrel |
| C–R to C–S–R | ○ | **Pt** | *Poa trivialis* | Rough meadow-grass |
| C | ● | **Pa** | *Pteridium aquilinum* | Bracken |
| R to S–R | ◊ | **Rf** | *Ranunculus ficaria* | Lesser celandine |
| C–R | ○ | **Rr** | *Ranunculus repens* | Creeping buttercup |
| S–C | ♦ | **Rfr** | *Rubus fruticosus* agg. | Bramble |
| (S) | | | *Sanicula europaea* | Sanicle |
| (C–S–R) | | | *Stellaria holostea* | Greater stitchwort |
| R | □ | **Sm** | *Stellaria media* | Common chickweed |
| C–S–R | △ | **Ts** | *Teucrium scorodonia* | Wood sage |
| C | ● | **Ud** | *Urtica dioica* | Common nettle |
| S | ■ | **Vr** | *Viola riviniana* | Common dog violet |

1200 plots sampled in 1991 were divided into 930 that were bearing forest when the first aerial survey of Prince Edward Island was made in 1935, and 270 that were not. Those plots which had been clear of trees in 1935 were dominated by softwoods, primarily white spruce *Picea glauca*. These were in the regeneration stage, carried a relatively low amount of timber, and were typical of an 'old-field' origin. Other characteristic species were early successional trees and shrubs, 'weedy' herbs and shade-tolerant mosses. In contrast, plots which were afforested in 1935 were typically dominated by hardwoods and positively associated with partial-cutting/thinning, though there was a marked gradation towards the softwood-dominated group in those plots which had by 1935 regenerated after previous clearance. Thus, the forest stands divide neatly into softwoods (conifers) and hardwoods, dependent upon their recent history.

Conifers encountered in the survey included balsam fir *Abies balsamea*, tamarack *Larix laricina*, white, black, and red spruce (*Picea glauca*, *P. mariana* and *P. rubens*), jack, red and white pine (*Pinus banksiana*, *P. resinosa* and *P. strobus*), eastern white cedar *Thuja occidentalis* and eastern hemlock *Tsuga canadensis*. There were 15 broadleaved hardwoods: red, striped and sugar maple (*Acer rubrum*, *A. pensylvanicum* and *A. saccharum*), yellow, white and grey birch (*Betula alleghaniensis*, *B. papyrifera* and *B. populifolia*), American beech *Fagus grandifolia*, white and black ash (*Fraxinus americana* and *F. nigra*), ironwood *Ostrya virginiana*, largetooth and trembling aspen (*Populus grandidentata* and *P. tremuloides*), red oak *Quercus rubra*, American mountain ash *Sorbus americana* and American elm *Ulmus americana*.

Sobey (1995a, b) employed two statistical methods based on reciprocal averaging (TWINSPAN and DECORANA, both devised by Hill, 1979a, b), to handle the very large amounts of multivariate data generated by this survey, which provided much useful information regarding successional processes and the influence of environmental variation. Figure 4.3 is a TWINSPAN classification of 1127 plots recorded in 1991. The first division was almost equal. Group 0, on the left side consisted of plots whose species were associated with broadleaved woodlands (hardwoods), those of Group 1 on the right were mainly characteristic of woodlands dominated by conifers (softwoods). Wood fern *Dryopteris spinulosa* was an indicator for the negative Group 0, while Schreber's moss *Pleurozium schreberi*, wild raisin *Vibernum cassinoides*, and blueberry *Vaccinium angustifolium* acted as indicators for the positive Group 1. An **indicator species** is characteristic of a particular habitat and in this case can be reliably used to decide or indicate which type of forest is being considered (see also Hill *et al.*, 1975).

The 595 hardwood plots of Group 0 then split unequally into the 165 of Group 00, that included the 156 plots belonging to the more nutrient-rich and wetter broadleaved woodlands of Group 0011 (**GP3** in Fig. 4.3) (delineated at the fourth division), and the 430 of Group 01. This was then split into the 79 plots of Groups 0100 **(4)** and 0101 **(5)**, and the 351 plots of Groups 0110 **(6)** and 0111 **(7)**. Group 1 was similarly divided into eight smaller groups as a result of three further divisions. Terms for the Stand Groups used in the figure follow the classical notation, those in bold type were used for the 15 end groups which Sobey finally employed. Of these, Groups 1, 2, 4 and 15 were so small as to be insignificant, the 21 plots of the rather oddly placed Group 11 were of a species-poor hardwood-related type, and the 12 plots of Group 12 were wet transition plot types. After a detailed review of the floristic and environmental features of all the plots involved, Sobey considered that major floral variations

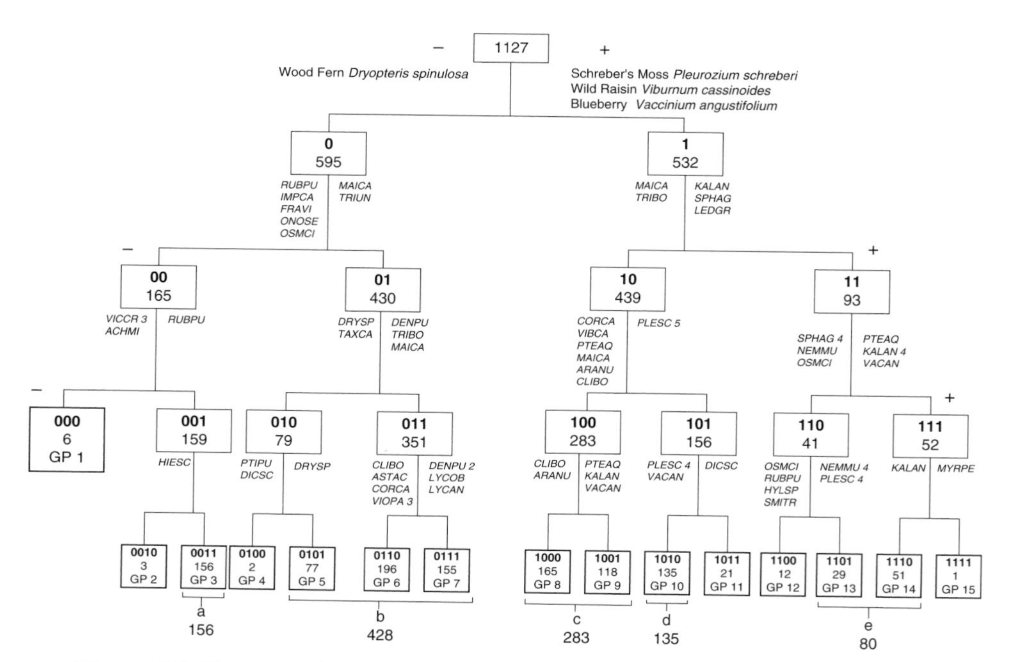


Figure 4.3 Two-way indicator species analysis (TWINSPAN) of 1127 of the circular ground flora plots (each with an area of $4\,m^2$) assessed in the 1991 Prince Edward Island Inventory. This dendrogram shows the classical notation to the third level of division; plot allocations for the fourth order division are shown together with the final Group Numbers used by the original author, who considered the major floral variations to be best considered under the five categories **a, b, c, d** and **e**. (After Sobey, 1995a. *Report to the Prince Edward Island, Forestry Division.*)

Scientific names of indicator species are given below.

within the remaining 1082 quadrats were best considered under the five categories of:

(a) Wet rich deciduous woodland (Group Number 3, of which 3B – not distinguished in Fig. 4.3 – had a ground flora associated with wetter and richer soils than 3A). Indicator species for the 71 quadrats of 3 A were starflower *Trientalis borealis* and lady fern *Athyrium filix-femina*: those for the 85 of 3B were sensitive fern *Onoclea sensibilis*, the moss *Mnium* sp., cinnamon fern *Osmunda cinnamomea* and jewelweed *Impatiens capensis*.

---

Caption for Figure 4.3 (cont.)

The five-letter short forms on the figure consist of the first three letters of the generic name and the first two letters of the specific name. Numbers following an indicator imply that a certain percentage cover has been exceeded.

| | |
|---|---|
| ACHMI *Achillea millefolium* Yarrow | MYRPE *Myrica pensylvanica* Bayberry |
| ARANU *Aralia nudicaulis* Wild sarsaparilla | NEMMU *Nemopanthus mucronata* False holly |
| ASTAC *Aster acuminatus* Whorled wood aster | ONOSE *Onoclea sensibilis* Sensitive fern |
| CLIBO *Clintonia borealis* Bluebead lily | OSMCI *Osmunda cinnamomea* Cinnamon fern |
| CORCA *Cornus canadensis* Bunchberry | PLESC *Pleurozium schreberi* Schreber's moss |
| DENPU *Dennstaedtia punctiloba* Hay-scented fern | PTEAQ *Pteridium aquilinum* Bracken |
| DICSC *Dicranum scoparium* Broom moss | PTIPU *Ptilidium pulcherrimum* Ptilidium moss |
| DRYSP *Dryopteris spinulosa* Wood fern | RUBPU *Rubus pubescens* Dewberry |
| FRAVI *Fragaria virginiana* Strawberry | SMITR *Smilacina trifolia* Three-leaved false solomon's seal |
| HIESC *Hieracium scabrum* Rough hawkweed | SPHAG *Sphagnum* Bogmoss species |
| HYLSP *Hylocomium splendens* Mountain fern-moss | TAXCA *Taxus canadensis* Yew |
| IMPCA *Impatiens capensis* Jewelweed | TRIBO *Trientalis borealis* Starflower |
| KALAN *Kalmia angustifolia* Sheep-laurel | TRIUN *Trillium undulatum* Painted trillium |
| LEDGR *Ledum groenlandicum* Labrador-tea | VACAN *Vaccinium angustifolium* Blueberry |
| LYCAN *Lycopodium annotinum* Bristly clubmoss | VIBCA *Viburnum cassinoides* Wild raisin |
| LYCOB *Lycopodium obscurum* Ground pine | VICCR *Vicia cracca* Tufted vetch |
| MAICA *Maianthemum canadense* Wild lily-of-the-valley | VIOPA *Viola pallens* Small white violet |

(b) Three different ground flora community variants associated with tolerant hardwood forests on dry upland soils (Groups 5, 6 and 7).
(c) Cut-over or disturbed conifer-dominated forest on dryish soils (Group 8 with a tree canopy dominated by softwoods, especially balsam fir, and Group 9 which is another similar successional type but on drier soils).
(d) Old field white spruce woods (Group 10).
(e) Black spruce 'bog' (Group 13) and 'heath' (Group 14) forests.

## 4.2 Regenerative strategies and vegetative spread

### *4.2.1 Pollination strategies*

Forest plants face a number of problems in respect of pollination, from the sheer size of the larger trees to the still, dark conditions inside a forest hampering pollinators. Trees in northern and temperate forests, especially the conifers, tend to be **wind pollinated (anemophilous)** but towards the tropics the trend is towards animal pollination, particularly by **insects (entomophilous)**. Wind pollination is often seen as primitive and wasteful of pollen (a single birch catkin can produce 5.5. million pollen grains) and yet is remarkably prevalent. The easy answer would be to say that in northern areas there are fewer insects and higher wind speeds, leaving wind as the best pollen transporter. While this is undoubtedly true, it appears to be so successful due to low species diversity of trees. Wind can transport pollen long distances but is unspecific as to where the pollen goes. In northern areas where forests are dominated by relatively few tree species this is not necessarily a problem since the pollen has a high chance of landing on a flower of the same species. The high species diversity of tropical forests, where individuals of a species are few and widely scattered, would make wind pollination too unreliable, and the higher costs of attracting animal pollinators are clearly worth the investment. Wind-pollinated flowers have no need to attract animals and so have dispensed with bright petals, nectar and scent. Various adaptations help to ensure the pollen goes as far as possible. Most deciduous wind-pollinated trees flower while the canopy is bare of leaves; conifers and evergreens improve pollen dispersal by placing the flowers on the ends of branches.

Since trees are so large and produce huge quantities of pollen, self-pollination is a likely problem whether wind- or animal-pollinated. Even if the tree is self-sterile, the female flower may still be swamped with its own pollen physically preventing pollen from other trees reaching the stigma. In wind-pollinated trees the most frequent way of avoiding this is to have separate male and female flowers on different parts of the trees, sometimes separated on the same branch (e.g. oaks) and sometimes on separate parts of the tree (as in many conifers, males at the bottom of the canopy, females at the top).

Plants buried in the dark, still conditions of the forest have their own share of pollination problems. As outlined in Section 1.4.2, many herbs of temperate woodlands flower early in the year before the canopy closes. Those that flower after canopy closure are usually insect-pollinated and tend to have pale or white flowers which increase visibility to pollinators or are heavily scented. More information on pollination in trees can be found in Thomas (2000).

### *4.2.2 Regenerative strategies and methods of seed dispersal*

In 1942, Salisbury put the study of plant reproduction on a firm footing in his classic work *The Reproductive Capacity of Plants*. Five strategies are widespread in terrestrial vegetation, and all occur in woodlands and forests. **Vegetative expansion (V)** is found most often where disturbance is at a low level. In habitats subjected to seasonally predictable disturbance by climatic or biotic factors that reliably creates openings **seasonal regeneration (S)** is common, seeds or vegetative propagules being produced together in a single cohort. In contrast, species with **persistent seed** or **spore banks ($B_s$)** enjoy a selective advantage in places where the timing of disturbance is unpredictable: the seeds are present whenever an opening occurs (see Section 4.6.2). Numerous **widely dispersed seeds or spores (W)** are common in species growing in habitats that are relatively inaccessible or subject to spatially unpredictable disturbance. Such species are often opportunists (or r-species – see Section 4.1.1), indeed both fireweed *Chamerion angustifolium* and the moss *Funaria hygrometrica*, which frequently develop on burnt areas, are widespread invaders. **Persistent juveniles ($B_j$)** are seedlings or sporelings which persist for long periods in unproductive habitats in which levels of disturbance are low. This last strategy is that of the **seedling bank** (Section 9.2.2), and is employed by a number of shade-tolerant trees such as the Norway spruce *Picea abies* whose dwarf trees persist for many years in primeval Scandinavian forests. These seedlings can quickly grow into a gap in the canopy ahead of a seedling that starts from seed. It is a risky strategy, however, since many seedlings will eventually die before a suitable gap appears.

The majority of the strategies outlined above depend upon seed dispersal. Consequently, the evolutionary pressure to find effective means of transport has been strong, resulting in diverse and fascinating mechanisms. Methods of seed dispersal vary considerably; many fruits are spread by wind, animals or explosive mechanisms. This last system, **ballistic dispersal**, is employed by violets and geraniums, including Herb Robert *Geranium robertianum*, a

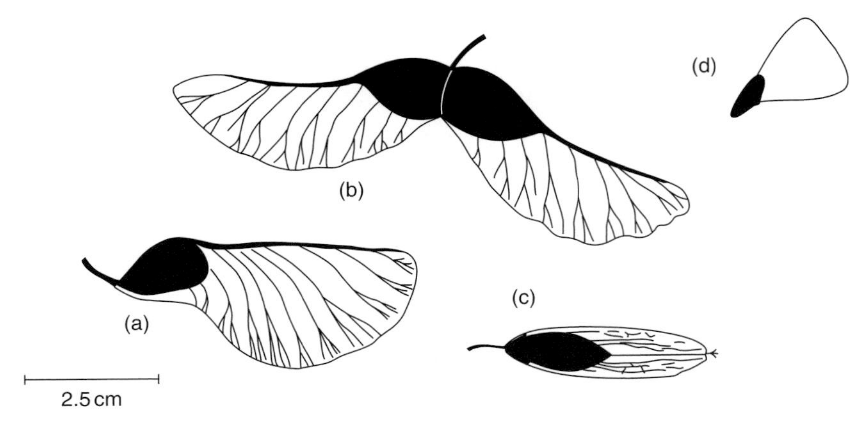


Figure 4.4 Wind-dispersed fruit (samaras) of (a) the Pride of Bolivia *Tipuana tipu* (b) sycamore *Acer pseudoplatanus* and (c) common ash *Fraxinus excelsior*. Convergent evolution has resulted in remarkably similar samaras in the Fabaceae (a), the maples (Aceraceae – b), and the ashes (Oleaceae). All spin as they fall, carrying them away from the parent tree when winds are strong. The samara of sycamore is double, and sometimes even treble; the line of separation can be seen in the figure. Many gymnosperms are also spread by winged seeds and those of the noble fir *Abies procera* (d) fly particularly effectively. (Drawn by John R. Packham.)

widespread woodland and coastal species in Britain and a widespread introduction in North America, especially western forests. Many legumes (Fabaceae) have pods each of which splits into two halves which dry, twist and expel the seeds explosively. Relying on wind, members of different families have, by a process of convergent evolution, produced similar winged fruits (Fig. 4.4). Many **animal-dispersed fruits**, such as those of enchanter's nightshade *Circaea lutetiana* and goosegrass or cleavers *Galium aparine* are covered in hooks, others are succulent and the seeds they contain may be transported considerable distances before being wiped off a bird's beak or egested. Some plants use two or more seed dispersers (**diplochory**), which increases the chance of successful germination. For example, a number of Mediterranean plants have explosive fruits and an elaiosome (an oily body) on the seed which attracts ants to carry the seed away. Pines with big seeds such as the coulter pine *Pinus coulteri* and sugar pine *P. lambertiana* of California, have wings on the seeds which aid spread but being big they tend to fall near the parent. But they are then gathered by secondary dispersers such as chipmunks, mice and corvids, are carried away and buried (**scatter hoarded**) in the soil (Box 4.1). Some will be eaten by the animals, but others will escape unharmed and germinate.

Vander Wall and Longland (2004) generalize that the first phase of dispersal moves seeds away from the parent plant and so reduces losses to seed

### Box 4.1 Seed-caching birds

The **Corvidae**, a widespread group of rather aggressive birds including the ravens, crows, rooks, jackdaws, magpies, jays, choughs and nutcrackers, show considerable intelligence and the rooks in particular seem to communicate by their calls. The group as a whole has a rather mixed diet, some feeding on carrion, but a number of corvids also form caches of seeds as a winter food supply. In doing so, they are crucial to the establishment of a number of tree species from seed. Most prominent are the 20 or so northern hemisphere pines that are reliant on corvids to disperse their large wingless seeds (Lanner, 1996). These include the pinyon pines, of which there are six in the USA alone. Islands of arolla pine *Pinus cembra* in the kampfzone of the European Alps, for example, arise from the activities of the nutcracker *Nucifraga caryocatactes*, which buries heaps of 10–30 seeds as winter food and either doesn't need them all or fails to relocate them. The next generation of trees grows from these ready-planted and undamaged seeds. The numbers of seeds buried by individual birds may be very large. The Clark's nutcracker of the mountains of western North America (*N. columbiana*) makes small caches of 1–15 seeds of the whitebark pine *Pinus albicaulis*. Over a season it may bury as many as 98 000 seeds in over 30 000 caches; relocating these is a remarkable feat of memory and it is not surprising that some seeds go unneeded or forgotten. Vander Wall and Balda (1977) found that a population of around 150 Clark's nutcrackers could cache 3.3–5.0 million seeds of the pinyon pine *P. monophylla* in one autumn, weighing between 658 and 1028 kg!

Corvids are also important in the dispersal of other species. For example, successful colonization by the European oaks is strongly influenced by the size of jay *Garrulus glandarius* populations; when these were low in the English Midlands in the 1970s oak seedlings were far rarer than they are now, when these birds can often be seen burying acorns in grassland and lawns.

predators by distributing seeds widely. The second phase often moves seeds to safe sites, which usually means below ground, where they are relatively protected from seed predators, and have a good chance of successful germination. Several techniques are showing promise in helping to track seeds as they disperse. These include radioactive labelling of seeds, attachment of fluorescent microspheres (particularly good at following seed dispersal in faecal material from birds), stable isotope analysis (since these naturally vary geographically, it might allow the origin of long-distance seed dispersal to be worked out) and molecular genetic markers, which would allow the matching of seedlings with their parents giving novel information on spatial patterns (see Wang and Smith, 2002 for more details).

A variety of nut-producing trees also depend upon animals (rodents such as squirrels, and birds, especially those in the crow family – corvids) for scatter hoarding, including walnuts *Juglans*, hickories *Carya*, oaks *Quercus*, beeches *Fagus*, chestnuts *Castanea*, chinquapins *Castanopsis*, tanoaks *Lithocarpus*, hazels *Corylus*, horse chestnuts *Aesculus*, and almonds, plums and cherries *Prunus*. Vander Wall (2001) puts forward evidence that this mutualistic relationship originated as early as the Paleocene, about 60 million years ago. Most nuts appear to have evolved from ancestors with wind-dispersed seeds to become highly nutritious often with a calorific value ranging from 5.7 to 153.5 kJ per nut, 10–1000 times greater than most wind-dispersed seeds.

Despite these mechanisms, many woodland plants, especially herbs in the field layer, are poor dispersers; they tend instead to be clonal (**V**) plants, have little recruitment from seed and possess no obvious mechanism for long-distance seed dispersal. This has led ecologists to wonder how they managed to migrate towards the poles after the last ice age. It has been assumed that given long periods of time, even small dispersal distances would allow woodland herbs to keep up with the migration of the forests. Cain *et al.* (1998) examined this assumption using mathematical models based on the seed dispersal of wild ginger *Asarum canadense* in north-east North America. Field studies showed that ants are responsible for dispersing wild ginger seeds (attracted by the nutritious elaiosome) and can move the seeds up to 35 m, the largest distance ants are known to move the seeds of any woodland herb. However, by this mechanism, over the last 16 000 years, wild ginger should have travelled only 10–11 km from its glacial refugia. In reality, wild ginger has moved hundreds of kilometres during this time. Their work strongly suggests that for wild ginger and other woodland plants occasional long-distance dispersals were important in allowing their spread. This has important implications for forest fragmentation since seeds may occasionally disperse further than expected and bridge gaps. There are, of course, any number of assumptions behind whether such dispersal would be effective in today's landscapes and against the current speed of loss. Trees have, throughout their lives, many difficulties to overcome, initial establishment in a suitable environment being one and the repeated onslaught of herbivores another. Woodland herbivores vary immensely in size; contrast the larva of the moth *Cydia fagiglandana* (Fig. 4.11) which invades and kills the embryos of beech seeds with the African elephant which attacks adult trees. Herbivores are important with regard to both **defoliation** and the **destruction of seeds**. Some trees, especially oak, are very resistant to defoliation. The seedling of an English oak depicted in Fig. 4.5, for example, developed from an acorn cached in a lawn at the margin of Cantreyn Wood, Bridgnorth, Shropshire. Its shoots were cut repeatedly

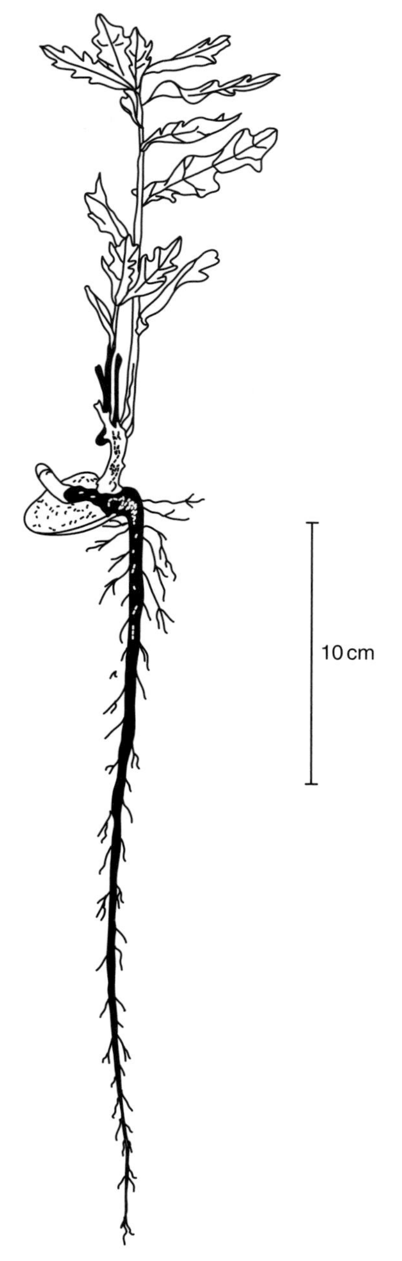


Figure 4.5 Caching and cutting. Young oak seedling developed from an acorn cached in a lawn, probably by a jay or other corvid (see Box 4.1), and repeatedly cut to the ground. (Drawn by John R. Packham.)

and it remained unnoticed until the lawn was left uncut for a month, at the end of which time the radicle had reached a length of 28 cm and the shoot a height of 16 cm. The large food reserves contained in acorns, which enables oak seedlings to survive such adverse circumstances, make them a valuable food source for a variety of herbivores including grey squirrels. Healthy young oaks readily arise in gardens and the trees produce an adequate rain of acorns, yet young oaks are a rarity even beneath the many mature oaks in Cantreyn Wood, which is changing its nature and has an abundance of young ash.

The tropics are renowned for the numerous tree species that occur at low densities with individuals more evenly spread than expected through the forest. A number of explanations have been suggested but the **Janzen–Connell hypothesis** has caught the imagination of many ecologists. The hypothesis, named after two ecologists, Daniel H. Janzen and Joseph H. Connell, who published on the subject in 1970 and 1971, respectively, states that tree seedlings are more likely to survive a certain distance away fom the parent tree. Figure 4.6 shows how this works. The number of seeds on the ground falls with distance from the parent. Seeds and seedlings nearest the parent are more likely to be found by species-specific predators or diseases so the probability of survival increases with distance from the parent. Thus the number of successful seedlings will peak at a certain distance away from the tree. Connell suggested that since seed mortality is always so high, this effect would be most easily seen in seedling numbers; it is with these that species-specific predators would have most effect. The argument goes that this does not happen in temperate forests because most predators and diseases are not species-specific so the chance of being killed is roughly equal with distance from the parent. This hypothesis has important ramifications in that it helps explain how so many rare species continue to exist for long periods of time. No one species, or small group of species, is so common as to gain dominance, so many rare species persist together. Certainly, Rees *et al.* (2001) suggest that this is one of the few ways that the co-existence of 1000+ species growing together in a tropical forest can be explained (see Section 2.2.2 for the role of soils in this). Much research has been invested in testing this hypothesis, particularly whether distance enhances survival, in plants and animals, but with mixed results. Hyatt *et al.* (2003) took an overview of these studies using meta-analysis, a statistical tool that helps synthesize the results of a number of studies. While greater seedling survival with distance from the parent was found in some species, this was balanced by cases where the hypothesis is not supported. Thus, while this mechanism is undoubtedly important in some species, it does not look to be a general underlying principle in all species.

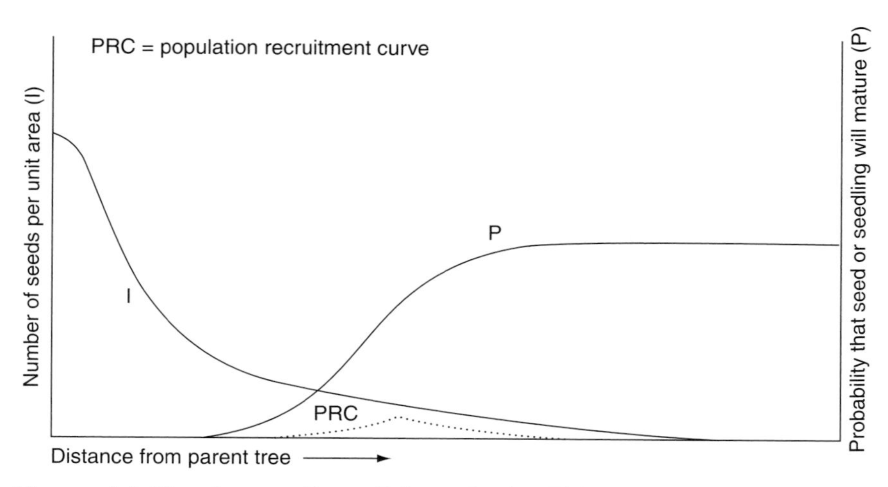


Figure 4.6 The Janzen–Connell hypothesis. This states that in tropical forests, tree seedlings are more likely to survive a certain distance away from the parent tree. With distance from the parent (left to right) the number of seeds on the ground declines (line I) but the probability of a seed or seedling being missed by species-specific predators or pathogens (which will be more numerous near the parent) increase with distance from the parent (line P). Most surviving seedlings will therefore be found in the dotted triangle, labelled PRC – the population recruitment curve. The distance from the parent tree at which this peaks will obviously vary with tree species and conditions. (Redrawn from Janzen, 1970. *American Naturalist* 104, University of Chicago Press.)

### *4.2.3 Early life of mangroves in relation to their adult distributions*

**Mangroves** are a group of phylogenetically unrelated trees (i.e. not from the same immediate ancestors) that live in very adverse conditions along the margins of tropical river estuaries and sea coasts, usually supported over a muddy substrate by **stilt roots**, most of which are frequently out of the water, or roughly horizontal roots bearing **pegs** or **knees** which project above the waterline. As mentioned in Section 2.3.3, in all these three types of pneumatophore, lenticels and aerenchyma (wide air passages) in the above-water roots and trunk allow gas exchange in submerged roots with oxygen diffusing in and waste gases passing out. Mangrove roots and lower stems are regularly bathed in salt water and some have root filters that effectively prevent the entry of salts. Other plants in similar environments, like tamarisks (or saltcedars *Tamarix* spp.), excrete briny water containing excess salt through special salt glands, while some species allow salt to accumulate in leaves which are then dropped. The adult distribution of various mangrove species shows marked

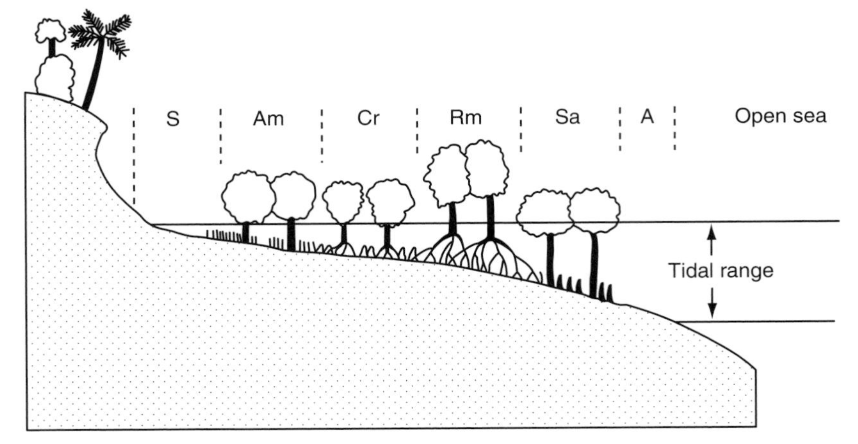


Figure 4.7 Zonation of mangroves on the coast of East Africa. S, barren sand; Am, *Avicennia maritima;* Cr, *Ceriops tagal*; Rm, *Rhizophora mucronata*; Sa, *Sonneratia alba*; A, algal zone. Horizontal lines indicate extreme high and low tide levels. (Redrawn from Jenik, 1979. *Pictorial Encyclopedia of Forests*. Hamlyn.)

zonation patterns (Fig. 4.7); these are now widely regarded as being largely due to a sorting of species along a salinity gradient as a result of competitive interactions. The mangrove and freshwater swamps of the tropics are commonly inhabited by crabs (true crabs belonging to the infraorder Brachyura); these excavate large corridors in the soil and so improve drainage and root oxygenation. Crustaceans of all sizes live in forests and in drier situations, such as the oak woods of Europe, woodlice are common, consuming plant remnants.

Clarke *et al.* (2001) investigated the dispersal potential and early growth of 14 tropical mangroves on the northern coast of Australia in an attempt to discover whether these correlated with their patterns of adult distribution. The species investigated belonged to eight different families; the Avicenniaceae, Combretaceae, Fabaceae, Meliaceae, Myrsinaceae, Plumbaginaceae, Steruliaceae and Rhizophoraceae. Some were **viviparous** (having seeds which could develop while still attached to the parent tree) and others were not, but both groups had broadly similar seed weights, buoyancy and rates of root and shoot initiation. **Diaspores** (reproductive units), which were either seeds or one-seeded fruits, were immersed in solutions containing 100%, 50%, 10% and 0% sea water, and then stranded on surfaces with the same salinity (trays with 5 cm depth of damp sand). Some floated while others sank even in fresh water, and there were also differences in predominant orientation of the diaspores and the time before roots were initiated. In all the species investigated the diaspores were relatively widespread; dormancy was found in only one species. Though dormancy is an advantage in many widely dispersed seeds – because it allows more time for dispersal – this is not the case with mangroves as swift development is an advantage upon stranding in a favourable habitat before they can be washed away. There seems to be a clear trend against the production of small dormant diaspores that are not likely to develop into saplings in such an unstable environment.

When early life-history traits of 12 of the 14 species were compared with patterns of adult distribution, correlation was poor. Traits concerning establishment were, however, better predictors of occurrence than those associated with dispersal. The proportions of diaspores germinating after being soaked for 15 days in various salinities varied widely amongst the species tested, as did the heights reached after 15 weeks in early growth trays. *Xylocarpus mekongensis* and *Avicennia mariuna* were amongst the species which grew well and rapidly under all four treatments, while *Aegiceras corniculatum* was adversely affected by high salinity and achieved optimal growth in 5% sea water.

### 4.2.4 Angiosperms, conifers and ferns: tree regeneration and dominance in South Island, New Zealand

The angiosperms rapidly assumed dominance of most terrestrial ecosystems after their origination early in the Cretaceous period (Fig. 1.1). While they diversified and spread, coniferous trees and other gymnosperms were virtually eliminated from the tropics and gymnosperm abundance in many other terrestrial ecosystems was reduced. Coniferous trees now generally dominate only at high latitudes, in subalpine forests, in arid regions and on nutrient-poor or poorly drained soils. Although they grow less rapidly than angiosperms during the early regeneration phase, they are better able to resist severe cold and can persist in less productive habitats where their tough, long-lived needles are advantageous on nutrient-poor soils. Thirteen co-authors from many parts of the world co-operated in investigations of the forests of the Waitutu Ecological Region of the Fiordland National Park, New Zealand, in an attempt to throw light on the mechanisms influencing the relationships between angiosperm and coniferous trees (Coomes *et al.*, 2005).

The forests investigated developed on a **soil chrono-sequence** in which soil nutrient availability became less and the drainage poorer on the older sites. Angiosperm trees were dominant on the 'recent' alluvial terraces which have been dated at less than 24 000 years old, while coniferous trees dominated the older marine terraces whose age varied between 79 000 and 121 000 years. The oldest marine terraces were 291 000 years old and were dominated by coniferous shrubs. The three main habitats investigated were the alluvial forest (most productive), terrace forest and shrubland (least productive). Soil phosphorus becomes increasingly limiting along this sequence and its scarcity is very marked in the shrubland. The ferns characteristic of the alluvial forest are *Dicksonia squarrosa*, *Cyathea smithii* and *Blechnum discolor*, while those of the terrace forest are *Blechnum procerum* and *B. discolor* (again). The dominant ferns of the shrubland are more varied consisting of *Hymenophyllum multifidum*, *Gleichenia dicarpa*, *Grammitis billardierei* and *Schizaea fistulosa* (Coomes *et al.*, 2005).

The title of their paper ('The hare, the tortoise and the crocodile: the ecology of angiosperm dominance, conifer persistence and fern filtering') reflected the nature of the three groups; the very active angiosperms dominating the better sites, the more conservative conifers coping well with less nutrient-rich and poorer areas while the ferns, a very ancient group, influence the regeneration of the trees. Tall ferns (crocodiles) and deep shade restrict regeneration opportunities in the relatively productive forests of New Zealand, thus diminishing the opportunities for conifers (tortoises) to escape competition from fast-growing angiosperms (hares). Less than 1% of total light reaches the floor of the alluvial forest; transmission of PAR is largely prevented by dense groves of tree and ground ferns, and by large-leaved subcanopy trees. Few young tree seedlings of any species were found on the forest floor, though angiosperm trees were particularly successful in colonizing rotting logs and tree fern trunks.

## 4.3 Reproduction and fruiting

Most woodland herbs produce **chasmogamous flowers** (CH) that open and expose the reproductive parts to wind, insects or other pollinating agents, encouraging cross-pollination (allogamy), so enhancing biological diversity in the next generation. But the flowers of some species remain entirely closed: **cleistogamous flowers** (CL) are structurally modified for self-pollination (autogamy); bud-like, never open, and are much reduced compared with the more familiar chasmogamous flowers. Petals are absent or greatly reduced, nectar and scent are not present, and stamens are frequently reduced in number and size. The advantages and disadvantages of the two forms of flower are reviewed by Berg and Redbo-Torstensson (1998), who investigated the sexual reproduction of wood-sorrel *Oxalis acetosella*, a perennial of European woodlands that produces both types. Figure 4.8 shows the seasonal distribution of the two types of flower. Chasmogamous flowers were produced in spring; ramets (individual members of a clone) with one or more CH flowers left unfertilized generally produced more CL buds than those in which all CH flowers were fertilized.

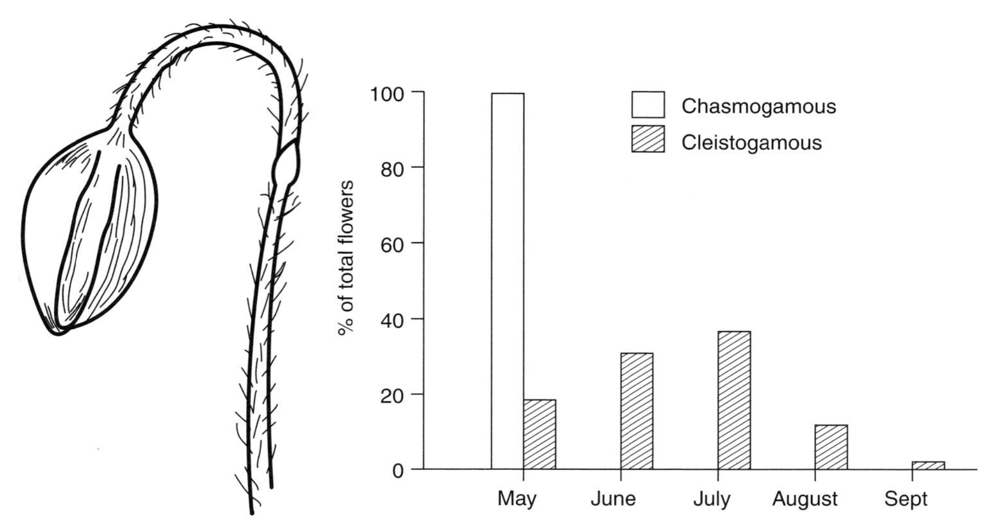


Figure 4.8 Left: Cleistogamous flower of wood-sorrel *Oxalis acetosella* borne on a flower stalk (pedicel) with two small modified leaves (bracteoles). The flower remains closed during development of the seeds until, as in normal opening chasmogamous flowers, they are discharged explosively through slits in the carpel walls to a distance of about a metre. (Drawn by John R. Packham.) Right: Pooled data for seasonal percentage distribution of chasmogamous and cleistogamous flowers of wood-sorrel in three woods near Uppsala, Sweden dominated respectively by birch/oak, Scots pine and Norway spruce. (From Berg and Redo-Torstensson, 1998. *Journal of Ecology* 86, Blackwell Publishing.)

Detailed analysis of the numbers of CH and CL flowers produced, the season in which they grew and the numbers of seeds maturing in each capsule at the three sites over a period of 3 years, demonstrated the impact of climatic variation and the resources available to the plants concerned. The general view of cleistogamy is that it is a fail-safe mechanism for back-up seed production unaffected by changes in resource supply and environmental conditions. In contrast, these studies showed wood-sorrel *Oxalis acetosella* to be a perennial with a bet-hedging strategy optimizing reproductive output in fluctuating environments.

Trees near the end of their lives sometimes fruit very heavily, as was the case in a common beech at the Patcham Place site (see Fig. 4.9), which became so dangerous that it had to be cut down in the autumn of 2002. The stump had a hollow centre, evidence of spalting (fungal rot) and a thin zone of active tissue adjoining the bark. The crown, however, bore an immense number of nuts, scoring the top grade 5, while the other trees in the Patcham Place site averaged

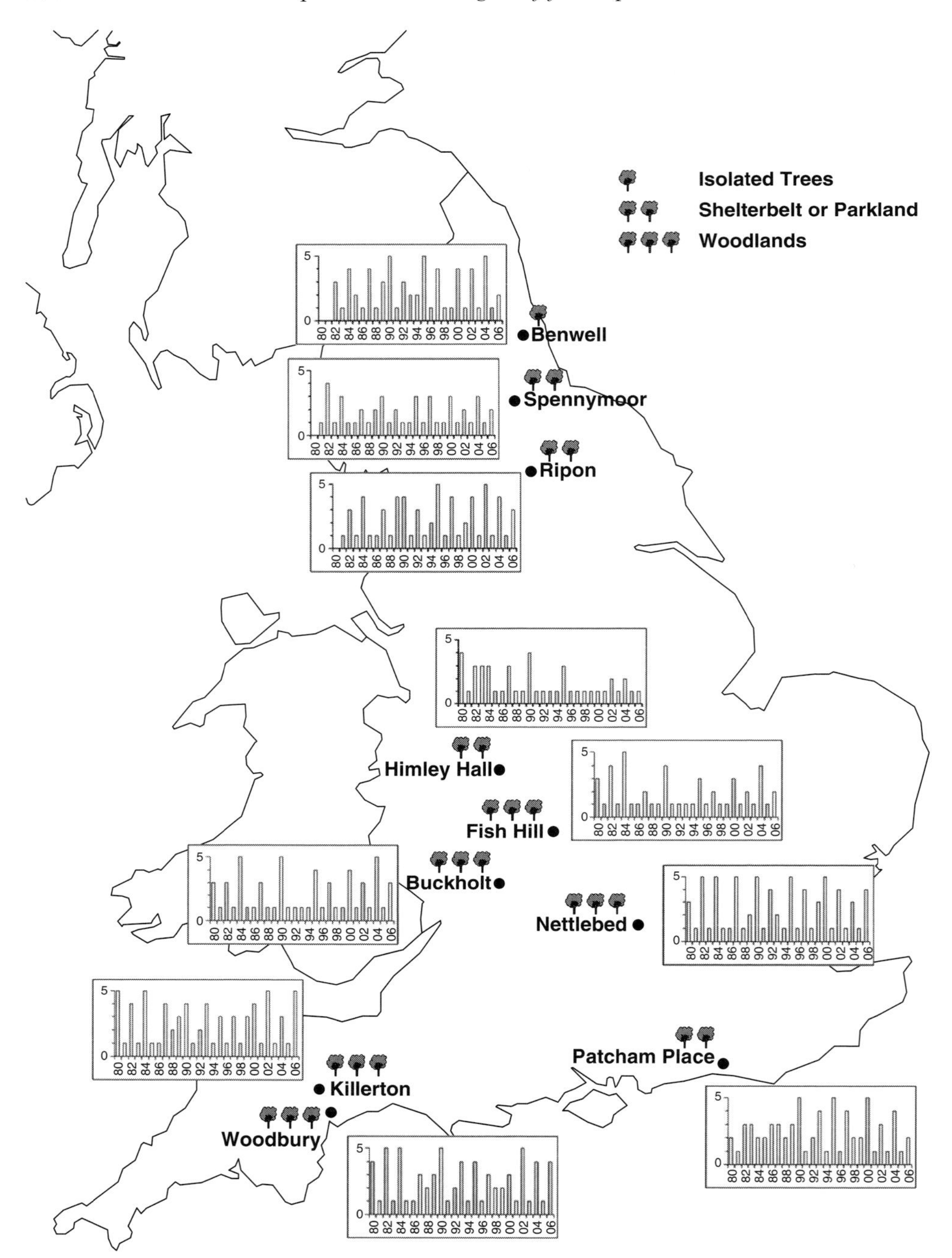


Figure 4.9 Positions of the main beech sampling sites used by the English Beech Masting Survey. Performance for each group of trees is expressed on a five-point scale, using the mean number of full nuts collected in a seven-minute sample, for the years 1980–2006. 1. <10 nuts collected; 2. 10–50; 3. 51–100; 4. 101–150; 5. >150. (Data of Packham, Hilton, Thomas and Lageard. English Beech Masting Survey.)

only a high grade 3 (Packham, 2003). A similar case occurred in a garden at Church Stretton, Shropshire, in the following year.

## 4.4 Masting

Masting is the irregular periodic synchronous production of large seed crops in perennial plants such that the majority of trees produce heavy seed crops in certain years (the mast years) while in other years the crop is small or non-existent. The pattern of reproduction involving the production of a superabundance of seeds became known as mast seeding from the German word for fattening livestock on abundant seed crops, so that years of high abundance became known as mast years. In many species, including those described below, good seed years are unequivocal and obvious. It should be borne in mind, however, that variability in seed production is a long continuous gradient with strongly masting species at one end and those that produce roughly equal number each year at the other. This does not detract from the interest of masting species but does explain the large number of weakly masting species that fall in between.

Masting occurs in a wide range of temperate and northern forest trees such as beeches *Fagus*, oaks *Quercus*, ashes *Fraxinus*, firs *Abies*, spruces *Picea*, pines *Pinus* and the Douglas fir *Pseudotsuga menziesii*. In the southern hemisphere masting occurs in a range of woody species. Norton and Kelly (1988) list several in New Zealand: three species of southern beech *Nothofagus* in the beech family, Fagaceae, four species in the podocarp family (Podocarpaceae) including rimu or red pine *Dacrydium cupressinum* a needle-leaved evergreen, the tawa tree *Beilschmiedia tawa* in the Lauraceae and the southern rata *Metrosideros umbellata* in the Myrtaceae.

### *4.4.1 Beech masting in England and continental Europe*

Masting is particularly well seen in European beech *Fagus sylvatica* and Douglas fir. The masting phenomenon is essential to the regeneration of common beech; in many forests the nuts of particular years have given rise to a complete new generation of young trees. This is because in mast years seed predators are completely satiated (but see Section 4.4.4), so many seeds survive uneaten and germinate the following spring. In a truly natural situation (virgin forest, called *urwald* in German) beech seedlings tend to develop most rapidly where old trees have fallen or died *in situ*, often leaving a magnificent erect **hulk**. The age mosaic found in virgin forest is to some extent imitated when continuous cover forestry is practised, gaps in the canopy encouraging natural regeneration when masting occurs so that the planting sequence associated with extensive clear felling becomes unnecessary.

Continuous systematic records of masting in common beech have been made at a number of widespread sites in England since 1980 (Fig. 4.9). They show an underlying biennial masting pattern of the kind described below, but in practice really heavy masts in England and many other places are, for various reasons, several years apart and result in very effective **predator satiation**. Considerable variations occur both between sites and even on a tree to tree basis within a single site. Less precise records of major masting years in England are known for a much longer period. These, together with other records of masting in northern Europe for the past two centuries, are summarized by Hilton and Packham (2003). Their fig. 2, which was derived from amalgamating the records for Denmark, England, Germany, the Netherlands and Sweden, gives the extent of beech masting in northern Europe for the years 1897 to 2001.

In his classic study of Swedish beechwoods, Lindquist (1931) discusses forestry records for four sites from 1895–1928. They show a clear pattern of masting every 2 or 3 years, broken only once. The suggestion is that there is an inherent periodicity which excludes significant flowering every year, but permits it every other year. Failures of the biennial pattern were related to the weather. There is a requirement for high temperatures in the previous June and July, when the flower buds are initiated, together with the absence of severe frost in late April and May in the year of masting when both male and female flowers would be damaged. Very hot summers resulted in good masting the following year, unless they coincided with good masting in the current year.

Maurer (1964) collated results for Lower Franconia, southern Germany, for many years, noting deviations between regions, though the large and famous masts – *die grossen und bekannten Masten* – of 1811, 1823, 1888, 1909, 1946 and possibly 1958, seem to have been common to most regions. Trees in many German beechwoods date from the 1888 *Jahrhundermast* mast of the century. The very heavy mast of 1946 did not have the same result as 80% of the nuts were collected for oil production. Figure 4.10 is a model to account for the periodicity and variability of masting in beech. A larva of the moth *Cydia fagiglandana*, which is responsible for most damage to beech nuts in Britain, is shown in Fig. 4.11 along with typical damage caused both by it and the birds which open the nuts to get at the larvae.

Damage by **mould and insect attack** has been carefully observed since 1980 as part of the English Beech Masting Survey. During the whole of this period, neither insect damage nor mould seem to have resulted in any significant effects on recruitment after the overwhelmingly important mast years which lead to the majority of seedling development. In 1995, for example, when both sites were high Grade 5 (see Fig. 4.9), 2.4% of full nuts at Patcham Place were

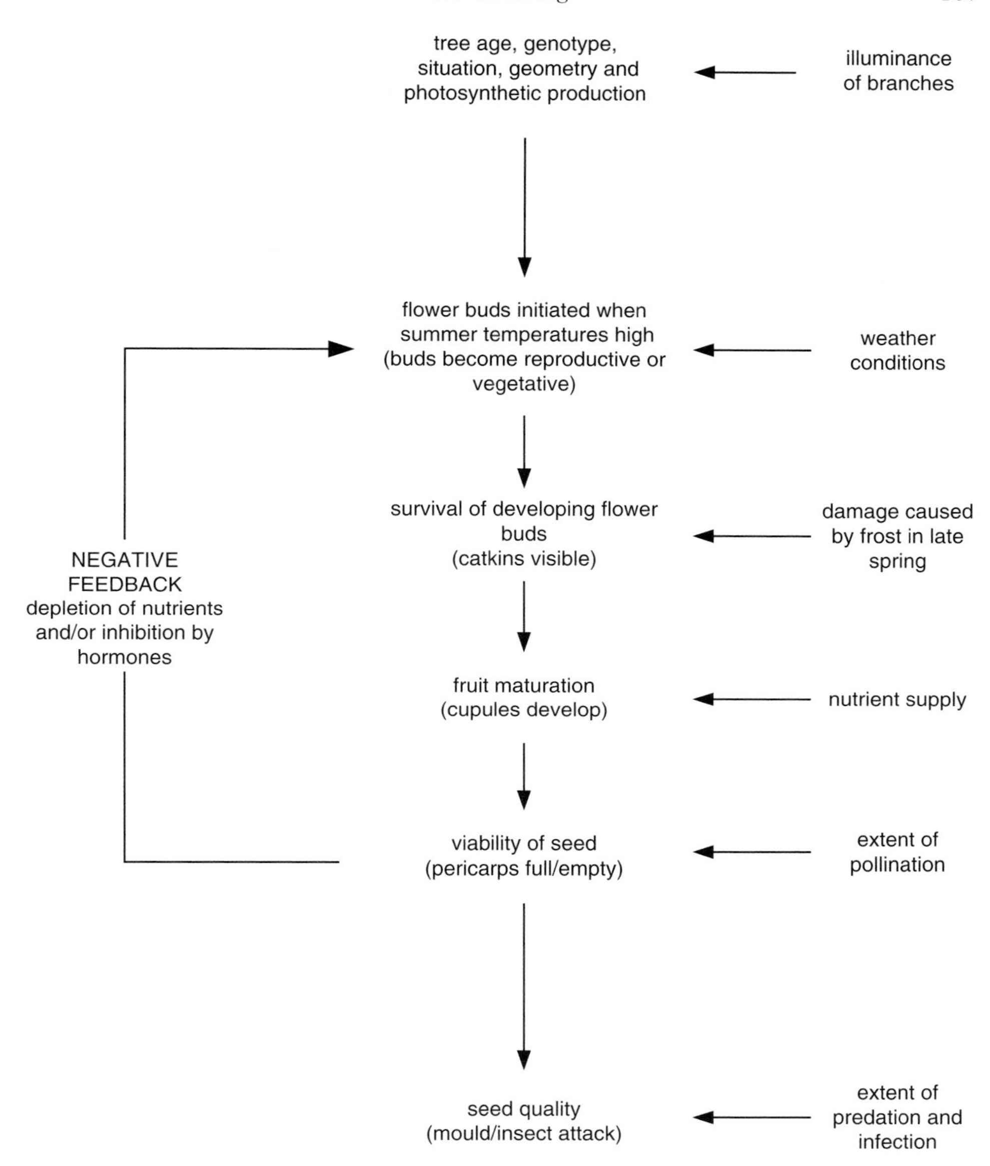


Figure 4.10 Model to account for the periodicity and variability of beech masting. Operation of the negative feedback loop depends on the extent of masting in the previous year. (Packham and Hilton, 2002. *Arboricultural Journal* 26.)

mechanically damaged or mouldy while less than 1% of full nuts of the single tree sampled at Withdean were non-viable. Equivalent figures for the mast year of 2000 are 2.1% for Patcham Place and zero for Withdean. The proportion of damaged full nuts is much higher in non-mast years. Regional variation

Figure 4.11 Four beech nuts damaged by larvae of the moth *Cydia fagiglandana*. A perfect exit hole caused by the departing adult moth is shown by the nut, bottom right. The other three show exit holes that have been enlarged by birds endeavouring to reach a larva, one of which is shown bottom left. Fungal mycelium is present on the two nuts on the right hand side. Note the millimetre scale. (Photographs by Malcolm Inman. From Packham and Hilton, 2002. *Arboricultural Journal* 26.)

remains important. At Buckholt in the mast year of 2000 the proportion of damaged nuts was much higher, being 14.9% for *Cydia* damage and 2.9% for mould, but this still left a mean of 113.7 viable full nuts per tree in each 7-minute sample. Assessments are made shortly after sampling, so the figures obtained give minimum estimates of such losses. Some seeds left on the ground will be consumed or rot during the winter, and there have been a few occasions when *Cydia* larvae have later bored their way out of seeds judged to be viable.

Empty pericarps are likely to result from a lack of cross-pollination. Nielsen and de Muckadeli (1954) enclosed branches on 28 beech trees to allow only self-pollination and recorded just 8% full nuts, whereas 58% were full on control trees allowed to cross-pollinate. At the Himley site (Fig. 4.9) in 1987, 29 well-spaced parkland trees bore 32% full nuts, while 20 adjacent woodland trees bore 77% full nuts.

Studies of masting in American beech *Fagus grandifolia* at a site in Michigan showed a similar basic pattern, broken by failure in pollination which led to large numbers of empty pericarps. As in the English beech mast survey, insect predation of the nuts was irregular, being most noticeable in poor mast years when a higher proportion of the already very small crops were destroyed.

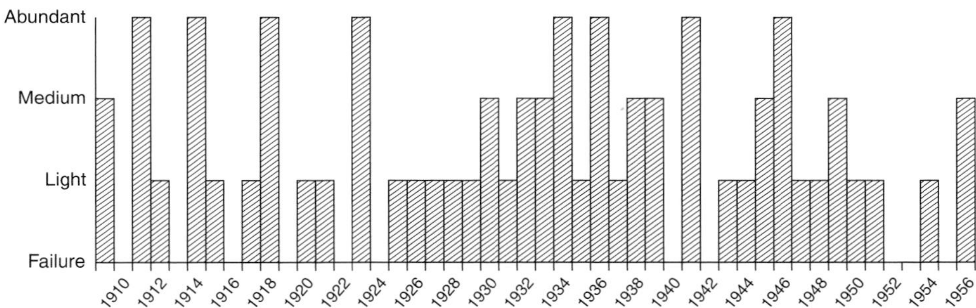


Figure 4.12 Rating of cone crop abundance for Douglas fir *Pseudotsuga menziesii* in western Oregon and Washington, USA, for the years 1909–1956. This species is the most abundant tree in the USA and has been much planted elsewhere since its seed was first collected by the Scot David Douglas in 1827; its timber is extremely valuable. The small drawing shows the distinctive cone. (Drawn from data from Lowry, 1966. *Forest Science* 12. These ratings were obtained from the US Forest Service and the Weyerhauser Timber Company. Cone drawn by John R. Packham.)

### *4.4.2 Apparent influence of meteorological conditions on cone crop production in Douglas fir*

Perhaps the most comprehensive attempt to investigate the meteorological requirements for masting is that made by Lowry (1966), who took the 48-year record of cone crop abundance in Douglas fir illustrated in Fig. 4.12 and correlated it with records of mean monthly temperature and mean monthly precipitation at Salem, Oregon, USA, which is central to the region concerned. He concluded that in this region an abundant cone crop in a given October required a warm January in that year, a high precipitation in March–April the year before, and a cool July 2 years before the harvest.

As with European beech, Douglas fir may go for several years without appreciable seed production; in 1924–1929 coning was never more than light. If October 1923 is taken as an example of a time of abundant coning, the theory is that the January of that year was warm, there was high precipitation in March–April 1922 and that July 1921 was cool. How could these meteorological conditions influence coning? A cool July 2.5 years before coning may be thought of as being too early to influence the reproductive physiology of the trees. The clear statistical link led the author to speculate that other long-term processes might be involved, possibly an adverse effect of a cool summer on pest build-up. The positive influence of high March–April precipitation in the year before coning may be related to the imposition of a suitably optimal intermediate temperature range in the meristematic tissue during a period when the cone primordia (the first cells of the cone) are being laid down. A warm January in the year of coning would avoid frost damage and might enhance formation of microspores and megaspores (pollen and seeds).

### *4.4.3 Masting in other tree species*

Oaks, like beeches and ashes, are also masting trees, though in some species their acorns take 2 years to develop, although those of the white oak group develop in one. Their mast years do not necessarily coincide with those of beech. In Britain, moderate acorn crops of pedunculate oak *Quercus robur* and sessile oak *Q. petraea* occur at intervals of 3–4 years compared with 2–3 years in beech; even in years of general failure there can be abundant seed in some areas. Years in which there are uniformly and exceptionally heavy crops over considerable areas are not more frequent than every 6 or 7 years in southern England (Jones, 1959), but still more frequent than the 5–12 year periodicity seen in beech. The common pattern in masting species is for seed production to be more uniform between trees in years of high seed production. Abundant crops produced by isolated trees suggest that cross-pollination is not essential to oak. Although the underlying reasons are not fully understood, they are probably related to climate; in many continental parts of Europe 20–25 years may pass without appreciable production of acorns, though a few trees may fruit heavily. Masting can have wide ramifications on the forest fauna. For example, Jones *et al.* (1998) highlight a link between oak masting, gypsy moth outbreak and Lyme disease in eastern USA. Mast years in the oaks, particularly the red oak *Quercus rubra* with its large acorns, led to increases in the number of white-footed mice *Peromyscus leucopus* the following summer since acorns are an important food source for them. Mice are also important predators of the pupae of the gypsy moth *Lymantria dispar*, a serious pest that defoliates millions of hectares of oak forest, so poor-acorn years may be an important factor in controlling gypsy moth outbreaks. The abundance of acorns also attracts white-tailed deer *Odocoileus virginianus* to the oak forests; in the autumn of mast years the deer spend more than 40% of their time in oak stands compared with less than 5% in non-mast years. Both the mouse and the deer are the main hosts of the black-legged tick *Ixodes scapularis* which also increases in density during and immediately after mast years. This tick is the vector of the bacterium *Borrelia burgdorferi* which causes Lyme disease in humans (see Section 5.7.1). Since deer numbers lead to high tick densities in the autumn of mast years, and the mice, which are primarily responsible for infecting the ticks with the disease, are abundant the following summer, the probability of humans being bitten by a tick, and of that tick carrying Lyme disease, increases dramatically 1–2 years after a mast year. Control is difficult since removing acorns would aid the gypsy moth and leaving the acorns would promote Lyme disease.

Not all mammals respond equally to a mast year, depending upon their food supply. In the hardwood forests of north-eastern USA, mast crops of red oak resulted in increased numbers of white-footed mice and deer mice *Peromyscus maniculatus* but did not affect red-backed voles *Clethrionomys gapperi*. In contrast, large red maple seed crops led to higher numbers of the vole but not the mice (Schnurr *et al.*, 2002).

The amount of fruit produced by the European small-leaved lime *Tilia cordata*, like that of many forest trees, varies greatly from year to year. The direct influence of climate is most marked at the northern limit of distribution, which is reached in northern England but extends further north in Finland and Sweden and is likely in future to be modified by the effects of global warming. At its northern English limit the species flowers in July and August, but produces only small quantities of fertile seed and then only in very warm summers such as that of 1976. This is caused by a failure of fertilization in the present cool oceanic climate of Britain. The more continental climate of Finland allows fertilization in most years, but the seeds of northern trees fail to complete their development by autumn. In central Europe fruiting is prolific and much more frequent, commencing when the trees are more than 25 years old. In Russia, small-leaved lime produces the best quality seed in the upper part of the crown; initiation of inflorescences and formation of fruit occur only on unshaded branches. This is also true even of trees in the Polish Białowieża Forest (pronounced Bee-ow-a-vey-sha), where July and August are normally warm and sunny with mean air temperatures of 19–20 °C. In the unbroken forest, production of fertile fruit was in 1973 almost confined to the emergent crowns of old trees, around which quite high densities of seedlings were visible even in August.

Growth and behaviour interact with the environment and there are considerable genetic differences within species, so studies of the flowering, fruiting, seed losses and germination of tree species in particular sites are of great value.

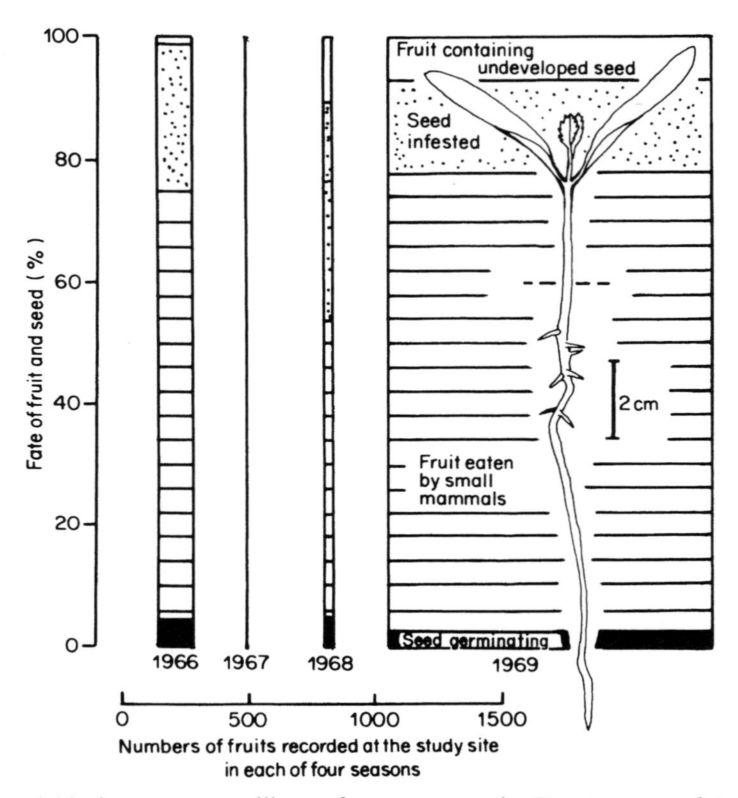


Figure 4.13 A young seedling of common ash *Fraxinus excelsior* and diagrams showing fates of ash fruit and seed up to the time of germination (numbers per $m^2$) in Meadow Place Wood, Lathkilldale, Derbyshire, UK, for the crops of 1966–1969. Most of the infested seed was spoilt by caterpillars of the moth *Pseudargyrotoza conwagana*. The largest seed loss was due to small mammals, notably wood mice *Apodemus sylvaticus* and bank voles *Clethrionomys glareolus*. A very small crop was produced in 1967, of which no seedlings survived, and frosting diminished the crop of 1968. Following these two very poor years, the crop of 1969 was very large. Seeds usually germinated in April or early May, about 20 months after falling in autumn (see Section 4.6.2). The values for seeds germinating are therefore based on the seedlings present in the quadrats two years after flowering. (Drawn from the data of Gardner, 1977. From Packham and Harding, 1982. *Ecology of Woodland Processes*. Edward Arnold.)

Observations of common ash at Meadow Place Wood (Fig. 4.13) demonstrated the very large proportions of the seed infested by caterpillars or eaten by small mammals. Although the proportion of seed that germinated was not high in the crop of 1969, a strong mast year, it was so prolific that the number of seedlings that resulted in 1971 was very great. Gardner (1977) found that fruit fall was evenly dispersed throughout the wood and continued from September until the following August. Each ovary contains four ovules of which one usually develops; some fruits contained no seed, 1.1% contained two seeds, and in one case all four ovules were seen to develop.

Common ash is a pioneer species whose seedlings have been recorded 125 m distant from the parent tree. Each tree may produce 10 kg of winged fruits, containing over 100 000 seeds in a mast year. Observations showed an alternation of good and bad years, a basically biennial fruiting pattern. This tree flowers on the wood of the previous year, so lack of carbohydrate or other nutrient reserve during that previous year could prevent initiation of flower buds. In 1968 flowering was considerably less than expected because the flowers of some trees were destroyed by frost and the drain of carbohydrate reserves was low. The fruit crop of 1969 was exceptionally good, presumably because photosynthate for 1969 was supplemented by reserves saved from both 1967 and 1968. Geographical position may also affect masting: seeding of Scots pine *Pinus sylvestris* is more irregular in the north of Sweden than in the south, where mast years are hardly discernible.

Different species of trees in temperate woodlands may mast together in the same season, but mast fruiting at the community level is especially spectacular in certain evergreen rain forests on nutrient-poor white sand soils in Borneo and Malaysia. Here various species of dipterocarp fruit synchronously over areas of several square kilometres at intervals of 5–13 years. Masting is normally rare in the tropics, presumably because seed predators can move and change diets in high enough numbers to eat excess seeds in a mast year. But the dipterocarp forests of south-east Asia are animal-poor and the extraordinary seed falls in mast years undoubtedly overwhelm the meagre seed eaters (see Section 4.4.4).

### 4.4.4 Causes of masting and its extent in perennial plants

Production of huge numbers of seeds by woody plants anywhere is always a notable event, seldom more so than in the tropics where complete areas of bamboo suddenly come into flower, or entire forests are swamped with seeds, as in the dipterocarp forests of Malaysia. Discussion so far has centred around intermittent synchronous production of large seed crops in particular species, but a number of authors have examined masting records, some collected over long periods of time, for a wide variety of trees. The intention has been to test the validity of the concept, ascertain the factors causing it, and establish the range of seeding patterns present in perennial plants in general and trees in particular. It is also important to realize that the resources released in mast years have effects throughout the forest ecosystem.

There is an enormous literature on masting, but a good modern summary is provided by Kelly and Sork (2002), who set out the major hypotheses very clearly, provide 570 masting data sets worldwide and give coefficients of variation (**CVs**; estimates of how variable masting is over periods of time) that throw light on the relevance of the various models put forward. Masting is best developed in species whose year on year fruit production shows high CVs.

As Fig. 4.10 indicates, weather and resources are both clearly involved in mast seeding, though the extent of their involvement is less clear. It might be that plants simply respond to variable weather by flowering more in years when it is favourable. Alternatively there may be selective advantages to masting (**ultimate factors**) that modify plant responses to weather and internal resource levels (**proximate factors**) that enhance interannual variation in seed production. The oldest model for mast seeding is the **resource matching hypothesis (RM)** which states that in the absence of selection for or against masting, seed crops vary in response to environmental variation. If this theory is accurate, years with good weather should involve good growth and high seed production. Jack pine *Pinus banksiana* growing in Quebec fits this pattern with strong correlations across years among reproduction, growth and good growing conditions. The reproductive pattern of this tree, however, is unusual in that it is strongly serotinous (see Sections 3.7.1 and 4.6.2) and the lifetime seed crop remains on the tree until fire occurs. In such a case the exact year of seed production is irrelevant and resource matching is thus an appropriate strategy. In most species, however, this does not occur and there is evidence of 'switching' of resources from growth to reproduction in mast years giving a negative correlation between growth and reproduction (good years for reproduction are bad for growth), thus refuting the RM hypothesis. Such results have been widely reported for many trees including firs *Abies*, maples *Acer*, birches *Betula, Dacrydium* (Rimu from New Zealand), beeches *Fagus*, spruces *Picea*, pines *Pinus* and Douglas fir *Pseudotsuga*. Koenig and Knops (1998) reviewed 298 data sets for northern hemisphere conifers and found negative correlations between growth and reproduction to be widespread. The conclusion is that switching, and hence masting, is much more common than pure resource matching. So the next question is, what triggers masting?

Masting requires synchrony amongst plants; this is usually in response to a weather cue. In tropical dipterocarps it is triggered by night temperatures dropping 2 °C over 3 nights (Ashton *et al.*, 1988), while in other species flowering may be triggered by drought or fire; none of these factors indicates an increase in resources. Masting is a reproductive strategy requiring a type of resource allocation strategy that exaggerates variation between years. If the plant has a physiological mechanism that alters flowering effort in relation to an environmental signal, why does it do so in this way? The prediction made by Janzen (1971) that in masting species selection should lead flowering to be hypersensitive to both weather variables and levels of nutrient reserves in the plant has been fully justified. It is now clear that proximate mechanisms integrating weather cues and resource utilization can produce more variable patterns of reproduction than expected from simple resource matching (Kelly and Sork, 2002). Why is this an advantage?

If mast seeding is to be selectively advantageous an **economy of scale (EOS)** is required (Janzen, 1978). The three EOS hypotheses with the most experimental testing are wind pollination, predator satiation and animal dispersal. These will be defined and reviewed after a description of three hypotheses that apply only in specialized situations.

(1) The **environmental prediction hypothesis** is that trees will reproduce strongly in years that will be favourable for seedling establishment. If this occurs the next question is, what is the cue that triggers masting? Unlike trees, which are **polycarpic** (= iteroparous) and accordingly fruit many times, many species of tropical Asian bamboos are **monocarpic** (= semelparous); these grasses flower and fruit only once and then die. The length of time between germination and flowering ranges from 3–120 years, according to species, and synchronized flowering occurs not only in local populations but also in widely transplanted individuals of the same genetic stock. What can have favoured the evolutionary development of such a genetic pattern, so different from that of masting trees where masting is largely environmentally induced?
(2) The **bamboo fire circle hypothesis** postulates that the synchronized death of bamboos after masting encourages fire, which prevents trees from out-competing the bamboo whose seedlings are able to establish rapidly on open sites.
(3) The **predator cleansing hypothesis** points to the way in which the synchronized death of bamboos reduces the densities of herbivores feeding on leaves of adult plants.

The three major EOS hypotheses are:

(a) The **pollination efficiency hypothesis** that masting should be strongly selected in species that are able to reach greater pollination efficiency through synchronized above-average flowering effort. If this hypothesis is true the percentage fruit set should be higher when flower density is higher. Pollination is also likely to be more sensitive to flowering density in obligate outcrossers (must cross-pollinate) such as dioecious (male and female on separate plants) and self-incompatible species. Masting is more likely to occur or evolve in trees growing in unproductive habitats where reserves are expended in occasional large efforts, while trees in much more productive sites might well be able to make a large flowering effort year on year. Kelly and Sork (2002) conclude that wind pollination often provides an EOS but that animal pollination does not.

(b) The **predator satiation hypothesis** (Janzen, 1971) concludes that seed predators cause selection for masting when larger seed crops synchronized among individuals experience lower percentage seed predation, as indeed occurs in beech (Section 4.4.1). As long ago as 1942 Salisbury noted that in beech and oaks the only seeds escaping predation were those produced in mast years, and that if a species had a constant seed crop its natural enemies could increase their numbers to the point where all seeds were destroyed every year. The mobility of the seed predator species is an important factor in the equation. Birds and mammals can move very considerable distances whereas some invertebrates, including *Cydia fagiglandana* (Fig. 4.11) normally do not. The high CVs of predator-dispersed plants were consistent with the trees benefiting from predator satiation rather than dispersal.

(c) The **animal dispersal hypothesis** is that masting should be selected against in plants dispersed by frugivores (animals eating only a proportion of the seeds they cache) that are saturated by large fruit production, creating diseconomies of scale (Janzen, 1971; Silvertown, 1980). An oak, for example, dependent upon birds and rodents to carry the seeds away and cache them underground (see Box 4.1), could have many seeds go to waste if they overwhelm the dispersing animals. Indeed scatter-hoarding birds collected 89% of seeds from the canopy of singleleaf pinyon pine *Pinus monophylla* in Arizona in a low-seed year but only 43% in a high-seed year (Vander Wall, 1997). Herrera *et. al.* (1998) analysed 296 data sets describing annual variation in seed output by 144 species of woody plants. They concluded that seed production was slightly more variable amongst wind-pollinated trees than those pollinated by animals. Plants dispersed by **frugivores** (animals that consumed some seed directly but concealed many more, some of which gave rise to seedlings later) showed less variation in annual seed production than those that were dispersed by natural processes such as the wind, or by animals that were predominantly seed predators. Moreover, oaks may benefit by drawing animals such as squirrels from considerable distances as decribed in (b) above. In the case of the pinyon pine, the birds responsible, the Clark's nutcracker *Nucifraga columbiana*, cache such large numbers of seeds (see Box 4.1) that the needs of the tree to reproduce are undoubtedly met even in mast years when more than half the seeds go to waste. The extra seeds not cached may act as a signal to attract the birds to feed.

Which of these hypotheses are true for plants? Sork (1993) finds support for the pollination efficiency hypothesis for several wind-pollinated temperate trees (e.g. beech) but not for oaks. She suggests that in their case, predator satiation more adequately explains the selection pressures driving masting behaviour. It is important to recognize that the three main hypotheses are not mutually exclusive; elements of one or more may apply to any particular masting plant.

Behaviour comparable to that of the masting seen in trees is extremely rare in animals, though there are a few insect species whose reproduction is both synchronized and **semelparous** (i.e. they breed just once and then die). The six species of cicadas are an excellent example of this (Heliövarra *et al.*, 1994). Synchronized **iteroparity** (repeated breeding) in which a long-lived animal will synchronously breed in some years but not in others, even given an adequate food supply, is very unusual indeed, though known in the cases of the kaka and kakapo *Strigops habroptilus* parrots. Both breed only in response to a masting food crop, regardless of supplementary artificial feeding, so perhaps they respond to the same environmental cue as the masting plants they are associated with.

Finally, Kelly and Sork (2002) note that the global pattern of masting shows the highest seed crop variability at mid latitudes and in the southern hemisphere, with a decrease towards the equator. Seed production is markedly affected by rainfall, whose pattern of annual variability is similar. Koenig and Knops (2000) examined 443 data sets showing patterns of annual seed production by northern hemisphere trees, finding that global patterns of annual rainfall and summer temperatures are generally similar to those exhibited by annual tree seed production but are more normally distributed. This, together with the inverse relationship between growth and reproductive effort shown by so many trees, supports the hypothesis that variability in annual seed production (masting) is indeed an evolved strategy.

### *4.4.5* Tachigali vasqueszii*: a successful monocarpic tree*

**Monocarpy**, which involves the death of the parent plant immediately after it has fruited just once, is a successful reproductive strategy for short-lived plant species growing in ephemeral habitats. There are, however, only four genera and around 30 species of long-lived monocarpic trees, amongst which are those of the genus *Tachigali*. Some authors have taken the view that members of this genus show a masting behaviour strongly influenced by seed predation. The fact that Kitajima and Augspurger (1989) found that the majority of the seeds of *T. versicolor*, a masting Panamanian species, were consumed by insects and vertebrates, counters this view.

*Tachigali vasqueszii* is extremely successful in the rain forests of the northern Bolivian Amazon where it can be the tenth most common species despite dying immediately after the production of the first fruit crop. This canopy tree is also known from the evergreen and semi-deciduous forests of Ecuador and Peru. Detailed investigations by Poorter *et al.* (2005b), who also analysed essential features of a number of polycarpic species growing in the same area so that it could be compared with an idealized polycarpous tree, revealed that there were a number of reasons for its success.

Seed production is very high. However, while seedling mortality of *T. vasqueszii* is less than that of an average polycarp, that of its saplings is higher. The most important point is that *T. vasqueszii* grows very rapidly, increasing its stem diameter at a rate far higher than that of its polycarpic competitors. This is facilitated by the production of a large leafy crown and a low wood density. As a result of this strategy the tree reaches reproductive maturity in only 49 years, rather than the 79 years for an average polycarpic species. This short life-span also reduces the risk of dying before reproduction; 50% more seedlings survive to maturity compared with polycarpic species.

## 4.5 Roles and influences of animals

### *4.5.1 Masting and reproduction of the edible dormouse*

Long-term studies of this alien species (see Section 9.3.4) have been made in the Chilterns of southern England by Morris (1997) and colleagues. This animal, together with its fellow alien the grey squirrel, consumes so many beech nuts that it may compromise normal methods of counting beech nut production. While the grey squirrel eats nuts on the ground, the edible dormouse *Glis glis* normally feeds on those in the tree. Interestingly, reproductive success in this animal, at least in the beech–spruce–larch woodlands near Tring, Hertfordshire, appears to be closely related to masting success in beech. In mast years the edible dormouse breeds well in this area, but if viable mast is not formed in the following year no, or very few, young occur. This corresponds with similar reports from continental Europe. Most recently there was a very great production of young edible dormice in the mast year 2002. In the following non-mast year there were very few, with another exceptionally good year in the mast year of 2004.

The patterns involved can best be illustrated by quoting from the records of juveniles weighing more than 100 g found in nest boxes kept at Hockeridge Wood in the Chiltern woodlands (Morris, pers. comm.). He used a mast grading scale running from 0–5. The number of juveniles present in September of the 7 years running from 1998 to 2004 were 1 (0), 137 (4), 167 (5), 1 (0), 323 (5), 7 (0), 319 (4) respectively, the figures in parentheses indicating the local mast grading for the year concerned. The connection between masting success in beech and breeding success in the edible dormouse is one of which the traditional dormouse hunters of Croatia and Slovenia were well aware. We have yet to pin down the key factor that leads these animals to have their young in the August of mast years, well before the nuts are properly developed.

### *4.5.2 Grazing by wild pigs and bears*

Wild pigs are important members of the woodland fauna in many parts of the world (Fig. 4.14). At the time of the Norman Conquest, swine still fed in quantity on acorns and beech mast in England. Emphasis on the right of **pannage** for

Figure 4.14 Wild boar *Sus scrofa*. Note the tusks formed by the lower canine teeth; these are not present in the sows. In earlier centuries this animal enjoyed the oak acorns in the New Forest, England, just as the local domestic pigs do today. Wild boar became extinct in England 300 years ago but has now escaped captivity to form feral populations. (Drawn by Peter R. Hobson.)

these domestic pigs suggests that the climate favoured oak and beech more than it does now; it became even better in the thirteenth century, the Golden Age of British agriculture.

The influence of bears on herbivore numbers is discussed in Section 5.8.2; here we are largely concerned with the herbivorous giant pandas, which modern DNA studies have shown conclusively to be bears, after many years of indecision. Those recently studied live their largely isolated lives in the remote Qingling mountain range of north-west China and are seldom seen. These were the earliest of the still-existing bears, and together with the spectacled bears of the Andes that are the next most ancient, they feed virtually exclusively on plants – bamboos in the case of the giant panda. Giant pandas *Ailuropoda melanoleuca* feed on two of the nine species of bamboo found within the gullies of the precipitous mountain reserves where they now flourish, and unlike other bears do not hibernate, solving the problem of winter by descending to lower altitudes, though even here they often have to knock the snow off the bamboo shoots. They lack enzymes with which to digest cellulose so the value of this food to them is very low, indeed they spend 14 hours a day in consuming 20% of their own weight of bamboo shoots. They are relatively inactive creatures that sleep most of the time they are not eating. Their territories are strictly marked out and individuals seldom meet.

All bears have a good sense of smell; the omnivorous and carnivorous species often locate potential prey and carrion in this way over great distances. Giant pandas use this ability to delimit their territories, having scent sites within their territories that they mark regularly using a scent gland near the anus, thus conveying information about the age, sex and condition of the marking animal. Remarkably, they urinate as high up the marker tree as they can reach from the ground, regarding the animal that can spray the highest as dominant. The only time they are likely to meet other giant pandas is in winter, and this is also almost the only time they show any rapid movement, sometimes chasing off their inferiors with remarkable elan.

Bamboos also form around 70% of the diet of some spectacled bear *Tremarctos ornatus* populations in Venezuela, though this is unusual. Bromeliad hearts (outer leaves are torn off and discarded) are usually the most important part of the diet, though other leaves, fruits and nuts are collected from the tree canopy by these agile climbers. They often make nests high in the trees and sleep in them. Spectacled bears are the most caring and affectionate of mothers, taking care of their young in dens and staying with them for 3 whole years. Both these bears eat carrion at times and the spectacled bear occasionally kills small calves, but clearly their most important influence is on the population size and reproduction of the plants they live on.

Though many grizzly bears still live in Alaska and north-west America, their range and numbers have been greatly reduced since 1900 when they occurred as far south as Mexico. Life for the polar bear has also become more difficult. Global warming is reducing the extent of the Arctic ice cap and the annual period of extensive sea ice is diminishing. Even the polar bear, which is the largest carnivore now existing and travels further than any other bear, spending much of its time hunting seals on the polar ice, makes long journeys over land in summer during which time it eats small rodents such as lemmings as well as berries and leaves from the tundra and the taiga. Its size is very different from that of the Malaysian sun bear, which is smaller than a wolf. Though it is the world's second largest carnivore, most individuals of the grizzly avoid direct human contact. Nevertheless, it causes at least a few deaths every year. The cardinal rule is never to run from a wild bear; not only will it chase you, its speed over short distances is incredible (Liddle, 1997).

Mature males of polar, grizzly and black bears are savage animals which young and female animals do well to avoid, indeed male grizzlies and black bears, though omnivorous, kill and eat the young of their own species. Young North American black bears are agile climbers and often ascend to the treetops to escape, yet more of them are killed by adult males than die from any other cause. There are more than twice as many of these black bears as the combined numbers of all bears in the remaining world population and their numbers are still increasing. Very intelligent, many live in urban situations and are a considerable nuisance as they invade private properties, eating anything that appeals to them and overturning and emptying trash cans.

## 4.6 Time constraints

### *4.6.1 Phenology: leafing, flowering and seed dispersal*

The time at which various natural events, such as the first opening of oak tree buds or the flowering of bluebells, varies from year to year. This area of science is called **phenology**, the study of recurring natural phenomena, especially in relation to climatic conditions. Further consideration of phenology, including leafing dates and the effect of climate change, can be found in Section 11.3.2.

Deciduous trees of various species commence leafing and leaf fall at different times, so the lengths of the annual periods during which they photosynthesize differ considerably. In Cantreyn Wood, Bridgnorth, Shropshire, England at the beginning of November 2003 ash had long since lost its leaves, and those of beech and hazel were following fast, while many of the oaks retained much of their canopy. The now yellowed leaves of the larch had yet to fall, while the dark green outlines of the Norway spruce were more prominent than ever, along with the ivy that grows in much of the wood. By the middle of the month many of the oaks and larch had lost a high proportion of their leaves, while the male catkins of hazel were prominently developed, though pollen discharge did not occur until mid January. The autumn of 2004 was very mild and some hazels were discharging pollen by 1 December, at which time a view over the Severn gorge at Bridgnorth showed that many deciduous trees, particularly oaks, had still retained a fair proportion of leaves though these were now falling.

Observations (JRP) of three English oaks at the southern margin of Cantreyn Wood, Shropshire, in the English Midlands reveal a similar story. A photograph taken on 4 May 2003 shows, as is usual, the western tree leafing first, followed by that to the east. Leaves of the central oak did not expand fully until the end of the month; some other oaks in the wood were even later. The same sequence occurred in both 2004 and 2005. Leaf fall does not necessarily follow the same order. In autumn 2003, after a very hot summer, the tree to the west retained the highest proportion of its leaves, while that in the centre had lost the most and the tree to the east was intermediate.

Phenology is also an important factor in **tropical forests**; indeed Richards (1996) has a whole chapter on the topic, devoting a considerable space to studies of leaf change. Tropical trees may be evergreen, develop their leaves continuously over the whole year, or expand them in conspicuous flushes. The total number of leaves in many evergreen species is considerably reduced before the new leaves flush. Deciduous trees may show marked lack of synchrony; in Ghana *Ceiba pentandra* may have expanded leaves on one half of the crown while the other is completely bare. The phenology of fungi is also interesting; in Malaya the fruit bodies of the larger fungi appear after dry spells and in normal years a succession develops over a 3-month period with each species having a season of a week or so.

A recent study by Hamann (2004) of the influence of climatic factors upon periods of flowering and fruiting in 5800 trees of the North Negros Forest Reserve on the island of Negros, Philippines only 10° 41′N of the equator, is of considerable interest. The study area is a transition area between the lowland evergreen rain forest and lower montane forests. Elements of both are found here on volcanic soils at a height of 1000 m on the north-west slope of Mount Mandalagan. Though daily maximum and minimum temperatures remain relatively constant throughout the year, in which day length varies by only an hour, typhoons are frequent during the winter monsoon from July to November.

Figure 4.15 shows average monthly rainfall and daily maximum and minimum temperatures, which sometimes differed by well over 15 °C, at the study

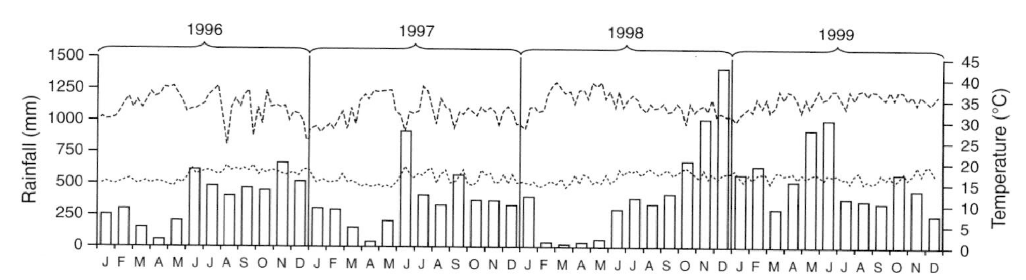


Figure 4.15 Average monthly rainfall (bars) and daily maximum and minimum temperatures (upper dashed and lower dotted lines, respectively) at 1000 m on the north-west slope on Mount Mandalagan, Island of Negros, Philippines. (From Hamann, 2004. *Journal of Ecology* 92, Blackwell Publishing.)

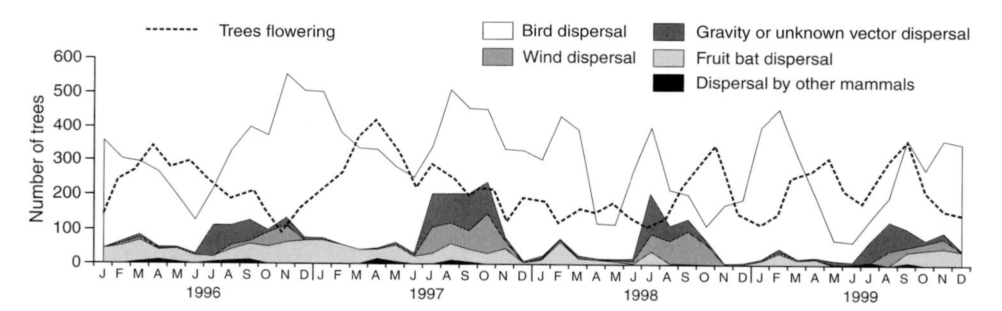


Figure 4.16 Number of trees flowering in the submontane rainforest at 1000 m on the north-west slope on Mount Mandalagan. The figure also shows the number of trees fruiting, partitioned by how the fruits/seeds are dispersed. Fruit production fluctuates almost twice as much between years as flowering. The proportion of trees with animal-spread fruits increases around February and September, and was lower during 1998 and 1999. Bat-dispersed species fruited throughout the year during 1996 and 1997, but was more seasonal in 1998 and 1999. Fruiting of wind- and gravity-dispersed species was restricted to July to October and did not appear to be influenced by climatic fluctuations (Fig. 4.15). (From Hamann, 2004, *Journal of Ecology* 92. Blackwell Publishing.)

site for the 4-year period 1996–1999. This should be compared with Fig. 4.16 that gives the number of trees flowering and fruiting during the same time period, and the general phenological patterns displayed by trees whose fruits are dispersed by birds, gravity or an unknown vector, wind, fruit bats or other mammals. The general similarity between the year on year fruiting patterns was influenced by the major El Niño that peaked in November 1997, causing a warming of the equatorial eastern Pacific Ocean, and a subsequent La Niña climate anomaly (cooling of the waters) that developed throughout the summer of 1998 with a peak in September. Such ocean temperature anomalies influence the western Pacific with Philippine weather patterns showing several months delay. Rainfall from February to May 1998 was exceptionally low.

The typhoon season (July to November) coincided with the extended periods of fruiting of the wind- and gravity-dispersed species. Fleshy-fruited trees showed peaks matching those of solar irradiance, either flowering at the beginning and fruiting at the end of the first peak (April), or flowering during the first peak and fruiting during the second (September). Some 95% of the intraspecific variation (i.e. variation within a species) of flowering and fruiting dates during the El Niño of 1997 and the La Niña climate anomaly of 1998 was ascribed to the delayed or advanced flowering and fruiting of a limited number of species. Mast fruiting in dipterocarps could not be correlated with the climatic events of 1997–8.

### *4.6.2 Seasonal climates, seed banks and dormancy*

Persistent seed banks can be found in the soil or in the tree canopy. The largest of these is the soil seed bank, created by seeds falling to the ground and becoming buried under litter and humus, or, in seeds with an elaiosome, being abandoned underground once the fatty body has been eaten. Salisbury (1961) emphasized the differences in germination patterns, dormancy and longevity of buried seeds. These features, together with the development of **soil seed banks**, are important in the reproduction of vascular plants. Figure 4.17 shows the four types of seed bank detected by Thompson and Grime (1979) by sampling the surface soil (0–3 cm depth) of ten ecologically contrasted sites in the Sheffield region. Sampling took place at 6-weekly intervals; the experiment was designed to detect both the transient accumulation of germinable seeds during the summer and persistent seed banks. Some species were in several types of habitat. The results suggested that seasonal variation in seed number is related to the species concerned rather than to the environment involved. Type IV seed banks are the most important in woodlands, where disturbance is likely to occur at long and very irregular intervals.

**Dormancy of seeds** stored in the soil can be caused by a variety of mechanisms including an impermeable seed coat (as in acacias and other legumes), a chemical inhibitor, a chemical lack, a need for winter chilling or (in a few woody species) a need for light. It can also be caused by immaturity of the embryo at the time of seed shedding. Fruit of the European ash *Fraxinus excelsior* falling in autumn is unable to germinate immediately as a period of embryo growth, in which its length almost doubles, is required. Embryo growth ideally needs a temperature of 18–20 °C. Only after this takes place

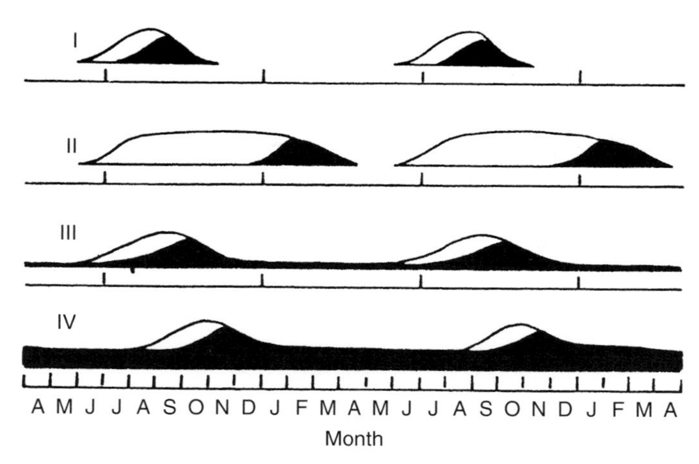


Figure 4.17 Four types of seed bank described by Thompson and Grime (1979). Shaded areas: seeds capable of germinating immediately after removal to suitable laboratory conditions. Unshaded areas: seeds viable but not capable of immediate germination. Type I, annual and perennial grasses of dry or disturbed habitats. Type II, annual and perennial herbs colonizing vegetation gaps in early spring. Type III, species mainly germinating in the autumn but maintaining a small but persistent seed bank. Type IV, annual and perennial herbs and shrubs with large persistent seed banks. (From Grime *et al.*, 1988. *Comparative Plant Ecology*. Unwin Hyman.)

can prolonged **stratification** (exposure to low temperature) at about 5 °C, overcome dormancy. Because of this, ash seeds usually germinate in April or early May 2 years after the flowering season in which they were formed. A very small proportion germinates after the first winter, and when foresters wish to germinate ash quickly seeds are picked green in August and sown immediately.

Seeds can remain dormant and viable for centuries but, as described in Section 9.2.2, the bank rarely includes tree species (although there are notable exceptions such as cherries *Prunus* spp.). The reason for this lack is probably due to the comparatively large size of many woody plant seeds: they are less likely to be missed by rooting herbivores and will also be more eagerly sought after as a worthwhile meal.

The most effective seed bank in trees is found in the canopy of more than 500 different species of tree growing around the world in fire-prone areas. In these **canopy seed banks** the seeds are stored in fireproof fruits that open after they are burnt or heated by a fire. This is referred to as **serotiny**. A number of pines in the northern hemisphere are serotinous (such as the lodgepole pine *Pinus contorta* and jack pine *P. banksiana* of North America, and the Aleppo pine *P. halapensis* and maritime pine *P. pinaster* of Europe – see also Section 3.7.1) but the majority of serotinous trees are found in the southern hemisphere among evergreen hardwoods (such as the proteas, banksias and eucalypts of the Cape area of South Africa and particularly the shrublands (bush) of south-west Australia). The development and characteristics of such seed banks in post-fire stands of four species of pine growing in south-east Spain were studied by Tapias *et al.* (2001). Aleppo pine started to fruit at 5 years of age and more than 86% of the cones were serotinous (sealed by resin); they had opening temperatures between 49.3 and 51.3 °C. The canopy seed bank of this species was much greater than in maritime pine, in which the proportion of serotinous cones was lower (66.7%). Only one population of this species was studied, in which first fruiting was later (12 years) and cone-opening temperature was almost 5 °C lower than in the Aleppo pine.

Serotinous cones help Aleppo and maritime pines survive fire; early flowering is also essential in such species, which grow in areas where crown fires are frequent. It is suggested that the late flowering and lack of serotinous cones found in the black *P. nigra* and Mediterranean stone pines *P. pinea*, where flowering is insignificant even 15 years after fire, results from evolution in areas where ignition led only to low-intensity ground fires.

This chapter has examined the responses made by forest plants to the environments in which they live. It pays particular attention to their regenerative strategies, the influence of animals, time constraints and the role of seed banks. The strategic responses of all living components of the ecosystem are of particular relevance to forest change and disturbance (Chapter 9) and to climate change (Section 11.3).

# 5

# Biotic interactions

## 5.1 Producers and consumers

**Primary production** is undertaken by **autotrophs**, which in forests are green plants (**photoautotrophs**) that produce complex compounds from simple raw materials using the energy of light in the process of photosynthesis. **Chemoautotrophs** do this using the energy of chemical reactions (chemosynthesis), but do not play an important role in woodlands. **Heterotrophs**, by contrast, consume other organisms and so are dependent on the uptake of energy in organic materials synthesized by these other organisms. Herbivores, carnivores, parasites and decomposers (saprotrophs) are all heterotrophs; they vary in size from microorganisms and insect larvae to elephants and all play important roles in woodland, forest and related ecosystems. The increase in biomass of heterotrophs is known as **secondary production**. Heterotrophs that exploit autotrophs directly are called **herbivores** or **primary consumers**. These are consumed by **secondary consumers, the carnivores**, and some of these may in turn be eaten by **tertiary consumers** to form **food chains**. It is rare for an animal to feed on just one other species, so in reality food chains become a **food web**, a network of interconnected food chains (see Fig. 1.10). Many of the consumers forming this plant-dependent web influence green plants adversely, often by feeding or trampling. Others are positive, acting as pollinators and dispersers of fruits and seeds, and even more significantly, promoting nutrient cycling (see Section 8.3).

Food webs are often very complex and it is easier to appreciate the overall biotic interactions in a forest by breaking the food web into three main subsystems as in Fig. 1.9. The **plant subsystem** is utilized by the **herbivore subsystem**, which in forests commences with the living tissues of plants eaten by primary consumers (including directly herbivorous animals – **grazing** the whole plant or **browsing** parts such as shoots, twigs and leaves – exudate feeders, parasitic plants and fungi) and eaten in turn by secondary and tertiary consumers. The **decomposition (or detritus) subsystem** involves the consumption of dead organic materials. The herbivores within this subsystem are **detritivores** and decomposer organisms (see Section 1.5 and Chapter 7).

Certain carnivores, which consume both herbivores and detritivores, link the herbivore and decomposition subsystems. There are many variations on these basic themes. Omnivorous forest bears eat plants, hunt live animals and eat carrion. Similarly some fungi are parasites (living on their hosts but not killing them), others are pathogens (rapidly killing their hosts) and others are saprotrophs (decomposing dead hosts), and yet others are combinations of these. A fungus might invade a tree without causing undue harm, but if the tree is weakened in a drought year, it may rapidly succumb to the same fungus which can then use the dead tree as a food supply while exploring for another host. Eventually the dead remains of autotrophs and heterotrophs, together with faeces, are mineralized and the whole process, which began with the exploitation of mineral nutrients in the soil, can start again.

## 5.2 The interdependence of producers and consumers

Figure 5.1 gives an insight into the relationships between producers and consumers. It shows a scheme predicting the changes in abundance of trophic (mineral nutrient) elements along a gradient from low to high plant productivity. At very low productivity (a) where the vegetation consists entirely of **stress-tolerators** (see Section 4.1), defences against generalist herbivores work so well that the herbivores are restricted to a small biomass of specialist feeders. The relations of eucalypts (Section 1.3.1) with the koala approximates to this model. Predators are also likely to be extremely sparse and specialized. At higher productivity with less effective defences (b), generalist herbivores may be expected in greater abundance as long as they are insufficient to provide a large and reliable food source for predators which would otherwise keep their numbers low. At the highest levels of productivity (c) high rates of population growth by herbivores are possible, but predators effectively suppress herbivore numbers thus protecting the palatable vegetation.

Plant litter produced by **competitors** tends to be copious but not usually persistent. Litter production in plants with the other two primary strategies is sparse, being persistent in **stress-tolerators** but not in **ruderals**.

It is clear from the above that soil and environmental conditions limit plant form and growth – which is why we have different biomes around the world (see Section 1.6.1), ranging from tough evergreen open woodlands in dry areas to dense rain forests in wet areas. The climate and the vegetation both in turn

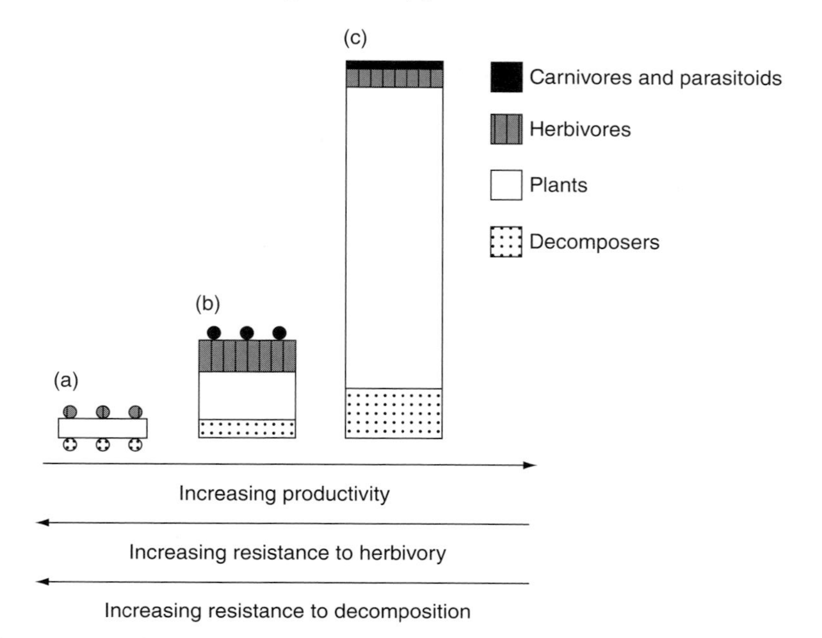


Figure 5.1 Scheme predicting the changes in abundance of trophic groups along a gradient between low to high productivity (a–c). The area of each section corresponds to its abundance. (Redrawn from Grime, 2001. *Plant Strategies, Vegetation Processes and Ecosystem Properties*. John Wiley & Sons.)

affect how many and what types of herbivorous animals are able to live. Of course, this is not a one-way process: the herbivores also affect the vegetation. Some effects are adverse for the plant – grazing and trampling – while others are positive – pollination, dispersal of fruits and seeds, and even more significantly, promoting nutrient cycling (see Section 8.3). These mutual influences result in an evolutionary stalemate where plants evolve defences which prevent decimation by herbivores, and the herbivores evolve ways of breaching these defences.

An interesting question to ask is whether these **herbivore–plant interactions** are sufficient to regulate the numbers of herbivores: does the quantity or quality of available food limit herbivore reproduction or survival? Or, are herbivore populations regulated by parasites and/or carnivores (**predator–prey interactions**). Such questions, whether control is bottom-up or top-down, have greatly exercised ecologists over the last century and, as yet, no real consensus exists (see Lindström *et al.*, 2001). However, of clear practical importance is what happens when the balance between plants and herbivores is disturbed by unusual conditions such as dry years or a sudden genetic change in one party (such as the

Dutch elm disease fungus) or the introduction of an exotic species that has left its controlling factors behind: these are discussed later in the chapter.

Defoliation by insects and other animals has complex side effects; it increases light penetration through the canopy, reduces competition for abiotic resources such as water and nutrients, leads to alterations in plant species composition (discussed later), increases the rate of nutrient leaching from foliage through damage, and accelerates the rate of fall of nutrient-rich litter into the decomposition subsystem. The cycling of nutrients (abiotic flux) is enhanced because defoliation stimulates redistribution of nutrients from boles and branches to components with high turnover rates such as leaves and buds. It can thus be argued that defoliation, paradoxically, promotes more consistent plant production in the long term than would be expected.

It is clear that defoliation affects not just the forest but also the plant itself. A number of deciduous trees show a burst of new **lammas growth** in late summer (named after Lammas day, the first day of August when the festival of first fruits was formerly celebrated). This is usually ascribed to the need to replace leaves destroyed by insects, and allow extra growth once the insects have pupated. Whether subject to insect attack or not, deciduous trees form new leaves throughout the summer, either continuously as in ash, or in the

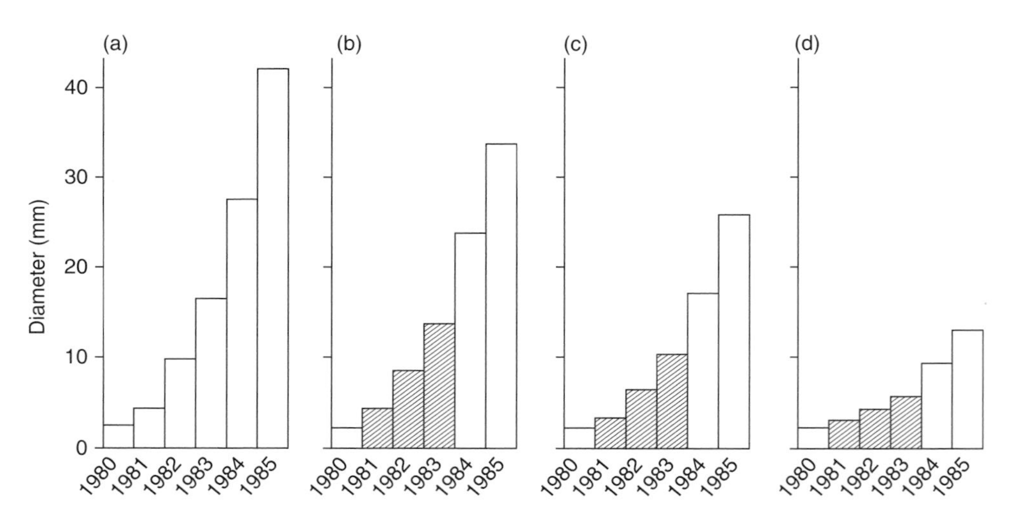


Figure 5.2 Mean diameters of young trees of pedunculate oak *Quercus robur* measured at the ends of the years 1980–85. They were grown in the open at a spacing sufficient to avoid competition but were subject to artificial defoliation. Trees (b), (c) and (d) were defoliated in the years 1981–3 (shown hatched) but not subsequently. (a), control, not defoliated. (b), one-third defoliated. (c), two-thirds defoliated. (d), totally defoliated. (From Hilton *et al.*, 1987. *New Phytologist* 107. Blackwell Publishing.)

distinct determinate flushes found in oaks, whose second flush of yellowish leaves is usually produced towards the end of June, and may be followed by a third at the beginning of August. Hilton *et al.* (1987) investigated the effect of defoliating oak seedlings at three different levels in mid-June for 3 successive years when the leaves had fully expanded. Defoliated plants differed from the controls in four major respects having: (1) earlier production of the lammas growth, together with the formation of more lateral branches which were susceptible to winter frost damage; (2) production of smaller but more abundant leaves; (3) reduced diameters of main stems (Fig. 5.2), from which were calculated relative growth rates (RGR) which varied both with defoliation treatment and growing conditions over the year and which returned to normal values as soon as defoliation ceased; (4) formation of wood with many vessels and a lower than normal proportion of xylem fibres, similar to the wood normally produced in spring.

## 5.3 Insect defoliation and damage

Insect larvae are effective herbivores that can defoliate – and kill – entire trees and stands. In North America, insects and pathogens affect an area almost 50 times larger than fire with an economic impact nearly five times as great (Dale *et al.*, 2001). Insect activities in this respect have been observed and studied in many parts of the world. In Britain, forests suffered severely in 1979, 'the year of the caterpillar', when defoliation of oak and other hardwoods affected whole hillsides in North Wales, mid-Wales, Cheshire, Lancashire, Devon and Dorset. Harding (1992) conducted a long-term study of the oak canopy of Chaddesley Woods National Nature Reserve (NNR) near Kidderminster, Worcestershire. This followed severe defoliation of broadleaves in 1979 and especially 1980. At the end of May 1980, following a very warm month with virtually no rain, numerous oaks, hazels and even ash were completely defoliated, though some oaks which had only just begun to flush escaped attack (Harding, 1992). Falling **frass** (insect faeces) could be heard hitting the dry litter, while trunks were festooned with silken skeins produced by caterpillars descending to pupate or find food. The insects involved were mainly 'looper' caterpillars (Geometridae, including the winter moth *Operophtera brumata*, and the mottled umber *Erannis defoliara*), which together with the green oak roller *Tortrix viridana* are well known for causing severe devastation, often for several successive years (Evans, 1984).

Defoliation in oaks is greatly influenced by the order in which individual trees produce their first leaves. At Roudsea Wood NNR, Cumbria, for example, Satchell (1962) found that certain individual trees retained relatively undamaged canopies in years of generally severe defoliation; his conclusion was that this resistance to attack was based on the phenology of oak bud flushing relative to insect egg hatch. Early trees can produce more foliage so they can 'grow away' from insect attack, while buds of late flushers open too late for the young larvae. Severe defoliation occurs most commonly on trees that flush at the same time as the larvae emerge. Such phenological sequences reflect genetic variation between individual trees, a phenomenon confirmed by Crawley and Akhteruzzaman (1988) who reported that the flushing order of 36 oaks at Silwood Park was consistent from year to year of a 7-year study.

Leaves of the dominant trees of temperate woodlands become tougher as they age, and also tend to possess increased concentrations of such compounds as resins and tannins which are distasteful to herbivores. Figure 5.3 illustrates both the phenology of insects that feed on oak and beech in the UK and the way that the tannin concentration of oak leaves develops during the season. Most forest Lepidopteran pests feed on new foliage in spring; interestingly winter moth caterpillars (b in Fig. 5.3) reared on oak leaves picked at the end of May developed into smaller – and thus less fertile – adults than those fed on younger leaves which had less tannin and so more available nitrogen. Figure 5.4 illustrates the caterpillar counts in 24 frass traps under six of the Chaddesley oaks during a period which included the major outbreak of 1990. Defoliation was total in 1980, and ranged from 30–78% between 1982 and 1990. Subsequent values failed to reach 10% until 2002, and rose to 60% in 2005. The large peak expected in 2002 failed to materialize.

Harding (1996, 2000, 2002) provides data regarding defoliation in eastern Europe. In Hungary, where oaks form 35% of the total forest area, forest owners report occasional major defoliation by Lepidoptera, notably by winter moth *Operophtera brumata*, mottled umber *Erannis defoliaria* and *E. aurantiaria*. The cyclic nature of these attacks which, during the period 1961–1998, showed really major peaks in 1962 and 1963 and lesser ones in 1972, 1983, 1993–5 and 2004, is compared with those in Poland and at Wytham Woods, Berkshire, UK in Fig. 5.5. In the Strict Reserve of the Białowieża Forest, Poland, there are in some years areas of lush nettles *Urtica dioica*, one reason for this being nutrient enrichment from geometrid caterpillar frass. This increased growth of herbs affords shelter to small mammals, leading to adverse effects on songbirds subjected to more than usual attacks by birds of prey attracted by the small mammals. Defoliation of broadleaves, of which hornbeams were a major component, was near 80% in 1992/3 and though no significant damage was reported in the period 1996–99, the cycle then resumed and peaked in 2003. As in Hungary and Wythan Woods (a and c in Fig. 5.5),

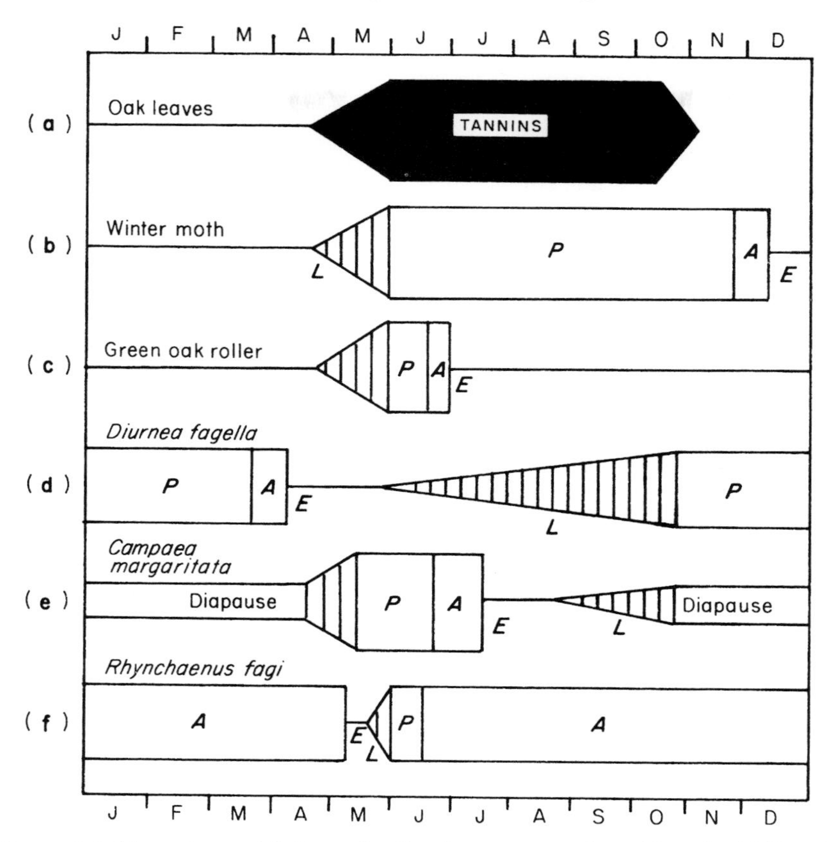


Figure 5.3 Phenology of foliage-feeding insects on oak and beech. (a) Tannin content of leaves of pedunculate oak *Quercus robur*.(b)-(e) Phenology of four species of Lepidoptera on oak (from Varley, 1967). (f) *Rhynchaenus fagi*; beech leaves are mined by the larvae and skeletonized by the adult weevils (data from Nielsen, 1978, *Natura Jutlandica* 20, 258–72.) E, egg; P, pupa; A, adult; L, larval growth periods, shaded. (From Packham and Harding, 1982. *Ecology of Woodland Processes*. Edward Arnold.)

there were approximately 10-year intervals between peaks. Only a small proportion of temperate forest insects go through these regular cycles: what causes them to do so? Berryman (1996) and others conclude that regular cycles of forest Lepidoptera, including winter moth, are most likely the result of interactions with parasitoids, predators or diseases, rather than through delayed induced defences in the tree host, but there is still much to learn about this complex and important phenomenon.

Not all cycles are regular, however. The eastern spruce budworm *Choristoneura fumiferana*, a lepidopteran native to North America, normally occurs at low densities but periodically (at approximately 35-year intervals) it

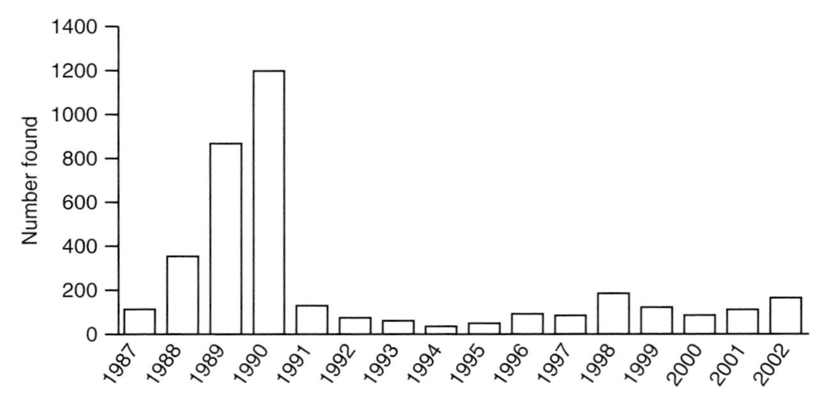


Figure 5.4 Numbers of geometrid caterpillars found in a 3-week peak activity period at the end of May/mid-June in 24 frass traps under six oak trees in Chaddesley Woods, Worcestershire, UK, over a 15-year period 1987–2002. (Drawn from data of Harding, 2002. *Quarterly Journal of Forestry* 96.)

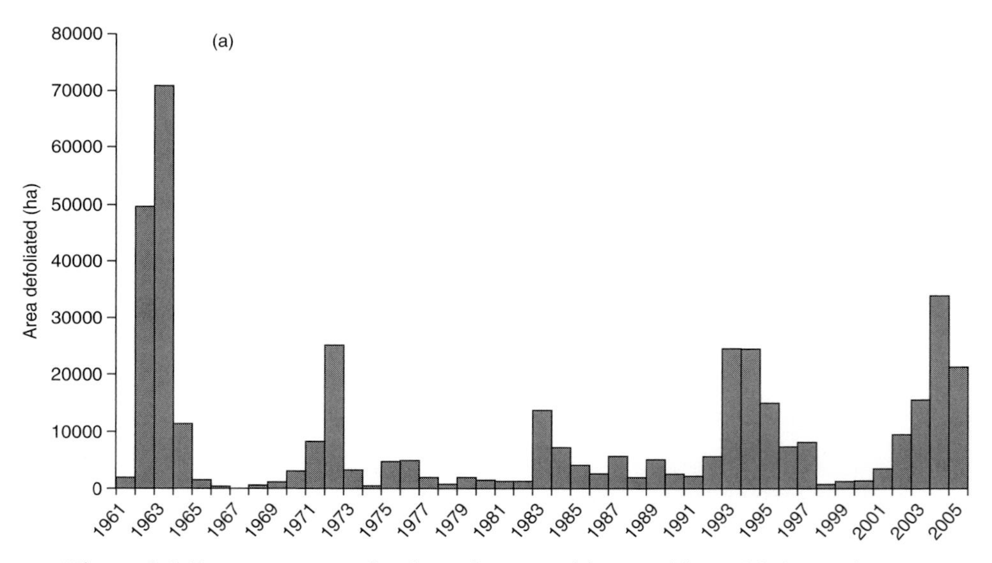


Figure 5.5 Long-term monitoring of geometrid caterpillars. (a) Annual areas (ha) of Hungarian forests defoliated between 1961 and 2005. (Csoka, pers. comm.); (b) Index of abundance of geometrids – *Operophtera fagata, O. brumata, Erannis defoliaria, E. marginaria, E. aurantiaria, Phigalia pedaria,* – on hornbean *Carpinus betulus* in May, Białowieża Forest, Poland between 1975 and 2005 (Jaroszewicz and Wesolowski, pers. comm.). This figure also shows a caterpillar descending to pupate; (c) Numbers of winter moth *O. brumata* caterpillars trapped beneath five oaks at Wytham Woods, UK (mean no. $m^{-2}$) from 1952–2005 (Cole, pers. comm.). No data for 1974, 1977–82. (Data compiled by David J. L. Harding.)

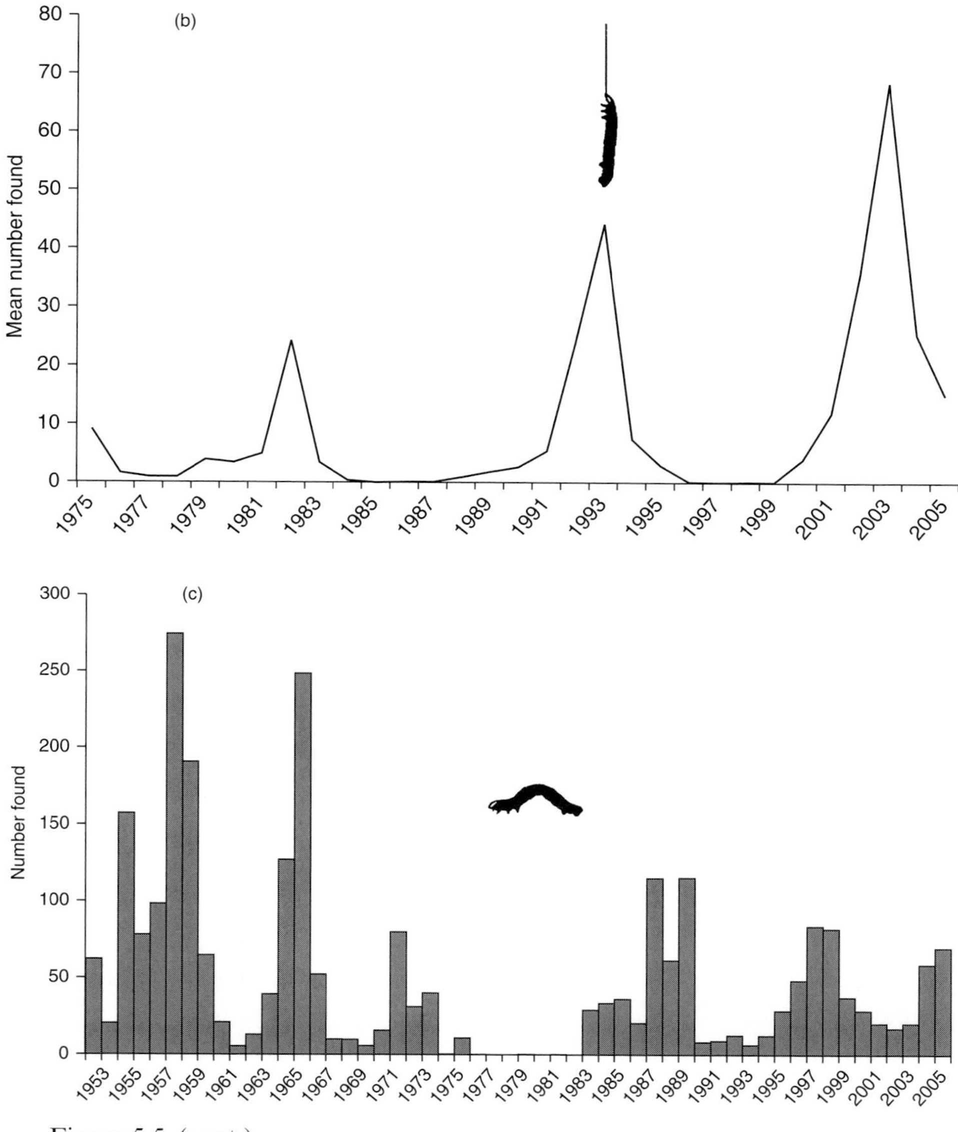


Figure 5.5 (cont.)

erupts in epidemics. These defoliate the local conifers (mostly firs and spruces) for periods as long as 5–10 years; in Quebec these epidemics are known to have occurred back to 1704.

Global warming is also likely to have an effect. In Holland investigations of the response of winter moth *Operophtera brumata* egg hatching and oak *Quercus robur* bud burst to changing seasonal temperature patterns during the period 1975–1999 appear to show a disruption of the synchrony of their phenology (Visser and Holleman, 2001). This was ascribed to the pattern of spring temperatures and winter frosts. The first stage in oak **bud burst** is initiated by the build-up of a 'chilling sum' by the trees in a time period beginning on 1 November, a form of **stratification** similar to that seen in seeds (Section 4.6.2). After this first threshold value has been reached they commence to build a 'warmth sum'; the buds burst only after this second threshold has been reached. In years when the mean winter and spring temperatures were high in relation to the number of frost days, bud burst in oak is earlier, but egg hatching is up to 3 weeks earlier than bud burst and many caterpillars starve. Warmer springs also lead to mistimed reproduction in Dutch great tits *Parus major* (Visser *et al.*, 1998), where caterpillars developed earlier but the time tit eggs were laid did not. During egg laying great tits forage in larch *Larix decidua* and downy birch *Betula pubescens*, whose dates of bud burst are much less temperature-dependent than that of oak and so have not advanced in the same way. In consequence, many more geometrid caterpillars feeding on oak are likely to survive the crucial early period.

Monitoring of a site in Surrey (Sparks and Smithers, 2002) shows an advance of nearly a month in oak leafing dates since records began in 1950, a greater response by the oaks than that expected if air temperature was the sole trigger. Harding (2002) found no evidence of such an advance at Chaddesley and thought this might be due to a shorter observation period of two decades. Is global warming the cause of the recent decline in episodes of severe defoliation by oak-feeding caterpillars in England but not in Scotland (interestingly, also seen in such things as declining cycles of vole numbers in Scandinavia)? Whatever the reason, although earlier leafing out, when it occurs, seems desirable to foresters as active photosynthesis continues for longer and **tree-ring growth** is less disrupted so timber production is greater, it does have negative effects: fewer caterpillars means less food for rearing the young of woodland birds.

Insects not only defoliate trees, they can damage shoots and cause damage by burrowing under bark (Fig. 5.6). They also often act as **vectors of disease**, spreading fungi through the wounds they make. As can be seen in Table 5.1, a number of serious pests in eastern USA are insects and many of the major fungal diseases are aided by insects, as discussed later in this chapter. Non-native insects inadvertently introduced may be even more devastating than their native counterparts if their normal population regulators (predators, climate) are left behind or their new food source is particularly suitable. Examples of widespread and rapid damage by introduced insects are legion. Two examples are given below; the first is of global concern, the second of a more localized but severe insect introduction.

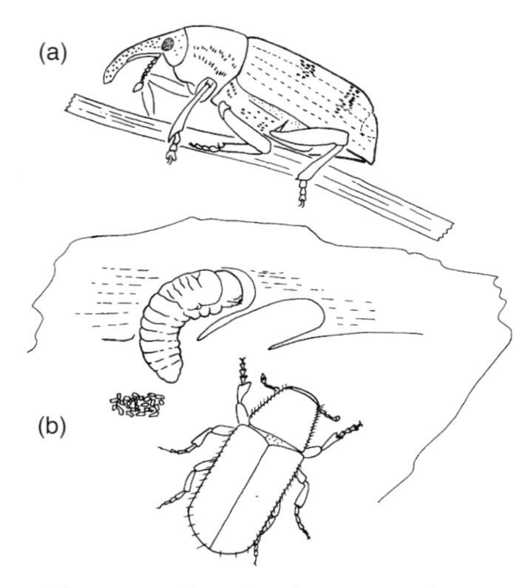


Figure 5.6 Common European beetles that cause damage to trees. (a) Large pine weevil *Hylobius abietis*: larvae develop in stumps and logs; adults damage bark of young pines. (b) Pine-shoot beetle *Tomicus piniperda*: larva and adult live under bark, a potential source of fungal infection of weakened standing trees. (Drawn by David J. L. Harding from Cannock Forest, Staffordshire, UK. From Packham and Harding, 1982. *Ecology of Woodland Processes*. Edward Arnold.)

### *5.3.1 Asian longhorned beetle*

The Asian longhorned beetle is a large (up to 35 mm long) and remarkably attractive black and white beetle with blue feet (Fig. 5.7), aptly named the starry sky beetle in its native Asia. In China, however, it is a particularly troublesome native pest of hardwood trees such that 40% of all poplar plantations have been damaged and around 50 million trees were felled in a 3-year period just in Ningxia Province in central China (an area twice the size of Belgium or equal to West Virginia). The larvae burrow under the bark and girdle branches and trunks (i.e. cut through the phloem around them, depriving the roots of sugars from above – see Section 1.3.2 for details of tree morphology) leading to death over a few years. Unfortunately, the Asian longhorned beetle is becoming a world pest. In 1996 it arrived in the USA from China carried as larvae/pupae inside solid-wood packing material. A year later and populations were established in New York City and Chicago. Since then it has been found in warehouses at ports around North America, and small populations have been found in Austria (probably arriving in 1998) and in Germany (2004). The beetle has also been found in numerous ports around Europe, including the UK, with consequent worries about it escaping into the wild.

Table 5.1. *List of exotic pests and pathogens that have caused or threaten to cause dramatic tree declines in forests of eastern USA.*

| Pest/pathogen | Type | Date introduced | Species affected |
|---|---|---|---|
| Gypsy moth (*Lymantria dispar*) | Insect | 1869 | *Quercus*; other hardwoods |
| Beech bark disease (*Nectria coccinea* var. *fraginata*) | Fungus | 1890 | *Fagus grandifolia* |
| Balsam woolly adelgid (*Adelges piceae*) | Insect | 1900 | *Abies balsamea*, *A. fraseri* |
| Chestnut blight (*Cryphonectria parasitica*) | Fungus | 1904 | *Castanea dentata* |
| White pine blister rust (*Cronartium ribicola*) | Fungus | Early 1900s | *Pinus strobus* |
| Dutch elm disease (*Ophiostoma ulmi*) | Fungus | 1930 | *Ulmus americana* |
| Hemlock woolly adelgid (*Adelges tsugae*) | Insect | 1950s | *Tsuga canadensis*, *T. caroliniana* |
| Butternut canker (*Sirococcus clavigigenti-juglandacearum*) | Fungus | 1967 | *Juglans cinerea* |
| Dogwood anthracnose (*Discula destructiva*) | Fungus | 1978 | *Cornus florida* |
| Asian longhorned beetle (*Anoplophora glabripennis*) | Insect | 1996 | *Acer*, *Betula*, *Populus*, *Ulmus* spp. |

*Source:* Based on Orwig, 2002. *Journal of Biogeography* 29.

Figure 5.7 Female Asian longhorned beetle (*Anoplophora glabripennis*). (Photograph by Linda Haugen, USDA Forest Service, www.forestryimages.org.)

So far outside Asia, the Asian longhorned beetle has been confined to urban and ornamental trees, particularly maples. Along the eastern seaboard of North America there are worries about the maple syrup industry and autumn colour tourism if the beetle becomes common. Moreover, there is a real threat that the Asian longhorned beetle could also become a serious global landscape pest, so much so that it has been included as one of the world's worst 100 invasive species in the Global Invasive Species Database. Visit the database (www.issg.org/database) for further details of this and other world pests.

### 5.3.2 Hemlock woolly adelgid

The hemlock woolly adelgid *Adelges tsugae* is a small aphid-like insect, less than 1 mm long, wrapped up in a white, woolly mass when laying eggs (Fig. 5.8a), that as the name suggests, infests hemlock (*Tsuga*) trees. Native to Asia, it was first found in western North America in 1924 and then spread through the continent, probably carried on infected hemlock trees. The hemlock woolly adelgid became a problem when it reached the east coast in the early 1990s because the two hemlock species here – eastern hemlock *T. canadensis* and Carolina hemlock *T. caroliniana* – proved much more susceptible to the pest than western or Asiatic hemlocks. The hemlock woolly adelgid is now seen as the single biggest threat to these hemlocks with potential impacts every bit as devastating as Dutch elm disease (Section 5.4.5) and chestnut blight (Section 5.4.6).

The small insects cluster on young branches and inject a toxic saliva as they suck the sap. Within a few months of infection, the needles and most buds die leading to the loss of major branches and often eventual death of the tree within as little as 4 years of infection, aided by other insect and fungal attacks on the weakened trees. Mortality in an area is high (50–100% of trees) and, since hemlock doesn't sprout from the base nor are its seeds stored in the soil (Brooks, 2004), death of the tree leads to the loss of the hemlock and conversion of the forest from a dense and dark conifer-dominated forest to hardwoods dominated by black birch *Betula lenta* with red maple *Acer rubrum* and oaks – a radical change with many ecological consequences (Kizlinski *et al.*, 2002). As with Dutch elm disease, hemlocks were devastated around 5000 years ago, taking 1000–1500 years to recover.

The hemlock woolly adelgid is currently found in 15 states along the eastern seaboard (Fig. 5.8b), covering around half of the native range of the hemlocks. The pest is spreading at the rate of 12–20 miles (20–30 km) each year, blown by the wind and easily carried by birds and animals. Added to this it has two

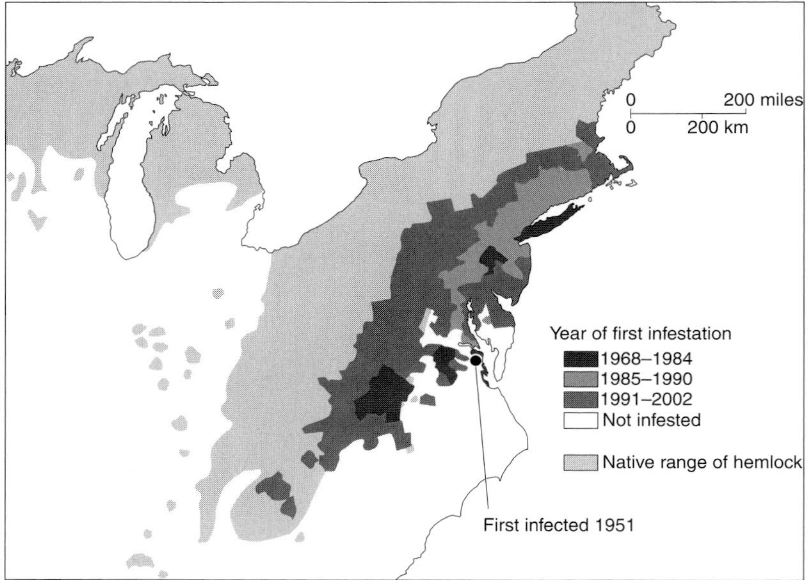


Figure 5.8 (a) Masses of the hemlock woolly adelgid (HWA) on the underside of an eastern hemlock shoot, (b) infestations of the HWA in eastern USA between 1951 and 2002 (a: Photograph by Dr Mark McClure of the Connecticut Agricultural Experiment Station [Retired]; b: redrawn from data of USDA Forest Service, Northeastern Research Station.)

generations a year and they reproduce parthenogenetically (without the need for males), so every new individual is female and capable of laying up to 300 eggs twice a year. In the short term, control is problematical. It is difficult to spot in the tree until it is well established; insecticides and other forms of sanitary control can be used on individual trees but are not practical on whole forest stands. Trees in wet areas usually survive for longer so irrigation during dry spells may help, but again this is difficult on a large scale. The main hope is biological control; a number of insects are possible contenders, the most hopeful being a ladybird (ladybug) beetle native to Japan, *Pseudoscymnus tsugae*. This feeds almost exclusively on adelgids and millions of these beetles have already been released in 15 eastern states in an effort to control the hemlock woolly adelgid.

## 5.4 Forest fungi

### *5.4.1 Fungal friends and foes: saprotrophic, parasitic, mycorrhizal, edible and poisonous species*

Fungi resemble animals in having **heterotrophic nutrition**. Unlike the **autotrophs**, which fix their own energy by photosynthesis (green plants) or chemosynthesis (certain colourless bacteria), fungi are dependent on living or dead organic matter for their sustenance. **Saprotrophs** obtain their energy from dead organic matter, which they digest. Many are completely harmless and some are edible, like the ubiquitous field mushroom *Agaricus campestris*, but others cause a great deal of damage such as the dry rot fungus *Serpula lacrymans* which establishes in damp timbers and then uses its mycelial strands, which can transmit water and nutrients, to invade drier parts of a building. **Parasites**, on the other hand, attack living organisms. Fungal parasites such as those causing Dutch elm disease (DED, see Section 5.4.5) and chestnut blight (see Section 5.4.6) have caused very serious damage to tree populations and have become increasingly difficult to control with the development of global trade. As mentioned in Section 5.1, many fungi can continue to live saprotrophically on the dead tissues of plants they have parasitized and killed.

The white-spotted red fruiting body of the highly poisonous fly agaric *Amanita muscaria* var. *muscaria* can often be seen growing on sandy soils. This fungus receives photosynthate from the birches with which it forms a **mycorrhizal association**. A **mycorrhiza** is an association between a root and a fungus (or fungi) to the benefit of both (**symbiosis**), found in over 80% of the world's vascular plants. Mycorrhizal fungi, such as the highly poisonous fly agaric, live in association with higher plants from which they receive photosynthate, while supplying in return water and important nutrients (particularly phosphorus but also nitrogen and any others for which competition is intense), derived from the soil. The main role of the fungus appears to be in increasing the effective absorbing surface area of the roots; Rousseau *et al.* (1994) found that fungus contributed 80% of the absorbing area of pine seedlings. Several forms of mycorrhizas exist (see Box 5.1). Of these, **ectomycorrhizas (ECM)**, in which fungal sheaths envelop the root, are of particular interest to forest ecologists (judged by the number of published studies) since they are confined almost entirely to woody plants. Ectomycorrhizal fungal species frequently live on more than one host plant and this makes possible the formation of common mycorrhizal networks between understorey and canopy trees, as Kennedy *et al.* (2003) have demonstrated in mixed evergreen forest stands in northern coastal California. Douglas fir *Pseudotsuga menziesii* dominated the canopy and tanoak *Lithocarpus densiflora* the understorey trees. Of the 56 ECM fungi found in this forest, 17 were found on both hosts, 27 solely on Douglas fir and 12 on tanoak. Although the number of fungal species found on both hosts was fairly low, their abundance was sufficiently high that 13 of the 17 multi-host fungal species were present on both hosts within at least one of the 18 soil samples taken. This suggests that it is very likely that tanoak is connected to the Douglas fir trees above by the mycorrhizal net. Do such common mycorrhizal networks result in a transfer of photosynthate from the roots of the canopy trees to those of the understorey? This problem was not investigated in the study discussed above, but Simard *et al.* (1997) found that significant amounts of carbon were transferred from paper birch *Betula papyrifera* seedlings growing in full sun to the roots of experimentally shaded Douglas fir seedlings. This suggests that carbon may move preferentially to plants in shaded environments; in any event individual higher plants tapping into ECM networks will have access to larger nutrient pools. Evidence of such **facilitation** was provided by Horton *et al.* (1999), who showed that Douglas fir seedlings shared many ECM fungi with mature *Arctostaphylos* spp., and that in the area concerned Douglas fir seedlings would establish successfully only when near *Arctostaphylos* individuals. Similar studies in a volcanic desert in Japan have shown that ECM fungi associated with pioneer willow *Salix reinii* were essential in facilitating the establishment of subsequent Erman birch *Betula ermanii* and Japanese larch *Larix kaempferi* (Nara and Hogetsu, 2004).

The importance of ectomycorrhizas to seedling success was demonstrated by Dickie and Reich (2005). They looked at how the ECM community changed with distance from the forest edge into two abandoned agricultural fields in

## Box 5.1 Types of mycorrhiza

Many plants require the symbiosis to prosper while for others, such as maples and birches, it is not essential. Others, notably members of the Proteaceae, including the proteas, banksias and grevilleas of the southern hemisphere, rarely if ever form mycorrhizas. The fungi vary from **obligate symbionts** that can obtain carbon only from the host plant (as in endotrophic mycorrhizas) to **facultative symbionts** that can be free-living (as in some ectotrophic mycorrhizas (ECM) fungi). Endotrophic mycorrhizas are typically more common where phosphorus is limiting but nitrogen is not. Ectotrophic mycorrhizas become most important with increasing latitude as nitrogen becomes more limiting (see Chapter 8).

**Ectomycorrhizas (ECM)** Fungi: usually Basidiomycetes, rarely Ascomycetes. The fungal partner forms a pale mantle around the outside of the root with hyphae penetrating between the cells of the root. Found in 3% of the flowering plants but including 90% of temperate and boreal trees of the northern hemisphere including most conifers. Also found in southern hemisphere trees such as southern beeches *Nothofagus* and eucalypts *Eucalyptus*, and some tropical families including the dipterocarps (Dipterocarpaceae).

*Endomycorrhizas*

Three main groups are recognized. In every case the fungal hyphae penetrate into the cells of the root.

***Arbuscular* (*AM*)** Fungi: usually Glomales (formerly included in the Zygomycetes but now assigned to a new group, the Glomeromycetes).

Found in a wide variety of families including some trees (Aceraceae: maples; Juglandaceae: walnuts, hickories; Ulmaceae: elms; Oleaceae: olives, ashes; Magnoliaceae: magnolias, tulip trees; Hamamelidaceae: sweet gums, witch hazels; Cupressaceae: cypresses, junipers; Araucariaceae: monkey puzzle; Ginkgoaceae: ginkgo; Taxodiaceae: redwoods, swamp cypress.

***Ericaceous*** Fungi: Ascomycetes.

Three types exist, all growing on members of the *Erica* family or close relatives, such as the heathers, rhododendrons and wintergreens *Pyrola* on very acid soils but also includes Indian pipes/yellow bird's-nest, *Monotropa* which are non-green and entirely myco-heterotrophic (living off fungi, see Box 3.1).

***Orchidaceous*** Fungi: Basidiomycetes.

Orchid seeds have almost no food reserves and so are totally dependent upon the fungus until the orchid becomes photosynthetic which may be a number of years after germinating. Some are non-green and permanently dependent upon the fungus.

Based on information from Calow (1998), Thomas (2000) and Deacon (2006)

Minnesota, USA. Seedlings of the bur oak *Quercus macrocarpa* were planted in the fields at different distances from a northern pin oak *Q. ellipsoidalis* (focal tree) growing at the edge of the forest. Use of a focal tree of the same genus but different species reduces the possibility of root grafting which is common in oaks, while host specificity of fungi tends to be at the genus level. When seedlings were harvested after a year they found that mycorrhizal abundance and species richness were high near the focal tree, declined rapidly around 15 m away (probably where the roots of the focal tree ended) and were uniformly low at 20 m. All seedlings up to 8 m from the tree were ectomycorrhizal but many at 16–20 m were uninfected. This demonstrates that ECM infection of seedlings is spatially complex and dependent on existing trees for infection.

The discovery that the mycorrhizal agaric fungus *Laccaria bicolor* traps springtails and supplies nitrogen to the trees with which it is symbiotic is important; in some cases 25% of the nitrogen used by the eastern white pine *Pinus strobus* is derived from these soil animals (Klironomos and Hart, 2001).

### *5.4.2 Fungal pathogens and death*

While mycorrhizal fungi are beneficial, there are a large number of formidable fungal pathogens that inhabit the world. Given their abundance it might seem a simple case of pathogenic fungus kills plant whenever the two meet. However, as is usual in the biological world, it is rarely quite this simple. Various examples could be used but perhaps one of the most widespread global fungal pathogens is the **honey fungus**, really a set of parasitic root and butt-rot fungal species that can live saprotrophically once its host is dead, using the dead stump as a base for further infection. There are numerous of these ***Armillaria*** species, geographically overlapping and varying greatly in their pathogenicity: *A. mellea* is often a virulent pathogen especially in lowlands, and is most dangerous in the absence of other fungi. Having killed a tree in a garden situation it may spread, devastating vulnerable trees and shrubs such as rhododendrons. It appears not to be so dangerous a parasite in forests where, in the opinion of a number of mycologists, the presence of other fungi diminish its effectiveness as a parasite. *Armillaria ostoyae* affects mostly conifers; *A. gallica* is normally a fairly innocuous saprotroph; and for many others little is known of their virulence. It has been suggested that no tree in the UK ever shows its full potential growth because of the stultifying effects of honey fungus even where it doesn't kill. This is undoubtedly a gross overstatement but demonstrates the legendary ability of some species to infect. But even here, the condition of the potential host can be of considerable importance in determining whether it will or will not become infected. **Water stress**, caused both by dry soils and waterlogging, has been shown to influence infection by honey fungus of susceptible woody species. Popoola and Fox (2003) used isolates of *Armillaria mellea* and *A. gallica* to infect healthy blackcurrant, strawberry, Lawson cypress and privet. Previous to this the host plants had been watered normally, subjected to drought or waterlogged for a period of 4 weeks. At the end of this period chemical analysis of the roots showed that levels of protein, lipids and carbohydrates were higher in both groups of stressed plants than they were in those watered normally. The increased nutrient levels found in both the droughted and waterlogged plants were sufficient to enhance the virulence of both *A. mellea* and *A. gallica*.

Pathogenic fungi are also known to affect hosts in different ways. A considerable number of mountain pines *Pinus mugo* in the Swiss Central Alps which had been killed by root fungi were investigated by Cherubini *et al.* (2002), in an attempt to evaluate the causes of the destabilization of the stands involved. In trees infected with honey fungus the decline in width of the annual rings after infection was relatively slow, whereas in *c.* 60% of cases involving Fomes root rot *Heterobasidion annosum* (an important circumboreal root pathogen) it was abrupt. It seems that honey fungus is, at least in this site, a secondary pathogen which tends to attack trees already weakened by competition, while *H. annosum* infects and kills trees relatively swiftly. Although tree rings can be used to indicate the history of decline they do not enable the date of infection to be determined. It should also be noted that in some trees, for a period prior to death, tree-ring growth occurred in only part of the trunk circumference.

A major difficulty in the investigation was that of **determining the year of tree death**. This has sometimes been taken as the year in which the tree no longer possesses any green needles; alternatively the year in which the outermost tree ring was formed may be taken as the last year of life. The problem is that green needles usually remain on a conifer for more than a year after the last annual ring is formed. The most striking discrepancy between the two criteria was in the case described above, where the variation was a remarkable 1–31 years.

Fungal attacks on seeds and young seedlings can play an important role in tree establishment. Evidence is building up that in tropical forests and more recently in temperate forests (O'Hanlon-Manners and Kotanen, 2004) the abundance of seedlings below canopy gaps is not just due to high light levels but also to temperature and moisture regimes less favourable to fungal pathogens. Below the canopy of trees loss of seeds and seedlings to pathogens is much higher than in warmer, drier gaps.

### *5.4.3 Long-distance transmission of fungal disease*

Fungal spores and pieces of mycelium are known to be spread by a variety of animals, particularly insects. The felted beech coccus *Cryptococcus fagisuga* spreads the beech bark disease, now causing damage in English pole-stage stands, through the wounds it makes. Endemic on *Fagus sylvatica* in Europe, *Nectria coccinea* is also spreading on American beech *F. grandifolia*, the minute sap-sucking felted beech coccus having been accidentally introduced into Nova Scotia around 1890.

For long-distance spread, abiotic agents such as water and wind are also important. Spores, being so small and light, are readily carried high into the atmosphere and moved long distances. This undoubtedly helps explain why the same or similar species of fungi are found on different continents. Nevertheless, spread by spore is not always the main agent of infection. Fomes root rot, mentioned above, will readily infect new host trees by spores enabling it to spread over long distances between forests but honey fungus species rarely spread this way, relying instead on vegetative spread by **rhizomorphs**, concentrated mycelial strands looking like black bootlaces ramifying through the soil.

The long-distance transmission of pests and diseases hidden in timber and infected planting stock is a major concern of foresters, and quite rightly so given the catalogue of major diseases released into new areas: Dutch elm disease (see Section 5.4.5), chestnut blight (see Section 5.4.6) and the important root rot *Rhizina undulata* introduced from northern areas to the tropics and southern hemisphere (Wingfield *et al.*, 2001) to mention a few. However, measures taken to prevent spread, particularly quarantine and specific regulations regarding treatment, have considerably improved, although there have been notable exceptions! The military have not always adopted equivalent standards for timber which they export, indeed a number of tree diseases that took a major toll in Europe during the twentieth century, including chestnut blight, seem often to have started near military bases. A particularly clear-cut case has recently been described from the Presidential Estate of Castelporziano 24 km south of Rome, long famous for its rich forest and exclusively Italian flora. The US Army was briefly encamped in this forest during its drive up Italy in 1944, a brief incident directly related to the death of large areas of stone pines *Pinus pinea* in the 1980s. This was caused by Fomes root rot *Heterobasidion annosum* which spreads from the root of one tree to the next by spores. When samples of the fungus were collected, all turned out to have DNA signatures typical of eastern North American variants of this worldwide species. The samples also differed amongst themselves, indicating that the fungus had gone through many generations since its introduction to the forest. The fungus may well have originated from untreated lumber brought in as crates or pallets by the army during 1944 and existed unnoticed for decades (Gonthier *et al.*, 2004). Once spread, control is often almost impossible; prevention is infinitely preferable to cure.

### 5.4.4 Some serious fungal pathogens

Many serious problems facing trees are due to fungi, as a number of diseases are sweeping through various areas of the globe. *Phytophthora cinnamoni*, introduced into Western Australia in 1921, is killing very large numbers of trees across southern Australia. The closely related **sudden oak death** *Phytophthora ramorum* causes serious problems in southern Oregon and California, where it has already killed more than 100 000 tanoak *Lithocarpus densiflorus*, Californian black oak *Quercus kelloggii* and other related species. The fungus thrives in warm, moist conditions and its spores can be carried in water or in mud on shoes or vehicle tyres; so far there is no concrete evidence that it is spread by birds or wild mammals. In Europe it is a threat not only to alien oaks and the native oaks *Quercus robur* and *Q. petraea*, but also to beech, sweet chestnut, maples and a variety of shrubs. It was found in England on established *Viburnum, Magnolia, Camellia, Rhododendron* and *Pieris* in the summer of 2003 by inspectors from the Department for Environment, Food and Rural Affairs (Defra). These were destroyed and the soil disinfected. In the autumn of that year came the first record of it infesting an oak in Britain; a bleeding canker was discovered in Sussex on the trunk of a 100-year-old exotic southern red oak *Quercus falcata* from central and southeastern USA. The two English *Quercus* species are members of the white oak group and not thought to be so vulnerable to the fungus. Moreover, despite urgent and detailed searches, the disease has not so far been discovered in the wild. Many of the trees infected have been in nurseries, growing next to infected viburnums and sharing the same watering system. The problem is made more difficult because Europe has been suffering from **oak decline** for decades, linked to a variety of causes including root-infecting fungi, including other *Phytophthora* species, drought and various insect attacks; it will be difficult to discern what sudden oak death is likely to contribute. Oak is a valuable timber tree and also important to many historic sites, for example the Boscobel oak in Shropshire in which Charles II is said to have found refuge after his defeat by Cromwell at the Battle of Worcester. The UK Government is treating this fungal threat with the highest priority; it may well be contained by vigorous inspection and the immediate elimination of infected trees. The danger remains that it will be spread on ornamental plants imported from continental Europe that have bypassed the statutory health checks required under legislation introduced in 2002.

Some fungal diseases, for example ***Dothistroma pini*** **needle blight** which afflicts radiata pine in New Zealand plantation forests, are relatively easy and cheap to control in large monocultures. In this instance the fungicide copper oxychloride applied at the rate of 2.5 kg in 20 litres of water per hectare is very effective; not more than three applications are needed in a 30-year even-aged rotation. This pine develops some natural resistance after the age of 15 years.

Root diseases are a common problem since the soil provides excellent conditions for fungal growth. The susceptibility of the highly adaptable species Port Orford cedar or Lawson's cypress *Chamaecyparis lawsoniana* to the root disease caused by ***Phytophthora lateralis*** was first noted in a horticultural nursery in Seattle, Washington in the 1920s. Its long-lived spores have since been carried far to the south by motor vehicles, on the feet of grazing animals and in floods of water. The disease also moves from tree to tree via root grafts. Infection at the root tip rapidly spreads to the whole root system; even large trees often die within a year of infection. Quarantine remains the best defence, though research into genetic resistance continues (Greenup, 1998). More recently this disease has also been shown to infect the yew *Taxus brevifolia* which commonly grows with Port Orford cedar in California and Oregon.

The basidiomycete ***Ganoderma*** causes root- and **butt-rot** (i.e. at the base of the tree) in a wide range of trees in many parts of the world. *Ganoderma applanatum*, which has a bracket-type fruiting body and is widespread in the northern hemisphere, attacks mainly deciduous trees including maples, beeches, limes, poplars, planes, oaks, horse chestnuts, birches, alders, ashes and willows, but is also found on conifers including firs and spruce. It is primarily saprotrophic rather than parasitic, being largely associated with trees whose roots already have large wounds. Trees infected by fungi often respond by producing **chemically modified barriers**, termed **R-zones** which help to arrest the spread of fungal hyphae and the consequent xylem dysfunction. Tests of this important ability against *Ganoderma* were made by Schwarze and Ferner (2003) who used sterilized wood blocks taken from a single London plane *Platanus x acerifolia*. The sapwood used as host material already contained naturally occurring R-zones. Five sides of each test block were sealed with paraffin wax while the sixth was exposed to one of the fungal species concerned. This novel approach demonstrated considerable differences in invasive ability and speed of decay between the species involved.

### 5.4.5 Dutch elm disease

Dutch elm disease (DED), one of the more serious tree diseases in the world, acquired its name because so much early research on it was done in the Netherlands (Gibbs *et al.*, 1994), rather than this being its source. Dutch elm disease, which may have originated in the Himalayas, is widespread throughout the natural distribution of the elm apart from China and Japan; it has even attacked exotic elms in New Zealand. Discovered in a number of European countries shortly after World War I, the first English record was made in Hertfordshire in 1927. This outbreak, which caused widespread deaths of elms in southern England, was caused by the insect-borne fungus *Ophiostoma ulmi*, probably brought into the country on infected logs from North America. In the 1940s the numbers of trees affected and the severity of the damage declined, though there were local 'flare-ups' from time to time.

In the late 1960s a new and far more severe epidemic developed. By 1980 it was estimated that it had killed over 25 million of the UK's estimated 30 million elms, causing losses throughout England and Wales and extending northwards into the central Scottish lowlands. The 1970s epidemic of DED in Britain resulted from the introduction of a very pathogenic form of the fungus (*O. novo-ulmi*) on diseased elm logs imported from Canada. This fungus exists as two subspecies, each of which has a different history of spread within the northern hemisphere. The outbreak also caused enormous damage in other European countries, south-west and central Asia and North America where more than half of the native *Ulmus americana* were killed. The 'Princeton' and 'Valley Forge' variants of this species are so resistant to DED that they are being planted again.

Much is now known about the onset of the disease. After infection, both species of fungus exist in the living tree as a yeast-like stage. This is spread through the tree in the sap of the vessels of the outermost xylem ring (the vascular tissue) and the fungus quickly blocks the water-conducting tissue of the tree. Soon after this elm twigs commence to wilt (hence why the disease is classified as a 'vascular wilt'). This directly relates to the activities of the fungus: fungal toxins are produced, some vessels fill with air as they are punctured by the mycelium, and tyloses (gummy extensions of the xylem walls) bulge into the cavities of the vessels giving rise to the dark streaks which can be seen in transverse sections of infected twigs. This is particularly damaging to the elm because it is a **ring-porous tree**, conducting all its water up the tree through just the outermost ring directly under the bark (in contrast **diffuse-porous trees** such as beech use the xylem of several years for water transport; see Thomas, 2000 for more detail). The virulent *O. novo-ulmi* can spread through a mature tree in

2 weeks. The fungus also produces mycelia in the bark of dead and dying elms and these give rise to two types of asexual fruiting structure. The sexual flask-like perithecia are formed in similar places but are fertile only if both the A and the B mating types are present. It is a combination of these spores that are carried between trees by the main disease vector – a beetle.

Where trees are connected by grafts, and in suckering elms, the fungus is transmitted directly from tree to tree. The main agents that transmit DED over any distance in Britain are, however, the large European elm bark beetle *Scolytus scolytus*, which reaches a length of 5–6 mm, and the small European elm bark beetle *S. multistriatus* that is about half as long and normally has only one generation per year. The larger of these two scolytids is the more important vector of the disease; the adult carries more spores than *S. multistriatus*, it can fly up to 5 km and is the only species that breeds successfully in Scotland and northern England. It also makes wounds that are more conducive to infection of the xylem. In the north of the UK it has a single generation each year, but in the south there are two and sometimes even a partial third generation. A very large number of mature beetle larvae can often be seen when the pupal chambers are revealed by paring away the outer bark of a recently dead tree. Figure 5.9 shows the progressive decline of a tree from initial attack by the vector beetle to severe infection by the fungus.

Wych elm *Ulmus glabra* (notably the variant 'Clusius') although not immune to DED, does show considerable **resistance** for a number of reasons, including the fact that it does not sucker (produce new shoots from the roots) as do most British elms and so is probably less liable to infection via root grafts, and it reproduces from seed and so each individual is genetically different. Its twigs are not as palatable to the beetles which spread the disease as those of the very vulnerable English elm *U. procera* (see Section 10.5.2), and the fungus *Phomopsis* rapidly invades the bark of newly dying wych elms providing a natural biological control of elm-bark beetles. The smooth-leaved elm *U. minor* ssp. *carpinifolia* of East Anglia is also resistant, as are some of the many regional forms of elm found in England.

Individual elms are usually capable of regrowth from root suckers or from epicormic/adventitious buds (pre-existing/brand new buds) at the base of the trunk (see Fig. 5.10). However, once these shoots are large enough to attract the attention of the beetles, they are inevitably reinfected. Native elms in the UK are probably doomed for the foreseeable future to go through repeated cycles of growth and above-ground death, remaining as understorey shrubs. However, bear in mind that the widespread elm decline of Neolithic times may well have been due to Dutch elm disease (Girling and Greig, 1985; Perry and Moore, 1987) and the elms did eventually recover!

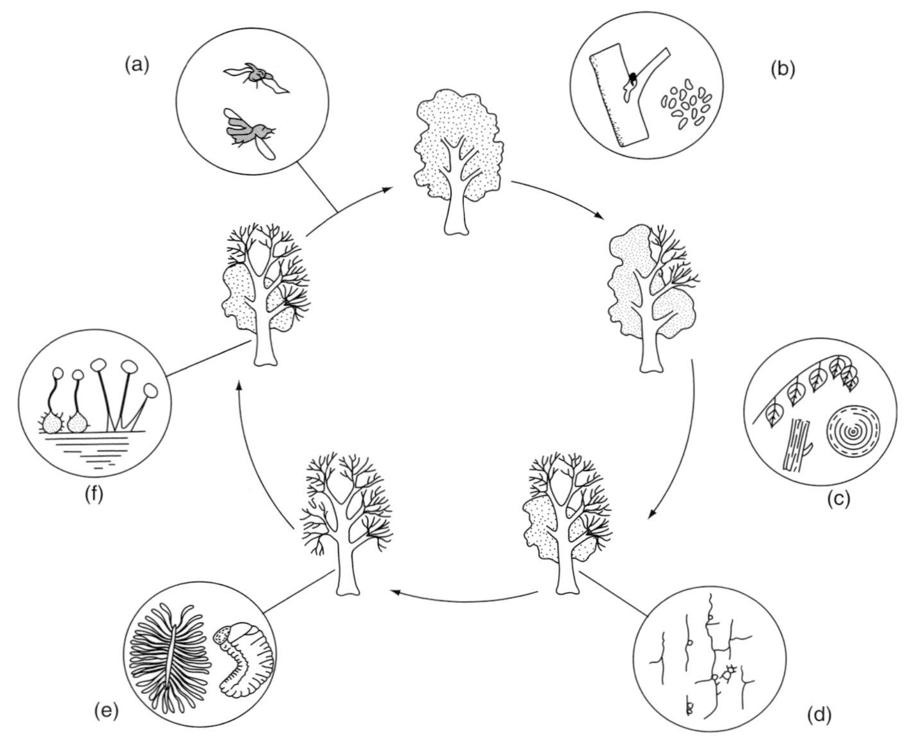


Figure 5.9 Stages in the infection of an elm tree and the life cycles of Dutch elm disease (DED) fungus *Ophiostoma ulmi* or *O. novo-ulmi* and its scolytid beetle vector.

(a) Adult beetles emerging from the bark of dead and dying elms in spring or summer carry spores of the causal fungus.
(b) Beetles feed in the twig crotches of healthy elms and introduce fungal spores into the wood.
(c) The infected regions wilt and diseased twigs show characteristic streaks or spots.
(d) Trees weakened by DED become breeding sites for beetles.
(e) The larvae form galleries by burrowing beneath the bark.
(f) The fungus fruits in the pupal chambers, discharging spores.

(From Gibbs, Brasier and Webber, 1994. *Research Note 252.* Forestry Commission. © Crown copyright material is reproduced with the permission of the Controller of HMSO and Queen's Printer for Scotland.)

Several different approaches, involving both the fungi and the beetle vectors, have been employed in attempts to counter DED. Though advances have been made and much learnt, none have been very successful. The use of insecticides to protect healthy trees from beetle attack was ineffective. Use of attractants in beetle-trapping operations which could have led to biocontrol

Figure 5.10 Part of the complex mosaic of a climax woodland. Black 'bootlaces' (rhizomorphs) of honey fungus *Armillaria mellea* run over the trunk of an elm killed by Dutch elm disease *Ophiostoma novo-ulmi*, which lies surrounded by dog's mercury *Mercurialis perennis*, common nettle *Urtica dioica*, lesser celandine *Ranunculus ficaria* and occasional cleavers *Galium aparine*. Suckers derived from the dead tree grew nearby. Buff Wood, Cambridgeshire, UK 1982. (Photograph by John R. Packham. From Packham *et al.*, 1992. *Functional Ecology of Woodlands and Forests*. Chapman and Hall. With kind permission of Springer Science and Business Media.)

using fungal or bacterial preparations that have worked with other insects was abandoned as the beetles are not susceptible to such agents, although **pheromone** ('external hormone') traps proved to be useful in monitoring beetle populations in sanitation control areas. Injection with a soluble formulation of the fungicide thiabendazole that proved to be successful in combating DED in valuable trees in the early stages of infection is too costly for commercial application. Investigations continue into the virus-like '**d-factors**' (mycoviruses) that have a deleterious effect upon both *Ophiostoma ulmi* and *O. novo-ulmi*. It now seems that hybridization between *O. ulmi* and *O. novo-ulmi*, and the consequent rapid genetic diversification, may have helped the disease at its peak by impeding the spread of these d-factors through the fungus population. Moreover, *O. novo-ulmi* is made up of two subspecies, ssp. *novo-ulmi* and ssp. *americana*, with the latter being the more virulent, but hybrids may be more virulent still due to enhanced resistance to d-factors (see Konrad *et al.*, 2002).

(Indeed, the newly discovered 'alder' *Phytophthora* may be a product of this hybrid vigour in pathogenicity.)

Some Asian elms, such as *Ulmus pumila* and *U. parvifolia*, although quite different in appearance to British elms, have high resistance to the fungus and may eventually be saved by the use of genetic engineering to introduce disease resistance. Complete immunity to DED is, however, unlikely ever to be achieved as the disease keeps changing; this is a battle where constant watchfulness and ingenuity will be required in the centuries to come.

### 5.4.6 Chestnut blight and the possibility of its control with fungal antagonists

The timber of the sweet chestnut of Europe (*Castanea sativa*) and the American chestnut (*C. dentata*) is an excellent decay-resistant hardwood, and the edible nuts produced by both species are consumed by wildlife and people. The damage caused to them by the chestnut blight fungus (also known as Asiatic blight), the ascomycete *Cryphonectria parasitica*, formerly known as *Endothia parasitica*, is thus all the more regrettable (Tattar *et al.*, 1996). In the native range of the American chestnut along the Appalachian Mountains, 99.9% of the trees, or about 3.5 billion trees, have been killed in just a century. The chestnut formerly made up around 40–50% of the trees in the canopy over an area of some 800 000 km$^2$ and by all accounts was a vigorous, quick-growing tree (Jacobs and Severeid, 2004). The loss is staggering and has led to very large changes in these forests: at the very least, squirrel populations crashed and seven species of moth that fed exclusively on the chestnut are now extinct.

The fungus, a canker disease, was unknowingly imported into northeastern USA on Asian chestnut trees at some time between 1882 and 1904, being noticed in Bronx Zoological Park, New York in 1904. It did not seem to cause extensive disease symptoms on Asian chestnuts, but the American chestnut *C. dentata* proved to be highly susceptible. Within two or three decades this tree was destroyed as a commercial crop in eastern North America. At the same time, the disease was first noted in European chestnuts (Pridnya and Cherpakov, 1996).

The fungus spreads when sticky orange spores ooze from its pycnidia and are carried long distances by insects (especially the two-lined chestnut borer *Agrilus bilineatus*) and birds, including migratory woodpeckers. Wind-borne ascospores, shot into the air from perithecia, also assist in dispersal. Much of the international spread has been in infected timber, resulting in a global spread, even reaching Australia (Cunnington and Pascoe, 2003).

Chestnut trees in which vigorous recovery shoots develop immediately below the lesions, or where cankers heal when the fungus fails to girdle the shoot, indicates the development of a degree of **field resistance** to the fungus. However, the American chestnut persists mostly as sprouts from mature trees or suppressed seedlings that become infected once big enough and so go through a repeated cycle of growth and death (Paillet, 2002). In the USA major attempts have been made to create disease-resistant hybrids between the American chestnut and Asiatic species such as *Castanea crenata* and especially *C. mollissima*. Attention is also being given to sweet chestnuts in several parts of Europe which became infected but which had subsequently recovered. This recovery has been associated with less virulent (hypovirulent) strains of the fungus that are infected with a **hypovirulent agent**, thought to be a fungal virus (hypovirus), a cytoplasmically transmitted double-stranded RNA, that may have entered *Cryphonectria parasitica* as a result of interaction with a related European fungus. This agent was successfully transferred to wild chestnut blight populations in North America (Griffin *et al.*, 2004), but unfortunately the degree of protection offered by its hypovirulence to the American chestnut *Castanea dentata* has not been as great as in Europe with *C. sativa*.

In addition to being a source of the hypoviral equivalent of Asian 'flu, other fungi appear to have important interactions with chestnut blight, some negative, some positive. *Phytophthora* root rot is thought to have been an important agent contributing to American chestnut dieback prior to the arrival of chestnut blight, and it may now present a serious limitation to establishment of blight-resistant hybrid chestnut (Rhoades *et al.*, 2003). In the Caucasus Mountains of Russia, Pridnya and Cherpakov (1996) suggest that other fungal and bacterial infections are now preventing the recovery of chestnut populations.

## 5.5 Specialized heterotrophs: epiphytes, parasites and saprotrophs

### *5.5.1 Epiphytic plants and lichens and surface-living microorganisms*

A large number of autotrophic algae, lichens and higher plants such as bromeliads and orchids hitch rides on the external surfaces of living plants, especially in moist subtropical and tropical climates. They take nothing from the host except a safe anchorage and so are classed as externally attached autotrophs (**epiphytes**). These obvious large epiphytes are accompanied by a much more widespread and varied microflora of bacteria, yeasts and filamentous fungi. Few of these can be classified as true epiphytes; most are heterotrophs, deriving a living from their hosts. The **rhizosphere** community (in the sheath of soil directly surrounding, and influenced by, the root, including the root surface – the **rhizoplane**) is mainly derived from the soil (Wood, 1989). It can have up to one hundred times the density of microorganisms as the surrounding soil (see Section 7.2.1 for further discussion). By contrast, the **phyllosphere** community (the leaf surface – the **phylloplane** – and surrounding space) derives mostly from species found on or in seeds and buds (Dickinson and Preece, 1976). Leaves often carry $10^6$ to $10^7$ bacteria per $cm^2$ (Lindow and Brandl, 2003). Some are epiphytes but most are heterotrophs that feed on either plant exudates and animal products (including the sugary excreta – honeydew – of aphids) or on dead tissues, such as the sloughed cells produced in even the earliest stages of root growth.

Roots secrete or leak out a number of compounds. Sugars, amino acids, enzymes, growth factors, organic acids and cyst–nematode hatching factors have all been identified in root exudates. Smith (1976) estimated that root exudates from hardwood trees at the Hubbard Brook Experimental Forest (see Box 8.1) amounted to 4 kg of carbon per hectare, mostly as organic acids. These compounds are mainly produced from the elongating region a few centimetres from the tip, and also from lateral roots, root hairs and senescing or damaged tissues. Such exudates are believed to stimulate the germination of fungal propagules, young roots being especially susceptible to colonization by saprotrophs and pathogens. Particular bacterial species also build up on young root surfaces, living in the numerous crevices. Some invade and disrupt epidermal and cortical cells, causing the sloughing of organic debris. Although some rhizosphere microorganisms compete with roots for essential nutrients, others, including the bacteria *Rhizobium* (associated with legumes) and *Frankia* (found on alders *Alnus* spp.) benefit higher plants by fixing nitrogen. An extra dimension is added by certain bacteria recognized by Garbaye (1994) that act to promote mycorrhizal development – the **mycorrhization-helper bacteria (MHB)**. A similar symbiosis occurs in the case of nitrogen-fixing actinomycetes growing on tropical leaves.

The activities of microbes that penetrate the surface defences of leaves, using enzymes including pectinases, and then exploit living tissues, are in stark contrast. Besides the pathogenic bacteria and fungi, there are many weak parasites, some of which cause no visible disease symptoms; others invade only tissues that are already damaged or ageing. As tissues age they release more exudates, develop a rougher surface and lower their defences against attack, so species diversity amongst the microorganisms increases with senescence. The succession of organisms recorded during the lives of leaves probably reflects the changing availability of different nutrients. The hyphomycete *Aureobasidium pullulans*, which has been noted on the buds and leaves of many coniferous and broadleaved species, grows on and in sycamore leaves for the first 2 months after the buds open. It then survives as resting chlamydospores.

Another fungus in the same group, *Epicoccum nigrum*, is amongst those colonizing mature sycamore leaves, active until leaf fall; it does not invade internal tissues until senescence. Many of these epiphytes probably hasten senescence, particularly when they have the potential to be pathogens. The fact that decomposition starts in the seedling stage emphasizes the difficulty of separating the herbivore and decomposer subsystems.

### 5.5.2 *Saprotrophic, parasitic and hemiparasitic plants*

A number of plants have lost their autotrophic status by losing their ability to photosynthesize, and live instead as heterotrophs, either as saprotrophs or parasites. The saprotrophs essentially live by extracting food from decomposing leaf litter using fungi as an intermediary and so should be called myco-heterotrophs (living off fungi), as outlined in Box 3.1. It is often difficult to tell what the fungus receives in return from the plant; perhaps the plant is primarily parasitic on the fungus? The parasitic plants are simpler in that they have a direct root connection such that the parasite draws all its needs (water, minerals, sugars and other compounds) directly from the host. To do so the parasite has to overcome a good deal of host resistance which requires a high degree of biochemical specialization, so parasites tend to be specific to a small range of hosts.

One solution to host resistance is to be a **hemiparasite**, a plant that takes water and minerals from the host but remains green and capable of photosynthesis. Water is extracted via short modified roots that form **haustoria** in the host. Because these are extracting liquid from dead tissue (the xylem) the host resistance is more easily overcome and a wider range of hosts are available. (The term semiparasitic is avoided since this implies it is partly parasitic, i.e. gets some of its carbon from the host.) Hemiparasites include many of the mistletoes living on trees, including members of the family Viscaceae (notably the European mistletoe *Viscum album* and the broadleaf mistletoes *Phoradendron* spp. of the New World – see Fig. 5.11a) and members of the Loranthaceae in the southern hemisphere. These are obviously green and photosynthetic, and gain advantage by being able to grow high up in a tree canopy in the light while stealing water from the host. On the whole they do little damage although heavily infected trees can lose vigour. Just to add confusion, it appears that some mistletoes can also steal sugars, and it should be noted that species of *Arceuthobium* (Viscaceae), the dwarf mistletoes (Fig. 5.11b), are small, often yellow and are completely parasitic! They are notoriously destructive pathogens of conifers in western North America and Asia.

Other forest hemiparasites are ‘rooted’ in the ground and look like normal green plants. Many are members of the figwort family (Scrophulariaceae)

Figure 5.11 (a) Broadleaf mistletoe *Phoradendron* sp. on oak in eastern USA; (b) American dwarf mistletoe *Arceuthobium americanum* on pine in western USA (a: Photograph by Edward L. Barnard, Florida Department of Agriculture and Consumer Services, www.forestryimages.org; b: David W. Johnson, USDA Forest Service, www.forestryimages.org.)

including the Indian paintbrushes *Castilleja* of North America and the circumpolar cow-wheats *Melampyrum*. They gain a competitive advantage by not having to grow roots and can invest more in above-ground growth.

## 5.6 Exotic plants

Many species, which in their native ecosystems are in reasonable balance with others, exert a much greater, and often adverse influence, if introduced accidentally or deliberately elsewhere. In the USA alone, it is estimated that there are between 2000–4500 exotic insects (i.e. non-native), more than 200 alien plant pathogens and around 4000 exotic escaped plants (i.e. growing outside cultivation). Of these a third of the insects, more than half the plants and almost all the pathogens are known to have a harmful effect (Pasek *et al.*, 2000; Campbell, 2002).

Exotic or alien plants (**neophytes**) can be harmful in a variety of ways. In the UK, efforts are being made to eradicate the rapid-growing turkey oak *Quercus cerris* from stands of native oaks, as it hybridizes with them producing trees with poor quality timber due to 'shakes' (internal cracks along the rays or growth rings). The large woody shrub *Rhododendron ponticum*, native to Turkey and Spain, has become a notorious intruder, especially in the British Isles where it is extensively naturalized and spreads by seeding and suckering. Introduced as cover for game birds such as the pheasant, it spreads to form dense thickets under which little else can live. Japanese knotweed *Fallopia japonica* and Himalayan balsam *Impatiens glandulifera* are amongst the larger herbs creating similar problems. A number of garden escapes – Spanish bluebell *Hyacinthoides hispanica* and variegated yellow archangel *Lamiastrum galeobdolon* ssp. *argenteum* – are causing concern in the UK as they hybridize with the native species.

Virtually every country in the world has similar problems with invasive species. Any number of web pages can be found that deal with weeds causing problems in forests and many other habitats. This is especially true in North America and Australasia where European weeds were brought in by colonists both deliberately and accidentally.

The dangers of globalization and diminution or potential loss of endemic species are well illustrated by the Island of Madeira where a decision to replant many forest areas previously cleared by fire was taken as early as 1515, some exotic species being brought in at a very early date. Douglas fir, Monterey cypress, Japanese cedar, sweet chestnut, sycamore and the common fig are amongst the very many introduced plants listed by Sziemer (2000) as having become naturalized or showing a tendency to do so. Eucalypts are the most prominent of the trees introduced for silvicultural purposes, often occurring on steeply sloping hillsides beside the Laurel Forest, which they overshadow and tend to invade. More remarkably, native trees are in some places also being overgrown and killed by the climber *Passiflora* x *exoniensis*, one of the passion flowers.

Globalization and the displacement of native forest is by no means a one-way affair. Wax myrtle *Myrica faya*, from Madeira and the Azores, is much more vigorous in Hawaii, where this introduced plant is spread by a bird introduced from Asia and supplants the native forest.

## 5.7 Herbivorous mammals and birds

Mammals have played a major role in the world's forests over a very long period. Their numbers are small compared with insects and other invertebrates, but their individual biomasses are large so collectively they have made a major impact on forest ecosystems. In Pleistocene times, for example, when even Australia had a marsupial elephant, elephants even larger than those of today were widespread. These enormous animals, which could break down large trees (Rackham, 2002), appear to have been exterminated in North America around 11 000 BC by Palaeolithic people employing spears tipped with flint Clovis points.

The Asiatic elephant *Elephas maximus* is the less formidable of the two contemporary elephants, but at 60 a plains male African elephant *Loxodonta africana* can weigh 6 tonnes; many reach 4.5 tonnes at 30. Despite their size and the damage they can inflict on vegetation by trampling and feeding, studies have shown that they benefit trees in a number of ways, including spreading seeds. As an example of their more indirect effect, Goheen *et al.* (2004) working on African acacia savannas found approximately twice the number of tree seedlings in areas with elephants, despite higher seedling death due to desiccation. In areas where elephants were excluded, the greater number of rodents and invertebrates led to ever-higher seedling death.

Elephants, which have rather few sweat glands, maintain a steady temperature by losing heat from their ears. Typical African elephants living in the relatively open savanna absorb a great deal of heat from the sun's rays and have very large ears, while those of forest type African elephants (see Frontispiece), whose total height is no greater than the shoulder of a savanna elephant, are relatively smaller. The ears of Asiatic elephants, whose whole skulls are different, are considerably smaller than either type of African elephant.

Large forest animals exert a very considerable influence on the trees they live among; they are amongst the agencies which may have led to the evolved ability to sprout from cut or broken shoots. Windblow, self-coppicing and fire are others. Humans have utilized this sprouting ability in **pollarding** and **coppicing** (the deliberate felling of trees above or at ground level, respectively, to stimulate long, straight growth). The history of coppicing is of too short a duration to have caused an evolutionary response itself.

It has been suggested (Redford, 1996) that many forests, particularly in the tropics, while appearing intact are actually devoid or depleted of large animals and consequently face long-term problems of pollination, seed dispersal, a detrimental build-up of herbivore numbers, etc., and therefore face severe long-term problems. Conversely, increasing numbers of some herbivores are causing equally severe problems, as discussed in the next section.

### *5.7.1 Megaherbivores and ungulates: deer and others*

Without question, large herbivores have an important effect on forest dynamics. They are also often the most likely animals to go extinct when forests change rapidly such as when humans appear (this is because they are usually the longest lived and so less able to adapt quickly by natural selection – see Brook and Bowman, 2005). The **ungulates**, herbivorous mammals with hooved feet and a herbivorous diet, have a particularly important role in forests. Those with an even number of toes, the **artiodactyls**, include the insectivores, pigs, hippopotamuses, camels and giraffes. These tend to have a larger global impact on plants than the **perissodactyls** (with uneven number of toes) such as the horses and rhinoceroses. The most successful modern artiodactyls are the **ruminants** (Fig. 5.12), which include the deer (Cervidae) and the cattle, sheep, goats and antelopes (Bovidae), which have flourished only since the Miocene (5–24 Ma). Ruminants all possess a stomach with four chambers, of which the first two (rumen, reticulum) receive the food after it has just been swallowed. Here it is mixed with mucus and acted upon by a mixture of bacteria and ciliates which rapidly break up the celluose chemically. Food is then returned to the mouth when the animal ‘chews the cud’. Following this it passes to the two latter chambers (omasum, abomasum) where useful material is absorbed. This feeding strategy enables the ruminants to feed very rapidly and then digest their food after retiring to a place of safety. It also enables them to utilize materials that are intrinsically difficult to digest, making them very effective grazing machines.

**Deer** are a most successful group of ruminants whose keen eyesight and sense of smell largely enable them to evade their enemies. They often graze extensively in small areas, yet being very mobile can spread their grazing over large areas, including even isolated woodlands. In the northern hemisphere, deer populations

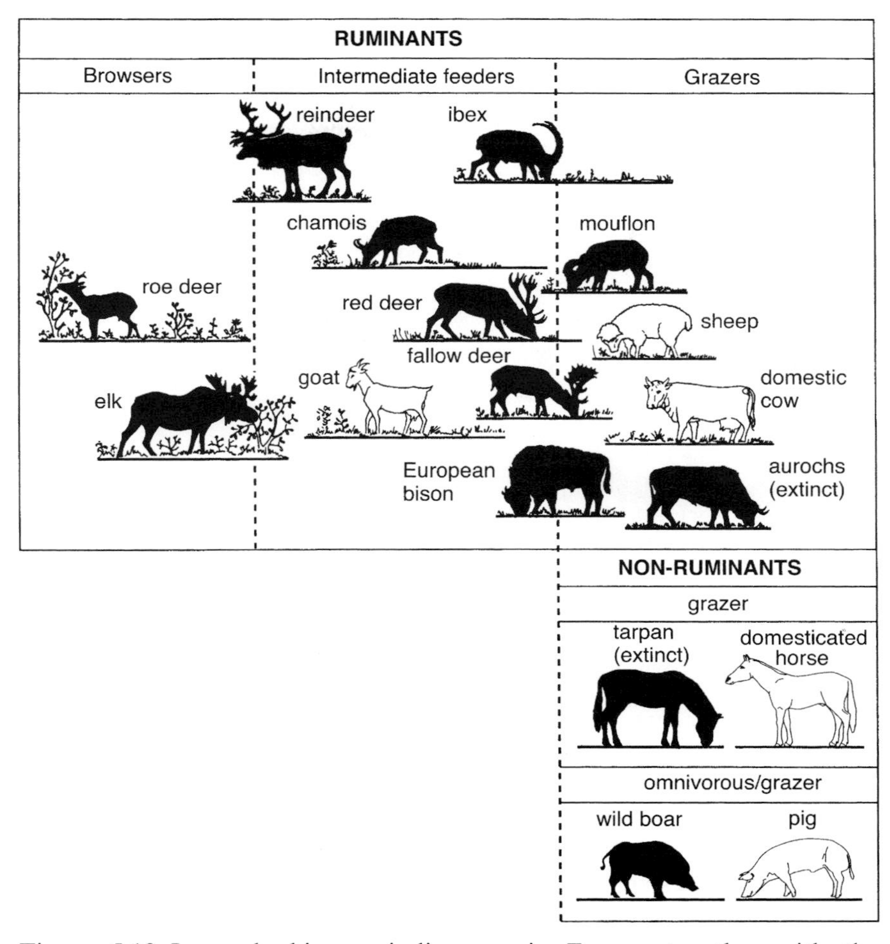


Figure 5.12 Large herbivores indigenous in Europe together with the omnivorous wild boar, classified according to their feeding strategy. Domesticated species are merely outlined. Indigenous species of the lowlands of central and western Europe include the aurochs, tarpan, European bison, red deer, elk, roe deer and wild boar. (Redrawn from Hoffman 1973, 1976, 1985, and Van der Veen and Van Wieren 1980. From Vera, 2000. *Grazing Ecology and Forest History*. CABI Publishing.)

have been expanding in size and distribution over recent decades. Red deer *Cervus elaphus* (the red deer of Britain and the elk of north-western American forests – Fig. 5.13) is a case in point. Although the claims are disputed, the population of red deer in the Scottish Highlands is thought to have risen from 300 000 in 1989 to 450 000 in 2002. If heather cover and woodland regeneration are to be protected here it will be necessary to reduce numbers of sheep as well as deer, and to protect sensitive woodland with fences. Similar problems are being experienced through most of its North American range as well.

Figure 5.13 The carnivore/herbivore relationship. The wolf *Canis lupus* and the skull of a red deer *Cervus elaphus*, known as elk in North America, which it had killed earlier. These deer defend themselves with their antlers, which the males also use in competitive battles with rivals. (Drawn by Peter R. Hobson.)

In Eurasia *Alces alces*, the largest deer in the world, is called moose or elk, or just moose if you are in North America. It is superbly adapted for snowy climates with long legs (males stand around 2 m tall at the shoulder, females 1.7 m) and wide shovel-like antlers on males used to help dig through snow to reach food. Males can reach 2.8 m in length and their antlers commonly have a spread of 120–150 cm. Much of its summer fodder comes from aquatic and wetland plants, but in the winter it spends most time feeding in forests. Inevitably, an animal that can reach 630 kg (in the male) requires a lot of food and its impact can be high through grazing, trampling, thrashing shrubs to remove velvet from the antlers and debarking trees. In Sweden, where it damages commercial forests, numbers of moose rose to very high levels from 1970 onwards. In the 1980s approximately 1.5 million were killed in Sweden alone (Cederlund and Bergström, 1996).

As well as the red deer, one other species of deer is native to Britain – the roe deer *Capreolus capreolus*, of which the largest bucks weigh only 20–30 kg. However, a number of deer have been introduced, adding to the deer problem. The fallow deer *Dama dama*, a native of southern Europe and Asia Minor, was introduced into Britain probably by the Romans. Males are renowned grazers rather than browsers. When kept in parks fallow deer are gregarious, but in the feral (wild) state they go about in small parties.

Deer which have reached Britain much more recently include the sika *Cervus nippon*, the Siberian roe *Capreolus pygargus*, the black-tailed deer *Odocoileus columbianus* from western North America, the Chinese water-deer *Hydropotes inermis* and the Indian muntjac *Muntiacus muntjak*. The Chinese muntjac *Muntiacus reevesi*, or barking deer has spread from Woburn Park. It stands less than half a metre tall and the antlers are very short, ending in a single spike. The upper canines of the males project from the mouth as small tusks. Though it is so small and primarily a grassland animal, the foraging of this species, like those of its larger relatives, is causing a great deal of damage, particularly to the herbs of the woodland floor.

Reindeer *Rangifer tarandus* are the only deer to have been truly tamed by humans, and are unusual in that both sexes have antlers. Although they have only recently been introduced into North America, there are several domesticated races in northern Europe. Within the Arctic Circle they are used in much the same way as cattle are further south. The caribou of North America is so similar that it is now placed in the same species, though it has never been tamed. Many caribou spend their lives on the vast treeless arctic tundra, while others are found mainly in the taiga (northern boreal forest). They gather in late winter in groups of 10 000–100 000 before the spring migration. In winter they feed mainly on lichens, in summer they have a much wider diet and consume mushrooms, fruit, green plants, twigs of trees and even discarded antlers.

Deer are often regarded as being **keystone species** in forests, able to have larger effects on their habitat than their numbers would suggest. The problems caused by an almost worldwide expansion of the deer population are considered in detail by Rackham (2003), who has first-hand experience of the damage they cause in Europe, America and Japan. Large populations of deer graze trees so thoroughly that the height of the **browse-line** they can reach up to is an indication of the dominant species present. Ivy and holly are particularly palatable, as are elm and hawthorn. Under severe conditions forests may be devoid of seedlings and young saplings; even resilient species like ash are often deformed if grazing is later lessened. Rooney and Waller (2003) point out a strong linear relationship between increasing deer browsing pressure and decreasing seedling density of various conifers and deciduous trees in the forests around the North American Great Lakes, leading eventually to complete regeneration failure. White-tailed deer *Odocoileus virginianus* have so changed the Allegheny National Forest in Pennsylvania that black cherry *Prunus serotina* dominates much of its 800 square miles (its leaves contain hydrocyanic acid that releases hydrogen cyanide – cyanide gas – and so are little grazed) and every hemlock *Tsuga canadensis* has a strong browse-line, and none are younger than 70 years. Deer grazing also tends to lead to an increase in grasses, sedges, rushes and ferns at the expense of the herbs that produce the spectacular spring show. Rackham states that most Japanese woods are free of deer browsing, but Mount Odaigahara is an exception. This mountain has zones of different tree communities, but its top is a graveyard of dead nikko fir trees *Abies homolepis*. The fir is killed by the resident deer whose population exploded after 1970. The deer eat the tree bark which they prise off with their antlers. By 1998 the ever-expanding circle of dead trees had reached a diameter of some 14 km and the deer were working their way through other conifers. Their main food is probably the dwarf bamboos forming carpets or thickets in Japanese woods, but even the rhododendrons show a browse-line.

Changes induced by deer go beyond just the plants since other mammals or insects dependent upon the same plants will also be affected, creating a **trophic cascade** (i.e. effects cascading through other trophic levels). As an example, a study in the pinewoods of Scotland by Baines *et al.* (1994) showed that grazing by red deer reduced the height of bilberry *Vaccinium myrtillus* by half, which in turn led to a fourfold reduction in Geometrid moth larvae. These larvae form an important food supply for insectivorous forest birds such as the capercaillie *Tetrao urogallus* and black grouse *Tetrao tetrix* and so the red deer may indirectly be a major factor in affecting their numbers via this trophic cascade. Similarly, Flowerdew and Ellwood (2001) reviewed how deer affected small mammals in British lowland woodland and concluded that deer can have strong impacts on most small mammals through modification of habitat structure and food supply. However, not all effects are bad since the more open habitat maintained by deer can be beneficial: in the authors' words, 'the presence of some deer is a good thing, too many deer and too few are not'.

Deer are also host to the ticks that transmit **Lyme disease**, which is caused by bacteria that bring about skin changes, flu-like symptoms and joint inflammation in humans. Though normally found on deer (see Section 4.4.3), these ticks can also infest dogs and humans. Though most prominent in the USA and first recognized at Old Lyme, Connecticut, in 1975, Lyme disease is also a problem in many parts of Europe.

Amongst the other ungulates, **sheep**, **goats** and **cattle** are renowned for their grazing ability. Fortunately, feral populations with access to wooded areas are fairly limited around the world, but in these areas their effect can be equally if not more deleterious as deer. The forests of the world contain many species of **wild boar** *Sus* and their relatives. European wild boar *Sus scrofa*, from which the domestic pig was derived, is common through Europe, North Africa and through Asia to Japan, and is being seen again in southern England (after an absence of four centuries) following its escape from captivity (Fig. 4.14). In 2003 it was estimated that a population of over 1000 animals had built up. The wild boar is a highly intelligent and fast-moving animal which can weigh up to 180 kg and be 2 m long. Its lower canines rise straight up outside the mouth and point backwards towards the eyes; these formidable tusks can rip open the body of an adversary. Males join the herds, each of which is led by an old sow, in December and January. They leave again after the rutting season and the attractively striped young are born 3–4 months later. Though individual animals are good at concealing themselves, wild boar leave unmistakable traces in the forest. They are omnivorous (eating anything with food value), consuming seeds, fruits, bulbs, insects, earthworms, reptiles, small mammals, and the eggs and young of birds. Their rooting activities are important in creating areas of bare earth on which tree seedlings can establish. Venturing from the forest they attack many farm and garden crops, rooting through the ground with their formidable snouts. They also leave very distinct disturbances in adjacent muddy lakes or swamps, where they wallow to rid themselves of parasites, later making muddy rubbing marks on tree trunks and branches.

### *5.7.2 Other grazing mammals and birds*

Small herbivorous mammals and birds play an equally important role in forests as the ungulates, and conservationists concerned with particular groups such as dormice (e.g. Briggs and Morris, 2002) have done much to investigate this interaction. Their effect is based not so much on the overall amount they eat but on what they eat, especially when it comes to seedlings and seeds. A 5-year study in New York State by Manson *et al.* (2001) used $40 \times 40$ m partial enclosures to manipulate the density of meadow voles *Microtus pennsylvanicus* and white-footed mice *Peromyscus leucopus* along the junction between a hardwood forest and a set of abandoned fields. They found seedling predation by meadow voles played a more important role than seed predation by white-footed mice in the invasion of trees. At the end of the study, there were an average of 19 tree seedlings $ha^{-1}$ under low vole densities and 8 $ha^{-1}$ under high density. The authors conclude that the main effect of the vole is to slow down tree invasion considerably, both directly by herbivory of seedlings, and also indirectly by reducing the establishment of shrubs which act as bird perches so the number of bird-defecated tree seeds reaching the site is less. This also means that the species composition of the developing tree cover can be changed as different tree species vary in their susceptibility to herbivory.

Other rodents can have even more significant effects on woodland. The ability of **squirrels** to climb and their continuously growing, extremely sharp incisor teeth, whose front surface consists of hard enamel and the rear of more easily eroded dentine, enable them to reach the seeds and cones of trees more effectively than any other group of mammals. In Virginia, eastern USA, native grey squirrels *Sciurus carolinensis* feed on acorns of white and red oaks, which are normally produced in very great quantities every year. The acorns of the red oak are slightly darker in colour than those of the white, but the squirrels are colour-blind and differentiate between the two by smell. Acorns of white oaks germinate immediately they are shed, whereas those of red oak remain dormant until the spring. In normal years the squirrels feed on white oak acorns in the autumn and store those of red oak for consumption during the winter. If only white oak acorns are available the squirrels bite out the embryo of each acorn before storing it, preventing the growth of any that might be forgotten. Since the death rate of young seedlings is high, this extra loss might have long-term consequences for the oak.

The importation of the eastern grey squirrel *S. carolinensis* into Britain in 1876 and elsewhere in Europe may be one of the main causes for the very great losses of the smaller and less aggressive Eurasian red squirrel *Sciurus vulgaris* through much of the British Isles and northern Italy (Gurnell *et al.*, 2004), as this species is badly affected by a virus (parapoxvirus) which the grey squirrel carries though itself immune to it (Tompkins *et al.*, 2003). Fortunately, the red still exists in the Scottish Highlands and parts of Wales, besides persisting in such outposts as the Isle of Wight. As importantly, grey squirrels were estimated in 2003 to cause about £50 million worth of damage to UK trees annually where they often occur at much higher densities than in North America. Major damage is mainly caused by juvenile animals and involves bark-stripping in mid-July or a little later. Maples and beech are very badly affected, though very few tree species are immune from attack. Complete ringing of the upper parts of the main trunk of young trees may make them useless for forestry purposes. The most effective method of control is by using specially designed poison feeding-hoppers.

This is a clear example of **globalization**, the introduction of organisms into territories from which they are normally absent. In other cases such organisms may cross-breed with native species, as when the Russians cross-bred the American and European bison. Unfortunately they then released some of the resulting hybrids into the wild Caucasian herds.

Seed movement by small herbivores is also important. For example, seeds of the yew tree (*Taxus baccata*) in Europe are dispersed primarily by birds (mainly thrushes, various *Turdus* species) that void the seeds once the fleshy red aril has been consumed. Seeds end up concentrated under the bushes that the birds perch in, giving the seed a favourably moist microhabitat and protecting the seedling from large grazing animals. However, the seeds have to run the gauntlet of small predators attracted to the food supply. Of the 2–6 million seeds $ha^{-1}$ produced by the yew, at least 60% that end up under bushes are normally eaten (Thomas and Polwart, 2003), predominantly by small rodents but helped by rabbits, squirrels and birds (mainly finches, great tits and woodpeckers). On the other hand, some of the left-over seeds carefully hoarded by rodents may be well placed for subsequent germination.

The part played by herbivores in spreading seeds is one of the most important of all biotic interactions. The cassowary, a very large flightless bird of ancient Gondwanaland lineage (Fig. 5.14), and said to distribute the seeds of some 250 species of understorey plants in its droppings, is probably a record breaker in this respect. Cassowaries belong to the ratites that all possess a flat sternum and have powerful legs for running. Cassowaries are now confined to New Guinea, parts of Polynesia, and the remaining tropical rain forest of North Queensland, the latter now largely cleared and used for growing bananas, amongst other crops. The cassowaries of New Guinea stand as tall as a person and are the largest animals in an extremely rich rain forest that contains roughly 20 000 species of trees and flowering plants. The forest is populated by agile marsupials like the spotted cuscus *Phalangista maculata*; there are no squirrels or monkeys.

Smaller mammals may also play a key role in herbivory of foliage. In Fig. 5.15 a browse-line can be seen that would be normal for a deer except this picture was taken near the treeline in Arctic Canada. In this case the cause was snow-shoe hares *Lepus americanus*, able to browse progressively higher as snow pack builds up over the winter, and able to remove as much foliage as an ungulate.

**Beavers** *Castor canadensis*, with their prominent orange incisor teeth and paddle-like tails, are the largest rodents in North America and have an intense if localized effect on forests by damming forest waterways and flooding low-lying areas, and by gnawing down trees. They construct stick-and-mud lodges with underwater entrances and inside platforms raised above water, emerging at dusk to forage for succulent plants or cut down trees and shrubs. Cuttings stored in late summer and autumn in the mud at the bottom of the ponds are eaten in winter and the young are born in spring. The European beaver *C. fiber* does not create large ponds, instead it burrows along the water's edge and makes small dams to slow streams that are too narrow or fast-flowing for its purposes. They also tend to be less aggressive in tree felling and less invasive of surounding waterways. These animals, which were in England hunted to

Figure 5.14 Cassowary (*Casuarius*). These birds all have a prominent horny helmet or casque on their heads; this is thought to protect the bird as it moves through the tangled undergrowth. Their feathers are highly modified and look like fur, those of the wings being reduced to quills hanging over the flanks. The birds defend their feeding territories and the males are particularly aggressive when tending their young; if aroused they strike out with both feet, which are each armed with three large spiked toes, and they sometimes disembowel human intruders. (Drawn by Peter R. Hobson.)

Figure 5.15 Browse-line on willow vegetation at head-height on the people, caused by snow-shoe hares able to browse to almost head-height by standing on the winter build-up of snow. Photograph taken north of Inuvik, Northwest Territories, Canada. (Photograph by Peter A. Thomas.)

extinction in the late Middle Ages, are now making a comeback after the European population had been reduced to very low levels after World War II, having been hunted for its meat, fur and the castoreum produced from its sex glands. Ten beavers released in Brittany in the 1960s have given rise to a population of 50, while individuals are now established in the UK at the enclosed Ham Fen nature reserve, near Sandwich, and a private estate in the Cotswolds. It remains to be seen whether free-living natural populations will again be allowed to flourish in the UK, although it is thought they are unlikely to develop into a widespread pest given their more sedentary behaviour compared with their North American cousins.

The above are examples of mammals that have significant effects on forests, but most mammals are less influential and often appear to have little day-to-day impact on their food supply. A prime case is the koala *Phascolarctos cinereus*, a marsupial which evolved in parallel with the eucalypts and is unusual in its ability to digest their nutrient-rich, but hard and chemically unusual leaves. These animals can live for 15 years; they sleep for up to 20 hours a day while they digest eucalypt leaves with the aid of a bacterium which is passed from generation to generation, the babies consuming some of their mother's excrement. The young animal is very small when born and is carried in its mother's pouch for 6 months, being later transported on her back until half-grown. So poor is the food supply that koalas can support only a very small brain (a very energy-demanding organ); offer it leaves lying flat rather than hanging and it has no idea what to do with them (Martin and Handasyde, 1999).

Other **marsupials** flourish on the floors of forests. The smaller ones exploiting the canopy include several arboreal tree kangaroos that flourish in the forests of Papua New Guinea where they spend half their time feeding on leaves, as they formerly did throughout Australia when it was still largely covered by trees.

## 5.8 The impact of woodland carnivores and omnivores

### *5.8.1 Roles of carnivores past and present*

Carnivores and the larger omnivores play an important role in nutrient cycling and, as illustrated in Fig. 5.13, are probably crucial in regulating the size and nature of herbivore populations. In most European countries, and many other parts of the world, including much of North America, large carnivores have either been very greatly reduced in number or entirely eliminated. A number of factors such as change in habitat, food supply and competitors may partly account for the resultant highly undesirable increase in deer numbers seen around the world, but lack of predators is certainly a major contributory factor. **Deer control** now involves either fencing off the habitat to be protected, or shooting animals whose populations have become excessively large. This was not a problem when large carnivores were common to almost all forests. Those of earlier times were often quite remarkable. In the early Tertiary the fossil sabre-tooth tiger *Smilodon* of the northern hemisphere, which struck with its whole head, had a marsupial equivalent, *Thylacosmilus* in South America, whose enormous stabbing teeth continued growth, unlike those of true cats, and whose lower jaw had a large flange to protect them. Amazingly, many of the other marsupial hunters which fed on both placental and marsupial herbivores, dominating the South American forest for some 30 million years, were superseded by very large, fast-running ground birds which used their massive beaks to stab and rip their prey as did the extinct carnivorous dinosaurs.

It had been considered that the early mammals living in the Mesozoic era, when the dinosaurs flourished, were insect-eating or herbivorous animals no larger than modern rats, mice and shrews. Very recently two much larger and carnivorous early mammals, 'Mesozoic dogs' as they have been termed, were discovered in the Yixian formation of China. Both belong to the genus *Repenomamus*, lived 130 million years ago, had powerful jaws, and fed on the young of small dinosaurs. At a metre long and weighing around 14 kg, the largest, *R. giganticus*, was a quite formidable member of the forest fauna and in effect a forerunner of the hunting dogs, a few of whose troops still cooperatively hunt antelope in African forests.

The reptiles in their time gave rise to the early mammals, the dinosaurs, and two types of flying animal, the pterosaurs and the birds. In contrast modern reptiles, though including tortoises, turtles, lizards, alligators, crocodiles and caimans, are a much reduced group, of which the most interesting are the limbless snakes. All these animals occur in the forests of the world, often near to waterways. The crocodiles and their immediate relatives remain powerful hunters though showing many early features. O'Shea and Halliday (2001) provide an excellent recognition guide to the major reptiles of the world and also the amphibians, including the newts, salamanders, frogs, toads and the limbless caecilians which also frequently occur in forests and often fall prey to reptiles, especially snakes.

There has been more than one trend for reptiles to lose their limbs in the past, but the snakes, many of which live in forests, have their origin in a single line which began around 100 million years ago and has given rise to some 2800 named species which vary greatly in size and ability. The giant green anaconda *Eunectes murinus* can reach a length of 10 m and weigh 200 kg, 7000 times as much as the tiny Brahminy blindsnake *Ramphotyphlops braminus*, while various snakes, despite their lack of limbs, can run, jump and swim. Many snakes kill their prey by means of the poison produced by their modified salivary glands, others by crushing and suffocating their victim. Both methods are extraordinarily efficient. A bite from a black mamba *Dendroaspis polylepis* can kill a human in 30 minutes, while a giant green anaconda, the largest living snake, can wrap itself round an animal as large as a caiman and crush it to death. The problem of swallowing comparatively large prey has, in the course of evolution, been solved in a quite remarkable way. Not only can the lower jaw be displaced from its point of articulation with the upper jaw, but the two sides of the lower jaw also come apart at the front when the animal swallows a large object. The green anaconda, an aquatic arboreal species that lives along forest waterways, can swallow and digest a complete caiman, bones and all, resuming its activities after a period of rest.

Fossil evidence regarding the ancestry of the various groups of snakes is almost non-existent, so relationships between and within some groups have been investigated by genetic studies of their DNA. Results from three groups of highly poisonous bushmaster snakes *Lachesis* spp. in central and southern America yielded results of considerable interest. Though all three groups are now geographically isolated, they show major resemblances. The southernmost bushmasters, however, show considerable differences from the others, from which they have been separated for 12 million years. Differences between the two more northerly populations, which have been isolated for each other for a shorter period, were considerably less. Snakes can live in a wide variety of places, even the sea, but being poikilothermic find northern winters very difficult. In North America red-sided garter snakes *Thomnophis sirtalis parietalis* have solved this problem by hibernating deep underground in limestone sinkholes. In spring up to 15 000 have been seen emerging from a single locality, the greatest concentration of snakes ever known.

### *5.8.2 Controlling the herbivore/hunter balance*

Modern-day forest large carnivores notably now consist of a variety of big cats at low densities (from the Siberian tiger *Panthera tigris altaica* of Eurasian conifer forests weighing up to 300 kg to the diminutive 2–3 kg kodkod *Oncifelis guigna* of the temperate mixed forests of Chile and Argentina) and the wolf. The wolf *Canis lupus* was formerly the most common predator in the woodlands and grasslands of the world. Large populations are now largely restricted to a few relatively remote territories, such as Alaska, northern Canada and eastern Europe, although the natural range of the grey wolf is probably greater than that of any other living mammal, apart from humans. Wolves are highly intelligent social animals that travel very great distances in search of their prey, often migrating with them. Their packs have a well-defined hierarchy with the alpha-male, to which all the other individuals defer, at its apex. These large animals co-operate to bring down and kill animals that are often much larger than themselves. Once a potential victim, typically a young, old, sick or wounded animal, has been identified the pack will continue to harry it even if it manages to rejoin its companions. Wolves can eat enormous amounts of meat in very short periods, and at frequent intervals, but are capable of going without food for several days.

Radio-tagging has recently thrown remarkable light on the early life of wolves, which often roam alone for at least 2 years after weaning. An 11-month-old grey wolf cub was found injured beside a ring road west of Parma, Italy, in February 2004. Park rangers were called and it recovered under the care of a veterinary scientist. It was released on a snowy day in March weighing a healthy 28 kg and its subsequent progress traced by means of a global positioning satellite unit on its collar. Whereas previous radio-tagged wolves had rarely travelled more than 50 km, this one followed a wandering path, crossing the Italian Maritime Alps to a nature reserve in Cuneo Province and staying there for a time with resident wolves. It eventually crossed into France and continued west and by September 2004 had travelled a total distance of over 1000 km and reached a point 350 km in a straight line from its point of release.

There are a number of varieties of wolf, of which the grey Mexican wolf *C. lupus baileyi*, now found in the wild again, was at one time reduced to just five animals, all of them in captivity. The task of gradually returning them to the wild was a long and complicated one, involving the mating of two healthy animals which became the alpha-male and alpha-female of the new pack, getting them used to a much wider territory and finally releasing them in an area where they taught themselves the art of tracking and killing appropriate prey.

In the absence of large carnivores, elk populations in Yellowstone National Park became excessive. The re-introduction of wolves was very successful in regulating the deer here. The carnivores, which were sedated, flown in from a distance and then released, soon adapted to their new habitat and prey. After initially feeding only on deer, they also learnt how to pull down and kill bison *Bison bison*, formidable animals able to kill badly placed wolves by kicking or goring them with their horns. In former times, relationships between wolves and the European bison *B. bonasus* (Fig. 5.16), a more long-legged animal than its American cousin, were much the same, with the wolves concentrating their attention on the young and sickly members of the herd.

The existence of **antipathy** between particular pairs of carnivore species, such as foxes and pine martens, is well established. In most cases, as with foxes and wolves, there is no direct competition for food or other benefit, but the result may well be conclusive as far as the less powerful predator is concerned. Foxes feed on rodents, fruit and invertebrates, and sometimes scavenge the remains of wolf kills; whereas wolves consume large herbivores, particularly moose. Nevertheless, foxes are killed by both wolves and lynx in the Białowieża Forest, Poland, and wolves are recorded as killing them elsewhere (Yalden, 2003).

Of the larger omnivores, **bears** are formidable animals and can be dangerous if disturbed, but are certainly much less ferocious than the considerably smaller wolverine *Gulo gulo*, which is capable of driving even a bear or cougar *Felis concolor* from its kill. The latter is believed to be the most widely distributed carnivore in the New World. Bears arose in the Miocene (5–24 Ma) as an offshoot of a central stock which gave rise to dogs, wolves and foxes. Though they have many colour variations, the present world population consists of

Figure 5.16 European bison *Bison bonasus*, which was once a common forest herbivore. Though now comparatively rare, populations exist in the southern mountains and Białowieża forest of Poland, in Russia, Lithuania, Belarus, Ukraine, eastern Slovakia and the Romanian Carpathians. Substantial remnants of American bison *B. bison* herds still survive. Both animals greatly influenced forest ecosystems in earlier times. (Drawn by Peter R. Hobson.)

only eight true species. Bears tend to be wide-ranging, spending only part of their time in forests, exemplified by the North American grizzly bear *Ursus arctos horribilis* which also spends time in the tundra and alpine zones. This often has an overall length of more than two metres; if disturbed it often rears up on its hind legs to view the situation, and can run as fast as a horse for short distances. Although capable of killing caribou and moose it generally feeds on smaller animals, carrion and plant material. The black bear *Ursus americanus* has a much wider distribution, being found in forests, swamps and mountains from Alaska to the subtropical forests of southern USA. Males are much larger than the females, but both are considerably smaller than the grizzly. Even so a sow will weigh 140 kg, so the weight of the average baby (about 240 g) is unusually small though it grows rapidly on the rich diet of milk it receives in the winter den. It is closely related to the Asian black bear *Ursus thibetanus*. Extinct in Britain long before the Norman Conquest, bears were hunted by the King of Spain in the first half of the fourteenth century and still survive in other parts of Europe. Sloth bears *Melursus ursinus* are found in forested areas south of the Himalayas.

The role of the **primates** in forest ecosystems, and in the Holocene especially, as agents of change, is of very great importance. They are remarkable for their range of size, from the tiniest monkeys adapted to exploit the very ends of slender branches to the huge gorillas, and for their very high intelligence and adaptability. It is indeed their remarkably large brains and ability to devise new strategies that makes the whole group so formidable, and the importance of their social attitudes so great, which can vary from the geniality of the huge orang-utans of Borneo to the sometimes vicious behaviour of the chimpanzees of Uganda. Social behaviour in the great apes with well-defined dominance ruled by the alpha-male, and bonds reinforced by grooming, is complex. Orang-utans normally live alone because food is scarce, but come together when it is not. As many as 20 may feed simultaneously on the fruit of a single masting tree. They have extraordinarily good memories, know just where particular tree species grow and also when they are likely to fruit. In captivity they have learnt to use tools such as the hammer and saw. The ruthlessness and efficiency with which chimpanzees, which eat meat as well as leaves and fruit, hunt other primates including Colobus monkeys through the tree tops is awe inspiring.

In recent times the rise of the hominids and modern humans has had an overwhelmingly important influence on forest structure, and all too often forest elimination. Amongst other features it was the development of relatively shorter and lighter forelimbs and the bipedal habit, which enabled them to run faster and to see further in tall grassland, that marked out the hominids. It is interesting to observe, however, that the great apes walk on their hind legs if wading in lakes and rivers when their body weight is largely supported by water. Humans are the only animals in which the thumb is fully opposable, so enabling them to use tools very efficiently. Together with the use of fire, this was to enable them largely to master many wooded environments.

## 5.9 Herbivores and the Holocene: did the lowland European forest have a closed canopy?

The steppe-tundra present at the end of the last ice age was open and dominated by grasses and herbs. When the temperature rose, animals adapted to very cold conditions, like the mammoth *Mammuthus primigenius*, moved away. As they did so a more familiar fauna of large ungulates, including the auroch *Bos primigenius* and the tarpan *Equus przewalski gmelini* which were, respectively, the wild ancestors of cattle and the domestic horse, took their place. Fossil remains show that these two species remained prominent well into the Holocene, roaming and grazing primeval vegetation along with other wild indigenous ungulates, including the European bison *Bison bonasus*, red deer, elk, roe deer and wild boar (see Fig. 5.12). These facts have long been known, it is the size of their populations and the nature of the forests in which they lived that are the critical issues raised by Frans Vera (2000, 2002). Until very recently it was widely assumed that until large-scale human intervention started, the lowlands of central and western Europe, which have a temperate climate, would have been covered with **closed canopy forest** wherever trees could grow. This view is largely based on studies of modern abandoned fields and pastures, which in the absence of heavy grazing pressure give rise to just such a wood or forest. Once established, such closed canopy forests are maintained by a variety of mechanisms, including patch dynamics, small regeneration gaps and regeneration waves, discussed in Section 9.2 and 9.3.

Vera, however, has posed an interesting re-interpretation of the available evidence, suggesting that much of Europe was a more **open and park-like wood pasture**, not unlike that now found in much of the New Forest, southern England and, to a certain extent, in the Oostvaardersplassen Reserve in The Netherlands (Vines, 2002 provides a good review). Vera suggests that these have developed in response to the grazing of a considerable population of large herbivores, particularly aurochs, horses and bison, keeping the woodland more open. Had populations of these animals become even greater, extensive areas of grassland or heathland would have formed as a result of the **retrogressive succession** of woodland.

Vera's view is that many of the lowlands of central and western Europe and eastern North America were covered by a grazed park-like landscape containing groves of trees *c.* 7000 years ago, before the arrival of humans (Hodder *et al.*, 2005). Individual groves enlarged their boundaries at the speed with which shrubs such as blackthorn could spread by means of their underground rootstocks (around 0.1–0.5 ha in 10 years). Isolated individual shrubs such as hawthorn would also appear as a result of animal dispersal of seed, helped by reduced grazing pressure in the summer (rapidly growing grass means less extensive grazing). These prickly shrubs gave shelter which allowed the seeds of animal- or wind-dispersed trees to develop. Jays, for example, transport acorns considerable distances, bury them and often fail to return to their food stores. On poorer soils trees could grow up protected by shrubs such as juniper or bramble. Thus, open park areas would be gradually invaded by forest. The groves would be dense, producing tall unbranched trees in the centre (as found in preserved bog oaks) with more branched trees at the edge. Within the groves, mature oaks were not replaced by young shade-tolerant trees because these were destroyed by large grazing herbivores wherever a gap began to form and grasses moved in. Eventually gaps became larger as older trees were windblown or died as a result of fungal infection, and individual groves degraded into grassland. Thus the presence of a considerable population of

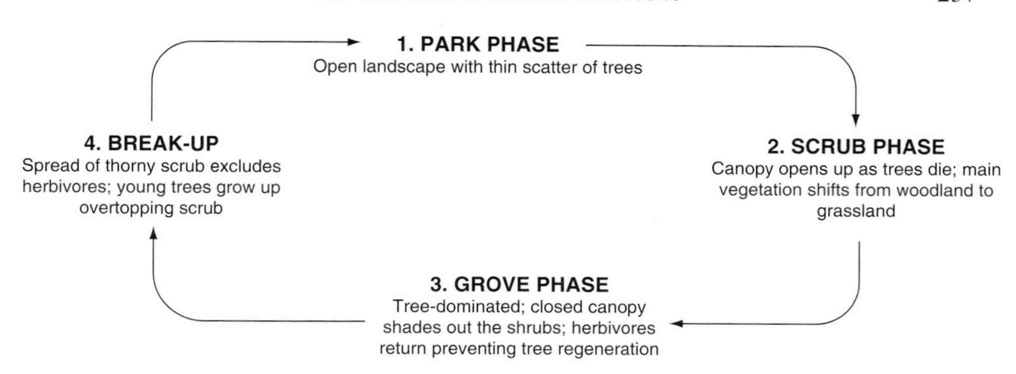


Figure 5.17 Essentials of Vera's model of primeval forest succession, together with the scrub phase added by Kirby (2003).

large grazing herbivores (probably helped by fires in dying areas) encouraged a shifting sequence: grassland – thorny shrubs – grove – grassland – thorny shrub – grove. . . (Fig. 5.17). This process, which Vera terms the **cyclical turn-over of vegetation**, may well have been important in the long history of many continental European forests. The case, however, must not be pressed further than the evidence warrants. Closed canopy forests do exist and are discussed later (Section 9.4).

Kirby (2003) suggests that most grove trees would be expected to live for 300–400 years but this does not preclude older veteran trees (> 450 years old) so typical of the British landscape. Kirby also suggests that some grove trees would survive as solitary trees until the young trees of the next grove phase over-topped the veterans and shaded them out.

Evidence bearing upon this problem is drawn from the pollen records of undisturbed prehistoric forests, population trends in forest reserves studied over a long period, and documentary evidence which in the case of forest charters of the Merovingian and Frankish kings goes back to the seventh century. Documentary evidence, particularly that provided by Forest laws and regulations in the various European countries, has to be treated with considerable caution as the meaning of various familiar words has changed with time, and the reasons for some prohibitions demand considerable under-standing. Pasture, for example, is now thought of as consisting of grassland affording food for cattle and sheep, whereas in the Middle Ages the concept applied to anywhere animals could obtain food, including such trees as the oak, wild apple, wild pear and wild cherry.

Vera's ideas require a re-thinking of the evidence which has been previously interpreted as showing a dense forest. His view is that the open parkland explains why hazel, pedunculate oak and sessile oak (and other light-demanding species) have been well represented in pollen records for thousands of years, along with that of shade-tolerant species such as limes, elms, ash, common beech and hornbeam. In closed-canopy forests and forest reserves where large gaps are not present, oaks tend gradually to diminish because their seedlings, unlike those of the shade-tolerant trees, cannot grow at the low light levels present in the limited gaps which do form. He also contends that a partial explanation for the very high proportion of tree pollen dating from this period is that grazing may have been so efficient that production of grass pollen per unit area was greatly reduced. Svenning (2002) counters this by pointing out in a review of north-west Europe that in many studies non-tree pollen correlates well with other measures of openness such as beetle, snail and plant macrofossils and concludes that forested conditions were the norm with open vegetation being restricted to floodplains or poor soils (sandy or calcareous) and in the continental interior of north-west Europe. It is also possible that the frequent pollen from light-demanding species may have come from relatively few trees growing in a limited number of openings in an otherwise dense forest. Some of the debate is hampered by just what a landscape driven by large herbivores would have looked like. Was it open savanna with less than 30% cover of trees or mainly a wooded landscape (> 70% tree cover) but with larger, shifting glades?

An insight into a developing situation of this type can be gained by measuring the trunk diameters of the main tree species present. When results of this kind were plotted in 10-cm diameter classes for trees at Dalby Soderskog in Sweden, a very clear picture emerged (Fig. 5.18). Here the number of small trees of pedunculate oak was very low; together with the high proportion of large trees, this indicated an increasing recruitment failure. In closed-canopy sites these light-demanding species are being replaced by shade-tolerant trees including ash, beech, elm, lime and sycamore. In such situations these species have a high proportion of smaller (and younger) individuals, so the curve of their diameter classes forms a reversed J, indicating successful regeneration.

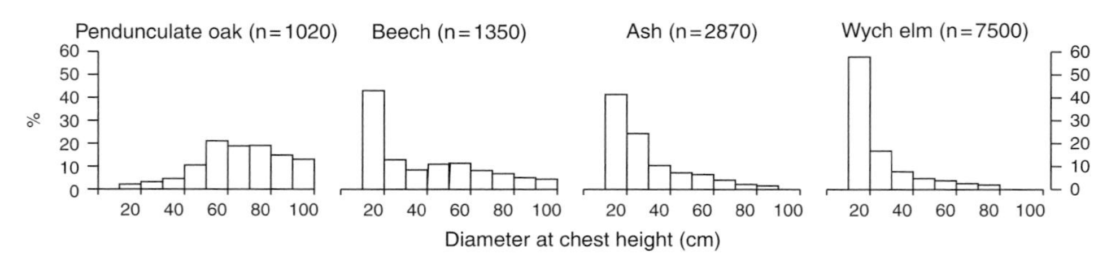


Figure 5.18 Percentage distribution into diameter categories for the main tree species growing at Dalby Soderskog, Sweden. Only trees with a trunk diameter of > 10 cm at chest height are included. (Redrawn from Malmer *et al.*, 1978. From Vera, 2002. *Arboricultural Journal* 26.)

Vera's ideas raise many interesting questions and may help explain a number of features of European woodlands. For example, does the existence of large open spaces explain why green woodpeckers *Picus viridis* feed on the ground, favouring ants? Or why there are so many common bird groups that feed on open-ground insects, such as starlings *Sturnus vulgaris*, wagtails *Montacilla* spp., pipits *Anthus* spp., and many corvids like crows and magpies (P. R. Hobson pers. comm.).

How Vera's ideas apply in the British Isles has been discussed in detail by Rackham (2003), Kirby (2003) and Hodder *et al.* (2005), who welcome many of Vera's insights though questioning a number of important points. In addition to the openness of the forest, a second area of debate is whether the openness here could have been driven solely by large herbivores. Conditions in the British Isles were markedly different from the continent, in that the number of large animals surviving the last glaciation was very restricted. Native horses and bison have not been present in this interglacial. Elk and bear probably died out in the Mesolithic (Middle Stone Age) and the auroch, which lived more in fens than forests, in the Bronze Age. Even the remaining red and roe deer, wild boar and beaver were uncommon by historic times. England was also more densely populated than continental Europe and more than half the original wildwood had gone by the Iron Age or even earlier. Oak may not always behave as Vera supposes, indeed in some English woods from which deer were largely excluded, occasional thickets of sessile oak resulted from the great mast year of 1976 and similar events. It is suggested that in the UK abiotic disturbance such as fire and windthrow may have had important roles in maintaining open spaces (Hodder *et al.*, 2005). Certainly the open landscape in the New Forest is currently maintained by such means as heather burning, and the numbers of grazing animals are managed. Hodder and Bullock sum up the debate as follows:

> The difference between the closed forest hypothesis and Vera's alternative of cyclical dynamics may be a matter of degree. While there is general agreement that the *original-natural* forest [in the UK] may have been more open than was previously thought, this is not equivalent to saying that a wood-pasture landscape would necessarily dominate the landscape. The balance of opinion is towards predominance of closed forest with localised, longer lasting openings.
>
> (*Hodder* et al.*, 2005, p. 31*)

On the whole it would seem that the British Isles landscape probably consisted of large tracts of wooded pasture with groves covering expansive areas. Rivers and floodplains would have acted as corridors for the large herbivores which migrated from one feeding ground to the other. This would certainly help explain the persistence of plants restricted to old woodlands (the British **Ancient Woodland indicators**) that are estimated to require many hundreds of years to invade isolated woodlands. These species and their dynamics do not fit a landscape made up of shifting, patchy groves in a sea of grass and scrub. To envisage genetically viable metapopulations of woodland plants in discontinuous, patchy landscapes would be almost impossible, particularly for species such as toothwort *Lathraea squamaria*. It is possible, however, that these plants have previously persisted in small woods on steep ground or islands inaccessible to large grazing animals.

Mitchell (2005) discusses the openness of European primeval forests using palaeoecological evidence from Ireland where only two native large herbivores, the wild boar and the red deer, were present in the early Holocene. This was in marked contrast with England and Scotland, which at that time also possessed roe deer, elk, reindeer, horses, aurochs, and beaver, and continental north-west Europe where bison and fallow deer were also present. This gives a quite exceptional opportunity of testing Vera's view that many large herbivores were required to create forest gaps and an open park-like landscape in which the high abundance of oak *Quercus* and hazel *Corylus* pollen found in European pollen diagrams could develop. The comparison is conclusive: the similarity between data sets from Ireland and continental Europe is so close as to make it clear that large herbivores were not required to maintain large oak and hazel populations in the primeval landscape. Mitchell goes on to demonstrate that percentage tree pollen is a reliable indicator of canopy openness in both Europe and eastern USA, where the palaeoecological data demonstrate that open canopy forest has only ever been maintained by human exploitation.

Rackham inclines to the view that in cases where beech is not involved, the landscape was less dynamic than Vera supposes. Beech was in any case mainly found in south-east England and concentrated in wood pastures in historic times. Moreover, topography and substrate conditions play a major role in the long-continued distribution of trees and grassland the world over. In surviving wood pastures the woods tend to be on hilltops and the pastures in the valleys, while individual trees favour rocky places, the shelter of thorny shrubs and grassland soil.

Vera's ideas do not answer all our questions about how woodlands used to look, and indeed have resulted in even more that still need to be answered. Whatever the eventual outcome, it is this type of innovative thinking that helps ecological science to develop.

# 6
# Biodiversity in woodlands

## 6.1 Genetic variation in populations and its implications

Genetic variations exist in virtually all forest organisms; they are particularly important in tree species and in the pests and pathogens that attack them. The genetic basis of forest pathology and its influence right up to the landscape level was the basis of the 1999 Montreal symposium of the American Phytopathological Society. Much of what was discussed then has been updated (Lundquist and Hamelin, 2005) in a volume that emphasizes how rapidly long-lived trees and forests can be devastated by pathogens and microscopic organisms such as fungi, bacteria, phytoplasmas and viruses. Pathogens have important effects on biodiversity, greatly influencing plant populations of both natural forests and plantations. The numbers of important indigenous species may be greatly reduced; some may be eliminated altogether. On the other hand pathogens themselves contribute to the biological diversity of the ecosystems they inhabit.

Severe devastation of the cypress *Cupressus lusitanica* by the cypress aphid *Cinaria cupressi* in East Africa appears to have been facilitated by the **narrow genetic base** of the planting stock. Similarly, Dutch elm disease in the UK (Section 5.4.5) was particularly bad because many of the 25 million trees killed were English elm *Ulmus procera* that all derived from a single clone brought to the UK from Italy by the Romans 2000 years ago for use in supporting and training grape vines (Gil *et al.*, 2004). Considerable care is taken to provide an adequate genetic base in forests of radiata pine *Pinus radiata* in New Zealand where particular clones are often employed. Trees of the same clone are genetically identical, those of different clones may vary quite widely in features such as growth and immunity to pests and diseases. Though these forests are monocultural they are not usually monoclonal; it has been argued that such forests should contain at least 7–25 genetically unique clones or else seedlings from seed lots with more than 16 parents.

The fact that individuals of a **self-incompatible species** such as the primrose *Primula vulgaris* are unable to breed with other genetically identical individuals has the advantage of ensuring that reasonably large populations remain genetically diverse. This perennial herb of moist, shaded habitats has a wide North Atlantic and Mediterranean distribution; most English individuals grow in woodlands but primroses can be quite abundant in hedges and old grasslands protected from drought. The distylous nature of this species, whose flowers have either short or long styles, is controlled by a super-gene. Plants with pin-eyed flowers are double recessive (ss) and those with thrum-eyed flowers are heterozygous (Ss). In thrum-eyed flowers the stigma is half-way up the centre of the perianth tube and the stamens are at the top in the eye of the flower, while in pin-eyed flowers the reverse is true.

Fewer alleles (alternative forms of a gene) can be physically contained in smaller populations, and some even of these can be progressively lost by small random changes in gene frequency (called **genetic drift**) leading to even lower genetic variation in the population. The effects of this on small populations can be severe. Habitat fragmentation and the resulting decline in population size have adversely influenced the reproductive success of primrose in Belgium, where it is a rare and declining species. Brys *et al.* (2004) examined 16 populations near Bruges and found that population size strongly affected reproductive success; plants from small populations produced significantly fewer fruits per plant and seeds per fruit.

## 6.2 Selection pressures and biodiversity

### *6.2.1 Influence of herbivores*

The size and vigour of particular herbivore populations at a given time is a major influence on forest trees and other plants. Many examples could be chosen to illustrate this point; here we consider deer and elephants.

Deer have increased in number and expanded their range in many parts of the world over the last few decades, becoming a particular problem (Section 5.7.1). This is partly due to our previous management. Historically game managers strove to augment and protect deer populations, and hunters learned to limit takes and favour bucks. Today, such precepts are outmoded, but unlearning old lessons and reversing this cultural momentum has proved difficult (Côté *et al.*, 2004). The result in many places is **overbrowsing** with consequent reductions in plant cover and implications for carbon storage and nutrient cycling. As importantly, selective browsing by deer can greatly influence the plant species composition of forests with consequences for animals needing different species, and affecting the species composition of young trees growing up to form the future canopy.

Elephants exert a strong pressure on the ecosystems in which they live, as is evident in the Chobe National Park, northern Botswana, which together with adjacent parts of Zimbabwe and Namibia, is home to more than 100 000 animals, the largest population of elephants in the world. Every day during the dry period they have to visit the Chobe river. In the wet season they spread out and disperse southwards drinking from temporary water pans. They browse heavily on the trees and shrubs of the river forest and also break down trees, few of which regenerate. Studies of the many ecological effects and history of this increasing population of **megaherbivores** are of great importance: indeed Skarpe *et al.* (2004) conclude that while the present population level is ecologically acceptable it may be necessary for social and economic reasons to cull elephants which persistently invade areas used for agriculture and housing.

A hundred years ago elephants here were rare due to excessive hunting, primarily for ivory. Moreover, rinderpest, a disease spread by cattle that had come to Africa in 1887, reached Botswana in 1896–1897 causing heavy mortality in both wild and domestic ungulates (hoofed animals). After restrictions were imposed on hunting, elephant numbers began to increase during the 1930s. There is now concern at the loss of scenic acacia woodlands along the riverfront. Many of these huge old acacia trees probably started life around a hundred years ago when populations of elephant and several other browsers were low due to the combination of hunting and sickness already mentioned. These trees are known to be unable to regenerate when impala densities are high. However, browsing by large herbivores is obviously not the only ecological factor at play: trees such as *Croton megalobotrys* and *Combretum mossambicense* have increased in density in recent years while *Faidherba* (formerly *Acacia*) *albida* and *Garcinia livingstonei* have decreased. Smaller and middle-sized browsers/grazers, particularly impala *Aepyceros melampus*, also consume tree seedlings and saplings, especially at sites close to the river. Grazing by ungulates leads to dominance of either fast-growing palatable species, which enhance the rate of nutrient cycling, or of slow-growing unpalatable species – often with physical or chemical defences – which diminish it. Small mammal populations varied greatly over short distances, being influenced by trampling, digging, defecation and urination by large herbivores.

Studies of many other facets of this ecosystem, including that of the lion population – several of which were radio-collared – are of considerable interest. In one 4-year study an incoming pride male killed the cubs of some of the female groups it took over; it was the only immigrant lion observed.

### 6.2.2 Beak size, guilds and resource partitioning

The evolution of herbivores is strongly influenced by the food available to them, a point well illustrated by comparing relative beak sizes of the birds of temperate forests with those of tropical forests, whose beaks are considerably larger. The diversity of tropical birds is also larger, a fact related to the greater diversity of foods available, particularly of large insects (Terborgh, 1992). Species utilizing the same pool of resources are referred to as a **guild**. These are often (but not always) made up of closely related species, such as the parrots in the Amazonian forests that form a guild of arboreal seed eaters. The smallest of these parrots is the size of a sparrow – the green-rumped parrotlet (*Forpus passerinus*), with the rather larger parakeets, the full-sized parrots, and finally the majestic macaws, following in a sequence of birds able to deal with successively larger seeds. The diminutive parakeets feed on fig seeds, discarding the pulp, whereas the beaks of the larger birds enable them to get at resources, such as hard nuts, that are much better protected. The evolution of these birds seems to have been influenced by Hutchinson's 'law of limiting similarity', that implies that two species cannot co-exist and utilize a common pool of resources unless they differ by some significant degree.

The packing of species in tropical forest guilds, with many species with only slightly different ecological requirements, is really quite remarkable. Such guilds routinely contain twice as many species, often many more, than in temperate forests. The neotropical **antwrens** (*Myrmotherula* species), which are basically similar to the warblers of the temperate zone, provide a good example of how specialization within the towering height and structural complexity of the tropical rain forest has led to speciation. As many as ten species of antwren can be found in a single locality. Similar in morphology and superficial appearance, all of them feed on insects but search for them in different ways. In the Amazonian forest the foraging zones of *Myrmotherula* species can be arranged vertically above each other with relatively little overlap; *M. haematonata, M. axillaris, M. menetriesii* and *M. brachyara* occupy successively higher zones, finally reaching the tops of the emergent trees.

### 6.2.3 Evolution, diversity and activities of insects

The insects are by far the most diverse group of animals that have ever existed, accounting for well over half of all the species of living things so far named and described (Fig. 6.1). The next largest group of living organisms is that of the vascular plants, which Grimaldi and Engel (2005) list as amounting to 248 400 species. Although this is certainly an underestimate (there are at least 400 000 species in the angiosperms alone), there are far fewer vascular plants than insects. Insect variation and evolutionary history over the past 400 million years are considered in detail by Grimaldi and Engel (2005), who trace their development and relationships with other organisms, many of which are components of forest ecosystems.

Anatomically, insects are divided into three parts, the head, thorax and abdomen, and have six legs. The greatest number of extant insects belong to the Holometabola (83%; Fig. 6.1 – insects that have distinct larval and adult stages). Of these by far the largest group is that of the beetles (Coleoptera)

Figure 6.1 Proportions of species in present-day insect groups. See the text for an explanation of the insect groups. The total number of named insect species now existing is believed to be over 926 000. (From Grimaldi and Engel, 2005. *Evolution of the Insects*. Cambridge University Press.)

whose forewings are modified to form protective elytra that meet along the mid-dorsal line and protect the membranous hind wings which fold beneath them. They have biting mouth parts and many of them, especially the death-watch beetle *Xestobium rufovillosum*, cause considerable damage to timber. The weevils (Curculionoidea), of which there are 11 families, are by far the largest major group of Coleoptera. One species, the spruce bark beetle *Ips typographus*, has killed millions of Norway spruce *Picea abies* trees. Next numerous are the Lepidoptera with less than half the number of species. Many of these four-winged moths and butterflies are extremely beautiful, but their larval stages often cause severe damage to plant populations. The hornets, wasps, bees, and ants of the Hymenoptera follow with 13% of species, while the two-winged flies of the Diptera have 12%. Various members of the Hemiptera (true bugs), whose mouth parts are used for piercing and sucking, attack both flowering plants and animals to obtain either sap or animal juices including blood and in so doing create entry paths for disease organisms. They form more than half of the remaining species, which account for 17% of the grand total. Amongst the remaining small groups are the wingless hexapods including the springtails and silverfish of the Collembola, the mayflies and dragonflies of the Paleoptera, the grasshoppers and crickets of the Orthoptera, the stoneflies of the Plecoptera, and the termites (Isoptera). The Paraneoptera include not only the bugs, but also thrips (Thysanoptera) and book-lice (Psocoptera). See Box 7.6 concerning the activities of colonial insects in attacking undecayed wood.

Jones and Elliot (1986) provide an excellent description of pests, diseases and ailments of Australian plants, together with a diagram (their page 100) showing the regions of trees affected by pests in the tropics. The importance of insects in causing and transmitting disease, as pests of trees particularly in the tropics, in defoliating trees, and in pollinating the flowers of a high proportion of the flowering plants (angiosperms) with which they co-evolved, is outlined below:

(a) Insects transmit fungal and viral diseases of herbs, shrubs and trees. They and other arthropods of medical importance also cause or transmit many important diseases of humans and other animals (Grundy, 1981). The two-winged slender flies of the mosquito subfamily (Culicini) that spread malaria, yellow fever, dengue fever and filiariasis, are amongst the greatest dangers of tropical forests.
(b) Insects attack many trees, particularly in the tropics (Speight and Wylie, 2001). The adverse influence of *Hypsipyla* shoot borers belonging to the Lepidoptera on the plantation growth of valuable timber trees in the Meliaceae is described in Section 10.6. Attacks by the sap-feeding yellow scale *Aonidiella orientalis* on the exotic neem tree *Azadiracta indica* growing in the Lake Chad region of sub-Saharan Africa were countered by use of residual contact insecticide, employing

older transplants for re-establishment and using previously uncultivated sites. These measures greatly reduced losses.

(c) The role of insects in defoliating trees is considered in Section 5.3. Defoliation of pines in Vietnam and China by the pine caterpillar *Dendrolimus punctatus* is often very severe, being aggravated by the fact that this Lepidopteran can have up to four generations a year and that each female lays an average of 300–400 eggs with a possible maximum of 800. Other species of *Dendrolimus* are also involved and every year about 3 million ha of forest are infested. Records of these destructive attacks on Masson pine *Pinus massoniana* and also slash pine *P. elliottii* and other pine species go back to 1530 AD. On a recent occasion almost complete defoliation was followed by the death of almost 25% of the trees and the volume growth of the survivors was reduced to 31% of normal.

(d) Pollination by insects is far less wasteful than that by wind, particularly when individuals of a species are far apart (see Section 4.2.1). In return the pollinators are rewarded by nectar and often use a small proportion of the pollen, so this is a very effective symbiosis.

## 6.3 Biodiversity at organism, population and habitat levels

### *6.3.1 The nature of biodiversity*

**Biodiversity** can be defined as the variety and abundance of species, their genetic composition, and the communities, ecosystems and landscapes in which they occur (Maclaren, 1996). The term is sometimes used very narrowly to refer to species-richness, but should be employed to describe the full complexity of life on Earth. Such descriptions still have far to go; as Box 7.1 points out, some 200 000 fungi have so far been described but these represent only a fraction of those believed to exist. Localized species-richness at a particular place is called **alpha diversity**, but the biotic community often changes in a traverse of the landscape as soil, slope or disturbance such as fire changes, frequently creating locally different habitats within a forest: **beta diversity** measures the extent of such change along a gradient and can be thought of as the diversity within a landscape. **Gamma diversity** is similar to alpha diversity but is a measure of species richness across a range of habitats within a larger geographical area and is often used to show regional diversity, which may include forests and other types of vegetation. **Temporal diversity** refers to the change of species over time. In practice alpha diversity is often calculated as a diversity index (such as the Simpson Index or the Shannon–Weiner Index), incorporating the number of species within an area and the evenness of spread of individuals across those species. Whittaker *et al.* (2001) gives a good review of the subject. Kimmins (1992) found that many temperate forests have low alpha but high beta diversity, while the reverse is true for many tropical rain forests, although seasonal rain forests can have both high alpha and beta diversity (Zanne and Chapman, 2005). Thus even within forests biodiversity varies greatly in different parts of the world.

Biodiversity within particular populations and habitats may be very important for their continued success and survival in the face of both competition and disease. A species or population showing a considerable range of genetic makeup may well have at least some strains able to resist incoming disease, while others may respond better to particular competitive or habitat changes. Such genetic variation within a species and the continuing evolution within particular populations can eventually culminate in forms so different that they are recognized as new and different species. The value of genetic variation is discussed by Booth and Grime (2003) in the introduction to a paper on the effects of genetic impoverishment on plant community diversity. They point particularly to comments by Harper (1977, p. 707) that 'Diversity of a plant community is inadequately described by the number and abundance of the species within it' and that 'a major part of the community diversity exists at the intraspecific level' (i.e. within a species). Their own experiments were concerned with 11 long-lived herbaceous species growing on an area of ancient limestone pasture in North Derbyshire. These were used to create a number of model communities identical in species composition but widely contrasted in genetic diversity. These communities were allowed to develop in microcosms containing natural rendzina soil (see Section 2.2.1) and exposed to a standardized regime of simulated grazing and trampling. Though a gradual loss of *species* diversity occurred in all three treatments employed it was, at the end of the 5-year experiment, highest in the most *genetically* diverse communities, a result of significance to all ecosystems including woodlands. Evidence of the type shown here and in Section 6.1 demonstrates that while we should be concerned over the loss of species, we should also be highly concerned over the more insidious and often largely unnoticed loss of genetic diversity.

### 6.3.2 Geographical patterns of old and young species in African forests

Given the high biodiversity of the world's forests (an estimated 90% of all terrestrial species) and the fact that over 8000 tree species are thought to be threatened with extinction (see Section 1.1.1), plus the role forests play in human welfare (Section 1.2), forests are obviously an important priority for conservation. However, forests are more than mere repositories of species; they can also be an important source of new species, vital to the evolutionary development of forests in the future. In their consideration of this problem

Figure 6.2 Map of central Africa showing geographical variation in the ratio between numbers of species in young and older groups of bird species. Slanting lines interrupting the signature indicate that the number of species is too low for a reliable calculation. (From Fjeldså and Lovett, 1997. *Biodiversity and Conservation* 6, Fig. 3. With kind permission of Springer Science and Business Media.)

Fjeldså and Lovett (1997) concentrated on the evolution of both birds and plants, which has been studied by many workers over a wide area. DNA evidence has been particularly valuable in determining when different species first evolved. There was, for example, a speciation burst in the bushshrikes (Malaconotinae) in the Miocene (5–24 Ma), and an explosive radiation of starlings (Sturnidae) in the Plio-Pleistocene (1–4 Ma). Figure 6.2 illustrates avian results in part of the area covered. Close examination of montane forest habitats in East Africa shows that they have high concentrations of species with restricted distributions. Peak concentrations of **neoendemics**, recently evolved species of local origin, occur in the same mountains as clusters of distinctive old species with relictual distributions.

Africa, in common with the rest of the world, has experienced many changes in its climate over the years, and an early theory was that speciation occurred as a result of temporary fragmentation of the main rain-forest blocks in Pleistocene forest refuges. The combined results of many workers in diverse fields have led to a re-interpretation of the spatio-evolutionary pattern of forest-adapted species in tropical Africa. The current view is that there are a number of montane areas of particular importance as evolutionary centres (or 'evolutionary fronts') from which large numbers of young species have migrated (Fig. 6.3). Vertical shading on the main diagram indicates 'museum areas' of lowland forest where species of potentially diverse origins accumulated

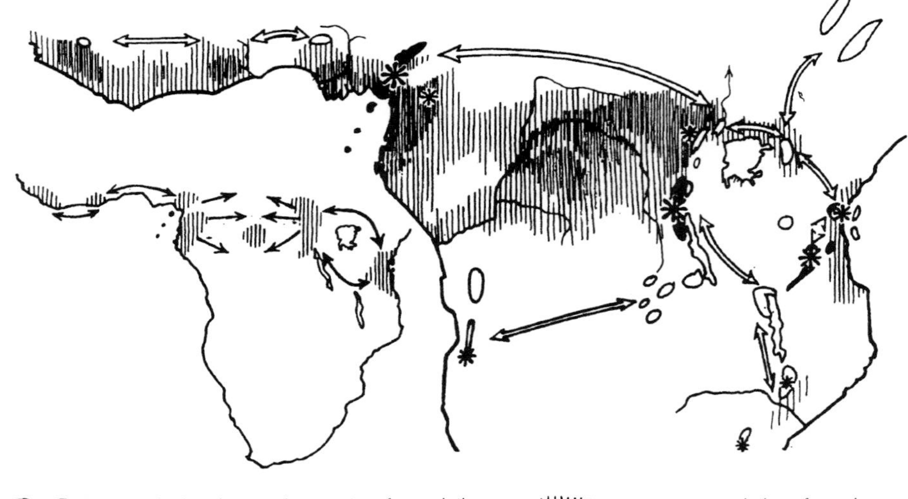


Figure 6.3 A re-interpretation of the spatio-evolutionary pattern of forest-adapted species in tropical Africa, with a smaller diagram illustrating the traditional interpretation of speciation on the left. (From Fjeldså and Lovett, 1997. *Biodiversity and Conservation* 6, Fig. 6. With kind permission of Springer Science and Business Media.)

during the upper Tertiary (1.8–23 Ma), with little speciation in the Pleistocene (the last 1.8 million years), and with species-richness patterns that may reflect rainfall and habitat heterogeneity. The asterisks show places where the existence of several relict species indicates long-term diversity. In contrast, black areas are centres of endemic species representing strong radiations in the Plio-Pleistocene; open areas represent less important centres. This illustrates the vital importance of these lowland African forests for the conservation of biodiversity, above and beyond just the number of species they contain.

### *6.3.3 Rediscovery of lost species: The Wollemi pine and the Gondwana bequest*

The excitement generated by the discovery of a modern representative of a group long thought to be extinct – a 'living fossil' as Charles Darwin put it – is always considerable. Perhaps the most striking in the twentieth century was that in 1938 of the coelocanth fish *Latimeria chalumnae* which had four limb-like fins on its underside and most striking resemblances to fossil forms from Devonian rocks at least 380 million years old. To botanists the discovery in

1945 of living trees of the deciduous dawn redwood *Metasequoia glyptostroboides* in the Hupeh and Szechuan provinces of China was just as important. *Metasequoia*, of which 10 fossil species are now known, had been described from fossil material in 1941 and was widespread in the northern hemisphere in Cretaceous times more than 65 million years ago. The insights which the living dawn redwoods gave into the evolutionary development of the evergreen giant sequoia *Sequoiadendron giganteum* and the coastal redwood *Sequoia sempervirens* were particularly important.

In 1994 came the discovery by David Noble of what is now called the Wollemi pine *Wollemia nobilis*, in the desolate canyons of Wollemi National Park in the Blue Mountains about 100 km north-west of Sydney, Australia (Woodford, 2002). The 'canyon band' is some 6 km wide and stretches for nearly 200 km; some of its estimated 500 canyons have never been visited by non-indigenous people and the danger of being caught in rising waters caused by torrential rains is such that the name wollemi is believed to be derived from the Darkinjung word *wollumnii* – aborigine for 'watch your step'. Desolate though these deep water-carved sandstone gorges are, it is almost certainly the exceptionally moist nature of their microclimate and isolation which have led to the survival of these unusual trees.

Noble was leading a group through an area new to him when he suddenly came upon some very tall strange trees growing in a slightly more open area. Their apices were well above the rest of the warm temperate rain forest dominated by coachwood *Ceratopetalum apetalum* and sassafras *Doryophora sassafras*. The stiff yellow-green adult leaves were unusually arranged in four rows along the tops of the branches. In contrast, the soft, green juvenile leaves found on the topmost branches were arranged in two rows along the branches and were dark green on the upper surfaces but had white waxy undersurfaces. The mature leaf outlines had a distinct resemblance to those of *Agathis jurassica* which had flourished over 150 million years before. Even the bark was unusual in having a bubbly surface which, with its subtle gradation of colour, has been described as resembling a settled swarm of bees, quite unlike any existing member of the Araucariaceae, in which family it belonged as subsequent collection of reproductive material made clear. Both male and female cones are borne high up at the ends of very long thin lateral branches of the same tree. Seed cones were first collected by a forest ranger suspended at the end of a cable beneath a helicopter. They are now caught in fine mesh nets below the canopy and many plants have been raised from seed (Fig. 6.4) while others have been grown from cuttings. In its natural environment it appears that dense shade and competition with other plants make development of a seedling into a tree most unlikely, though seedlings grow well and rapidly in cultivation.

Figure 6.4 Seedling of Wollemi pine *Wollemia nobilis* growing on the forest floor amidst leaf debris of coachwood and sassafras in Australia. Note the two cotyledons and double row of juvenile leaves. Very few seedlings survive the low light levels and generally unfavourable conditions on the floors of their native canyons. (Drawn by John R. Packham.)

If this new species is in the Araucariaceae why does it deserve to be put into a new genus? The answer to this becomes apparent when the spiny female cones and their cone scales are examined. In *Agathis*, the exterior of whose female cone is smooth, the ovule rests on top of the scale and when the mature seeds fall they spin like rotors, because each of them has a wing on one side. In contrast the ovule of *Araucaria* is embedded in the scale and the seed is not winged, while the cone scales are sharply pointed. The Wollemi pine differed from both in that each seed was winged all round and rested on a pointed cone scale.

DNA studies by Hill showed that in the long-distant past *Araucaria* and *Agathis* diverged from a common ancestor, and that *Wollemia* branched off the *Agathis* line later still. Moreover, genetic constitutions of the wild Wollemi pines turned out to be remarkably uniform. When Peakall (1998) plotted the DNA peaks forming a considerable proportion of the complete **genome** – the total set of the genetic information within an organism – he was amazed to find that the DNA peaks from different trees exactly overlapped each other. This was true even if one of the two trees came from stand one and the other from stand two which, although they are only 3 km apart, are effectively completely isolated genetically. DNA peaks of the seedlings were also similar. This is a very unusual situation, especially as low genetic diversity normally implies that populations are especially vulnerable to infection. In contrast, plots of DNA peaks in various species of both *Araucaria* and *Agathis* showed that different trees of the same species had marked differences. This is not surprising: for example trees of the Moreton Bay or Hoop pine *Araucaria cunninghamii*, which with the Monkey Puzzle is an important source of timber, have continued to evolve against different environmental backgrounds in Queensland and New South Wales.

It turned out that the distinctive form of the Wollemi pine pollen had been seen many times before in geological borings. The presence of these tiny fossil pollen grains, which had received the name *Dilwynites*, showed that *Wollemia* and/or its close relatives had existed in the Cretaceous at least 91 million years ago and that its distribution had at one time extended to both the central Australian Desert and Antarctica. There are, moreover, two types of this fossil pollen grain, *D. granulatus* (whose pollen is almost identical to *Wollemia*) and *D. tuberculatus*; proof that the Wollemi pine was once part of a bigger and more diverse group of trees. It was also much more widely distributed within Australia; in 1986 cores from rocks 50 million years old in the central Australian desert yielded fossil pollen of both *Dilwynites* and of *Fischerpollis halensis*, a possible species of the Venus flytrap, the modern version of which (*Dionaea muscipula*) is now restricted to south-east USA. At that time annual rainfall in the area may have been as high as 1800 mm, easily enough for the Wollemi pine. Subsequently *Fischerpollis* pollen has also been found in the states of Victoria and south Australia.

The Araucariaceae first appear in the fossil record around 245 million years ago, growing in the northern hemisphere in the Triassic period. By the beginning of the Jurassic (205 Ma) its representatives were present in Australia as well as the northern hemisphere. The cosmic catastrophe which probably wiped out the dinosaurs at the end of the Cretaceous, 65 million years ago, seems also to have eliminated the Araucariacean forests of the northern hemisphere apart from a few minor populations in south-east Asia, leaving those of the southern hemisphere as a Gondwanaland survival.

Jones *et al.* (1995) played a major role in investigating and elucidating the nature of these strange ancient Wollemi pines, some of which live to an age of at least 400 years. The slender trees of today, the tallest of which currently reaches 38 m (125 ft), are in fact difficult to age as they throw up numerous basal shoots so producing a natural 'coppice' of shoots of varying age, a most unusual feature in a conifer. One fallen trunk has been dated at 350 years old, but as older trunks are gradually replaced with younger ones, the individual rootstock from which it came may have been much older. The Royal Botanic Gardens, Sydney, has subsequently assisted in the protection and conservation of the only three groups of Wollemi pines – with 23, 17 and 3 trees respectively – to be discovered. This species is of major value to conifer collectors and a major commercial launch was made in 2005/6 when many thousands of young trees were made available for sale. It has so far proved tolerant of temperatures down to $-5$ °C.

As mentioned above, the Araucariaceae as a whole form an important part of the flora contributed by Gondwanaland (see Section 2.5.2); *Araucaria* and *Agathis* are both of major value in forestry and as specimen trees. Some of the 18 species of *Araucaria*, whose native trees are now confined to the southern hemisphere and found in South America, Australasia and the islands of the South Pacific, have been very widely planted. In this genus male and female cones are usually on separate trees but are sometimes found on separate branches of the same tree. The rather bizarre monkey puzzle *A. araucana* with its sharply pointed leaves is found in many public parks the world over (Hora, 1981).

The Bunya pine *Araucaria bildwilli* has its present centre of population in the Bunya Mountains of south-east Queensland, but was far more widespread in the Jurassic (135–205 Ma) when it extended into the northern hemisphere. These trees were at one time widely planted in Australian parks but their seed-cones, which can reach 40 cm in diameter and weigh over 11 kg, proved to be a major health hazard. They fall suddenly and with an impact sufficient to have killed a dinosaur in former times! The seeds are an important article of diet for the aborigines. Tall and imposing Norfolk Island Pines *A. heterophylla* are commonly planted in parks and along seafronts. They are also very attractive when small and are often grown as pot plants.

*Agathis*, the most tropical of all the conifer genera, has 15–20 species that are found in the wettest tropical rain forests of the Malay archipelago, Sumatra, the Philippines and Fiji, with outliers in the subtropical forests of Queensland and northernmost New Zealand. Mature kauris are often very large – in New Zealand *A. australis* reaches a height of over 30 m with a trunk diameter greater than 3 m (Salmon, 1991) – and their timbers are amongst the most valuable softwoods in the world. That of most species is strong and remarkably free of knots because the trunk sheds its lower branches as it grows upwards. Fast-growing plantation forests of *A. dammara* in Java yield much Kauri timber. Though care is now taken to conserve them, the *A. australis* Kauri stands in North Island, New Zealand, were initially over-exploited both for their timber and the resin (kauri gum) which the trees of this and other species yield spontaneously and around sites of injuries. Fossil gum in peat bogs where the tree no longer grows is particularly prized; both forms are used as a basis for varnishes, linoleum and paints. The so-called Kauri of the East Indies *A. brownii* is one of many introduced trees growing in Madeira.

The Wollemi pine and its relatives are a mere fraction of the Gondwanaland bequest, indeed the long period that Australia – the driest of the continents – has been isolated has led to a most distinctive flora and fauna. Its climate varies immensely, from arid desert to rain forest and swamp, from baking heat to cold mountains. Of its flowering plant flora of some 20 000 species, 85% are endemic (found only in Australia), as are 500 (roughly a third) of the genera. Daytime temperatures can fluctuate by as much as 40 °C, and high evaporation rates have led to the development of scrub, **mallee** (small multi-stemmed trees) and open **sclerophyll** (leathery-leaved) woodlands often dominated by gum trees *Eucalyptus* spp. (Bopp, 2005). Cronin (1987, 1989) provides good illustrated accounts of some typical wild flowers, and of the palms, ferns and allied groups. Some species of the tree fern genera *Dicksonia* and *Cyathea* reach considerable heights in the understorey of eucalypt forests. Australian palms vary widely in their size, properties and appearance. Some have a crownshaft embracing the leaf bases at the top of the tall stem, others do not, while the leaves themselves may be pinnate or fan shaped. The slender stems of climbing palms, such as 'vicious hairy Mary' *Calamus radicalis*, are equipped with hooked prickles that enable them to climb over surrounding trees. The northern kentia palm *Gronophyllum ramsayi* of the open forests and adjacent islands in the far north of Australia has a trunk up to 35 m tall and 30 cm in diameter. Very many Australian plant communities are adapted to fire, and the grass tree *Xanthorrhoea australis* which commonly occurs in eucalypt forests, is often known as blackboy because of the blackened appearance of its trunk, from the top of which fresh leaves develop when the rains come. It is not a true tree since its 'stems' consist not of wood but of tightly packed leaf bases.

### 6.3.4 The Brazil nut: edible product of a complex ecosystem

In many ecosystems certain species can be lost without great impact on the way the ecosystem works. This is because several species may fulfil the same functional role (such as pollinators or food for carnivores) so if one is lost this **functional redundancy** allows the other species to fill the gap (see also Section 9.5). On the other side of the coin are **keystone species** which play a key role in an ecosystem, the loss of which affects many other species and can lead to severe loss or change in diversity. Often, however, the inter-relationships between species can be much more complex, forming a web where each relies on the other so that all are essential and no one species is the sole keystone. This is beautifully illustrated in forests by the Brazil nut tree *Bertholletia excelsa*. One of the tallest and most magnificent trees in the Amazonian rain forest, it produces its capsules – each of which is slightly larger than a cricket ball and contains 10–25 seeds (the nuts) – at the top of the canopy (Fig. 6.5a). These capsules are often 50 m above the ground so in January and February the nut gatherers are at some risk from the falling fruits as they compete with

Figure 6.5 (a) Fruit of Brazil nut tree *Bertholletia excelsa* sawn through to reveal the seeds within. Besides being common in the upland non-flooded regions of the Amazonian forests of Brazil, Peru and Bolivia, these trees are also found in Colombia and Venezuela. (b) An eighteenth century engraving of an agouti *Dasyprocta aguti*, a rodent which feeds on the nuts and whose neglected caches give rise to new seedlings. (Both these illustrations are from *Trees*, the journal of the International Tree Foundation – formerly known as 'Men of the Trees' – which does such a good job in planting trees in many places in desperate need of them, such as the Sahel south of the Sahara Desert.) (From Prance, 2003. *Trees* 63.)

the agoutis *Dasyprocta aguti* that both feed on and disperse the seeds (Fig. 6.5b). The agoutis are large rodents similar to guinea pigs that consume seeds and flowers on the forest floor. The fruit are technically known as pyxidia, which are capsules that normally open by a lid coming away from the top; in this case, however, the fruit do not open and seed release is dependent upon the very sharp teeth of the agouti that enable them to penetrate the wall of the capsule. The agouti then caches the seeds in the forest floor to eat later, just like the many corvid birds around the world do with pine seeds and acorns (Box 4.1). Being rich in vitamins, selenium, the amino-acid leucine, oil and protein, the nuts are very valuable to the agoutis and to humans. Attempts made to produce nuts from plantation trees have not been very successful because of the natural complexity of their ecology, as outlined by Prance (2003).

The trees flower from October to December and it takes 14 months for the fruits to mature after pollination. The large and complex flowers possess a hood that prevents most animals from reaching the floral organs; indeed only large bees are capable of lifting it and gaining access to the nectar on the inside of the hood. The commonest of the pollinating bees are the euglossine or orchid bees, whose males gather scent from the orchid flowers, pack it into pouches on their hind legs and use it to attract the female bees during mating. The success of Brazil nut production is thus linked with the epiphytic orchids that grow well on forest trees, but have not usually been present in plantations where the lack of suitable pollinators has been critical. The pollinating bees in the forest canopy, the presence of epiphytic orchids on neighbouring trees, and the agoutis scavenging on the forest floor, all play a vital role in the production of this valuable sustainably produced non-timber forest product.

## 6.4 Changes in species diversity over time

### *6.4.1 Plants, lichens and fungi indicative of old woodland*

Biodiversity may be related to the age and previous history of the ecosystem concerned. For example, the richness of epiphytic lichen communities is frequently related to the length of time the dominant trees have been established in the forest. *Fagus sylvatica* has a south-western distribution in Sweden where aged beech trees in old forest stands often have an extremely rich lichen flora including the rare *Lobaria pulmonaria*, a leafy green lichen found throughout the northern hemisphere. As well as the age of the individual trees, the age and 'continuity' of the forest stand are very important. In the county of Halland, Sweden certain lichen species, such as *Pyrenula nitida*, *Catinaria laureri* and

*Bacidia rosella* occur only in areas which have been covered by beech for many hundreds of years. In contrast, planted *Fagus sylvatica* forests on ground that lacks **'beech continuity'** are never inhabited by these species (Hultengren, 1999).

Such lichens are thought formerly to have been widespread and relatively common in virgin forests throughout Europe (Rose, 1988), forming part of the *Lobarion* assemblage of foliose and crustose lichens which develop as a late-succession grouping on the bark of large trees. They include species of the genera *Lobaria, Sticta, Pseudocyphellaria, Parmeliella, Pannaria, Nephroma, Peltigera* and *Parmelia. Lobarion* occurred through most of Europe well into the nineteenth century, but changes in forest practice and pollution have greatly reduced its area; it is now largely confined to montane forests and the lowland oceanic zone from south-west Norway to the Iberian Peninsula.

The diversity of lichens and fungi also varies with the antiquity of the forest; ancient woodlands such as Windsor Forest possess extremely rich fungal communities and are relict centres of diversity. Fungal indicator species have been used to indicate the conservation value of boreal conifer forests for some years, and in countries such as Sweden and Estonia are employed in countrywide surveys to locate **woodland key habitats**. Independent investigations in England and Denmark have shown their value with regard to beech forests. The computer database of the British Mycological Society is of great value in such investigations though its records have to be used with care, bearing in mind that levels of monitoring vary widely over the country. Moreover, the existence of records for many different fungal species for a woodland may not necessarily imply that the site is of high conservation value; some taxa are of greater value than others as far as conservation is concerned. The diversity of saprotrophic fungi in beech forests has been shown to be related to the age and continuity of woodlands in the same way as that of lichens by Ainsworth (2004, 2005), who surveyed saprotrophs growing on large-diameter beech logs in Windsor Forest for a decade. Standing and fallen wood, including **coarse woody debris (CWD)** is an important element in the biodiversity of forests (see Section 7.7); Peterken (1996) provides detailed information concerning the amounts present in a number of managed and natural woodlands.

The above work was the basis for generating a list of 30 fungal species used to define British beech forests of high conservation value. The 30 indicators now used on bulky beech substrata such as logs and branches in Britain, belong to several fungal groups and are also found on other species of tree. Besides the species depicted in Fig. 6.6, they include thick tarcrust *Camarops polysperma*, spiral tarcrust *Eutypa spinosa* (an ascomycete whose fruit-bodies spiral down beech trunks), toothed powdercap *Flammulaster muricatus*,

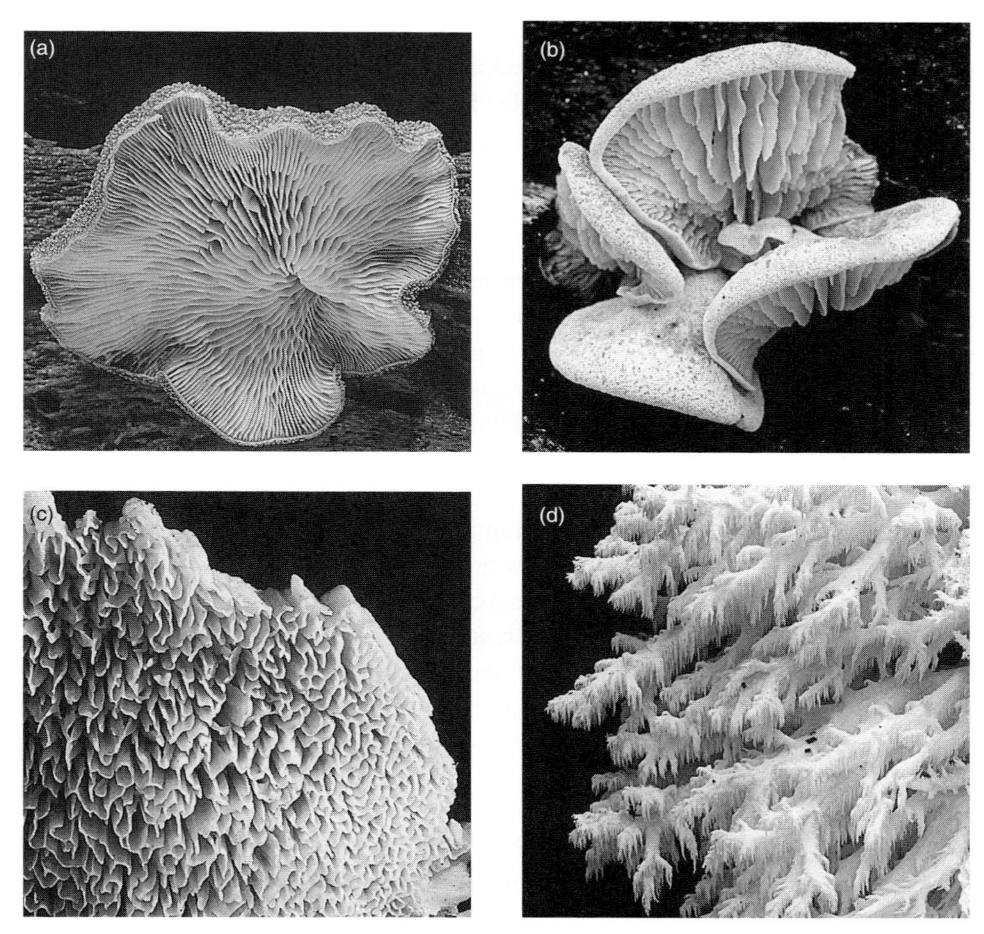


Figure 6.6 Four saprophytic indicators for bulky beech substrata. (a) Woolly oyster *Hohenbuehelia mastrucata*, a gill fungus without a stem. Species of this genus capture and prey on nematode worms. (b) Fox cockleshell *Lentinellus vulpinus*. (c) Spongy mazegill *Spongipellis delectans*, with its strikingly patterned lower surfaces. (d) Coral tooth *Hericium coralloides*, whose fruit-bodies are irregular masses of white branches bearing combs of downward-pointing spines beneath. (Photographs by Martyn Ainsworth.)

fragrant toothcrust *Mycoacia nothofagi*, and a number of gilled and poroid bracket fungi.

In Denmark a list of 42 species of potential indicators was proposed by Heilmann-Clausen and Christensen (2000) and subsequently used, with variations, in much of continental Europe. The use of a single European list for such indicators is not straightforward because of the differing geographic (and dynamic) ranges of the fungi involved. Nevertheless, results obtained when a list of 21 European indicators was applied to 126 European beech forests were extremely interesting. Two Slovakian communities come top, both with a score of 16 out of the 21 possible, followed by others from the Czech Republic (15), France (15) and Denmark (14). The Wood Crates (13) and Denny Wood (12) areas of the New Forest in the UK were respectively 6th and 8th. All of the 11 British beech forests were in the top 30, so the use of this relatively new method of assessment has emphasized their conservation value; one hopes that even with global warming beech forests will continue to grow in the Cotswolds, Chilterns, Wye Valley and North and South Downs. Box 6.1 deals with an example of the conservation value of ancient communities of a particular tree species.

Certain vascular species are also indicative of old forest and are consequently of value in the UK when attempting to get some idea of **woodland age** and particularly in determining which are ancient woodlands (see Section 9.1.4 for a definition). These **Ancient Woodland Vascular Plant Indicators (AWVPs)** are to some extent influenced by topography and geographical location, but many are useful over considerable areas. Peterken (1993, 1996) did much pioneer work on this subject in Britain, continental Europe and North America. The best indicators have a strong affinity for ancient woods, show little or no ability to colonize secondary woodland and are rarely found in recent woodland or other habitats. In central Lincolnshire smooth-stalked

**Box 6.1 Conservation value of aspen stands in maintaining the diversity of animal, plant and fungus groups in Scotland**

A considerable number of boreal fungal, lichen, bryophyte and invertebrate species occur on aspen *Populus tremula* in the boreal woods of Scotland. The mature form of the fungal parasite *Taphrina johanssonii* occurs only on female flowers; it persists during the long periods between flowering in a yeast-like state on the shoots. The aspen hoverfly *Hammerschmidtia ferruginea* has larvae which feed on the rotting cambial layer of recently dead aspen, while the bristle-moss *Orthotrichum gymnostomum* is an epiphyte previously thought to be extinct in Scotland, but recently found distributed vertically down a number of aspen trunks. These are just three of the many important species associated with aspen in the Scottish Highlands, and the Highland Aspen Group (HAG) is making strenuous efforts to document and protect natural stands of this valuable tree (Cosgrove *et al.*, 2005). Many of the associated plants and animals are dependent upon particular stages in the life cycle of aspen, so from the conservation point of view it is important to maintain adequate areas and genetic diversity of this clonal tree.

sedge *Carex laevigata*, pendulous sedge *C. pendula*, small teasel *Dipsacus pilosus*, wood horsetail *Equisetum sylvaticum*, herb paris *Paris quadrifolia* and small-leaved lime *Tilia cordata* provide a very strong indication of ancient woodland, while wood anemone *Anemone nemorosa*, wood-sorrel *Oxalis acetosella* and yellow archangel *Lamiastrum galeobdolon* are amongst the species which occur to a rather greater extent in other habitats and so are weaker indicators.

As already mentioned most **indicators of ancient woodlands** do not readily colonize new sites so their presence tends to show that the community is long-lived, although there are regional differences. In Shropshire, for example, dog's mercury *Mercurialis perennis* is found in gardens and recent hedgerows as well as base-rich woodlands, but it is an ancient woodland vascular plant indicator (AWVP) in the more continental climate of eastern Britain (Rackham, 2003). The AWVP list for Shropshire gives both strong indicators and others which are either less strong or relatively weak (Whild, 2003). Species such as herb paris are found only within ancient woodlands in Shropshire, but in general a really good indication is provided by the presence of at least half a dozen AWVPs of which bluebell *Hyacinthoides non-scripta*, early-purple orchid *Orchis mascula*, moschatel *Adoxa moschatellina*, primrose *Primula vulgaris*, sanicle *Sanicula europaea*, sweet woodruff *Galium odoratum*, toothwort *Lathraea squamaria*, violet helleborine *Epipactis purpurata*, wild garlic *Allium ursinum*, wild service tree *Sorbus torminalis*, wood anemone, wood fescue *Festuca altissima*, wood melick *Melica uniflora*, wood sedge *Carex sylvatica*, wood-sorrel, yellow archangel and yellow pimpernel *Lysimachia nemorum* are sound examples.

Rose (1999) points out that AD 1600, the defining date for ancient woods in Britain, was the time when reasonably accurate estate maps and also the first known widespread tree plantings began. He also makes a very strong case for the use of AWVPs in woodlands, pointing out that the flowering plants, native conifers, ferns and fern allies are by far the easiest of all groups, plant and animal, to observe and identify. In terms of biodiversity he emphasizes the old adage that the longer a habitat has been established the more species it is likely to contain. Of the 130 AWVPs he lists for four regions of southern Britain; a total of 23 are shown in Fig. 3.9, which illustrates the phenology of damp central European oak–hornbeam woods. These act in this way in all four regions including the south-east, while the grass wood barley *Hordelymus europaeus* is used only in central southern England and oxlip *Primula elatior* in the south-west (where Solomon's seal *Polygonatum multiflorum* is not used). Bird cherry *Prunus padus* and yellow star of Bethlehem *Gagea lutea* are AWVPs in East Anglia, as is spindle *Euonymus europaeus* (also used in the south-west).

Similar associations of species with primary forest occur in North America, though in such a large area these vary from one area to another. In the densely wooded Harvard Forest of Massachusetts, Gerhardt (1993) found that whorled aster *Aster acuminatus*, pipsissewa *Chimaphila umbellata*, bluebead lily *Clintonia borealis*, bog fern *Thelypteris simulata*, painted trillium *Trillium undulatum* and hobblebush *Viburnum alnifolium* had significant positive associations with old woodland. Species with an affinity for secondary woods in this area include the clubmosses *Lycopodium clavatum* and *L. obscurum*, bracken *Pteridium aquilinum*, the bramble *Rubus fragillaris*, Canada mayflower *Maianthemum canadense* and the hay-scented fern *Dennstaedtia punctiloba*.

### 6.4.2 Plants of secondary woodland

Some differences between the vascular and non-vascular floras of ancient and secondary forests result from changes inflicted on long-established communities that have considerable amounts of dead wood. In the virgin Norway spruce forests of Sweden five liverworts growing on logs (*Calypogeia suecica, Lophozia incisa, Odontoschisma denudatum, Riccardia palmata* and *Scapania umbrosa*), together with *Herzogiella seligeri* and *Splachnum rubrum*, did not survive initial logging and transition to a managed forest. Other secondary forests develop from the planting of formerly cultivated land (whose soils often have a high nutrient content), grasslands or heaths. Ruderals (weeds) are initially abundant on former farm soils, while characteristic grassland and heathland species may persist for a very long period, competing with the young trees until suppressed by shade. In North America secondary forest very often adjoins old forest, as it does in the Harvard Forest, Massachusetts. Conditions for the rapid migration of old woodland indicator species into such secondary woodlands seem straightforward, yet it does not seem to be occurring. A number of studies have followed the migration rates of plants into new woodlands that have been planted against old woodlands. The rates have been measured at 0–1.25 m per year in deciduous woodland in south Sweden (Brunet and von Oheimb, 1998), 0–2.5 m $y^{-1}$ in eastern North American hardwood forests (Matlack, 1994) and as slow as 0.18–0.38 m $y^{-1}$ from an oak–hornbeam woodland into a pine plantation in southern Poland (Dzwonko, 2001). But there can be surprises; wood anemone *Anemone nemorosa* can migrate at more than 20 m per year when seeds are carried by deer and wild boar. The usual slow migration seems to be linked to slow rates of seed dispersal and the harsh woodland conditions of low light and high humidity (fostering fungal diseases). Matlack (1994) observed that ingested seeds moved quickest, followed by adhesive seeds clinging to fur and feathers, and then wind-dispersed seeds, while the slowest were those moved by ants.

### 6.4.3 Small-scale diversity and variation: cryptogamic soil crusts and mats

Trees include the largest organisms known, but some of the greatest sources of diversity in forests are the small organisms that live on their outer surfaces as epiphytes or are present in or on the soil. **Cryptogamic (or microbiotic) soil crusts** are delicate symbioses of cyanobacteria, lichens and mosses that exist on the surface of arid and semi-arid soils. They increase the amounts of organic matter and available phosphorus in the soil, assist soil stability and rates of water infiltration, form favourable sites for the establishment of vascular plants including trees and assist in nitrogen fixation. They are also extremely fragile and on a world basis their area has been greatly reduced by trampling, particularly by livestock.

In contrast to cryptogamic soil crusts, the **cryptogamic mat** of lower plants developed over bare rock in many northern forests is a different and sturdier community; its most important components are mosses, liverworts and lichens, with larger vascular plants invading once a largely organic substrate has accumulated. **Lichens** are symbiotic associations between fungi and the photosynthetic algae, which are sometimes unicellular, that inhabit them. There are at least 14 000 lichens on the world list; many develop on bare rock or soil, while others are epiphytic on trees and shrubs. Their beauty and diversity is illustrated by the photographic plates in Moberg and Holmasen (1990), which also show their distributions in Scandinavia.

A very brief description of a cryptogamic mat has been given earlier; indeed Fig. 2.8 demonstrates how its existence assists the establishment of tree seedlings. On the other hand, erosion of the cryptogamic mat on the path along Goat ride, Fiby urskog, Sweden, illustrates its vulnerability to damage caused by visitor pressure, trampling by moose, roe deer and hare, animal faeces, sliding movements of mat units, tree fall, and naturally initiated fires. Investigations of the cryptogamic mat developed over the granite in Fiby urskog (Fig. 6.7), have demonstrated both its complexity and the patterns of cyclic succession involved. Change, in the form of **cyclic succession** (Section 9.2.2), results in a **mosaic structure** within the vegetation concerned and, because conditions differ in various parts of the mosaic, it promotes biodiversity. This is true at many different scales. In boreal forests patch size can vary from the gaps caused by fires or the fall of one or a group of trees due to gales or some other cause, to the much smaller scale involved here. In this instance the scale involved is so small that the species present were recorded in

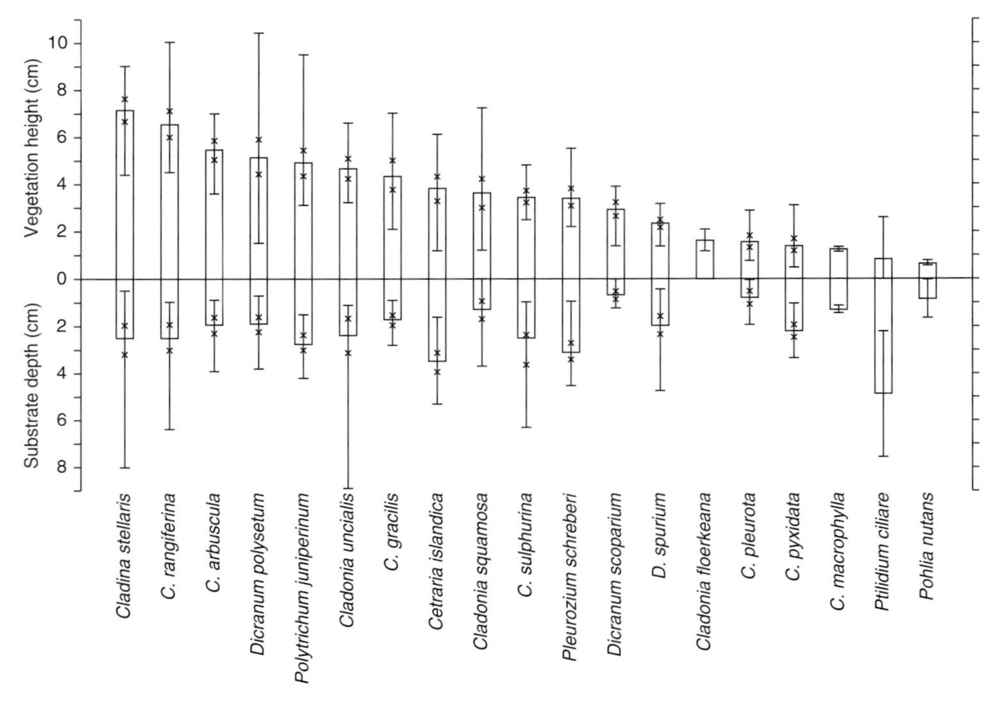


Figure 6.7 Heights of lichen and bryophyte species commonly occurring in the cryptogamic mat, together with depths of substrate resting on the underlying granite in Fiby urskog, Sweden. The main horizontal line represents the top of the substrate of soil and organic matter. The heights and depths of the columns represent mean values. Vertical lines show maxima and minima; standard errors are indicated by x, apart from the four species where less than the standard 10 observations were made. *Cladonia floerkeana* rested directly on the granite. (Unpublished data of Moberg, Hytteborn and Packham.)

six hundred 0.01 $m^2$ quadrats which together formed an area $3 \times 1$ m. The resulting map of the cryptogamic mat on a small hillock away from the main paths in the Lichen Hill area, Fiby urskog (Area 1 in Fig. 2.7) was divided into nine zones. *Cladonia stellaris, C. rangiferina* and *C. arbuscula* were prominent amongst the lichens here; there was also quite a high proportion of bare rock, mosses, litter and heather *Calluna vulgaris*.

There is a scattered field layer of heather and cowberry *Vaccinium vitis-idaea* rooted in the cryptogamic mat at the top of Lichen Hill; other field-layer species here include wavy hair-grass *Deschampsia flexuosa*, bilberry *Vaccinium myrtillus*, narrow buckler-fern *Dryopteris carthusiana* and juniper *Juniperus communis*.

Apart from relatively small amounts of mineral nutrients delivered in rain and weathered from the acid granite, and nitrogen fixed from the air, nutrients available to plants on the granite ridges are of organic origin, particularly from plant litter. The small amount of mineral soil is confined to depressions in the rocks and rock joints. Commencement of the invasion of the bare rock surface is of particular interest with the root-like rhizinae of the lichens and the rhizoids of the bryophytes being fastened directly to the granite. The commonest bryophytes growing directly on the bedrock are *Andraea rupestris, Dicranum scoparium* and *Racomitrium microcarpon*. True mat-forming vascular species grow on a rather thin layer of humus and decomposing plant litter, which is only truly anchored where roots penetrate rock joints. Parts of the mat resting on sloping rocks occasionally move, leaving totally bare patches from which even algae and crustose lichens are absent. These gliding movements tend to occur after heavy rain in areas of the mat that have exceeded a certain critical thickness.

The heights reached by the various components of the mat vary considerably, as Fig. 6.7 shows. Substrate depths were determined by probing with a metal pin. The three species of *Cladonia* formerly placed in the genus *Cladina*, including *C. rangiferina* the so-called reindeer 'moss' and *C. arbuscula*, are the tallest plants present, while the much-branched mildly calcifuge (calcium-hating = acid-loving) liverwort *Ptilidium ciliare* and the more markedly calcifuge moss *Pohlia nutans* are the shortest. The latter is, in Britain, common on heaths and moors, and to some extent in woodlands, being an almost constant member of the limited moss flora of cut peat surfaces in the north-west. Of the three *Dicranum* species shown, *D. polysetum* is the tallest, slightly over-topping another conspicuous acrocarpous (upright) moss *Polytrichum juniperinum*. The red-stemmed and strongly calcifuge moss *Pleurozium schreberi*, is in contrast pleurocarpous so its shoots are horizontal and much-branched. Of the remaining lichens, the eight species of *Cladonia* vary greatly in height. *Cetraria islandica* is present throughout Scandinavia.

Broadly speaking the bryophytes dominate the slopes with lichens towards the top and the most exposed places. The dominating lichen species are, in order of height, *Cladonia stellaris, C. arbuscula* and *C. uncialis*. These and the mat-forming bryophytes are not generally fastened to the bedrock, the mat being stabilized primarily by its own weight and its fit against the uneven rock surface.

When trees and large shrubs establish in the mat they reach mature size only if able to root in the rock joints, obtaining additional supplies of water and nutrients, as well as effective anchorage. Tree falls tend to pull up areas of cryptogamic mat leaving bare areas of granite, thus initiating another sequence of cyclic change, in which crustose lichens such as *Rhizocarpon geographicum* are amongst the first to re-invade.

### *6.4.4 Biodiversity in relation to current ecoclimatic stability: the use of remote sensing*

One of the major aims of large-scale studies of biodiversity is to try both to locate and understand the positions of biodiversity 'hotspots'. Fjeldså *et al.* (1997) used long-term GAC (Global Area Coverage) remote sensing data provided by satellite images in an attempt to determine whether such hotspots were related to how stable the ecoclimate (equivalent to microclimate: the immediate climate of individual organisms) has been over recent years. This subject is too complex fully to explain here, but some of the major points are of interest. The first is that this method gives an entirely new viewpoint concerning environmental variation. Light areas on maps based on daily scenes photographed from meteorological satellites for the period 1981–1991 indicate high inter-annual differences in surface conditions; the darker ones are more stable. 'Hotspots' of bird diversity and aggregates of endemic plants (see Section 6.3.2) are superimposed on the map.

Regions with the most stable climate (characterized by low ecoclimatic variability between years) are predominantly occupied by old species of forest birds. Those whose ecoclimatic variability is complex are dominated by birds of recent origin. Interestingly, however, the 'hotspots' themselves, which have the peak concentrations of endemic species, either have a local reduction in climatic variability, or are situated on the boundary of a stable region. Endemic plant species in these lowland tropical forests are usually associated with special soil conditions (such as being rich in heavy-metals or calcium). But even with these there seems to be a link with ecoclimate since endemics in the tropics are strongly associated with mist zones, and so presumably the more constant the mist zones are from year to year, the greater the number of endemics. Though this very broadbrush approach to biodiversity and the problems of endemic species has its limitations, it is clearly useful as a preliminary to much more detailed investigations of the impact of climatic variations on biodiversity and endemic species in tropical forests.

## 6.5 What allows species to co-exist in a woodland?

This question has taxed ecologists and foresters for centuries and has been the theme of thousands of publications. Despite this, we are still struggling to explain why one or just a few species do not come to dominate a forest by killing off the weaker species. Requirements for certain species are remarkably precise, as in the case of Abbot squirrels which have become so dependent on Ponderosa pine, whose seeds and shoot-tip bark they consume, that they are now found only in association with this tree (Attenborough, 2002). However, other more general considerations apply elsewhere. As has been noted previously, tropical forests tend to hold far higher numbers of species per unit area than any other forest. Tropical rain forests not uncommonly contain over 100 tree species in a hectare, and up to 283 per ha. Moreover, Pitman *et al.* (1999) found 825 tree species in 21 forest plots covering just 36 ha scattered through 400 $km^2$ of the Amazon forest in Peru. This compares with 620 tree species in the whole of North America covering almost 25 million $km^2$. The question of how species co-exist is thus even more pertinent in the tropics.

In general terms high diversity in the tropics is linked with 'productive habitats that have a benign, stable climate, a diverse physical environment, abundant biotic interactions, and long periods of time without major climatic or geologic disturbances' (Schemske, 2002, p. 167). Abundant biotic interactions, such as having specialized pollinators or prey to eat, reduce competition between species allowing them to co-exist (i.e. the ghosts of competition past), as exemplified in Section 6.2.2. If several species of tree, for example, were competing for the same pollinators, the best competitor would win driving the other extinct, but if they use different pollinators, that element of competition is removed. Such **specialization** is dependent upon a stable environment allowing a species to narrow its niche which, in turn, allows more species to be packed into the forest. So, a bird might eat the fruit of just one tree in the tropics while in temperate regions a similar bird would need a wider range of foods to cope with seasonal differences and year to year variations in fruit production. The main premise of this argument is that a reduction in competition between species allows more species to co-exist. However, it can also be argued that greater competition may help co-existence by suppressing the vigour of otherwise dominant species. Having stressed the value of stable conditions, **non-equilibrium mechanisms** (such as variation in habitat across an area, or variation in time by periodic disturbance or gentle changes in local climate) can contribute to species co-existence by giving spatial or temporary advantage to different sets of species, again preventing one set from driving others out. (Tropical forests, of course, show low stability in the face of large disturbances, so the emphasis above is on gentle or small-scale changes to enhance biodiversity.) The books *Foundations of Tropical Forest Biology* (Chazdon and Whitmore, 2002) and *Species Coexistence* (Tokeshi, 1999) are recommended for further development of these complex arguments.

A further solution has been borrowed from the evolution of genes. In the 1960s, when it seemed that the huge number of genes found in organisms was too many to explain just by natural selection, the neutral theory of molecular evolution came to the rescue. In this model, neutral, harmless mutations occur randomly, gradually increasing gene diversity beyond those selected for by natural selection. Stephen Hubbell (2001), who developed the **Neutral Theory of Biodiversity**, suggests a similar process may happen with species, in that diversity is increased by the accumulation of species by chance. If the environment is stable enough, for example, to allow a number of species to survive, then chance can lead to them accumulating into a rich forest. The mathematical models accompanying this theory do indeed often predict results very similar to those actually found in nature. Even this seems not to be the whole answer, since assemblages of species in the tropics build up more quickly than random accumulations could predict (see Nee, 2005).

The following sections explore some of the factors that affect biodiversity and co-existence in more detail.

### *6.5.1 Soil conditions and leaf litter as sources of diversity*

Along with propagule availability, historical factors, light conditions and grazing by animals, **soil conditions**, notably soil pH and fertility, drainage and water availability, greatly influence biodiversity in the tree and shrub layers and understorey floras. Many of the 30 or so species of oak found in south-east USA are closely associated with particular soil conditions. White and black oaks (*Quercus alba* and *Q. velutina*) are found on well-drained upland clay soils, shingle oak *Q. imbricaria* on calcareous soils, water and willow oaks (*Q. nigra* and *Q. phellos*) in moist river basins, turkey and blackjack (*Q. laevis* and *Q. marilandica*) oaks on dry sand hills and live and laurel oaks (*Q. virginiana* and *Q. laurifolia*) in maritime forests. Terborgh (1992) compares these American forests with those of the primary lowland forests of south-east Asia, which are some of the most valuable and best stocked of any in the tropics. Their commercial value is largely due to the high proportion and many species of Dipterocarpaceae present. Many of these dipterocarps have very straight trunks, are up to 60 m tall, and have wood with excellent working qualities which is often sold under the misleading name of 'Philippine mahogany'. The two or three most abundant dipterocarp species present in a stand often characterize its floristic composition (particular groups of species growing under different dipterocarps) and are related to the substrate on which the forest grows, but the number of associations present is far greater than in the oak forests just mentioned.

Soil conditions and light intensity have a strong influence on **bryophyte distributions and zonation**; indeed mosses and liverworts are in general very good indicators of small-scale mosaic patterns involving soil pH, levels of mineral nutrients, humidity and illumination. Coppicing causes the relative humidity of the air to decrease, while the light and temperature range are increased. This has a differential effect on the growth of those bryophytes which are either able to invade or are present already. Gimingham and Birse (1957) found that the life form sequence – dendroid forms (e.g. *Thamnium alopecurum* and *Mnium undulatum*) and thalloid mats (e.g. *Pellia epiphylla*): rough mats (e.g. *Eurhynchium striatum*): smooth mats (e.g. *Hypnum cupressiforme*): short turfs (e.g. *Ceratodon purpureus*) and small cushions (e.g. *Orthotrichum anomalum*) – occurred along a gradient in which light intensity increased and atmospheric relative humidity decreased. This sequence helps us to interpret the striking contrast in growth form distribution (see Section 3.1.1) which may be seen when tracing bryophyte communities along a stream which runs through both felled and unfelled regions of a wood.

The rugged bases of old ash stools in Hayley Wood in southern England bear species of moss and liverwort such as *Lejeuna cavifolia, Porella platyphylla, Homalia trichomanoides* and *Neckera complanata*, for which Cambridgeshire is otherwise too dry. *Hookeria lucens*, a moss of heavily shaded moist places, such as the deeply incised Seckley Ravine, Wyre Forest, English Midlands, is particularly susceptible to environmental change; exposure to direct sunlight kills it. Soil pH is also important. The liverwort *Pellia epiphylla* is favoured, like the moss *Mnium hornum*, by surface acidity, whereas *Pellia endivifolia* is a calcicole requiring alkaline conditions.

Some soil-dwelling organisms facilitate the presence of others. The nest mounds of ants (Fig. 6.8) are hotspots for litter-dwelling earthworms and 37 of the 369 ant-associated species of beetle recorded in Denmark and Fennoscandia are actually dependent on ants (i.e. are **myrmecophilous**).

Soil loss and deterioration can be very severe if forests are managed incorrectly; rates of erosion when tropical rain forests are felled in hilly country are very high. The influence of various trees on the degradation of forest soils is considered in Section 2.2.3, while recent declines in the floras of English ancient woodlands, often caused by nutrient enrichment (eutrophication), are discussed in Section 8.6.2.

### *6.5.2 Fungal hotspots of insect diversity in boreal forests*

The fact that many species of insect are found in association with wood-decaying macrofungi has long been known, but these have often been clumped together with other invertebrates associated with dead and decaying wood. Up to a thousand species of fly and the same number of beetles are believed to be associated with these two habitats in Fennoscandia alone. A review of recent major advances in our knowledge of species-richness, host specificity and the

Figure 6.8 Ant hill beneath Norway spruce *Picea abies* forest at Hall, near Ullanger, central Sweden. Bilberry *Vaccinium myrtillus* and seedlings of rowan *Sorbus aucuparia* are present in the field layer. (Photograph by Andrew J. Packham.)

rarity of the insects directly involved with wood-decaying fungi is provided by Komonen (2003). Species of boreal **bracket fungi** (Fig. 6.9) which have been investigated in this way include *Amylocystis lapponica, Fomes fomentarius, Fomitopsis pinicola, Fomitopsis rosea* and *Piptoporus betulinus*. All were associated with considerable numbers of insect species, 172 being recorded on the birch polypore *Piptoporus betulinus* in Canada. The largest number so far listed is for *Polyporellus squamosus*, which is associated with 246 species of beetles. The perennial bracket fungus *Fomitopsis pinicola* is more common on conifers, including Norway spruce *Picea abies*, than on broadleaved trees. This species of fungus is unusual in that clear droplets of fluid are exuded from the young fruiting body.

Figure 6.9 Left: Bracket fungi on the dead and fissured trunk of a common beech *Fagus sylvatica* in central Europe. Mosses cover the trunk base. Right: The hoof or tinder bracket fungus *Fomes fomentarius*. A specimen of this bracket fungus was found growing on a Bedford willow *Salix fragilis* ssp. *russelliana* by Shrawardine Pool, Shropshire, UK in 2004. Though often found on birch in Scotland and northern England, this fungus is rare further south and had been recorded only once before in Shropshire. Hoof fungus was at one time thought to be common in England owing to confusion with *Ganoderma applanatum*, which causes a serious and damaging heart rot in ageing beeches. (Drawn by John R. Packham.)

Some 36 insect species were found to be associated with this fungus in Norway, of which six were rare and five had a strong preference for this particular fungal host.

The collection of fungal fruiting bodies for laboratory rearing has enabled researchers to be much more certain about relationships between particular insects and the fungi with which they are associated. It turns out that the fruiting bodies of particular fungal species support a very similar insect fauna over wide geographical areas, whereas those of other fungal species are dissimilar. Even ecologically and taxonomically related fungal species growing on the same fallen log may host almost completely different faunas. It is also the case that most primary **fungivores** (fungus-consuming animals) are associated with parasitic wasps and flies, many of which are extremely host-specific. The combined host specificity of many primary fungivores and their parasitoids is significant; it is thought that microhabitat specialists are more prone to extinction than habitat generalists.

Old-growth forests favour high insect diversity; indeed some rare insects are absent from managed forests. This may be because, like the moth *Agnothosia mendicella*, they specialize on fungal species occurring primarily in old-growth forest, or because they require the common fungi in which they develop to be in mature managed or old-growth forest, as in the case of the beetle *Cis quadridens*. It is also important that the areas of old-growth forest available should be reasonably large. In terms of insect conservation it is better to have a few large areas of old-growth forest than many scattered fragments. Our increasing knowledge of the factors controlling forest biodiversity in general, and that of insects in particular, has practical applications. Normal forestry practice has traditionally regarded decaying wood as being insanitary, harbouring diseases and pests. It is now clear that its removal markedly decreases biodiversity. Such wood should be allowed to remain *in situ* if some of the rare insects and other organisms are to be saved for posterity (see Section 7.7).

### *6.5.3 Resilience of tropical rain-forest bird populations to habitat degradation: does complexity beget stability?*

Conservationists everywhere are faced with the problem of how to maintain biodiversity, and also prevent the extinction of as many species as possible, in the face of habitat loss and modification caused by humans. This problem is nowhere more acute than it is in the tropical rain forests, that are both more diverse and being more rapidly degraded than any other biome. In the mid-twentieth century MacArthur (1955) and Elton (1958) promoted the view that community complexity begat stability, that the more complex the ecosystem, the more resistant it was to change. By the 1970s it was becoming increasingly obvious that the opposite was the case; an increasing number of studies showed that the overall tendency was for inherent stability to decrease as complexity increases (May, 1972). In fact, as explained further in Section 9.5, stability begets complexity; it is the stability of the tropical environment that has led to the high species diversity by allowing each species to occupy a narrower niche safely. As Danielsen (1997) reminds us, the fact that communities in climatically stable regions are often both fragile and complex, while those where large and erratic climatic changes occur are relatively robust and simple, are in agreement with this. In view of this, one might expect communities in areas whose ecoclimatic histories were unstable to be more resilient to change than those that evolved in tropical rain forests of palaeoecologically stable regions.

Existing bird studies related to selective logging (and hence the removal of trees used by particular birds) in Asia and of habitat fragmentation in Latin

America did not enable this question to be answered with confidence. Studies of selective logging and habitat fragmentation in Africa, on the other hand, seem to confirm this assumption. This difference is not surprising as there are many problems in such investigations. The great diversity of the natural habitat (in which there are often relatively few common trees but many unusual ones), difficulties of access, lack of suitable controls in the majority of investigations, use of various census methods and inadequate descriptions of disturbance regimes, all make comparisons difficult. Two things in particular stand out from our existing knowledge. The first is the enormity of the damage to the biodiversity of complex tropical rain-forest faunas caused by habitat destruction and fragmentation, and the second is the way this contrasts with the simpler and more resilient communities of the temperate regions. Despite the widespread conversions of European landscape by humans, there have been comparatively few extinctions apart from some megaherbivores and large carnivores. However, these large animals have undoubtedly played an important role in forest ecology in Europe (see Section 5.9).

## 6.6 Conservation, biodiversity, population integrity and uniqueness

Biodiversity is, as we have seen, a measure concerned with the variety and abundance of living things, of which the number of species present is an important aspect. High biodiversity is seen as a desirable attribute, hence communities possessing it are more likely to be a higher priority for conservation than those which do not. The rationale of conservation, however, is far from simple. Communities with rare species facing possible extinction are often given more emphasis than more diverse communities composed of common species. Conservation issues often also involve questions of aesthetics, while many communities which much of the public regard as natural actually depend for their continued existence on continued human influence. The beautiful open countryside of the Lindis Pass, New Zealand, where little taller than tussock grasses occurs, for example, is dependent for its continued existence on regular burning, fertilization and grazing. In their absence woody vegetation would invade and presumably dominate.

Maintenance of the grassland and heathland on the Long Mynd hills, Shropshire, in the English Midlands is similarly dependent on human intervention in preventing the development of woodland. In time past, senescent heather *Calluna vulgaris* was periodically burnt off and the development of woodland prevented by intensive sheep grazing. In recent decades there has been a considerable spread of bracken *Pteridium aquilinum* and various trees and shrubs are also beginning to move in. Rowan *Sorbus aucuparia* is scattered here and there; a few of its trees are very large while many young trees are developing from seeds deposited in bird droppings. Young hawthorn *Crataegus monogyna* are developing in much the same way. Intriguingly some hill sides tend to have many young rowan and others hawthorn, rather than the admixture that might have been expected. The occurrence of wood-sorrel *Oxalis acetosella*, a shade species common in ancient woodlands, on the Long Mynd is also of great interest (Packham, 1979). The plant occurs on screes and grasslands on open hills and mountains where protection from exposure is much less than in forests. The slope and aspect of a variety of wood-sorrel sites on the Mynd were plotted on a slope-aspect polargraph in which increasing slope is represented by greater distance from the centre of a circle. Grasslands on north-facing slopes are relatively cool and moist, receiving little or no direct sunlight when steeply sloping, and often provide conditions suitable for the growth of woodland plants such as wood-sorrel, dog's mercury *Mercurialis perennis*, and wood anemone *Anemone nemoralis* as well as plants found in marshes such as glaucous sedge *Carex flacca* and marsh thistle *Cirsium palustre*.

New Zealand is unique in possessing large areas of natural forest in which there is a great deal of public interest. The indigenous biodiversity of the country is protected, and invasion of the original forests by exotic species is prevented as far as possible. A major problem here is that of **wilding trees** which arise from seeds carried from exotic plantations. Though many pioneer tree species are intolerant of shade and so cause little problem in southern beech forests, Douglas fir *Pseudotsuga menziesii* has considerable shade tolerance and its wildings invade canopy gaps in indigenous forest, while sycamore *Acer pseudoplatanus* is the most shade-tolerant tree in the country, is multi-leadered, coppices, and is very difficult to remove as European foresters know only too well. Fortunately, herbivores find it highly palatable and it is not commonly planted. Wilding trees, often of exotic species introduced by foresters in earlier times, also cause problems by invading open countryside. Lodgepole pine *Pinus contorta* causes the most difficulty, as it has enough shade tolerance to grow close to the base of grass tussocks and sets seed at 5–8 years of age when it may still be hidden by the grass. This tree can spread quite rapidly across the landscape. It also forms dense stands which suppress any understorey.

Some southern beech forests grow under such high harsh environments at higher altitudes that the trees themselves are the only higher plants (i.e. seed plants) present for hundreds of metres. Biodiversity could hardly be lower yet few would deny the value of these unique areas. In other cases southern beech forests with abundant rain, adequate nutrients, higher temperatures and sufficient light beneath the tree canopy have a dense and varied understorey.

A high diversity of indigenous understorey plants can also occur under monocultures of radiata pine *Pinus radiata* when conditions are good; this may be influenced by soil improvement due to mycorrhizal fungi. An extreme example of high biodiversity in an exotic stand was the occurrence of 30 different indigenous orchids beneath black pines at Itwatahi, while two old radiata pines near Fox Glacier bore 8 species of fern and 11 angiosperms as epiphytes.

The biota of conifer plantations and native broadleaf woodlands sometimes differ markedly. In Britain Tickell (1994) is reported as having shown that the latter contained 18 times as much insect life per unit area, yet the current conclusion is that conifer forests are not the 'deserts' they seem to some observers. Maclaren (1996) compares the biodiversity of pasture with that of natural and exotic forests in New Zealand, pointing out that the most important contrast is between pasture and forest plantation since these two land uses are interchangeable, whereas almost all indigenous forest is now protected. In practically every case planting exotic trees on pasture would increase indigenous biodiversity for understorey vegetation and birds, and probably aquatic species (including fish) also. **Hulks**, or **snags** as they are called in North America (large, standing dead trees – see Section 7.7.1), are uncommon in modern plantations and are a potential source of danger and disease. They do encourage a variety of other organisms, however, and a cavity in such a radiata hulk near Tokoroa once contained 20 long-tailed bats. Others have large and interesting populations of insects, mites and fungi; cores from them may furnish climatic information also.

Although bird populations of young conifer plantations are low, those of older exotic stands are usually as abundant, though different from, those of indigenous forests. In New Zealand, plantations miss out primarily on fruit-feeding and nectar-feeding guilds, from an avifauna which is relatively poor anyway: parakeets *Cyanoramphus* spp., kaka *Nestor merldionalis*, yellowhead *Mohoua ochrocephala*, native pigeon *Hemiphaga novaeseelandiae* and kingfisher *Halcyon sancta* are amongst the birds which tend to be absent. The endangered kokako *Callaeas cinerea*, however, feeds on insects, which are abundant in the pine plantations where it is frequently encountered. Exotic plantations can thus be useful storehouses of biodiversity (and, as in New Zealand, sometimes more diverse than native forests). Plantations also offer protection to the native forest by preventing their unsustainable exploitation by providing the necessary wood products; this allows natural assemblages of species to be more easily conserved.

# 7

# Decomposition and renewal

## 7.1 The vital key to a working forest

Without decomposition the dead material in a forest would physically swamp the field and ground-layer vegetation, prevent new seedlings establishing and, most importantly, lock up nutrients to such a degree that woodland processes would grind to a halt. **Decomposition** is the breakdown by physical (abiotic) and biological means of organic material in and above the soil. The organic substances involved include plant material ranging from large woody trunks to leaves and shed parts, dead animals and their excrement. The process of converting the organic matter into the final humus is called **humification**. Along the way, as complex organic components are broken down, the nutrients are **mineralized** (transformed from organically bound nutrients to simple gases such as ammonia and carbon dioxide or soluble (ionic) forms) and released into the soil water where they can be absorbed by decomposers and green plants. As described in Chapter 1, in any ecosystem nutrients are recycled usually with very little input from outside, so any major bottle-neck in the cycle will affect all future activity, including plant growth.

The major gaseous product of decomposition is $CO_2$ so decomposition is crucial in influencing the amount of carbon locked up in forests, a matter of increasing importance when looking at the implications of climate change scenarios and the large amount of carbon currently stored in the soil (see Chapter 11). This chapter will examine how litter and other dead material is decomposed and by what, what controls the processes, what is left at the end, and take a detailed look at the value and decomposition of woody material.

It is tempting when standing in a forest to see the vegetation as the main 'keystone' component, the most crucial aspect of the forest, and therefore the part that controls the running of the forest. However, ecologists are increasingly recognizing the importance of multiple links between the above- and below-ground communities (see Wardle, 2002), which can be crudely equated with the twin processes of plant growth (plus associated herbivory) and decomposition. Such linkages are demonstrated in the two extreme categories of humus normally recognized (Ponge, 2003) – mull and mor (see Section 2.2.2). **Mull humus** is associated with the most fertile soils (typical of grasslands and temperate deciduous woodland) characterized by a large number of burrowing animals, which intimately mix the decomposing organic matter into the mineral soil. This tends to lead to rapid decomposition and turnover of nutrients available to plants. Plants grow well and produce easily decomposed (labile) litter. So here we have a positive feedback loop where the nutrient-rich and easily decomposed litter leads to the development of mull by encouraging an abundant soil fauna and the mull in turn helps soil fertility and the growth of nutrient-rich plants. At the other extreme **mor humus** consists of deposits of largely undecomposed organic matter built up on top of the underlying mineral soil. This tends to develop under harsh cold climates or over poor acidic parent rock, where conditions discourage the larger soil fauna and most decomposition is carried out by fungi. This severely limits the speed of decomposition, leading to organic matter build-up and leaving the vegetation short of nutrients (maximum conservation of organic matter and minimum release of nutrients). In response to the harsh climate and lack of nutrients, the plants are usually evergreen and slow-growing and tend to put more investment into defences such as lignin and tannins (see Section 7.5.2 below), making the litter difficult to decompose. Moreover, before the leaves eventually fall, a greater proportion of useful nutrients are reabsorbed back into the plant. Both these factors combine to produce a very recalcitrant (decomposition-resistant) and often toxic litter which further decreases the diversity of soil fauna, further hindering the release of nutrients.

## 7.2 Decomposition

### *7.2.1 The process of decay*

When litter reaches the ground, or roots die, physical decomposition mainly involves rapid leaching of soluble substances (particularly dissolved organic matter and nitrogen – DOM and DON – see Section 8.4.2) by percolating water which can account for as much as 25% or more of weight loss from some litters (Fig. 7.1). Physical change also involves freeze–thaw and wetting–drying cycles, and the movements of animals which can contribute to the physical fragmentation of the litter. At the other extreme, fire can cause very rapid physical and chemical decomposition of litter and accumulated humus. In

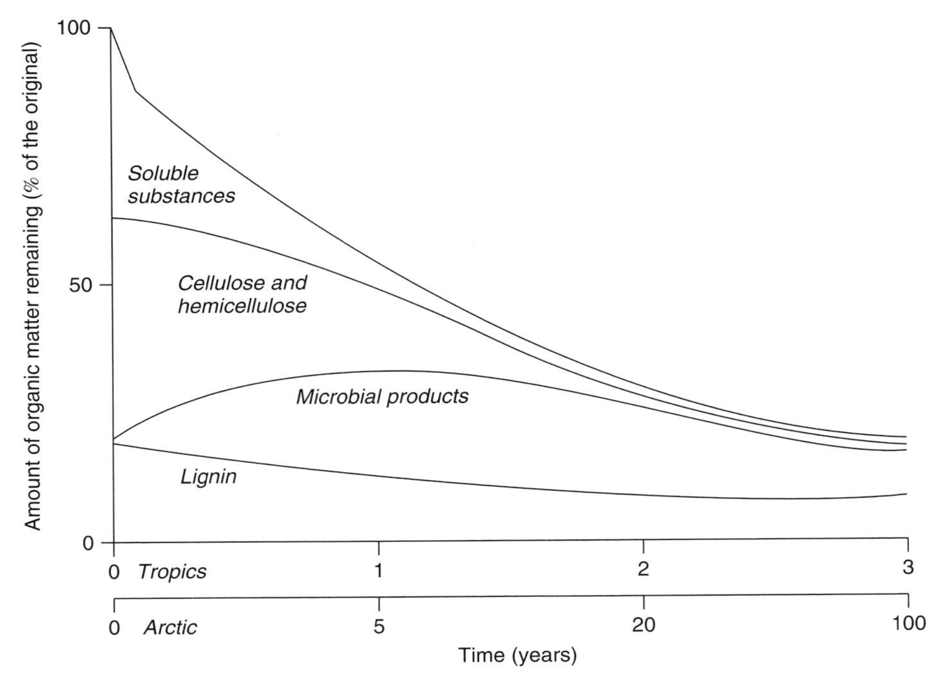


Figure 7.1 Changes in plant litter over time once on the ground and decomposition begins.

most soils, however, it is the 'biologically mediated' decay which accounts for most decomposition.

Decomposition of the dead material (**necromass**) by living organisms can be initiated by fungi living on the phylloplane of the leaf or the rhizoplane of the root (Section 5.5.1), such that decay has started before the leaf hits the ground or before the root is fully dead. Necromass is subsequently colonized by saprotrophic fungi and bacteria, some of which may be preyed upon by certain nematode worms, protozoa and rotifers, but it is this microbial conditioning which enables detritivores, such as various oribatid mites and springtails (Collembola), millipedes and woodlice, earthworms and potworms (Enchytraeidae) to exploit the dead remains. Molluscs, ants and termites are less dependent on such conditioning. The term **detritivore**, as mentioned in Section 1.5, is used here in the broadest sense to include decomposer animals that exist primarily on fungi or bacteria, as well as those which ingest necromass. The main role played by these animals is the breaking up (comminution) of the litter. Such fragmentation disrupts cell walls, exposing the more readily digested contents and considerably increasing the surface area of the litter, so making it more accessible to the microbiota (see below). Harding and Stuttard

(1974) estimated that the chewing up of a 60-mm-long conifer needle into 10 μm$^3$ fragments (μ, a micron, is a thousandth of a mm) by mites results in a 10 000-fold increase in surface area. Many detritivores are inefficient digesters, with as much as 80–95% of their food passing through without much alteration except being physically chewed up and coated with mucus which has a 'primer effect' encouraging the attachment of bacteria cells. Larger animals such as certain earthworms are also useful in mixing the litter into the mineral soil, helping it stay moist (and decompose quicker), and opening channels for air and water movement into soil, thus acting as ecosystem engineers (Bardgett, 2005). Paradoxically perhaps, their decay-resistant casts prevent all organic matter from being used up in the soil, which helps maintain fertility and structure. Detritivores are thus the millstones of the decomposer subsystem doing the heavy work of mixing and grinding, helping microbes that do the bulk of the actual decomposition by improving their food source and aiding spore dispersal (see Moore *et al.*, 1988 for a more detailed review). In these ways, the soil animals help to unblock bottle-necks in energy flow and nutrient cycling and so contribute much more to the functioning of the soil than is indicated by their own energy requirements (Macfayden, 1961).

The soil in forests contains a far greater diversity of organisms than those we notice above-ground. The role of soil animals in forest soils has long been recognized as being of fundamental importance (Wallwork, 1970). A single square metre of soil in temperate woodland may contain more than 1000 species of animals, from protozoa to earthworms. Some of these will be carnivores and herbivores feeding on roots but the majority are detritivores. Generally when large decomposers such as earthworms, woodlice and millipedes are absent, this is compensated for by higher numbers of smaller animals. Northern conifer forests, with their acidic mor humus, are renowned for the high numbers of mesobiotic organisms (see Fig. 1.11): a square metre may contain up to 6.3 million nematode worms and 400 000 springtails and mites (Sohlenius, 1980; Hole, 1981). Enchytraeid worms (macrobiota) are considered to be keystone detritivore species in such soils (Bardgett, 2005). Despite these large numbers of small animals, the total biomass of soil animals increases by a factor of six from northern conifer forests to temperate and subtropical areas, declining again towards tropical forests. Anderson and Swift (1983) point out that tropical rain forests generally have a lower weight of soil fauna than temperate deciduous forests, but the overall activity of soil organisms is higher. The effect of earthworms is being dramatically shown in many northern forests in North America which had no native earthworms after the last ice age and which are now being invaded by worms, mainly of European and Asian origin with the help of humans (such as worms carried on vehicle wheels, throwing away of unused fishing bait and disposal of horticultural waste) – see Bohlen *et al.* (2004). Their effect is still not completely known but they seem to shift the soil system from a slower cycling fungal-dominated system to a faster cycling, bacterial-dominated system with complex changes to nutrient and carbon cycling, invertebrate and vertebrate populations. The direction of these changes varies depending upon the system and the reasons are not always easy to unravel. Most strikingly, a number of studies have shown that the abundance and diversity of understorey vegetation and tree seedlings declines dramatically with the arrival of worms. A study by Hale (quoted in Bohlen *et al.* 2004) in hardwood forests of northern Minnesota showed that over 40 years a lush, diverse forest was reduced to only one species of native herb and virtually no tree seedlings by earthworm invasion.

**Mites**, of which the free-living oribatids are particularly interesting, play important roles in all soils from the poles to the equator. Soil mites are not only hugely abundant and diverse in soils everywhere, they are also most important in organic decomposition and nutrient recycling. Their feeding habits vary from the grazing of bacteria, and the consumption of all kinds of decomposing materials, to brutal carnivory.

Above-ground factors can have a large influence on the soil fauna. Wardle *et al.* (2001) looked at the effect of introduced browsing animals, including deer and goats in New Zealand (there are no natural large forest-dwelling herbivores there) on above- and below-ground flora and fauna using a series of exclosures. These browsing animals not surprisingly reduced the overall vegetation density and the number of palatable broadleaved species, and promoted other less palatable types and hence lower quality litter. Below-ground, the microbes and nematodes were largely unaffected but the larger animals, such as collembola, mites and snails were reduced in browsing areas with consequent (though variable) negative effects on decomposition. Dead wood lying on the soil has also been seen to influence the larger soil fauna (Spears *et al.*, 2003); Jabin *et al.* (2004) found that dead wood had a positive influence on the numbers of beetles, spiders, millipedes and centipedes in a 120-year-old oak–beech woodland in Germany. They found there were usually twice as many individuals of these groups in soil less than 10 cm from dead wood than in soil more than 5 m from such wood. This is because the wood supplies food, and also keeps the surface soil horizons moister (particularly in spring) and warmer.

The **microbiota**, and particularly the fungi and bacteria, are the real chemical decomposers, and account for about 80% of the energy flow within the decomposer subsystem. **Soil bacteria** are a very diverse group consisting of probably over 500 000 species, 95% of which do not show up in traditional culture methods and have only recently been found using modern molecular techniques (Fitter, 2005). Not only are they difficult to find, they are also remarkably numerous. There may be more than 1000 species and more than 200 million bacterial cells in one gram of soil! The cells tend to be anchored to soil particles particularly in the rhizosphere where they benefit from organic acids (such as malate, citrate and oxalate) secreted by the roots. The plants also benefit since the soluble nutrients mineralized by the bacteria in the rhizosphere are readily absorbed by the plant. Moreover, the rhizosphere and its microbiota have been implicated in many soil processes including the mobilization and uptake of less soluble nutrients such as phosphorus, and the detoxification of metals such as aluminium (see Section 2.3.4).

Ecological interactions between the multitudinous species occurring above the ground have been investigated for centuries. In contrast, many questions regarding the ecological significance of the biological diversity of the soil, and how it affects ecosystem function, have not been investigated until recently due to methodological difficulties. New technical developments, including isotopic and molecular methods, together with fresh experimental and modelling approaches, have thrown light on the ways in which the biodiversity of the soil influences the vital processes taking place within it (Bardgett, 2005; Bardgett *et al.*, 2005). The soil is no longer a 'black box' since increasingly we are learning which does what to, or for, whom (see Box 7.1 for an example).

**Box 7.1 Fungal diversity**

In the world as a whole there are far more fungal species (many of them still not described) than there are of vascular plants, which they outnumber by at least 6 to 1. Around 200 000 species have so far been described worldwide and it is thought that 1.5 million species exist (Hawksworth, 1991). The range of fungi is generally greater in the tropics and subtropics than elsewhere, particularly so in fungal groups dominated by decomposers, though pathogens and beneficial symbionts are also common (Lodge, 1997). The amount of food available, the diversity of habitats and the sheer host diversity (numbers of plants and animals) are thought to be the main contributors to high fungal diversity in the tropics, with diversity in certain groups of decomposer fungi being strongly related to host diversity.

Fungi are divided into the macro- and microfungi, the main difference being the size of the fruiting body. Macrofungi produce the familiar mushrooms while microfungi produce fruiting bodies hard to see with the naked eye and include such organisms as moulds, mildews, and many diseases such as rusts and smuts, including the many *Phytophthora* species (see Section 5.4.4) and the chestnut blight, *Cryphonectria parasitica* (Section 5.4.6).

The majority of decomposers in forests (as well as the majority of mycorrhizal fungi – Section 5.4.1) are macrofungi, readily seen in the autumn by the abundant fruiting

**Box 7.1 (cont.)**

bodies. Many of these species are showing a sharp decline in Europe (52% of basidiomycetes are on at least one country's red data book list; Arnolds and de Vries, 1993) and heading this way elsewhere. A variety of reasons have been put forward, including air pollution and decline of old forests but intensive collection of the delectable edible fruiting bodies for culinary use may possibly be playing a significant part.

Although less conspicuous, microfungi (especially vegetative or asexually reproducing forms of ascomycetes which produce no fruiting bodies aboveground) are nonetheless important in decomposing plant debris. For example, just 5 grams of litter from tropical forests in Costa Rica have yielded up to 134 different species of microfungi (Bills and Polishook, 1994). A number of the microfungi, such as species of *Aspergillus* and *Penicillium*, assist in the breakdown of clay complexes, helping to make the soil structure more crumbly and less clayey in nature.

The degree of host-specificity varies tremendously between different fungal groups. Agarics decomposing fallen leaves rarely show strong host-specificity. Even when they do it is normally very broadly so, with distinctions being between monocotyledonous and dicotyledonous hosts in the tropics, and gymnosperms and angiosperms in temperate forests. In contrast, larger ascomycete fungi of the family Xylariaceae are often restricted to the leaves and fruits of particular host-plant genera or families in the tropics. Of the 13 species of *Xylaria* ('dead men's fingers') encountered in Puerto Rico, over half were thus restricted. *Xylaria aristata, X. axifera, X, meliacearum, X. phyllocharis* and *X. stromatica* were host-specific on leaves, and *X. warburgii* and *X. palmicola* on fruits. Others showed no strong preferences: *X. apiculata, X. appendiculata, X. clusiae, X. ianthinovelutina* on leaves, and *X. mellisii* and *X. multiplex* on fruits.

The amount of fungal biomass varies between different types of soil but can be up to around 70% of all the living biomass (including animals) in the soil, and between 60–90% of the microbial mass in forests, especially on acidic soils where bacteria do less well (see Box 7.1 for a discussion of soil fungi diversity). Most **fungi** grow as thin white threads (**hyphae**), 2–10 μm wide, which group together to form **mycelia**. In addition to decomposition, fungi are also important in helping to glue together soil particles into larger 'crumbs' giving the soil a more open structure with more air spaces. Most woody material in forests is decomposed by a sequence of different fungi, discussed in Section 7.7, and the size and type of substrate involved has considerable influence on the decomposer organisms involved in particular degradative successions. At El Verde in the Luquillo Mountains of Puerto Rico, Lodge (1966) looked at the fungi found on six classes of substrate: logs (> 10 cm diameter), branches (1–10 cm), twigs < 1 cm diameter, leaves (including petioles), roots and soil. Of the 705 tropical decomposer fungi involved, 493 species (70%) were confined to one substratum, 246 only on leaves, 43 in soil, 13 on roots, 27 on twigs, 84 on branches and 80 on logs. A further 25% (173 species) were found only on two similar substrata, with a mere 5% on three or more substrata. These results confirm a degree of substratum specificity also demonstrated by others.

Decomposition tends to be rapid when it is primarily bacterial and much slower when fungal. However this does depend to some degree on the abundance of the meso- and macrobiota (worms, beetles, millipedes, etc.). Bacteria are fairly immobile and need fresh litter brought to them and so greatly benefit from mixing by the soil fauna. By contrast, fungal hyphae are very capable of growing into fresh litter and by the same token are readily broken up by the mixing of the litter. So litters with a low soil fauna, where there are summer droughts, very acidic conditions, or low nutritional status, tend to be fungal dominated. Fungi also directly suppress the abundance of bacteria by producing antibiotics (Alexander Fleming named his initial discovery of penicillin from the first fungus he saw with antibiotic properties, *Penicillium notatum*). This may be partly explained by it being an adaptation of mycorrhizal fungi (see Section 5.4.1) for protecting their partner plants from bacterial disease (Marx, 1969; Rasanayagam and Jeffries, 1992). Whatever its origin it is effective. Experimental elimination of ectomycorrhizal fungus from a Monterey pine stand in New Zealand resulted in a greatly accelerated activity of microorganisms and saprotrophic fungi resulting in more rapid litter decomposition, known as the **Gadgil effect** after the researchers, Gadgil and Gadgil (1975). This oppressive effect of ectomycorrhizal fungi has been attributed to their inhibition of other organisms, high competitiveness for nitrogen and possibly low soil moisture levels caused by high water uptake by the mycorrhiza (Bending, 2003).

### 7.2.2 The mechanics of decomposition

Fungi and bacteria work in similar ways: soluble substances such as sugars and low molecular weight compounds are directly absorbed from their surroundings, and larger compounds are broken down by extracellular enzymes secreted into the organic substrata. Some fungi are efficient in breaking down cellulose-containing material, such as plant cell walls (e.g. *Chaetomium*, *Trichoderma*) while others have enzymes capable of breaking down keratin in skin and hair (e.g. *Myxotrichum*) or lignin in wood (e.g. *Phanerochaete*). The nutrients released by decomposition and absorbed by the microbiota are immobilized in the fungal and bacterial bodies and remain so until the cells die or are eaten by other organisms (see Fig. 7.1 and Section 7.3). These nutrient-rich microbiota are fair game for being eaten just as plants are fair game for herbivores, and a number of animals feed directly on bacteria and fungi. To do this they need the appropriate enzymes to be able to digest their food. Thus a number of mites produce chitinase to attack fungal walls and trehalase to digest the fungal storage compound, the sugar trehalose (Luxton, 1972).

Soil animals that directly eat dead plant material are faced with a complex mass of cellulose and lignin since the more readily assimilated compounds such as sugars and proteins will soon have been leached or exploited by the microbiota (Fig. 7.1). Animals such as snails, wood-boring beetles and at least one species of termite, can produce their own cellulases although these may be supplemented by microbial enzymes. In other organisms there is a much clearer case of symbiosis where microbiota (particularly bacteria but also protista) are given a home in a gut and digest cellulose for the host. To be effective this normally needs a long gut to give time for digestion and absorption of the nutrients; those with a shorter gut can get round the problem by **coprophagy**, eating of their own faeces so that partly digested food has a second passage through the gut. This habit is well known in rabbits but is also found in certain millipedes. Feeding on the faeces of other species is also a well-used strategy in soil organisms where each successive organism will work over the microbially conditioned remnants of the resource until all usable energy and nutrients have been extracted. This explains the apparent paradox in soil that there often does not seem to be enough litter input for all the soil biota; much of it is reworked several times. For example, detritivores may work their way through one particular food source for long periods of time (months in the case of woody substrates such as beech cupules) until no more than a dung heap remains, but which is ripe for further exploitation by other organisms.

It is fairly easy in the laboratory to investigate what food each particular type of soil organism prefers. Under field conditions, however, it is much more difficult to observe what an organism is feeding upon. One way around this is to examine gut contents, while remembering that more detailed analysis is needed to establish what has actually been digested. The result is surprising. Although soil animals usually show a clear preference for particular microbes or leaves of particular species in the laboratory, allowing the construction of theoretical food webs, the contents of guts from organisms taken in the field are often remarkably similar, even across different types of organism. A similar story is seen when constructing soil food webs by following the journey of stable radioactive isotopes through different organisms. This apparent free-for-all (a lack of 'food **niche** differentiation') is in distinct contrast to the organization of above-ground food webs. Competition between species would normally force them to be specialized otherwise competition would drive one species to extinction. Possible reasons for this sharing of food sources below-ground may be attributable to litter arriving in bursts and swamping the numbers of detritivores so that food is not limiting. Or there may be a separation in space (by vertical zonation in the soil for example) or in time that helps prevent competition for the same food source. There may also be other factors which control populations, such as disease, predators, climate or limited egg-laying sites, that reduce competition for food. Wardle (2002) gives evidence that fungal-dominated decomposer systems are limited by food resources and bacteria-dominated systems are limited by predation.

## 7.3 Degradative stages

As decomposition proceeds, all the contained energy is eventually lost in respiration along with much of the carbon as $CO_2$ while the nutrients are released in an inorganic state, potentially available to the plant subsystem, ready to start the cycle again. But the availability of these nutrients to plants is not just a straightforward slow trickle.

When fresh litter arrives on the soil surface it has a high carbon to nitrogen ratio, that is, it is rich in carbon (C) but comparatively short on nitrogen (N), and usually other nutrients such as phosphorus (P). Living plant material has a **C/N ratio** of 50–100 but due to resorption of valuable nutrients, the litter produced has a ratio nearer 100–200. The microbes doing the decomposing, and to a certain extent the soil animals physically chewing up the litter, will soak up all the N and P being released by decomposition for their own use and lock up the nutrients in their cells. More than that, however, their need is often so great that they will also pull in nutrients from surrounding areas and lock these up as well. So rather than resulting in mineralization, the microbes at this stage actually cause immobilization. This is why wood chips (which have a very high C/N ratio) used as a garden mulch can result in plants showing classic nutrient shortage symptoms; they are being outcompeted for nutrients by microbes. (Note that bark chips have a much lower C/N ratio and do not cause anywhere near the same problem!) This explains why gross nitrogen mineralization rates are often an order of magnitude higher than net rates; most of the mineralized nitrogen is absorbed directly back into the microbial pool. Carbon is progressively used up in respiration and lost as $CO_2$, but N, P and other nutrients are conserved and reused. This cycling of nitrogen as it is taken up by microbes and released as they die to be taken up by other microbes is referred to as the **microbial nitrogen loop**. It is this loop that keeps the nitrogen in constant use and unavailable to plants. But as carbon is progressively used up, the ratio between C and the nutrients declines, reaching a C/N ratio of 10–20 in decomposed soil organic matter. Eventually the concentration of nutrients is sufficient to meet the needs of the decomposer organisms and at this point as microbial cells die, nutrients will be released from the loop and made available for uptake by plants. In any given **substrate–microbe complex** the processes of immobilization and mineralization occur simultaneously, with the difference between the two processes (**net mineralization**) determining when nutrients are released into the soil. From this it can be seen that when a fresh batch of litter arrives on the forest floor there is a variable lag before the carbon has been reduced sufficiently to allow nutrients to be freed for plant growth, the process being regulated by the microbial community (see Ågren *et al.*, 2001 and Attiwill and Adams, 1993 for more detail). Thus, the dead organic matter is the main bottle-neck controlling the availability of nitrogen to plants.

This immobilization when carbon is overabundant may go some way in explaining why N experimentally added to soils, especially where there is a lot of organic matter such as in podzols, has little effect on decomposition (see data from Prescott, 1995 below); extra N is soaked up by an expanding microbial community. Fogg (1988) points out that a large number of experiments have found that adding N to soils as fertilizer may actually slow decomposition. The reasons are not completely known but may well involve:

(a) changes in the composition of the microbe community through competition;
(b) N (as ammonia) suppressing the production of enzymes needed for breaking down lignin and other recalcitrant compounds; and
(c) ammonia and proteins reacting with organic matter to form recalcitrant material.

Part of the problem of interpreting the numerous nitrogen addition experiments results from the different forms of nitrogen added at different rates and frequencies, and complications caused by shortages of other nutrients.

The section above describes what happens to each input of litter. Decomposition can also be looked at over a longer period as conditions within a forest change, but it is much more difficult. As an example of this, W. W. Covington (1981) investigated what happens to forest floor organic matter over a 200-year period when a hardwood forest is felled by clear cutting. He looked at 14 hardwood stands that had been felled at various times through the past 200 years in the Hubbard Brook Experimental Forest in the White Mountains of New Hampshire and made the assumption that the pattern of soil organic matter changes between different-aged sites would mirror what would happen on one site over the same length of time, i.e. that

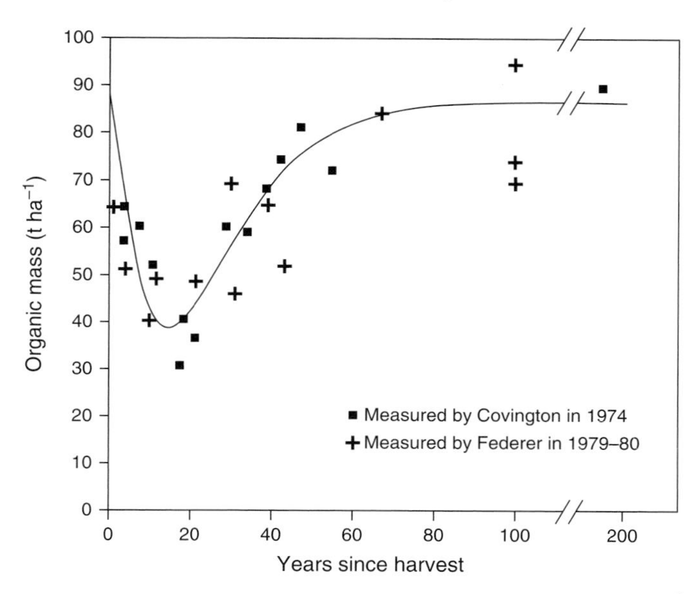


Figure 7.2 The amount of forest floor organic material (in t $ha^{-1}$) over a 200-year period following the felling of trees in hardwood stands in the White Mountains of New Hampshire. The graph is based on an original study conducted by Covington (1981) with additional stands added from Federer (1984). (Redrawn from Yanai *et al.* 2003. *Ecosystems* 6, Fig. 1. With kind permission of Springer Science and Business Media.)

the situation found on a site felled at 10 years occurred at the tenth year in a 200-year-old site. The resulting **Covington curve** (Fig. 7.2) suggests that the amount of organic matter in the soil declines sharply after clear-felling, with 50% (some 40 t $ha^{-1}$) lost in the first 20 years. This curve has become very influential in modelling carbon losses from forestry operations and in larger models of global carbon budgets, and yet may not be wholly true. Federer (1984) replicated the study and suggests that a more reasonable estimate of organic matter loss solely due to opening up the stands is 36% or approximately 30 t $ha^{-1}$. Yanai *et al.* (2003) in turn revisited Federer's plots 15 years later and found that individual plots had not moved along the Covington curve; some had lost organic matter and others had gained it but not in any pattern relating to time since clearing.

It was originally suggested that the loss of organic matter was due to (a) reduced litter input while the new trees that started growing were still young, and (b) increased decomposition of soil organic matter. The decomposition of surface litter would have been quite slow because although exposed to the sun and so warmer, it would also have readily dried in the open conditions. But the organic matter below would have been warm and moist (protected by the mulch of dry litter and with extra water that would have been taken up by the trees in an intact forest), excellent conditions for rapid decomposition. However, few studies have found these two assumptions to be reliably true. It seems much more likely that organic matter loss after tree felling may have been due to mixing of the organic matter into the mineral soil by mechanical disturbance of the felling operation, thus accelerating decomposition. If this is true, Covington's curve may be more of a reflection of changing forestry practices over time rather than natural changes in decomposition. This has very important implications for using Covington's data in predicting carbon budgets for natural forests where there is no mechanical disturbance. It also demonstrates the great difficulty in finding ways to accurately test and understand how whole forests work.

## 7.4 How much dead material is there?

The majority of studies that have looked at the rates of dead material production have concentrated on **litter** – small-diameter twigs, leaves, flowers, fruits and bud scales. But dead material also includes large-diameter dead wood (standing and fallen), deceased roots or material sloughed off roots below-ground (which, as described below, can be a considerable amount), dead animals and the excrement of their live relatives. Less is known about these because they can be harder to quantify since large dead bodies (tree or animal) and hidden roots are difficult to quantify and, in the case of animals, generally they make up a small proportion of the annual input of dead material to the decomposer cycle. Nevertheless, such components as animal remains can be locally important; a dead deer or an excess of insect bodies will provide a large and concentrated input of nitrogen and other nutrients to the soil (see Box 7.2). Using a wider definition of what constitutes dead material than just 'litter' leads to a problem of terminology. **Detritus** is often used but strictly this could also mean rock fragments. The most all-encompassing term is **necromass**, the total mass (weight) or volume of dead organisms or their parts in an area or volume at a given time. As this decomposes it merges into the humus remains found on and in the soil, at which point it is easier to talk about the **soil organic matter**, all the dead organic debris no matter what its stage of decomposition comprising a cellular fraction and amorphous humus (Swift *et al.*, 1979).

There is a major difference between the **total amount** of a nutrient stored in a soil and its **availability**. For example, in soils with abundant mor humus, such as in boreal forests, the total amount of nitrogen stored is high but because turnover is low, availability is low. In the tropics, the total nitrogen pool in the soil is low but availability can be high due to rapid turnover.

**Box 7.2 Pulsed resources**

Forest research tends to concentrate on normality, in this case how much dead material normally falls in a forest. However, there is growing interest in **resource pulses**, periodic large increases in food for organisms. This is seen in mast years when an unusually heavy seed crop is produced (see Section 4.4). In terms of decomposition, Zackrisson *et al.* (1999) observed that in a **mast year** of Norway spruce *Picea abies* in northern Sweden, an extra 5–10 kg of nitrogen was added per hectare to what is normally a nutrient-poor soil. Seeds that failed to germinate decomposed more rapidly than needle litter and released 65–75% of their nitrogen in the first growing season, which was linked by Zackrisson and colleagues to better growth of Scots pine *Pinus sylvestris* seedlings. They suggest that this nutrient pulse from the dead seeds is a significant factor in the strongly pulsed natural tree regeneration found in these northern forests.

**Insect defoliation** can also produce a concentrated pulse of nitrogen, redistributing it from the trees to the soil microbes in the droppings or frass (Lovett *et al.*, 2002) with potential long-term consequences for reproduction and growth as the trees compete with the microbes to regain nitrogen. In eastern USA there are a number of **periodical cicadas** which swarm in spectacular numbers for up to a month every 13–17 years (including the 17-year cicada *Magicicada septendecim*), most recently in 2004. Densities of adults can reach over two and a half million per hectare over hundreds of square kilometres. While the havoc these plant-feeding insects can cause is well known, it seems they may have benefits when they die. Yang (2004) provides compelling experimental evidence that adding 120 dead cicadas per square metre in hardwood forests led to an increase in soil bacteria and fungi within a month with increase in soil ammonium of 412% over 30 days and 199% increase in soil nitrate over 100 days. Moreover, he has shown that this results in 12% more nitrogen in foliage and 9% bigger seeds in the American bellflower *Campanula americanum*, a common understorey plant within the cicada's range. It may also explain the small increase in growth seen in pines after a cicada event, given that pines rarely suffer cicada damage (Koenig and Liebhold, 2003).

There is an overall increase in annual above-ground litter-fall from the poles to the equator (Vogt *et al.*, 1986); with broad averages of 0.1–5 tonnes per hectare (t $ha^{-1}$) in boreal forests, 3–10 t $ha^{-1}$ in temperate deciduous forests and 5–15 (even 30) t $ha^{-1}$ in tropical forests. There is an apocryphal story about a park keeper who wanted only evergreen trees in his park because they shed fewer leaves. Fortunately for him, that appears partly true. In temperate and northern regions, evergreen forests produce slightly less litter than deciduous forests (a smaller proportion of the leaves are replaced each year) but this difference is pronounced (and even then there is a good deal of variation

Table 7.1. *The fate of the biomass grown (net primary production) in one year in 23- and 180-year-old Pacific silver fir* Abies amabilis *stands in the Washington Cascade Mountains. In these stands the dominant fir is mixed with western hemlock* Tsuga heterophylla, *mountain hemlock* Tsuga mertensiana *and noble fir* Abies procera. *The understorey is dominated by broadleaved deciduous shrubs with a mostly deciduous field layer. Note that the figures for living material are after the detritus has been taken away, thus above-ground the vegetation produced 4.3 t ha$^{-1}$ of living material plus 2.1 t ha$^{-1}$ that subsequently died in the year. T = trace.*

| | 23-year-old stand | | 180-year-old stand | |
|---|---|---|---|---|
| | t ha$^{-1}$y$^{-1}$ | % of total | t ha$^{-1}$y$^{-1}$ | % of total |
| Above-ground biomass added | | | | |
| Tree | 4.3 | 23.3 | 2.3 | 13.8 |
| Shrub stems | 0.06 | 0.3 | T | T |
| Total living | 4.3 | 23.6 | 2.3 | 13.8 |
| Detritus production | | | | |
| Falling trees and litter | 1.8 | 9.9 | 2.2 | 12.9 |
| Field layer litter | 0.3 | 1.8 | 0.05 | 0.3 |
| Total detritus | 2.1 | 11.7 | 2.2 | 13.2 |
| Total above-ground | 6.5 | 35.3 | 4.6 | 27.0 |
| Below-ground biomass added | | | | |
| Total living (big roots) | 1.8 | 9.7 | 0.7 | 4.2 |
| Total detritus | 10.0 | 55.0 | 11.5 | 68.7 |
| Total below-ground | 11.8 | 64.7 | 12.2 | 72.9 |
| Total primary production | 18.3 | 100.0 | 16.8 | 100.0 |

*Source:* Data from Grier *et al.*, 1981. *Canadian Journal of Forest Research* 11.

and overlap) only at high latitudes away from the tropics. Note that the difference is between evergreen and deciduous rather than between coniferous and broadleaved; the amount of litter produced by a deciduous larch forest and a deciduous oak forest is of the same order of magnitude.

The next question to ask is how much of the forest's new growth in a year ends up as dead material. Grier *et al.* (1981) looked at annual litter-fall in a Pacific silver fir forest (Table 7.1) and found that in a young stand (23 years old) about one-third of the above-ground biomass grown in a year ended up as litter (2.1 of 6.5 t ha$^{-1}$). In an old stand, 180 years old, around half of the year's growth fell as litter (2.2 of 4.6 t ha$^{-1}$). Almost the same weight of litter was produced in both stands but the young stand was growing more vigorously and a greater amount of the year's growth was kept in living tissue.

Despite its importance, litter-fall is not necessarily the main source of necromass. Grier *et al.* (1981) also looked at the amount of 'detritus' (as they chose to call it) produced each year below-ground from dying roots. In the young and old stand, respectively, below-ground detritus was 10.0 and 11.5 t ha$^{-1}$ in the year, compared with around 2 t ha$^{-1}$ above-ground (Table 7.1). So in both stands more than three-quarters (around 83%) of the detritus was produced below-ground. Adding together these above- and below-ground figures gives a total detritus production of 12.1 and 13.7 t ha$^{-1}$ of dead material produced in the year in the young and old stands, respectively. This is two-thirds (66.6%) of total production that year in the young stand; in the old stand it was more than three-quarters (82%) – Table 7.1. Eventually, in a mature stand, that has reached its maximum accumulation of organic matter, decomposition will theoretically be 100% of the year's growth (see Chapter 8).

The exact amount of litter produced each year in the same stand can vary considerably due to factors such as weather, numbers of flowers and fruits produced (e.g. mast years), leaf production and factors such as pests and diseases or storm damage killing part or all of a tree. In temperate deciduous trees, around 75% of above-ground litter is composed of leaves, 10% flowers and fruits, and 15% branches and twigs although there can be tremendous variation within these figures.

There can also be considerable seasonal variation in litter-fall. The most conspicuous seasonality is in temperate deciduous trees dropping the majority of their leaves in the autumn. Some leaves can also be lost at other times of the year due to disease, insects or inclement weather. Added to this is the seasonal shed of bud scales in spring, flower parts (normally in spring or summer) and immature fruits in summer. Temperate evergreen species are less synchronous in litter shedding; some shed at irregular times through the year (e.g. Norway spruce *Picea abies*) while others peak in spring (e.g. holly *Ilex aquifolium* of Europe and holm oak *Quercus ilex* of the Mediterranean region), summer (e.g. cypresses *Cupressus* species) or autumn (e.g. many pines). Tropical evergreen trees are more complicated since individuals of a species may shed their leaves together or as individual trees or even a branch at a time; shedding may also be at irregular intervals or, if regular, not necessarily linked to the calendar year.

## 7.5 What controls the rate of decomposition?

Three major factors control decomposition: climate, quality of the litter, and the soil microbial and faunal communities, as shown in Fig. 7.3. Other factors

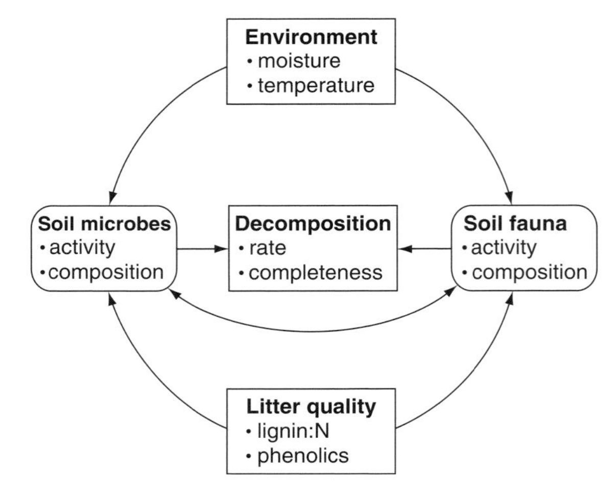


Figure 7.3 Factors that affect how quickly litter decomposes. (Reprinted from: Prescott *et al.* 2000. *Forest Ecology and Management* 133 with permission from Elsevier.)

can be important such as soil pH and aeration but tend themselves to be influenced by the three main factors. As with most ecological processes, environmental conditions such as temperature and moisture will affect how fast the process goes. With decomposition, the quality of the food supply – the material for decomposition – must inevitably be important, together with the nature of the soil biota discussed above. In most situations, the early decomposition is usually closely related to climatic conditions and concentrations of water-soluble nutrients while later decomposition rates are more influenced by the quality of the necromass, particularly the quantity of important nutrients and compounds that tend to impede competition such as lignin and other polyphenols (Prescott *et al.*, 2000).

### *7.5.1 Environmental conditions*

Providing other factors such as moisture are not limiting, decomposition rates increase with temperature, usually increasing by 2–3.5-fold in freshly fallen litter with a 10 °C increase (this is called the temperature coefficient or $Q_{10}$ value).

Moisture availability also regulates the rate of decomposition, especially summer precipitation when temperatures are warmest and decomposition is potentially most rapid. At the other extreme, under waterlogged conditions, oxygen availability is drastically reduced because diffusion is limited, and anaerobic conditions soon develop. In such cases, only limited decomposition is carried out by anaerobic bacteria, and with time peat can begin to form.

Indirectly, the moisture conditions of a soil can influence pH by affecting the oxygen supply, which in turn will influence the activity of decomposer organisms.

The influences of temperature and moisture interact such that under extreme moisture conditions, temperature has much less effect on decay rates. Furthermore, weather conditions influence the efficiency of decay. For example, the amount of nitrogen released per gram of carbon used varies when identical litters are incubated under different temperature and moisture conditions.

### *7.5.2 Quality of the necromass*

Animal necromass in the form of dead bodies and waste products may be relatively small in quantity and irregularly produced but its quality, especially in terms of its nitrogen content, can have very important effects in a forest. For example, the dung and urine produced by moose have been seen to have a larger effect through increased nitrogen mineralization and other soil processes in a willow forest (diamondleaf willow, *Salix planifera*) in Alaska than the more obvious effects of browsing (Molvar *et al.*, 1993). Some studies have shown that such mineralization tends to lead to nitrogen loss from soils by oxides being leached or escaping as gas. Sometimes, however, faeces can cause nitrogen immobilization at least for a time. For example, Christenson *et al.* (2002) found that the nitrogen in droppings (frass) of the gypsy moth *Lymantria dispar* in a hardwood forest of New York State was more quickly mineralized than from leaf litter (as predicted above) but was quickly immobilized by soil microbes stimulated by the readily available carbon in the frass.

As noted at the beginning of the chapter, plant litter quality varies greatly depending upon the plants it comes from; both in terms of what species are present and also how their growth is influenced by nutrient stress. Plants growing under nutrient shortage tend to produce harder, nutrient-poor, recalcitrant litter. While animals normally keep the elemental composition of their bodies within narrow bounds, plants are much more flexible so the nutrient content of their litter can vary enormously even within the same species.

A large number of different measurements of necromass quality have been implicated in limiting decomposition, including such considerations as toughness, thickness, carbon to nitrogen ratio, and the content of calcium, nitrogen, carbohydrates, lignin and other polyphenol compounds. As is often the case, different factors will be limiting under different circumstances. For example, in general, litters from tropical sites have higher **nitrogen** (N) values and lower **lignin to N ratios** (also written lignin/N) than litters from other regions. In the tropics, Mediterranean region, and temperate forests the lignin/N ratio is usually the best predictor of decomposability (i.e. litter with proportionately

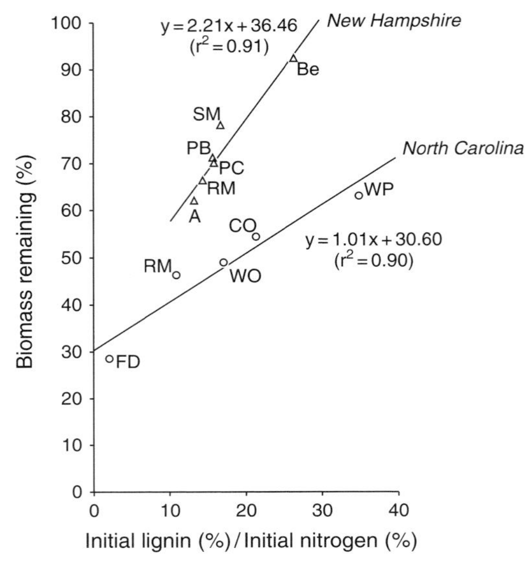


Figure 7.4 The percentage of litter remaining after 1 year plotted against the ratio of lignin to nitrogen in the litter at the start of the experiment. The experiment was conducted in a hardwood forest at the Hubbard Brook Experimental Forest, New Hampshire and in a similar forest in North Carolina, using litter of individual species placed in nylon mesh 'litter bags' which were then placed out in the forest and recollected for weighing to estimate weight loss (or biomass remaining). The equations are for the regression line fitted to each set of data: $r^2$ values run from 0 to 1; the closer the value to 1, the closer the data fits the line. The species used are: A white or American ash *Fraxinus americana*; Be Beech *Fagus grandifolia*; CO chestnut oak *Quercus prinus*; FD flowering dogwood *Cornus florida*; PB paper birch *Betula papyrifera*; PC pin cherry *Prunus pensylvanica*; RM red maple *Acer rubrum*; SM sugar maple *Acer saccharum*; WO white oak *Quercus alba*; WP eastern white pine *Pinus strobus*. (From Melillo *et al.* 1982. *Ecology* 63.)

more lignin to nitrogen decomposes more slowly), as shown in Fig. 7.4. Certainly the rate of N mineralization depends strongly on the lignin/N ratio in these areas (Rees *et al.*, 2001). However, N does not appear to be a reliable chemical predictor of decomposition in northern areas. For example, Prescott (1995) conducted a range of experiments looking at the effect of nitrogen on decomposition of litter of jack pine *Pinus banksiana* in the Canadian boreal forest. In some of these, N was applied externally as fertilizer (an 'exogenous' supply) and in others, litter was taken from trees grown with and without fertilizer which gave different amounts of internal N ('endogenous') – 1.56% N in litter from fertilized trees, 0.33% N without fertilizer. In all cases Prescott found that the speed of decomposition was the same or poorer with added N and she concluded that internal or external N does not control rates of litter decomposition in these northern forests. Attiwill and Adams (1993), reviewing published studies, go further and state that in warm-temperate areas and the tropics, and most likely also in northern boreal forests, available phosphorus (P) is much more likely to be the factor limiting decomposition.

The lignin/N ratio has been found to differ between species of trees by a factor of two or more. This should, and does, give quite a spatial variation in soil fertility within a forest depending upon species composition and arrangement of trees. Surprisingly, however, this pattern is not necessarily reflected in how well vegetation grows. For example, Finzi and Canham (2000) found that variations in N availability within a New England forest made little difference to sapling growth; light was far more important, explaining up to 79% of the variation in growth. Only the growth of sugar maple *Acer saccharum* and red maple *A. rubrum* was related to N and even here it explained less than 7% of the variation in growth. This is unexpected given the clear role that N plays in forest growth on a regional level but may be explained by the wide-spreading roots of most trees and nutrient movement across root grafts (see Section 1.3.2).

The nitrogen and phosphorus content of litter can be manipulated by the plant reabsorbing nutrients from the leaves before they fall. On a common-sense basis we might expect that (a) evergreen plants would reabsorb more nutrients than deciduous plants, and (b) that plants on more nutrient-poor soils should reabsorb a greater proportion of nutrients to reuse the next growing season. In both cases the consequence would be to make the C/N and C/P ratios of the litter poorer and so more resistant to decomposition. Aerts (1996) conducted a wide-ranging review of this using data from 44 studies primarily from the USA and Europe. Overall he found that an average of 50% of N and 52% of P was reabsorbed from leaves. Phosphorus reabsorption was similar across evergreen and deciduous woody plants but less N was reabsorbed by evergreen woody plants (47%) than by their deciduous counterparts (54%). But evergreen leaves started out with some 40% lower concentrations of N and P than did deciduous leaves (Table 7.2) so the litter of evergreen species is poorer in these nutrients. This agrees with a study by Killingbeck (1996) who collated published data on woody plants from around the world. He similarly found that the litter of evergreens was lower in P than deciduous species, and heading that way in N although the data were not significantly different (Fig. 7.5). It follows from this that the evergreen litter is likely to be more difficult (and so slower) to decompose than that from deciduous species. Importantly to the second intuitive guess above, the proportion of N and P recovered did not vary in any consistent way with the nutrient status of the plant. So it seems that species on poor sites are not reabsorbing higher amounts of nutrients from their leaves. In that case, how

Table 7.2. *Mean nitrogen (N) and phosphorus (P) concentrations in mature leaves from evergreen and deciduous woody plants (trees and shrubs) growing primarily in the USA and Europe, and litter from these plants (note, these last figures are calculated from the mature leaf concentrations knowing the reabsorption rates. The data can be converted to percentages by dividing the given figures by 10).*

| | Mature leaves (mg g$^{-1}$) | | Litter (mg g$^{-1}$) | |
|---|---|---|---|---|
| | N | P | N | P |
| Evergreen | 13.7 | 1.02 | 7.30 | 0.50 |
| Deciduous | 22.2 | 1.60 | 10.21 | 0.79 |

*Source:* Data from Aerts, 1996. *Journal of Ecology* 84.

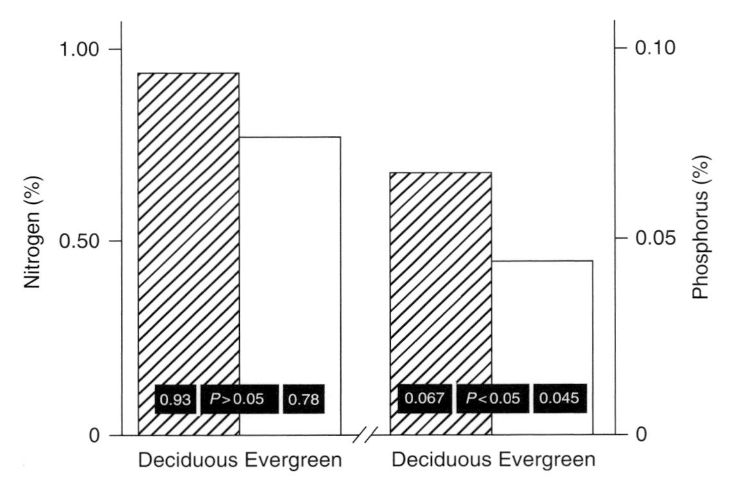


Figure 7.5 Mean nitrogen (N) and phosphorus (P) concentrations measured in the shed leaves of 45 species of deciduous woody plants and 32 species (N) or 31 species (P) of evergreen woody plants from around the world. These data can be compared with Table 7.2 by multiplying the figures by 10 to get mg of N or P per gram of leaf (mg g$^{-1}$). *P* values in the boxes indicate that the two nitrogen concentrations are not significantly different but the two phosphorus figures are. (Redrawn from Killingbeck, 1996. *Ecology* **77**.)

are nutrients conserved on poor soils? It appears that leaves live longer (this is as much to conserve carbon as nutrients) and they start with a lower initial nutrient concentration.

As well as the *absence* of nutrients, the *presence* of certain compounds has also been strongly implicated in slowing decomposition. In most cases these are **polyphenols**, a range of compounds found in many plants but here referring mainly to those of higher molecular weight such as lignin and tannins that are found almost exclusively in woody plants. Lignin and other phenolics are often referred to as anti-herbivore compounds but they may have just as large an after-effect on decomposition.

**Lignin** is a highly complex and variable compound (a macromolecule) that forms between a quarter and a third of the weight of wood (and so is the second most abundant organic material after cellulose). It gives strength and rigidity to the cellulose wall, is what makes a woody plant woody, and what makes brown paper brown and tough; it also protects structural polysaccharides such as cellulose from microbial attack, whether the plant is alive or dead. The lignin molecule is twisted into a random, highly heterogeneous and unordered structure. Since enzymes work by fitting themselves to the shape of their target compound, this makes lignin very difficult to digest and requires a variety of enzymes to do the job. For most organisms it is too expensive to produce all the required enzymes, so lignin is fairly immune to decomposition (see Fig. 7.1) except by specialists such as the white rots. Lignin concentration is thus often negatively correlated with decomposition rates in litter – the more lignin, the slower decomposition. It is usually the case that plants on nutritionally poorer sites produce more lignin, primarily as an anti-herbivore defence to protect the scarce nutrients. But, perplexingly, this does not always appear to be the case. Kitayama *et al.* (2004) looked at the amounts of lignin in leaves in tropical rain forest leaves in Borneo and found that lignin concentrations changed little with productivity of forests (as judged by the amount of litter-fall). Because lignin is resistant to decay, its concentration increases during the decomposition process and eventually forms a significant proportion of humus (Section 7.6.2).

**Tannins** are also polyphenols and are common enough to be the fourth most abundant plant compound after cellulose, lignin and hemicellulose. Leaves and bark may contain up to 40% tannin by dry weight, and in leaves and needles tannin concentrations can exceed lignin levels (see Kraus *et al.*, 2003). Tannins are useful in plant defence because they bind to proteins making these less digestible to herbivores, and may also inhibit fungal and bacterial activity. The ability of tannins to bind to protein and so inhibit decomposition has long been exploited in preserving leather. Their effect on decomposition in the soil can be as important as the nitrogen content of litter especially in their binding to proteins to form polyphenol-protein complexes (PPC). These are formed either in the dying plant cell or in the soil. The formation of PPC causes the brown colouration of dying leaves and can make up 20% of the dry weight of the leaf (see Hättenschwiler and Vitousek, 2000). Polyphenols and PPC are resistant to decay except by fungi with the appropriate enzymes (polyphenol oxidase) and to a lesser extent certain earthworms and millipedes. Polyphenols may have other roles in the soil such as helping to make P more available in acidic soils by preventing its adsorption by soil particles, but more study is needed to see how widespread such a mechanism is.

A good deal of research has been carried out on the effect of **mixing litter from different species**. The results have not always been predictable but in 67% of the cases looked at by Gartner and Cardon (2004), rates of decomposition were faster when the litter of several species was mixed together than when kept separate; the litter disappeared quicker, releasing more nutrients with a greater abundance and activity of decomposers. The reason behind this is that different litters complement the deficiencies or problems of each other. For example, nutrients released from easily decayed litter can stimulate decay in adjacent more recalcitrant litter. There may also be a number of indirect reasons as well. For example, a study by Hansen and Coleman (1998) and Hansen (1999) found that mixtures of yellow birch *Betula alleganiensis* and sugar maple *Acer saccharum* decomposed quicker if leaves of red oak *Quercus rubra* were added even though red oak litter is much harder to decompose. The underlying reason is that the oak leaves support a more diverse and abundant community of mites, and the activity of these mites in turn increased the moisture-holding capacity of the litter which itself helped further decomposition. In some cases decay can be slowed down when litters are mixed because of some antagonism caused by the release of inhibitory compounds such as phenolics or tannins. For example, compounds released in leachates from black oak *Quercus nigra* are known to slow the decay of leaves of sweetgum *Liquidambar styraciflua*. But, as with many aspects of soil biology, the bigger picture of what these beneficial mixtures contribute to the growth of whole stands is obviously complex and as yet less fully understood. A review of the nutritional status of mixed stands around the world by Rothe and Binkley (2001) found that the pool of soil nutrients available to plants generally did not differ between stands containing a mixture of trees and monocultures, despite the differences in decomposition rates observed above.

## 7.6 Rates of decomposition

### *7.6.1 k values and mean residence time*

With so many factors potentially affecting decomposition it is no surprise that decomposition rates vary tremendously around the world. These rates can be expressed in absolute terms as the weight of litter that disappears in an area

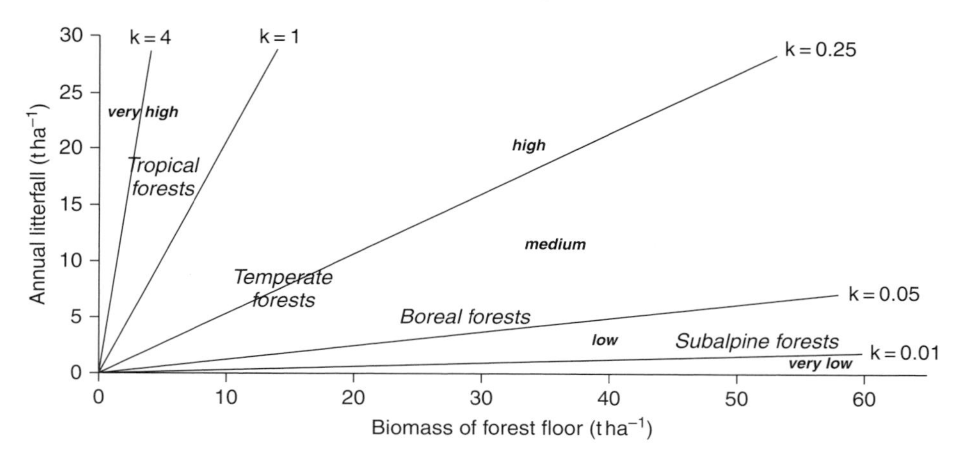


Figure 7.6 Estimates of the decomposition rate constant (k) for different forests, defined as the ratio of amount of annual litterfall to the amount of organic matter in the forest floor. Higher values of k indicate more rapid long-term decomposition rates. The inverse of k is called the mean residence time and gives an estimate of how quickly litter decomposes. (Redrawn from data from Olson, 1963. *Ecology* 44 and Rodin and Bazilevich, 1967. *Production and Mineral Cycling in Terrestrial Vegetation*. Oliver and Boyd.)

over a set time. Such figures can be useful in helping physiologists understand how much of a nutrient is available over time. For the forest ecologist an equally useful measure is to get a feel for the overall long-term decomposition rate by looking at the ratio of how much litter (or better still, total necromass) falls in a year relative to the amount of organic matter stored in the forest floor. This figure is referred to as **k** or the **decomposition rate constant** (implying that this figure stays constant in any particular forest). If a great mass of litter falls in a year and very little of the organic matter remains in the soil (so k is high) then the litter must be disappearing soon after it hits the ground and decomposition rates are high. Conversely, if the same mass of litter arrives but there is a good accumulation of organic matter in the soil (giving a low value of k) then decomposition rates, evened out over the years, must be low. Globally, the value of k varies from over 4 in some tropical forests to less than 0.01 k in subalpine forests with a cold, short growing season (Fig. 7.6).

These values of k reflect two general trends. The yearly production of necromass tends to decrease with increasing distance from the equator, and the amount of organic matter building up in the forest floor is, on the whole, the reverse: it increases with latitude. Thus organic matter contents should be low in tropical forest soils and high in boreal forests. As a gross generalization this is true but there is a very wide variation, including many organic-rich tropical soils, such as in cloud forests, deep tropical peat bogs and freshwater swamp forests merging to mangroves, and very sandy, impoverished northern soils resulting from severe fires. Typical approximate figures for the accumulation of organic matter in forest floors are 15–100 t $ha^{-1}$ in northern boreal forests, 7.5–12.5 t $ha^{-1}$ in temperate broadleaved forest and 1–2.5 (and up to 10–12) t $ha^{-1}$ in tropical forests (Vogt *et al.*, 1986).

Another way of looking at these types of data is to consider the **mean residence time**, defined as **1/k** or how long it takes for the equivalent mass of litter to disappear completely. In other words, if 1 tonne of litter falls into an area in one year, how long does it take for 1 tonne of weight to disappear from the forest floor (although not necessarily the same tonne that was added)? The material whose destruction is responsible for this weight loss will be made up predominantly of the softer and easily digested parts of the new litter but will be supplemented by some of the tougher components from older organic matter slowly decomposing lower in the soil. Since this is the inverse of k, in tropical forests with a k value of 4, it will take ¼ = 0.25 years for that amount of organic matter to decay and disappear. Estimates of the mean residence time vary from 0.25–2.5 years in tropical forests (longest in those with the most infertile soils) through 3–4 years in warm temperate deciduous forests to 60–350 years in boreal conifer forests. On a global scale, average residence time for litter is around 5 years and for coarse woody debris 13 years (Matthews 1997); see Section 7.7.

The time taken for an individual piece of litter to decompose will be longer than the mean residence time (as noted above, apparent decomposition of fresh litter is helped by other organic matter decomposing). The technical reason for this is that for a single piece of litter k is not really a constant (Fig. 7.6); decomposition slows down as the most easily digested compounds are used up and the more recalcitrant portions are left. Olson (1963) modelled this process in 1963 and suggested that since the rate of decomposition is negatively exponential it would take a period of 3/k to reach 95% total decomposition and 5/k for 99% decomposition. Thus in a forest where k = 0.5, the mean residence time of a new input of litter would be 2 years (1/k), but an individual leaf might take 6 years (3/k) to 95% decompose and 10 years (5/k) for 99% of it to decay.

Different types of litter will, of course, have their own k values which contribute to the overall forest k value. Vogt *et al.* (1983) record that the 99% disappearance time in a fir stand in Washington state was 2.2 years for species of *Allectoria* lichen and leaves of blueberries *Vaccinium* species, 33 years for Pacific silver fir *Abies amabalis* cone scales and 38 years for mountain hemlock *Tsuga mertensiana* cones. Similarly, different soil horizons have their cumulative k value. In the Vogt *et al.* (1983) study the mean residence time for the litter was 11–12 years and for the organic matter below the litter it was 32 years in the 23-year-old stand (3.2% annual turnover) and 69 years (1.5%) in the 180-year-old stand.

An interesting corollary of Olson's work is that if decay of a leaf becomes progressively slower with time (i.e. negatively exponential) such that it takes 5/k years for 95% to be lost, then the amount of organic matter in the forest floor will be slowly increasing through that time. So for a northern or subalpine forest where k values are low (0.01) the amount of organic matter in the soil may be slowly increasing for centuries after a disturbance ($5/0.01 = 500$ years for 95% decomposition) before it reaches a nominal steady state, and in the case of the subalpine forest Olson suggests that it may actually take 3000 years to reach that steady state. This is long after the time the vegetation above has reached maturity when most ecologists would have considered any change long finished; a salutary lesson not to ignore what happens in the soil!

The existence of different rates of decomposition for different fractions of the soil organic matter has led soil scientists, particularly those interested in climate change and carbon storage, to separate soil organic matter into three distinct pools of organic matter (but bear in mind that the 'pools' are conceptual artifacts to help understand a continuous trend). The **active carbon pool**, consisting of root exudates and the rapidly decomposing components of fresh litter, has a mean residence time of a matter of years and so is very susceptible to rapid change. The **slow carbon pool** is made up of the material coming from the active pool which is more resistant to decomposition and typically contains a higher proportion of components such as lignin. This pool has a turnover time that is measured in decades. Finally, the **passive carbon pool** contains the humus which, despite continual microbial attack, is the most stable organic matter with a turnover time of millennia.

### 7.6.2 The final product – humus

The amorphous and colloidal humus in the slow carbon pool is the most persistent part of the soil organic matter. This is partly because the humus is made up of the more inert residues of decaying organic matter (including a good deal of lignin) and complex microbial products synthesized during decomposition. Its persistence is also due to the humus tending to combine with clay to form stable clay–humus complexes which are more resistant to decay. It is perhaps not surprising then that humus may account for 60–90% of the soil organic matter. As well as being the major store of carbon in the soil, the humus is also important because it has a high cation exchange capacity (see Section 2.2.1) and so has a strong influence on nutrient cycling in a soil.

Humus can be rich in nitrogen but this is mostly locked up in complex molecules and so largely unavailable to plants.

Humus can be separated by chemical extraction into three distinct components: humins, humic acids and fulvic acids. Humins are black, insoluble and have the highest carbon content and the lowest oxygen content of all three. Humic acids are dark brown to black and are not soluble in acidic water; they are the most easily extracted components of humus. Fulvic acids tend to be yellow or yellow brown, are soluble in water at any pH, have the lowest carbon content but the most oxygen. The ratio of humic acids (HA) to fulvic acids (FA) changes predictably among soils. Forest soils often have an HA/FA ratio of less than 1 (i.e. comparatively low amounts of humic acids) compared with a ratio of more than 2 in grasslands.

## 7.7 Woody material

### *7.7.1 Quantities of dead wood*

Dead wood in a forest is much more variable in time and space than leaf litter and consequently has been the Cinderella of the forest, largely ignored for a long time but now seen to be crucially important. Dead wood appears in many forms, sizes and positions (and so is difficult to measure) including standing dead trees (snags), dead branches in the canopy, and trunks and branches on the ground. It should also include the dead heartwood of living trees but, as George Peterken (1996) points out, this is usually ignored. A useful term for this motley collection is **coarse woody debris (CWD)** defined as all dead wood over a minimum diameter, typically 2.5 cm although in some studies 7.5 cm is used; anything smaller than this minimum is included in the litter. Coarse woody debris thus includes everything from the largest fallen tree to small branches held in the canopy.

As well as being variable in size and location, CWD is also produced much more irregularly over time than litter. The timing of litter production tends to be fairly regular and predictable (with some exceptions – see Section 7.4), leaves, twigs and reproductive parts being shed in a process largely controlled by the vegetation, overlain by a less predictable amount caused by disease and the like. The production of large pieces of dead wood, on the other hand, is mostly controlled by factors outside the tree and tends to occur more irregularly over the years. In some cases this is fairly predictable over time. For example, in a young stand severe competition and suppression of growth will lead to partial or complete death of weaker individuals at certain points in the stand's history. Successional changes, such as when oaks take over from pioneer birch, will similarly produce a burst of dead wood. Other factors that create dead wood, such as disease, high winds, ice, snow and fire, are less predictable in occurrence. Some of these events produce very large quantities of dead wood such as the New England hurricane of 1938 (Fig. 7.7), the severe windstorm on 12 October 1962 which blew down an estimated 26 million cubic metres of timber in north-western North America and the high winds of 16 October 1987 in southern England which blew over 15 million trees. In the last case, high wind speeds just short of hurricane force were aided by wet soils and a late autumn (trees still had a large 'sail area' of leaves to catch the wind). Despite the external forces governing dead wood production, in some forests there can be a tendency towards seasonality of tree fall due to predictable high winds or the wet season; for example, in the tropical forests of Panama, tree fall is maximal in August through to September. Forest fires are particularly good at producing standing dead wood (snags). Cline *et al.* (1980) counted 815 snags per hectare after fire in the moist coniferous forest of western Oregon.

Despite the problems of measuring the amount of CWD, and its great variability in timing, estimates of the yearly production of CWD have been made: 0.12–30 t $ha^{-1}$ $y^{-1}$ in temperate coniferous forests, up to 16 t $ha^{-1}$ $y^{-1}$ in deciduous forests and 5 t $ha^{-1}$ $y^{-1}$ in tropical rain forests of Costa Rica (see Harmon *et al.*, 1986 for more detail). Generally, more CWD is produced in a year in coniferous forests than deciduous forests.

Typically, dead wood in a forest forms up to a quarter of all the above-ground plant biomass. Total accumulations of CWD are typically in the range of 11–50 t $ha^{-1}$ in deciduous forests. Conifer forests commonly hold more CWD than deciduous forests, from 10 to more than 600 t $ha^{-1}$ although intensively managed forests may have almost none; something around 100 t $ha^{-1}$ is typical. Tropical forests, with more rapid decomposition, usually have lower amounts of woody accumulation but levels up to 100 t $ha^{-1}$ are possible in more waterlogged areas of the Amazonian forest. The highest amount of CWD on record appears to be in a 63-year-old messmate eucalypt *Eucalyptus obliqua* forest in Tasmania regenerating after a fire containing 1089 t $ha^{-1}$ of CWD $>$ 1 cm diameter (Woldendorp and Keenan, 2005). The largest accumulations of CWD in any forest are usually in streams and rivers where waterlogging slows decomposition. In deciduous forests at least, there is a suggestion that accumulations of CWD are higher in cooler regions (Muller and Liu, 1991) with accumulations in eastern North America of 22–32 t $ha^{-1}$ in warmer old-growth forests of Kentucky and 34–49 t $ha^{-1}$ in cooler forests of New Hampshire. This would seem to be due more to slower decomposition rather than a greater annual input of dead wood. If 100 t $ha^{-1}$ of wood was

Figure 7.7 Pictures of the 1938 hurricane at Harvard Forest. On 21 September the 'Great hurricane of 1938' cut a 150-km wide strip through New England, starting at the Long Island coast and moving up through central Connecticut and Massachusetts to north-west Vermont, maintaining a wind speed of 220 km per hour (140 mph) with gusts up to 300 km per hour (186 mph). Along the way it destroyed 8900 homes and damaged another 15 000, making 63 000 people homeless and killing 600. Estimations of between

spread evenly over the forest floor it would amount to 10 kg in each square metre, but the bulk of the wood is in large pieces so the floor is not so cluttered. Typically, less than 5% of the ground will be covered by woody debris although this can rise to around a third cover in very heavily loaded coniferous forests.

Species composition of the CWD does not necessarily reflect the make-up of the canopy. Muller (2003) found in a mixed hardwood forest in the Appalachians of Kentucky that sugar maple *Acer saccharum*, yellow poplar/tulip tree *Liriodendron tulipifera* and black gum *Nyssa sylvatica* were 50% less abundant than their proportion of living biomass would suggest, and white oak *Quercus alba*, red oak *Q. rubra* and black locust/false acacia *Robinia pseudoacacia* were 50% greater. This is partly explained by ease of decomposition: the first group are rapidly decomposed. But since large woody pieces take so long to decay, the CWD tends also to reflect the past status of the forest and longer-term dynamics.

**Snags** (standing dead trunks, otherwise known as **hulks** – Fig. 7.8) are of particular wildlife interest. Up to 25% of dead wood in most forests is standing, sometimes more (up to 40%) but often much less even in old stands that have had time to accumulate dead trees. In the Białowieża Forest of Poland, one of the most pristine forests in Europe, Bobiec (2002) observed that standing dead wood varied from 3–21% of total CWD. Cline and colleagues (1980) looked at 30 conifer forest stands from 5 to 440 years old in the mountains of Oregon and found the densities of snags over 9 cm diameter decreased with stand age (from 35 to 18 per hectare in stands 120 and 200 years old, respectively) but the mean diameter of snags increased from 13 to 72 cm over stand ages of 35–200 years; larger snags survive longer. This is underlined in Fig. 7.9 which shows the progressive deterioration of a Douglas fir snag. Not all species collapse in such a graceful manner: western red cedar *Thuja plicata* in the same forests forms a bark-free grey 'buckskin' snag which remains essentially entire until the base is rotted through and it falls in one go 75–125 years after death. In the mixed spruce and fir forests of New Brunswick, Canada,

---

Caption for Figure 7.7 (cont.)
275 million and 2 billion trees uprooted or damaged have been made, accounting for more than 4 billion board feet of timber (6.5 million cubic metres). The first picture shows the damage in 1938 at Harvard Forest, central Massachusetts and the second shows some of the glut of timber being stored in water in Tom Swamp, Harvard Forest (March 1939) to reduce rotting until it could be used. (Photographs courtesy of Harvard Forest Archives, Harvard University.)

Figure 7.8 Dead 'snags'. The first is a common beech *Fagus sylvatica* snag (= hulk) in the virgin forest (urwald) of Kyjov, Slovakia and the second shows that fallen trees may not initially be on the ground (taken 1924, Harvard Forest, central Massachusetts). (First photograph by John R. Packham, the second courtesy of Harvard Forest Archives, Harvard University.)

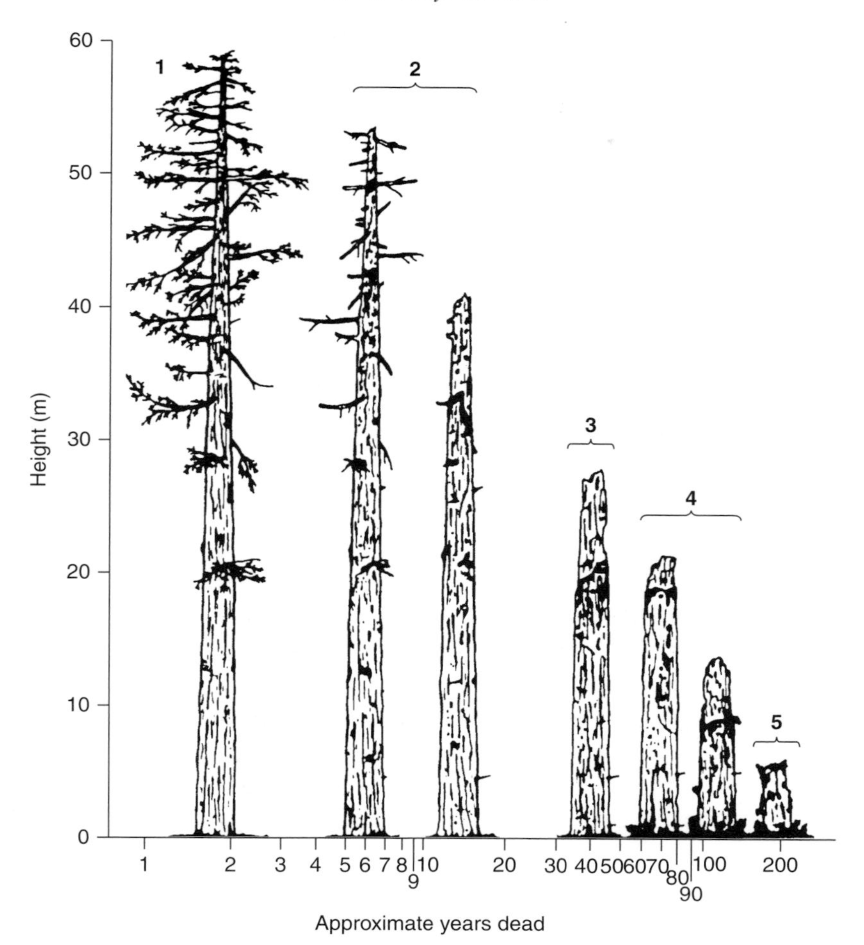


Figure 7.9 Stages of deterioration of large Douglas fir *Pseudotsuga menziesii* snags in the Coastal Range mountains of western Oregon. Stage 1 (0–6 years): much activity from wood-boring beetles and hence woodpeckers, some wood decay starting. Stage 2 (7–18 years): fine branches gone and big limbs and top breaking off; larvae of large beetles (e.g. long-horned beetles) penetrating deeply into the wood and increasing decay encouraging hole-nesting birds. Stage 3 (19–50 years): Sapwood extensively decayed and breaking off, heartwood extensively burrowed and decaying. Stage 4 (51–125 years): Sapwood gone and no sound heartwood remaining; vegetation begins to colonize the snag and the mound of broken wood accumulating at the base. Stage 5 (126 years and older): very rotten short trunk remains but stabilized by roots of invading shrubs and trees. Note this time sequence applies to large snags over 50 cm diameter; smaller snags progress through the stages more rapidly: snags 19–47 cm diameter are likely to reach stage 5 in around 60 years, snags 9–18 cm will reach stage 4 in less than 20 years and will fall before reaching stage 5. (From Cline *et al.* 1980. *Journal of Wildlife Management* 44.)

50% of dead trees go through this gradual standing deterioration with progressive loss in height while 23% are windthrown intact (Sarah Taylor, pers. comm.).

### *7.7.2 The role of coarse woody debris in forests*

The vital importance of dead wood in forests has been increasingly appreciated over the last decade (e.g. Kirby and Drake, 1993). It is important in carbon budgets of forests and also as an invaluable wildlife resource, especially in streams and rivers where CWD has a major influence in regulating sediment transport and storage. In large rivers, CWD provides a diverse array of habitats and in heavily wooded areas such as the Pacific Northwest is particularly important. Interest in this material is of long standing: some of the most interesting early work on CWD longevity was done in Sweden in the 1930s by Sernander (Box 7.3).

Coarse woody debris is a crucial habitat for terrestrial invertebrates such as bark beetles, wood-boring beetles, termites and carpenter ants, many of which are specialist **saproxylic insects** living on dead wood. They use it for food, hibernation, shelter and breeding (see Peterken, 1996). Bouget and Duelli (2004), reviewing the effects of windthrow on insects, highlight that the 'dead-wood islands' created are regional hotspots of insect biodiversity. Moreover, 20% of Swedish red-listed beetles are favoured by these extensive windthrows, and only 4% are judged to be harmed (Berg *et al.*, 1994). Large snags, particularly when over 50–60 cm diameter (Mannan *et al.*, 1980), are readily used by birds for feeding and nesting. One example of the many that could be chosen illustrates this: Cline *et al.* (1980) found that 20% of all bird species in the conifer forests of Oregon featured in Fig. 7.9 depend on snags. In Great Britain, around the same percentage of birds, including the willow tit and the spotted woodpeckers, depend upon dead wood.

Rotting wood is favoured by a number of plants. Mosses and liverworts (see Box 7.4) do particularly well because it is a firm substratum to grow on, holds water and is above the worst of the field layer and litter accumulation which would otherwise cover them up. Fallen **nurse logs** are renowned as nurseries for tree seedlings, especially in moist forests (such as the Pacific Northwest and the boreal forest) for similar reasons (see Fig. 1.7). Harmon *et al.* (1986) gives a good review of this phenomenon and notes that shade-tolerant species such as western hemlock and sitka spruce are particularly common on nurse logs. It is not just trees that benefit; in the moist forests of Oregon, red huckleberry *Vaccinium parvifolium* and salal *Gaultheria shallon* are commonest in, and dominate, the vegetation growing on broken stumps.

**Box 7.3 Change and decay: necrotization of spruce logs and tree regeneration in Fiby urskog, Sweden**

Compared with such trees as Scots pine, birch or aspen, Norway spruce *Picea abies* shows remarkably little resistance to wind, and in the boreo-nemoral (open boreal) primitive forest of Fiby urskog it is frequently blown down leaving **storm gaps** which facilitate tree regeneration. Sernander (1936) allocated fallen trunks to **six necrotization classes**, and used their condition in attempts to estimate the dates at which particular storm gaps were formed. Logs which were not penetrated when an iron spike was thrust at them fell into one of the first three categories. Logs in class 1 still had needles on the branches. Bark had begun to fall off logs in class 2, but no epiphytic bryophytes were present. These bryophytes first developed on class 3 logs, from which all bark has fallen. Logs which were penetrated to a depth of 4–6 cm belonged to one of classes 4–6. In logs belonging to class 4 the side branches had collapsed so the log was in contact with the ground; there was also a considerable cover of epiphytic bryophytes, mainly of species not common on the forest floor. Logs in class 5 had begun to disintegrate while those in class 6 were rotten right through and totally covered in bryophytes, particularly large species characteristic of the forest floor such as *Dicranum* spp., *Hylocomium splendens, Pleurozium schreberi, Polytrichum* spp. and *Rhytidiadelphus triquetrus*.

Figure 7.10a shows a 25 × 30 m area of Sernander's plot III of 1935 as it was when remapped in 1985 by Hytteborn and Packham (1987). Trees and fallen stems are shown, with fallen aspen *Populus tremula* being indicated by dashed lines. The necrotization condition of individual logs is shown by a number. The pecked line shows a storm gap, rather new in 1935, whose position has moved somewhat. Many young spruce exist for years as **dwarf trees** before the creation of storm gaps provides the light required for them to grow comparatively rapidly into adults (see Section 9.3.3). Growth rings of dwarf trees are very narrow and often incomplete (semi-lunar). The size of the symbols indicates the height classes of the live trees present.

Figure 7.10b shows the distribution of seedlings and saplings near the living spruce Z, whose position in the larger area of plot III is shown in (a). The tops of the young aspen and many of the rowan had been grazed by deer. Many aspen shoots subsequently die, with suckers regrowing from the base. Numerous small birch *Betula* sp. seedlings and saplings were also present, particularly so on the base of the large spruce B, which crushed the fallen log of spruce A which when living shared the same root system as the large living spruce Z (which showed 185 annual rings at a height of 0.35 m). Root grafting is very common in spruce and often occurs when the trees are very young.

Working on a large scale, Sernander allocated a single grade number to each log and thought that a period of 90 years would have to elapse before a log was so disintegrated that no trace of it was obvious on the forest floor. The later

**Box 7.3 (cont.)**

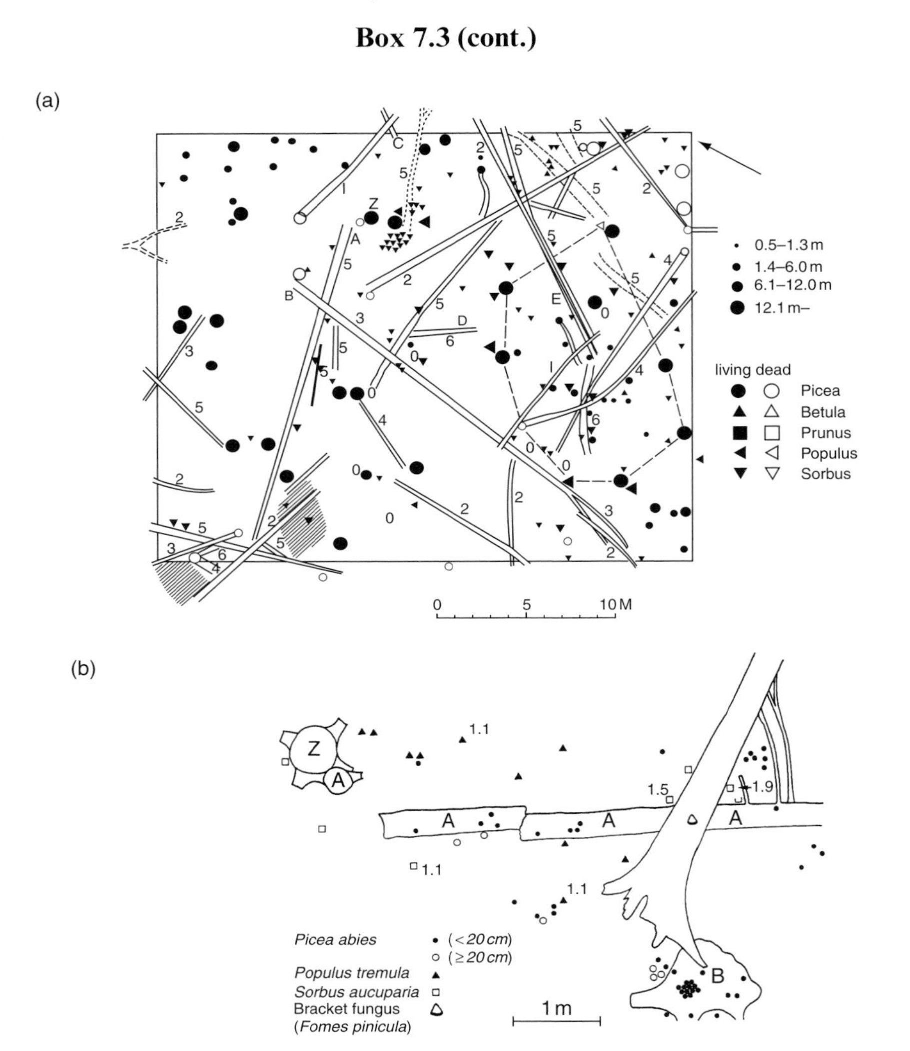


Figure 7.10 Change and decay: necrotization of spruce logs and tree regeneration in Fiby urskog, Sweden.

investigations of Hytteborn and Packham (1987) showed this to be an overestimate. Eight of the stems which had fallen since 1935 had reached class 5 in 50 years or less. Decomposition rates will in any case differ in particular cases and even within individual logs. Log A was a living tree in 1935, the basal portion of it was adjacent to its old tree base and its now utterly rotten wood entirely covered by a bryophyte mat in which *Ptilium crista-castrensis* was conspicuous. Regions of the same log on the other side of log B were much less decayed, the first five metres falling into class 4 and parts near the stem apex into class 3.

(Both figures from Hytteborn and Packham, 1987. *Arboricultural Journal* 11.)

### Box 7.4 Mosses and liverworts on rotting wood

Managed and unmanaged forests differ in many ways, particularly in terms of their biodiversity. Eighty plant species, of which 50 are mosses and liverworts (bryophytes), are acutely threatened by intensive forest management in Sweden, with another 72 that are vulnerable (40 being bryophytes). Andersson and Hytteborn (1991) investigated the bryophyte populations of an area of the natural forest of Fiby urskog dominated by Norway spruce, and of a managed forest about one kilometre away. The managed forest was clearcut for charcoal production in the 1930s, thinned in 1973/4, and fertilized with ammonium nitrate ($NH_4NO_3$) in 1979. Some 55% of the trees were Scots pine, the remaining being Norway spruce with a few birch; aspen occurred only in Fiby urskog. Eight $10 \times 20$ m sample areas were employed in each forest. Records were made from the decaying wood of the stumps and logs, but not the standing dead snags, in each plot. In the managed forest, decaying wood on the ground consisted mainly of stumps and tree tops from the thinnings, whereas Fiby urskog contained many fallen logs whose decay is described in Box 7.3.

Altogether a total of 54 bryophyte species, belonging to four different groups, were found on decaying wood. Of these 48 came from Fiby, compared with 33 from the managed forest (six of which were not found in the Fiby plots). The primeval forest had a larger amount of dead wood, much of it consisting of large-diameter logs that were absent from the managed forest. The presence of particular species was related to the type of tree involved as well as its degree of decay. Eight decay classes were used, starting with wood that was hard and parts of logs with intact bark and finishing with completely soft material. **Facultative epiphytes**, e.g. *Ptilidium pulcherrimum*, were the first to colonize decaying rotting wood, being followed by **epixylic** specialists (living on the outside of wood) and then by **epigaeics** (living at the soil surface), such as *Hylocomium splendens* and *Rhytidiadelphus triquetrus*. **Opportunistic generalists** including *Hypnum cupressiforme* and *Brachythecium rutabulum* showed a more irregular pattern of occurrence. There were 16 epixylic specialists, plants such as *Aulacomnium androgynum* dependent on decaying wood, in the Fiby plots, but only five of these were present in the managed forest.

Coarse woody debris has an important part to play in carbon budgets of forests (see Chapter 11) but much less of a role in nutrient budgets simply because wood is so poor in nutrients. Laiho and Prescott (2004), in a review of the subject in northern coniferous forests, point out that CWD contributes 3–73% of above-ground litter input, but $< 20\%$ of N, P, K (potassium), and Ca (calcium). Although up to 54% of accumulated organic matter (including that in the soil) is CWD, it contributes $< 5\%$ of the N, $< 10\%$ of the P, and

< 25% of the K, Ca, and Mg (magnesium). As discussed above with litter, CWD is initially a sink for N and P, becoming a source as decay progresses but it still plays a minor role in overall availability. Moreover, due to the low initial N levels, wood can be a net sink for N for many decades.

The importance of maintaining adequate amounts of dead wood in fully functioning and highly diverse ecosystems is emphasized in Section 3.8, which deals with veteran trees. The unquestionable value of CWD for wildlife has led to many recommendations for forest management to maximize the amount of dead wood (see Chapter 10), despite the human desire for order and tidiness in forests. These recommendations include such notions as leaving mature trees to die, rather than being harvested, because they provide the most wildlife-friendly CWD, and keeping forked trees as they are likely to have a higher biodiversity of fungi (Heilmann-Clausen and Christensen, 2003). This train of thought has led to suggestions that dead wood should not just be kept, it should be created by deliberate mutilation. Ways of mutilating trees are many and, to quote George Peterken (1996, p. 420) '. . . can be achieved by sawing rot holes and ring-barking, or by more exciting techniques such as fire, exploding the crown from the trunk or holding a vandals' convention'. These techniques do work. A major fire at Ashtead Common in Surrey, England seriously scorched and damaged a significant proportion of the 2000 oak pollards and many maiden trees. The subsequent aim was to retain as much standing dead oak as possible while rendering the standing trees safe, so chainsaws were used to remove dangerous branches, and replicate the shattered ends of storm-damaged limbs in a technique known as **coronet cutting**.

### *7.7.3 Ecology of wood decomposition*

Wood is difficult to digest not just because it is chemically tough but because it is so poor in nutrients. It is composed of 40–55% cellulose, 25–40% hemicelluloses and 18–35% lignin (conifers having a greater proportion of lignin than hardwoods). Wood is thus high in structural carbohydrates (which require specialized enzymes to break them down) but poor in such elements as nitrogen; wood has 0.03–0.1% N (by mass) compared to 1–5% in foliage.

Compared with litter, freshly dead wood loses little of its weight by initial leaching, partly because wood is low in soluble substances but also because of the low surface area to volume ratio. Physical fragmentation is, however, an important part of wood decomposition; breaking into smaller pieces increases the surface area exposed to the biological agents of decay. Snags are most prone to breaking up since they are exposed more to changes in temperature and high winds as they stand vertically. As shown in Fig. 7.9 the thinnest parts of snags are broken off first. Thus, twigs and small branches are rapidly lost (within a handful of years) followed by progressively fatter parts of the trunk. Most snags will eventually fall once the roots are weakened within a decade or so, usually precipitated by strong winds, although as outlined above, there are exceptions with western red cedar standing for over a century and bristlecone pines for millennia. When the snag hits the ground, if it is sufficiently rotted it will shatter upon impact. At the other end of the size spectrum, individual fibres are physically separated and washed off the outside of sun-exposed logs. **Fire** is capable of consuming large amounts of CWD but since larger logs can be almost completely fire-proof (due to their low surface area to volume ratio) its influence is not always as much as might be expected. Tinker and Knight (2001) looked at CWD consumption during a fire in Yellowstone National Park and estimated that only 8% of CWD (above 7.5 cm diameter) was consumed by the fire and another 8% turned into charcoal. Moreover, fires also produce more CWD by killing branches and whole trees, and Tinker and Knight concluded from a computer simulation that over 1000 years a fire-return interval of 100 years resulted in more CWD than a 200- or 300-year interval.

Wood decomposition proceeds at an impossibly slow rate without the intervention of biological agents. Of these the two most important groups are burrowing invertebrates and particularly fungal rots. Fallen trees go through a process of collapsing and settling. At first the trunk may be supported above the ground by the branches, but as these rot it settles into contact with the ground, keeping it moister. The bark is gradually lost and as the log begins to rot it settles further, increasing its contact with the ground, changing its suitability for microbes, invertebrates and vertebrates. Thus, as decomposition progresses, there is a succession of different organisms involved.

Heliövarra and Väisänen (1984) recognized four phases of **insect succession** on fallen wood in temperate forests. Phase A starts with short-term feeding on bark by bark beetles and longhorned beetles (Scolytidae, Cerambycidae). Phase B is composed of species living under the bark and in the surface layer of the wood. By this time the bark has fallen and Phase C follows, a long stage of several decades of wood-inhabiting species. In the final and longest Phase D, the wood-inhabiting species are replaced by animals living under the shelter of decaying logs, such as soil insects, snails and centipedes. During this stage other animals like frogs, salamanders (see Box 7.5) and snakes burrow under the log and moles and shrews tunnel in and around the log foraging for their prey. Finally the stem breaks up and merges into the soil organic matter. Burrowing insects, especially termites and the larvae of large beetles, can make sizeable channels into the wood and contribute greatly to decomposition (see Box 7.6). However, the main

### Box 7.5 Newts, salamanders and dead wood

Amphibians such as newts and salamanders (strictly speaking, newts are a type of salamander) are often ignored in forest ecology as interesting but minor players. Yet, in the hardwood forests of Hubbard Brook Experimental Forest in New Hampshire, Burton and Likens (1975) found around 2950 salamanders $ha^{-1}$, which may not sound many but this works out at more salamanders in the forest than either birds or small mammals. More importantly, the biomass of the salamanders was more than two and a half times that of all the birds during the peak of the breeding season, and equal to the biomass of all the mice and shrews.

The red-backed salamander made up almost 95% of the salamander biomass in these forests. Experimental removal of this species from similar forests by Wyman (1998) showed that the salamanders significantly reduced the decomposition of litter by 11–17%. The most likely cause is that they eat a significant number of the leaf fragmenters such as beetles, millipedes, snails and insect larvae.

Many forest salamanders are associated with the dead wood of streams and dry land. The red-backed salamander in particular is associated with a thick leaf litter and decaying logs or stumps. Thus, anything that affects CWD will affect the salamanders. Bury (2004) looked at whether salamanders and other amphibians and reptiles were endangered by prescribed burning and thinning for fire control since it reduces the amount of CWD and so reduces cover. Bury's evidence, however, suggests that gentle fires themselves have little effect because big pieces of wood are not consumed in normal fires, and during hot dry conditions suitable to fire, the animals are deep underground.

Figure 7.11 Newts, salamanders and dead wood. The red-backed salamander *Plethodon cinereus* of New England, USA. (Photograph by Brooks Mathewson).

**Box 7.6 Colonial insects and wood decomposition**

A number of termites, beetles and ants attack undecayed wood. Of these, termites account for the very rapid (and to us, potentially devastating) disappearance of wood in warm temperate and tropical forests. There are more than 2500 species of termites worldwide and the genus *Coptotermes* has 28 pest species, of which the Formosan subterranean termite or FST *C. formosanus* is the most widely distributed and economically important. Although probably endemic to southern China it was apparently transported to Japan prior to the 1600s and is now found in many other countries, having spread widely in the USA since the 1960s. As with other termites the FST has three primary castes: the reproductives, soldiers and workers. New colonies are each founded by two 'alates', one male and one female, from the huge swarms which fly on calm and humid evenings in early summer. The colony establishes on moist wood and takes 3–5 years to become large enough to cause severe damage. This termite attacks living plants and a number of structural materials as well as wood. Laboratory studies show that termites can consume up to 10% of their body weight in a day which, with colonies containing several million individuals, makes them significant and very difficult to control.

contribution of most burrowing insects is in producing galleries through the wood and in bringing fungal spores with them for direct inoculation inside the wood. Both of these help fungi to colonize more rapidly.

The larvae of the British stag beetle *Lucanus cervus* can reach a length of 12 cm and produce large and significant galleries in the decaying roots of deciduous trees which they help to break down. This is the largest beetle in Britain; adults can most often be seen from May to August, occasionally flying on warm evenings. The beetle, which causes no damage to garden plants, has become less frequent in recent years and it is now a UK Biodiversity Action Plan priority species.

Some of the wood-boring insect grubs found in fallen rotting logs, and for that matter tree trunks and limbs, of the forests of Australia and Papua New Guinea are surprisingly large and rapidly accelerate the decomposition process. Moreover, they are very attractive food items and the striped marsupial possum *Dactylopsida trivirgata* locates them by tapping the wood for the betraying hollow sound caused by the tunnels and then breaks in, further breaking up the log and skewering the grub with the sharp claw of the elongated fourth finger. This is by no means a modern strategy, indeed it may well have been employed by the now extinct *Heterohyus nanus*, which had two very long spindly fingers on each hand. These were probably used to extract insects from deep within rotting wood. Fossil remains of this animal which belonged to the apatemyids, a small group of insect-eating mammals known from the Palaeocene to the Oligocene, were found in the Messel pit near Frankfurt, Germany. The kaka *Nestor meridionalis*, a parrot which lives in large tracts of southern beeches in New Zealand, also uses its strong curved bill to dig into decaying trees for grubs and other insects.

Although insects are important, microbes are by far the most important biological agents responsible for wood decomposition. Of these, **fungal rot**, particularly by basiodomycetes, is responsible for the disappearance of most wood at moisture levels above 20%. Bacteria can break down woody cell walls but tend to be localized in their effect. They dominate in waterlogged wood over fungi, which is the main reason why submerged wood decomposes so slowly. As with insects, fungal decomposers go through a succession of types. The first invaders of freshly dead wood are normally the moulds and staining fungi which live on the easily digested cell contents (the **stain fungi**, mostly Ascomycetes, have pigmented hyphae which stain the wood, lowering its commercial value which is why foresters try to process cut wood as soon as possible). These early invaders fail to affect the integrity of the wood itself because they cannot digest **structural carbohydrates** or **lignin**, and in fact may slow subsequent decomposition by using up the free sugars and starches (the non-structural carbohydrates).

Three types of structural rot are often distinguished. A relatively small number of wood decay fungi break down just the cellulose leaving the brown-coloured lignin intact and so are called **brown rots**. These occur mainly on coniferous wood and cause the commonest type of wood decay in boreal forests. Other fungi break down both the lignin and cellulose components of wood leaving a weak white residue and so are known as **white rots**. Finally the **soft rot** fungi produce spongy pockets of decay in the outer layers of wet wood and are major players only in wet environments. Both brown and white wood rots attack wood using extracellular enzymes secreted outside the hyphae. Celluloses are very long unbranched strands of linked glucose units and, although insoluble, their repeating unit structure makes them readily degradable by cellulase enzymes. Hemicelluloses are more complex, being composed of branched polymers that contain several different sugars, and so are less easily decomposed. Lignin is particularly difficult to break down and requires around 12 enzymes, including ligninases and ring-splitting deoxygenases. The lignin is not itself used as a food source but its removal allows the attached cellulose to be digested.

The decomposition of a log usually involves several species of fungi working in succession, perhaps related to different complements of enzymes held by different species. However, the pattern is not normally clear and a large number of fungi may inhabit the same log at the same time. When this happens, the territories occupied by different species or different colonies of the same species are demarcated by black zone lines. These are double walls, built by the two touching colonies, composed of gnarled dark-coloured hyphae and functioning to reduce competition. Woodworkers enjoy this 'spalted' wood for its attractiveness.

A number of wood-decaying basidiomycetes that rely on wood as their only source of food are able to **forage** or search for new pieces of fallen wood on the forest floor. The hyphae of the fungus grow out through or over the soil in all directions from the wood it is currently living in. When the hyphae happen upon a new piece of wood all new growth of the fungus is concentrated on reinforcing the link between old and new wood with more hyphae to form a **mycelial cord**. This root-like structure transports water and nutrients to the new wood aiding the invasion process. A superb example of this military-like invasion is shown by the honey fungus *Armillaria* whose well-developed mycelial cords, called **rhizomorphs**, are so effective that they allow the fungus to invade living tissue (see Sections 5.4.2 and Fig. 5.10).

In most forests, wood will be colonized by fungi within a year and completely colonized within 5–10 years. However, **decay rates of wood** vary tremendously depending upon the climate, decaying organisms available and the size and type of wood. Small-diameter logs of Monterey pine left over from forestry operations in New Zealand were found by Ganjegunte *et al.* (2004) to be largely intact after 4 years, to have lost their bark by 9 years with wood still superficially sound, and to have lost 59% of their initial mass by 13 years. In tropical forests wood above 3 cm diameter takes at least 15 years to decompose; logs may disappear more slowly than might be thought (Anderson and Swift, 1983). In Fiby urskog, Sweden, Norway spruce *Picea abies* logs observed for 50 years were set to decay completely in less than 90 years. Conversely Franklin *et al.* (1981) found that 80-cm diameter Douglas fir logs in the Pacific Northwest lost around 40% of their original wood only after 150–200 years, and 90% after 480–580 years. Rates of decay are variable even within a single genus; the lifetime of eucalyptus logs ranges from 7 years in mountain ash *Eucalyptus regnans* in high rainfall mountainous areas to 375 years in the river red gum *E. camaldulensis* along water courses (Mackensen *et al.*, 2003). In general terms, pioneer trees invest less energy in protecting their wood from rot (they go for speed rather than defence and logs on the ground have a half-life of a few decades at most) compared with longer-lived late-successional trees which may persist as dead wood for a century or much longer. But again, environmental conditions play an important role in determining decay rates; balsam poplar *Populus balsamifera* logs which would decay away within 40–60 years on land have been found to last over 250 years when waterlogged in a beaver pond.

# 8

# Energy and nutrients

Growth and the limitations to it are of general interest to forest ecologists and of particular concern to foresters who grow trees as a crop. All tree species influence the nutrient content and other aspects of the soils in which they grow. Whether this results in an overall advantage or disadvantage to themselves, other plants, animals and the soil often depends upon local conditions and forestry practice. This chapter investigates how growth is measured, how it accumulates as biomass and the distinction between how much a tree can grow each year versus how much biomass the whole forest ecosystem can accumulate. The second part deals with nutrient flows and the problems of gains, losses and limitations. Much of what we know about temperate hardwood forests stems from the work carried out at Hubbard Brook (see Box 8.1).

## 8.1 Growth of forests

### *8.1.1 Biomass and productivity*

The sheer size of forest and woodlands is what people often comment upon. The weight or mass of organic material present is referred to as the **biomass** (or sometimes the standing crop). It should be borne in mind that this can be a somewhat loose term since it may or may not include dead wood or litter. Table 8.1 shows that the biomass above ground increases from the boreal forest towards the tropics, starting from very low levels at the Arctic treeline and reaching in excess of 940 t ha$^{-1}$ in the Amazon basin. However, there are exceptionally large forests outside the tropics, the record-holders being the temperate forests of the Pacific Northwest of North America including stands of huge Douglas fir (1600 t ha$^{-1}$) and coastal redwoods, the tallest trees in the world (trunk biomass of 3450 t ha$^{-1}$ with total **net primary productivity (NPP)** possibly approaching a staggering 4500 t ha$^{-1}$ y$^{-1}$). In most forests, more than

## Box 8.1 Hubbard Brook Experimental Forest

The name Hubbard Brook appears in many studies on nutrients and productivity in forests and these have provided some of the best information on the functioning of temperate hardwood forests. **Hubbard Brook Experimental Forest** in the White Mountain National Forest of New Hampshire, eastern USA (3160 ha) was established in 1955 as a centre for hydrological research. In 1963 the **Hubbard Brook Ecosystem Study** was set up within the forest by Herbert Bormann, Gene Likens, Noye Johnson and Robert Pierce specifically to study linkages between hydrology, nutrient dynamics and cycling in response to natural and human disturbances, such as air pollution, forest cutting, land-use changes, increases in insect populations and climatic factors. The forest is composed of a large watershed with all minor streams draining into the Hubbard Brook subdivided into a number of discrete watersheds underlain by impermeable bedrock of schist and granite (Fig. 8.1 below). The forest is second-growth hardwood forest dominated by sugar maple *Acer saccharum* (from which maple syrup is tapped), American beech *Fagus grandifolia* and yellow birch *Betula alleghaniensis*.

Much of the original work done here was based on inputs and outputs. It was assumed since the rock underlying the watersheds is impermeable, that looking at what arrived in precipitation and dry deposition and what nutrients came out in stream flow gave a good indication of what was happening in the forest in terms of nutrient budgets (mass balances).

The following are the notable uses of the different watersheds.

**W6** Untouched control. No. 6 has been used for energy and nutrient studies (see Sections 8.2 and 8.6.1) and has been undisturbed by fire or cutting since 1919. The vernal dam work of Muller and Bormann (1976) was carried out here (see Section 8.4.1).

**W2 (1965)** All trees were cut down but not removed, followed by herbicide use for 3 years to prevent regrowth. Used for nutrient loss studies (see Sections 8.4.1 and 8.6.1).

**W4 (1970)** Trees cut and removed in 25-m strips, every third strip being cut in 1970, second set cut in 1972 and the final strips cut in 1974. Used for nutrient studies (see Section 8.6.1).

**W101 (1970)** Trees cut and removed in blocks as a comparison to No. 4.

**W5 (1983)** All trees cut and removed to mimic commercial whole-tree harvesting conditions, using heavy machinery on the flat and chainsaws on inaccessible slopes. Partly used to study fine root biomass (Section 8.1.1) and nutrient loss studies (Section 8.6.1).

**W1 (1999)** Control untouched until 1999. Calcium has been declining in the forest over the last 50 years probably due to acid rain. In 1999 fifty tonnes of calcium was added to the watershed to bring calcium levels to pre-acid rain levels.

**Box 8.1 (cont.)**

Figure 8.1 Hubbard Brook. Above is a view of Watershed 2 and below is a view of Watershed 2, 4 and 5 (from left to right). Descriptions of the watersheds are given above. (Photographs courtesy of USDA Forest Service, Northeastern Research Station.)

85% of the biomass is usually contained in the above-ground portion of the woody plants. The extra biomass present in roots is often unmeasured but Jackson *et al.* (1996) in a comprehensive review show that root/shoot ratios increase from 0.19 in tropical evergreen forests (i.e. 0.19 tonnes of roots to

Table 8.1. *Above-ground biomass and net primary productivity (NPP) in four forest types. The data give the usual range encountered but inevitably examples of outliers above and below the figures given can be found. Figures from Reichle (1981) quoted in Packham* et al. *(1992) are considerably lower, but more recent work justifies the data given below.*

| | Above-ground biomass (t ha$^{-1}$) | NPP (t ha$^{-1}$ y$^{-1}$) |
|---|---|---|
| Boreal conifers | 10–160 | 1–10 |
| Temperate broadleaved deciduous | 150–300 | 7–12 |
| Temperate rain forest (Pacific Northwest) | 500–1000 | 15–25 |
| Tropical montane evergreen | 450 | 10–24 |
| Tropical broadleaved evergreen | 450–700 | 17–32 |

every tonne of biomass above-ground) to 0.23 in temperate areas to 0.32 in boreal forest. Although a greater *proportion* of the biomass is below-ground away from the tropics, above-ground biomass decreases along the same gradient, so there is still less biomass below-ground away from the tropics (49 t ha$^{-1}$ in tropical evergreen forests; 40–42 t ha$^{-1}$ in temperate and tropical deciduous forests; 29 t ha$^{-1}$ in boreal forests). Even fewer studies of fine roots ($< 2$ mm diameter) have been conducted: in the hardwood forest of Hubbard Brook (mostly in W5 – see Box 8.1) fine roots were calculated to add 4.7 t ha$^{-1}$ by Fahey and Hughes (1994). The **sclerophyllous shrubs** of dry Mediterranean environments, which are evergreens with small, hard, thick leaves, resistant to hot, dry conditions, put more of their growth below ground than above (root/shoot ratio of 1.2), presumably as a defence against frequent fire.

Although useful, biomass is a static measure of how much there is at any one time and gives no indication of how quickly new growth is being added or lost, and so gives little insight into how the forest is functioning. More useful are estimates of the **productivity** of the forest, how much new material is being added. Most primary productivity in a forest (the manufacture of organic complex compounds from simple inorganic substrates) is carried out when green plants use light in photosynthesis. By adding up all the photosynthetic activity in a forest the **gross primary productivity (GPP)** can be calculated in terms of the carbon fixed or sugars created. Plants have, however, to expend some of this energy in respiration (energy used to keep existing tissue alive). Gross primary productivity minus the respiratory costs gives the net primary productivity (NPP), the actual increase in growth that can be measured in tonnes of new material per unit area per year. For the forest to grow in size, GPP has to be bigger than the respiratory costs so that the NPP or **carbon balance** is positive. In practice, GPP is very difficult to measure since it is almost impossible to trace all carbon uses and losses in a complex forest. But estimates have been made which show that respiratory losses account for between a third and three-quarters of the GPP. Such estimates are useful since they give insight into how energy is flowing in a forest. For example, Kira (1975) found that a 25-year-old beech forest in Denmark which expended around 40% of photosynthate in respiration, had an NPP comparable to that of a mature lowland tropical rain forest, which, although having a GPP more than double that of the beech forest, had to support respiratory losses of 75%. Both GPP and respiration (and hence NPP) are dependent upon factors in the environment such as temperature, light and water, as described in Chapter 3. Nutrients are also vitally important to growth and so form the subject of part of this chapter.

Forest ecologists are particularly interested in the NPP since this is what other organisms have to feed upon, and it relates to how much carbon is being stored in a forest (discussed further in Chapter 11). Table 8.1 gives average rates of NPP for different forests, ranging from 1 t $ha^{-1}$ $y^{-1}$ in boreal forests to over 30 t $ha^{-1}$ $y^{-1}$ in tropical rain forests, and a maximum of 36.2 t $ha^{-1}$ $y^{-1}$ in a 26-year-old western hemlock *Tsuga heterophylla* stand in the Pacific Northwest.

There have been comparatively few studies that have investigated how the new material produced in NPP is distributed within the forest. Figure 8.2 shows such a study for a mixed-oak forest in Europe. It can be seen in this example that 85% of the new growth is above-ground and that only 15% is below-ground. If just the woody plants (trees and shrubs) are considered, the ratio is still 83% above-ground to 17% below-ground. This ratio may well be closer in some forests, possibly even reaching 50:50. Within the woody plants, 26% of the production goes into leaves, 65% into woody material (twigs, branches and trunks) with the remainder being such structures as flowers and fruits which, with dead wood and twigs, are absorbed into the non-leaf litter (Fig. 8.2).

It should be noted that as a tree gets bigger it becomes more or less fixed in size and so holds a maximum number of leaves. This creates an upper limit on GPP for an individual tree. At the same time, as the trunk gets thicker, the respiratory burden grows from keeping more living tissue alive. Thus, NPP of a tree tends to decline with age (see Section 10.1.3 for further information). If all the trees in a stand are of the same age, it can be expected that the overall stand NPP will also decline with age (Fig. 8.3). This decline would be less pronounced in a large uneven-aged stand where the proportion of different-aged trees stays more constant over time.

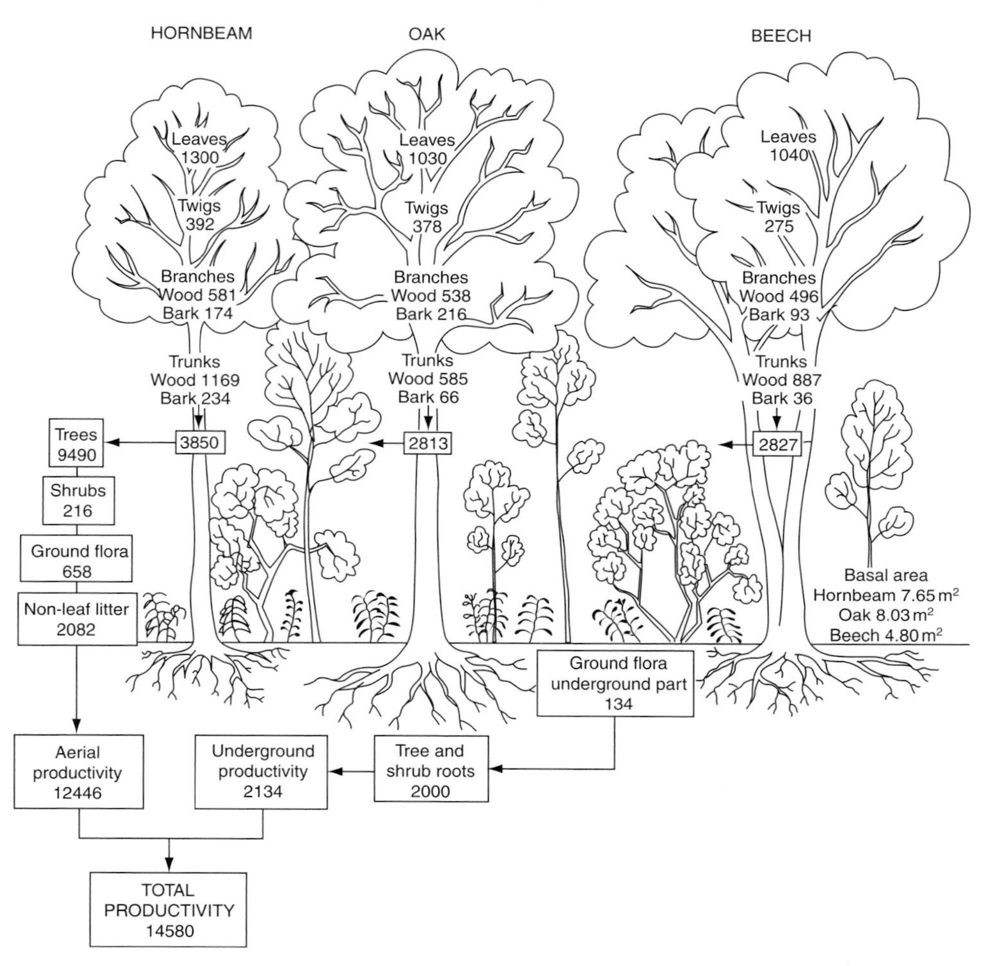


Figure 8.2 Annual net primary productivity (kg of dry weight $ha^{-1}$) of a European mixed-oak forest. Data can be converted to t $ha^{-1}$ by dividing by 1000. (Based on data from Duvigneaud and Denaeyer-De Smet, 1970. From Archibold, 1995. *Ecology of World Vegetation*. Chapman & Hall, Fig. 6.3.1. With kind permission of Springer Science and Business Media.)

Peter Attiwill (see Attiwill and Weston, 2001) describes three stages that a eucalypt forest goes through in growing to maturity which are undoubtedly widely applicable. In the first stage, a few decades long, the forest goes through a **building phase** where the major proportion of the NPP goes into building biomass, with a net uptake of nutrients from soil reserves. The second stage sees the rate of growth slowing and an increasing proportion of NPP stored as dead heartwood (see Fig. 1.2). Internal recycling of nutrients becomes more important and uptake from soil reserves decreases. In the final stage, the trees

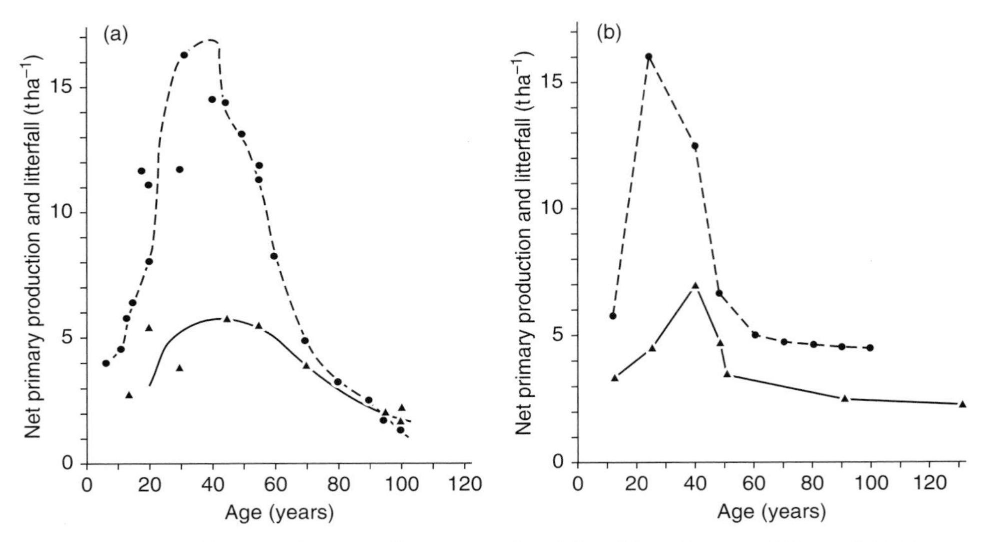


Figure 8.3 Changes in net primary productivity (•) and annual litter-fall (▴) with age in two Russian forest types, (a) southern taiga pine forests (see Section 1.6.1 for more information on taiga), and (b) oak forests near Moscow. The graphs show that net primary productivity reaches its highest levels at 40–50 years and 25 years old, respectively, in the pine and oak stands. (From Rodin and Bazilevich, 1967. *Production and Mineral Cycling in Terrestrial Vegetation*. Oliver and Boyd.)

reach **maturity** and the amount of added biomass approaches zero; the major proportion of NPP is shed as litter and coarse woody debris (CWD), and nutrients are increasingly immobilized in the litter.

### *8.1.2 Ecosystem productivity*

It is important to differentiate between the productivity of the component plants and the total productivity of the stand. Though the NPP of the individual trees declines with age, in a large uneven-aged stand young trees replace dead trees, so in the absence of any major disturbance, NPP should remain more or less constant with time. When all the components of a forest including the soil are considered, however, the total productivity of a stand will inevitably decline with age. As shown in Fig. 8.4, gross stand production increases rapidly with age, matching almost exactly the pattern of increase in biomass of leaves. As the forest stand gets denser, leaf biomass and gross stand production reach a peak followed by a slight drop as the stand self-thins to a sustainable level. Thereafter, gross stand production and leaf biomass settle down almost to a constant, reflecting the more or less constant NPP. Over time, however,

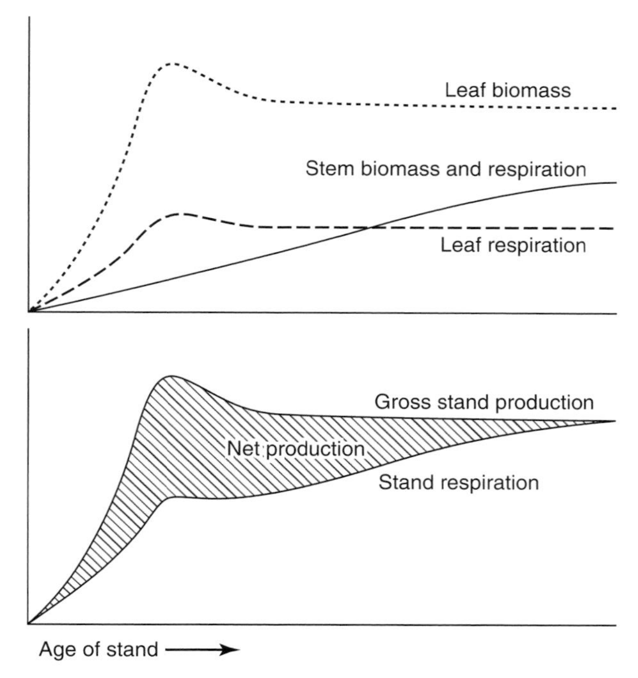


Figure 8.4 Changes in productivity in a hypothetical forest stand with age. (Based on data from Duvigneaud, 1971. In: *Symposium on the Productivity of Forest Ecosystems*. Ed. P. Duvigneaud. UNESCO.)

the respiratory cost of running the stand increases. This is partly because of the build-up of living biomass (as trunks get bigger and the flora builds to a maximum) and which forms part of the NPP calculation as discussed above. Also, however, there is a build-up of organic matter in the soil and dead woody material which decomposes, causing a build-up of respiration needs in the decomposer subsystem. As total respiratory losses of the stand increase for both these reasons, there will come a point where the total stand respiration equals the gross productivity (Fig. 8.4). At this point the *net* stand productivity (gross productivity minus respiratory costs) approaches zero; the stand stops getting bigger because much NPP is shed as dead material and decomposes. All the biomass produced in a year is matched by an equal loss of biomass used in respiration above and below the soil surface. To avoid confusion with NPP, this net stand production is referred to as the **ecosystem productivity**. As the stand reaches maturity the ecosystem productivity approaches zero.

As will be seen in Chapter 11 the distinction between NPP and ecosystem productivity is crucial when it comes to looking at locking up carbon in forests to mitigate climate change. The NPP figure suggests that a large amount of biomass (half of which is carbon) is being stored in a forest every year. However, the ecosystem productivity shows that if all parts of the stand are considered, the forest is growing significantly only in size (and thus soaking up carbon) until it reaches maturity. Thereafter no more carbon is stored because as much is lost each year as enters. As will be seen in Chapter 11, in practice this is a slight oversimplification but the principle stands.

## 8.2 Energy flow through forest ecosystems

Another way of looking at the dynamics of a forest is to investigate how energy flows through the system. This can give important insight into how a forest works as energy is always limiting and places stringent limits on the number of plants and animals a forest area can support. As discussed in Chapter 1, nutrients are cycled through an ecosystem while energy passes through, entering as sunlight and leaving as heat from respiration. Due to the complexity of forests, there have been few studies complete enough to give a good overview of energy flow through a forest.

One of the best examples comes from a study conducted at Hubbard Brook by James Gosz *et al.* (1978). In a typical year at this latitude in New Hampshire, the total energy input was measured at 1 254 000 kilocalories per square metre (kcal $m^{-2}$). Of this, just 0.8% was 'fixed' by photosynthesis (gross primary production – GPP), of which 55% was used in respiration leaving 0.4% of the sun's arriving energy to be used for producing new plant growth (net primary production – NPP). This may seem very inefficient but if only the 4-month growing season (June to September) is considered, solar radiation was 480 000 kcal $m^{-2}$, of which 10 400 kcal $m^{-2}$ were fixed by photosynthesis in GPP (about 2%) which after respiration left a NPP of 4680 kcal $m^{-2}$ or about 1% of incoming energy. This figure is a typical conversion rate of the sun's energy to NPP. Although only around 2% of the sun's energy is used in photosynthesis the rest is not wasted since another 42% is used in the growing season to heat up the environment and another 42% is used to evaporate water in transpiration, an essential element in a tree's growth. The remaining 14% of sunlight is reflected back to the sky.

Of the NPP (4680 kcal $m^{-2}$) in a year, 26% (1199 kcal $m^{-2}$) was stored in new growth. The remaining 74% went to tissues subsequently dealt with by decomposers with 65% (3037 kcal $m^{-2}$) and 9% (437 kcal $m^{-2}$) dying above and below ground, respectively. In most years less than an additional 1% (41 kcal $m^{-2}$) was eaten by herbivores although this can rise to as much as 44% when, for example, defoliating caterpillars peak in numbers and strip the trees of leaves, and a good deal of what would otherwise fall as litter is eaten by the

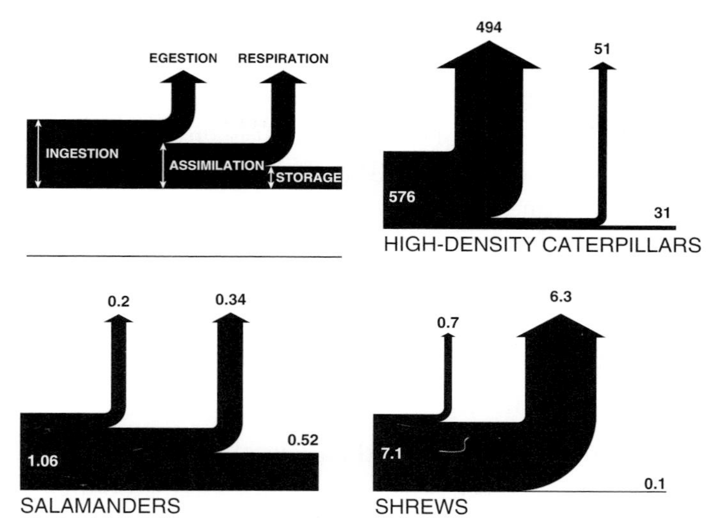


Figure 8.5 Energy budgets for three of the main consumer organisms in the hardwood forests of the Hubbard Brook Experimental Forest of New Hampshire. The sizes of the strips through the boxes are proportional to the amount of energy passing through each process in kcal $m^{-2}$ of ground in the forest. Energy ingested in the form of organic matter is either assimilated into the organism or lost in material egested from the body. This energy is then used for either respiration or put into new tissue through growth or reproduction. Much of the energy taken in by caterpillars is not absorbed but remains in the material that they egest as faeces. Shrews being 'warm-blooded' use much of their energy in respiration (in maintaining their high body temperature) while the 'cold-blooded' salamanders put proportionately more energy into growth and reproduction. (Redrawn from data in Gosz *et al.*, 1978. *Scientific American* 238.)

insects. When it is considered that of the 1% of energy transferred to herbivores normally only 1–10% of that (0.1–1% of NPP) is transferred to carnivores it can be appreciated why most food chains are fairly short. Figure 8.5 shows what happens to the energy consumed by different animals and reveals the great differences in energy use. Caterpillars can consume huge amounts of foliage but they are inefficient digesters and a large amount of the contained energy is lost in the faeces (egested) and, in the example given, only 14% of the energy ingested is assimilated; of that around 60% is used in respiration and 40% is incorporated into new tissue. Salamanders and shrews are important carnivores in the Hubbard Brook forests and show different strategies of energy use. The shrews being 'warm-blooded' (more accurately, **homoeothermic**: keeping a high constant temperature) and active through the year are efficient at extracting energy from their food (90%) but use roughly 98% of their assimilated energy for respiration and only about 2% for growth and reproduction. By contrast, salamanders consume only about a sixth of the energy of a shrew and are almost as good as shrews at extracting energy from it (81%) but being 'cold-blooded' (or rather **poikilothermic**: with a body temperature tending to follow that of the surrounding environment) they use less energy to keep warm and around 60% of their assimilated energy goes on growth and reproduction. Salamanders are thus very good at transforming energy into increased biomass (see Box 7.5).

## 8.3 Nutrient cycling

### *8.3.1 Losses, gains and cycling*

Plant and animal tissues are composed largely of carbon, oxygen and hydrogen which flow through an ecosystem as part of large global gaseous cycles. A number of other nutrients, making up at most 5% of a plant's bulk, are also crucial to growth. As described in Chapter 2 these essential mineral nutrients are divided into those required in larger quantity (the **macronutrient elements** – N, P, K, Ca, S and Mg) and those needed in much lower amounts (the **trace or micronutrient elements** – see Figure 2.3). The surface layers of the Earth hold large amounts of these chemical elements but a high proportion are locked up in minerals and sediments (or atmosphere in the case of nitrogen) as part of large **biogeochemical cycles**. Forests are part of these cycles and interact with them through the small losses and gains of nutrients made over time but the bulk of the nutrients are recycled within the ecosystem.

Most nutrients do not have a gas phase under normal conditions and so go through a strictly sedimentary cycle with inputs to living organisms being through uptake from the soil. The exception is nitrogen: this is usually the nutrient element needed in highest quantities (typically ten times greater than phosphorus and several hundred times greater than a micronutrient like zinc), is usually most limiting, and is the one element that has a large interaction with the gases of the atmosphere (sulphur can appear as gaseous hydrogen sulphide under certain circumstances and so is a minor exception). Much of the nitrogen used by plants has its origin directly from the atmosphere. The processes that control the availability of nutrients are of obvious interest to forest ecologists in understanding how the forest functions.

The main **sources of nutrients** in a forest are chemical weathering, biological fixation and atmospheric deposition. **Chemical weathering** of rocks releases a number of nutrients, particularly phosphorus, magnesium and calcium, the amounts available depending upon the rock type. Plants can make an important contribution by roots physically breaking up rocks and by root exudates accelerating the weathering process. **Biological fixation** primarily involves the conversion of atmospheric nitrogen into a form useable by organisms – this is discussed further under nitrogen below. **Atmospheric deposition** is the input of particles and gases carried or dissolved in rain, snow and fog as wet deposition, and by dry deposition as dust and fragments carried by the wind. Such inputs, especially from dust carried from other soils, can be very important in forests with poor, highly weathered soils. Chadwick *et al.* (1999) show that after four million years of weathering, the soils of the oldest Hawaiian islands are very nutrient poor and the rich rain forests flourish on cations (positive ions, such as calcium) supplied in sea spray carried by the wind and on phosphorus arriving in dust blown from central Asia over 6000 km away. Saharan dust is similarly important to parts of Africa. Lichens can make a considerable difference to atmospheric deposition. Knops *et al.* (1996) examined the lichen *Ramalina menziesii* on blue oak *Quercus douglasii* in the fog belt along the Californian coast and found 3.8 kg of lichen per tree, totalling 590 kg $ha^{-1}$ for the woodland, equivalent to over half of the oak leaf biomass (958 kg $ha^{-1}$). The extra surface area of the lichen intercepted significant amounts of many nutrients from wind and rain. It is estimated that the lichens intercepted an extra 2.85 and 0.15 kg $ha^{-1}$ $y^{-1}$ of nitrogen and phosphorus, respectively, against background levels of 0.88 and 0.06 kg $ha^{-1}$ $y^{-1}$.

Agriculture and industry are together putting increasingly polluting amounts of some nutrients into the atmosphere, particularly nitrogen and phosphorus. In areas of North America and Europe the atmospheric deposition rates of nitrogen are up to 40 times the normal background rate of less than 1 kg $ha^{-1}$ $y^{-1}$. The implications of such high levels are discussed in Section 11.4.

**Loss of nutrients** from a forest is usually by leaching to groundwater and consequent loss to streams, and gaseous loss to the atmosphere. Loss by leaching partly depends upon variations in nutrient supply and to demand; if a nutrient is available in excess to what organisms can immediately use (and so store), it is more likely to be lost by leaching. Loss also depends upon the form of the nutrient and its chemical reactivity with the soil. These both help explain why nitrate is more readily leached from soil compared with ammonium which is rapidly taken up by microbes and held by the soil cation exchange complex. Phosphorus, by contrast, is captured in the soil in an insoluble form and is less readily leached than some forms of nitrogen. Gaseous losses of nutrients are possible for those elements with a gaseous phase such as sulphur as hydrogen sulphide and nitrogen as methane and several other forms.

Once nutrients are in an ecosystem, they can be cycled a number of ways. In forests, the dominant path is in litter-fall, decomposition to mineral form and uptake by plants (discussed in detail in Chapter 7). Central in this is the role of microbes (fungi and bacteria). The **co-evolution of fungi and trees** over vast periods of time has resulted in extensive mutual interaction that influences this cycling; indeed Rayner and Boddy (1988) consider there is constant interaction between them via chemically communicated feedback systems and dynamic interactions between boundaries. Mycorrhizas (see Box 5.1 for more detail) protect the tree against toxins and forage for nutrient elements which they supply to trees; the trees in exchange provide the fungus with photosynthate. **Endomycorrhizal** or **arbuscular (AM)** fungi are able to garner extra supplies of phosphorus for their host principally by increasing the soil volume exploited by the fungal mycelium. **Ectomycorrhizas** become most important with increasing latitude as nitrogen becomes more limiting and are able to exploit extended soil volumes and by accessing nitrogen and phosphorus in forms otherwise unavailable to the host tree. **Ericaceous mycorrhizas** are beneficial due to their ability to utilize nitrogen and phosphorus from complex organic sources. Mycorrhizas are also capable of degrading proteins and transferring amino acids directly to the host plant. Mycorrhizal fungi additionally provide contact between the tree and the roots of adjacent members of the same species. Different species of mycorrhizas provide a link between trees and other plants. The success rate of tree seedlings is greatly enhanced when their roots form mutualistic associations with appropriate fungi (Section 5.4.1).

### *8.3.2 Deficiencies and proportions*

**Deficiencies** in any of the essential nutrients lead to characteristic symptoms such as discolouration of leaves, and changes in growth such as size, number of leaves and lengths of internodes. Such deficiencies can be related to availability in the underlying bedrock but more usually are a product of soil pH. As shown in Fig. 2.3, acidic soils tend to be deficient in nitrogen, phosphorus, calcium, magnesium and potassium. Equally important, acidic conditions lead to greater solubility and increasingly toxic amounts of aluminium and iron. There is increasing evidence that nutrients are used more **efficiently** on nutrient-poor soils (Paoli *et al.*, 2005): organisms grow with lower tissue nutrient concentrations.

As noted in Chapter 7 **nitrogen** is usually thought of as the nutrient most limiting in temperate forests. Outside these well-studied temperate forests, especially on soils of great age, **phosphorus** may well be the limiting nutrient. It may seem strange that this is not definitively known. The main problem is in knowing how much of the total amount of a nutrient like phosphorus (which may appear abundant) is actually available to the plant, and how the conditions in the rhizosphere can influence and change its availability at the site of uptake. It is certain that in most soils the amount of soluble P is a very small fraction of the total phosphorus content. To add further confusion, Carline *et al.* (2005) found that when red deer *Cervus elaphus* were excluded from birch woodlands for 14 years the limiting nutrient changed from nitrogen to phosphorus. This appeared to be due to increased nitrogen mineralization (see Section 8.4.1) making N readily available and so leaving P in short supply; quite why the exclusion of deer should cause this is open to speculation.

The relative proportion of nutrients available (covered by the term **stoichiometry**) is a more useful way of looking at nutrient supply than concentrating on a single deficiency. Plants and animals need a balance of nutrients and this goes a long way in explaining why adding nitrogen or phosphorus to a forest soil may have little effect. In a forest the addition of a single element may often elicit some response but usually it also reveals the existence of other limiting deficiencies. The interactions between different quantities of nutrients can be very complex as they interfere (perhaps competing for uptake sites on roots) or synergistically interact with each other.

The characteristic ratios between carbon, nitrogen and phosphorus (**C/N/P stoichiometry**) are widely used in marine ecosystems for investigating productivity, but less so in terrestrial ecosystems. Forests tend to have relatively higher levels of carbon than marine systems due to the need for high investment in structural and defensive compounds (see McGroddy *et al.*, 2004). C/N/P ratios of foliage and litter are globally variable but C/nutrient ratios in litter are consistently higher than in foliage (particularly C/P), suggesting that resorption of nutrients is a globally important mechanism of conserving nutrients that are in short supply (see Section 7.5.2).

## 8.4 Nitrogen

Nitrogen is often limiting and always a central nutrient in forest growth. As a consequence it has been extensively studied and thus makes a good example nutrient in which to delve more deeply into its dynamics.

### *8.4.1 The cycling of nitrogen*

Inert nitrogen gas ($N_2$, dinitrogen) is the most abundant element in the atmosphere forming around 78%. Unfortunately nitrogen in this form is not directly usable by plants. As Galloway *et al.* (2003) put it: '. . . more than 99% of this N is not available to more than 99% of living organisms'. The reason for this seeming contradiction is that plants and animals use nitrogen mostly in various inorganic forms: nitrate ($NO_3$), nitrite ($NO_2$) and ammonia ($NH_3$), usually found in the soil as ammonium ions ($NH_4^+$). Nitrogen is **fixed** by the reduction of gaseous nitrogen to an inorganic form by **nitrogen-fixing microbes** or lightning and to a lesser extent by cosmic radiation and meteorites, and from volcanic eruptions (cumulatively around 10% of all fixation). The nitrogen-fixing microbes, including free-living (asymbiotic) bacteria such as *Azobacter* and *Clostridium* and blue-green algae (cyanobacteria) such as *Nostoc* and *Calothrix*, produce ammonia by splitting the $N_2$ and combining it with hydrogen ions. These nitrogen-fixers can be free-living, often in the rhizosphere of roots or in symbiotic association with a plant species (such as *Rhizobium* bacteria found in the root nodules of legumes – see Section 5.5.1 – and *Frankia* actinomycetes in other families, notably alders). Ammonium ions produced by the microbes are rapidly oxidized to nitrite and then to nitrate in the process of **nitrification** by fungi and particularly the aerobic bacteria *Nitrosomonas* (ammonium to nitrite) and *Nitrobacter* (nitrite to nitrate). This happens particularly at neutral or higher pH. Lightning and the other physical processes combine N and oxygen (O) to form nitrates directly. The opposite process of **denitrification** happens when oxygen is absent (anaerobic conditions) and nitrate is converted back to gaseous nitrogen by different fungi and bacteria (e.g. *Pseudomonas*). Nitrogen is also held in organic compounds in the soil. These are converted into readily usable inorganic forms by the process of **mineralization** (see Chapter 7). Nitrogen is taken up by plants not only as nitrate and ammonium but also as organic compounds such as amino acids (see Section 8.4.2). Nitrate is readily stored or moved in a plant but ammonium has to be incorporated into organic forms in the roots because of its toxicity.

Nitrogen can be gained and lost by the forest. Nitrogen fixation in young ecosystems can be substantial (more than 100 kg of nitrogen per hectare each year) but over time it is the recycling of the accumulated nitrogen that supplies the majority of ecosystem needs. Nitrogen-fixing microbes are normally responsible for around 90% of all the **gains of nitrogen** in a forest. Of this, 70% comes from symbiotic bacteria (see Box 8.2) and only around 30% is fixed by free-living bacteria and actinomycetes, but can reach high levels in those trees that do not readily form symbiotic relationships. For example, in eastern white pine *Pinus strobus* this can be in the order of 50 kg $ha^{-1}$ $y^{-1}$ and in red alder *Alnus rubra* up to 320 kg $ha^{-1}$ $y^{-1}$. Experiments to measure this in eastern white pine (see Knops *et al*., 2002) were done using trenched plots to sever any roots coming into the plot to ensure there was no 'root mining', bringing in nitrogen from the surrounding area. The inferred input of the free-living microbes is confirmed by the fact that the trees would have had to mine

**Box 8.2 The 'benefits' of symbiotic nitrogen fixation**

Black locust (false acacia in Europe; *Robinia pseudoacacia*) is a native of the Appalachian uplands and has been so extensively planted through temperate North America and Europe that Boring and Swank (1984) estimate that it is now the second most abundant deciduous tree in terms of planted area in the world (after eucalypts). Like other members of the pea family Fabaceae, this tree fixes nitrogen using symbiotic bacteria, and can add tremendous amounts of nitrogen to the soil. A study by Rice *et al.* (2004) looked at what happened to a nutrient-poor pine-oak forest in New York state 20–35 years after the invasion of black locust. The upper organic layer (A horizon) of black locust soils had 1.3–3.2 times more nitrogen relative to pine-oak soils, with net mineralization rates 25–120 times greater. Litter falling from the black locusts had low lignin levels and was readily decomposed. Over the year the black locust stands produced double the leaf litter of pine-oak stands containing 86 kg of nitrogen per hectare compared with 19 kg $ha^{-1}$ in pine-oak stands. Thus, when black locust invades nutrient-poor pine-oak stands, it supplements soil nitrogen pools, increases nitrogen return in litter-fall and enhances soil nitrogen mineralization rates. This may be considered a distinct benefit if, for example, you are considering growing a crop of trees on the site but will inevitably cause a loss of diversity in the native flora as one or more species able to benefit from the extra nutrition come to dominate.

an area five times the plot size to produce this amount of nitrogen. Much of the nitrogen fixed, even that in root nodules, ends up in the soil pool. Conifers like Douglas fir *Pseudotsuga menziesii* show much better growth on nutrient-poor soils when associated with red alders. It looks highly likely that the rhizosphere does not just contain more bacteria but also a higher proportion are specialist bacteria that can break down protein to ammonia, and a lower proportion are of nitrifying microbes. Norton and Firestone (1996) found a 50% increase in ammonium mineralization and immobilization in the rhizosphere of ponderosa pine *Pinus ponderosa* compared with soil more than 5 mm from any root. The presence of the roots reduced microbial consumption of ammonium ions by microbes, and the pine roots were able to compete with microbes for limited nitrogen. The pine roots were also more successful at competing for nitrate than ammonium (they consumed 30% of available ammonium and 70% of nitrate), helped no doubt by nitrate's faster diffusion rates through soil which benefits relatively sparse roots over the ubiquitous microbes (Zak *et al.*, 1990). All this is in addition to the benefits of mycorrhizas discussed above.

**Losses of nitrogen** can happen strikingly due to fire (through direct volatilization and by leaching from ash) and by canopy removal by logging or wind damage (so there is less growing vegetation to take up the nitrogen being mineralized). Losses can be great following disturbance because there is less vegetation so less nitrogen uptake and more ammonia is available for nitrification. At Hubbard Brook, Bormann and Likens (1979) measured a large nitrate spike in streams following the clear-felling of a watershed – see Section 8.6.1. Nitrogen can be lost more slowly by gaseous emission from waterlogged (anaerobic) soils by denitrification but most commonly it is lost by the leaching of nitrate and dissolved organic nitrogen (DON; see below). The soil nitrate pool is generally small because of rapid uptake by plants and microbes but since nitrates are ions with a negative charge (anions) they are much more readily leached out of soil, especially sandy soils, than ammonium ions which, being positive cations, are strongly held by the cation exchange complex of the soil (Section 2.2.1). Ammonium is thus the major form of inorganic 'available' nitrogen in soils. The balance between nitrate and ammonium is affected by pH: nitrate is the main form available in alkaline soils, ammonium ions in acidic soils. Acidic soils have fewer nitrifying bacteria and so the ammonium remains largely unchanged. But there is still usually nitrate in some acidic soils, it is not all or nothing, and there is a wide variation in the ability of acid-loving (calcifuge) plants to utilize nitrate.

The loss of nitrogen (in whatever form) is also affected by the vegetation. If the seasonality of plant activity does not match the realizability of N supply from microbes, it can leave a pool of unused nitrogen that is readily leached. This is seen in early spring in northern hardwood forests, where microbial mineralization of organic matter and nitrification begin before uptake by plants becomes an important sink. Muller and Bormann put forward the **vernal dam hypothesis** in 1976 which proposes that spring ephemerals which grow before canopy closure take up nitrogen and other nutrients before they can be leached, and that these are subsequently made available to other plants as they die back from lack of light and decompose. They found that the yellow trout lily *Erythronium americanum* growing in the Hubbard Brook Experimental Forest saved almost half of the important nutrients from being washed away: they used 43% and 48% of the potassium and nitrogen, respectively, released in the spring, the rest being lost in stream water. Since then the hypothesis has had mixed support: some experiments have supported Muller and Bormann's data (e.g. Tessier and Raynal, 2003), others have shown that the microbe population itself is better at soaking up the spring burst of nutrients (e.g. Zak *et al.*, 1990), and still others have shown that while the dying back of vernal plants can indeed produce a burst of nutrients available to other plants in the summer (e.g. Anderson and Eickmeier, 2000), they are not very efficient at taking up nutrients in the first place (e.g. Anderson and

Eickmeier, 1998; Rothstein, 2000). Undoubtedly some of the differences come from looking at different plants in different forests.

The vernal dam hypothesis gives the impression that any peak or pulse of nutrients without a complete biological sink is unfortunate, but this is not always so. It is apparent that nutrient fluxes in tropical forests are in pulsed surges (see page 289 for further discussion on pulsed resources). It is becoming clear (e.g. Lodge *et al.*, 1994) that pulsed inputs of nutrients have different effects from a constant, gradual input. If the level of nutrient mineralization is synchronized with plant uptake it leads to tight nutrient cycling and, once competition has ousted the weakest, a reduced competition between microbes and plants for limited nutrients. If, however, mineralization happens in bursts (due to the arrival of abundant litter, or a change in environmental conditions) it will lead to losses of nutrients from the system (see Section 8.5 for the myth that tropical systems are tightly closed) but may also play a critical role in maintaining productivity in tropical forests with otherwise tight nutrient cycling.

Finally in this section, it should be remembered that animals can make a large difference to nitrogen gains and losses. Animals such as salmon, birds and termites that travel long distances can bring in significant amounts of nitrogen and other nutrients. Although salmon die in streams once spawned, their nutrients are available to those plants with roots in or near the water (such as those of willows or alders). Animals can also significantly reduce nitrogen input by eating nitrogen-fixing plants such as legumes. Moreover, ruminants that hold microbes in their stomachs can speed up nitrogen cycling by acting as living decomposition vessels.

### *8.4.2 Inorganic nitrogen and dissolved organic nitrogen (DON)*

As discussed in Chapter 7, nitrogen is readily taken up by microbes and immobilized in the **substrate–microbe complex** because plant roots cannot compete with the microbes. The traditional view is that only inorganic nitrogen in excess of microbial requirements would be available to plants. It is becoming apparent, however, that this picture is not quite right in two ways: some plants are now known to use *organic* nitrogen, and some of this organic nitrogen is thought to be immobilized without the involvement of microbes, with important consequences for plant nutrition and nitrogen limitation of growth (see Neff *et al.*, 2003).

Soils contain a large pool of organic nitrogen, mostly in the form of **dissolved organic nitrogen (DON)**, made up of a wide range of compounds from simple, easily digested amino acids to much more recalcitrant compounds of tannins

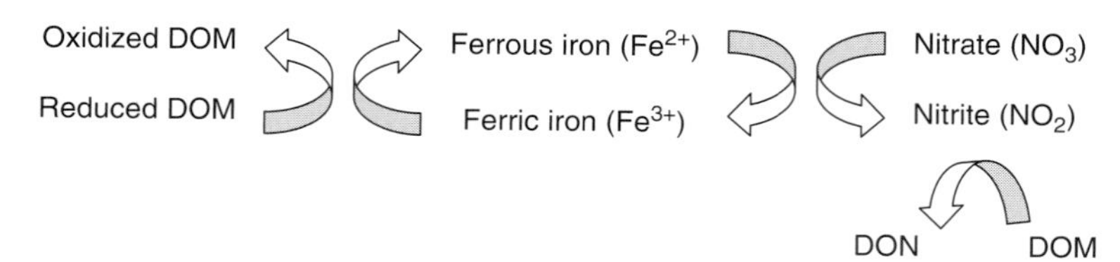


Figure 8.6 The 'ferrous wheel' hypothesis for immobilizing nitrogen in organic compounds without the use of living organisms. The process depends upon dissolved organic matter (DOM), made up of compounds from the organic matter in solution (see Section 7.6.2 for a discussion on the soluble fractions of humus). The carbon compounds in the DOM are usually in a 'reduced' form (chemically lacking in oxygen) capable of stripping oxygen from oxidized ferric iron ($Fe^{3+}$ also called Fe(III)) to create ferrous iron ($Fe^{2+}$ or Fe(II)). In turn, the ferric iron can gain an oxygen molecule back by changing nitrate into nitrite, and so creating a 'wheel' where the iron molecules circle between the two forms. Importantly for nitrogen immobilization, although nitrate is largely unreactive with soil organic matter, nitrite reacts readily, quickly and abiotically with dissolved organic matter (DOM) to form dissolved organic nitrogen (DON). It looks as if the phenolic compounds common in soil DOM solutions react with nitrite to form nitrophenols. This change from nitrite to DON may take just minutes to complete. (Adapted from Davidson *et al.*, 2003. *Global Change Biology*, 9.)

and other polyphenols. Dissolved organic nitrogen is brought to the soil in precipitation (collecting nitrogen as it runs over the foliage and drips to the ground as throughfall) and is also created in soils by a complex mix of processes. Some of these involve organisms and extracellular enzymes breaking up proteins and peptides in the soil organic matter. Soil microbial biomass typically turns over several times a year producing a pool of nitrogen that is rapidly degraded and reused. But it is now realized that a significant portion of the more complex organic nitrogen is created in soils by a set of chemical reactions not requiring living organisms (and so is **abiotic**). Recent evidence that abiotic immobilization of inorganic nitrogen into organic forms may be an important process challenges the previously widely held view that microbial processes are the dominant pathways for nitrogen immobilization in soils (see Davidson *et al.*, 2003). The most likely way this happens is encapsulated in the **ferrous wheel hypothesis** of Davidson and his colleagues – see Fig. 8.6. This abiotic pathway helps to explain why nitrogen added as fertilizer is rapidly immobilized even when the soil is sterilized.

Increasing importance is now being given to this pool of DON as ecologists find out more about its role in the nitrogen cycle. Dissolved organic nitrogen is the most abundant form of nitrogen in forest soils, making up to 50% of the total nitrogen. In boreal soils, amino acids make up to 10–20% of DON, compared with less than 10% in temperate forests. A small proportion is dissolved in the soil water (most commonly as the amino acids aspartate and glycine, and the amide glutamine) but these are quickly absorbed by plants and microbes. The majority is linked to soil organic matter and especially soil minerals, particularly clays. The degree to which this fixed DON is available to plants is still open to speculation but it seems that the majority of compounds in DON are labile and readily decomposed or used directly.

It is becoming clear, however, that many plants and bacteria are capable of absorbing **amino acids** directly from the DON pool, and may do so preferentially over other nitrogen sources, and are thus able to **short-circuit the nitrogen cycle** by not having to be dependent upon the mineralization of nitrogen by microbes (the same may also be true for organic phosphorus). Certainly plants in boreal, arctic and alpine forests seem capable of direct absorption (see Lipson and Näsholm, 2001 for a review), generally preferring amino acids over inorganic nitrogen, perhaps because of the larger amounts in northern soils (see above) and because the soils tend to be wet and cold in these northern and upland areas, microbial mineralization is slow. There is also evidence that pines have evolved polyphenol-rich litter which increases their nitrogen uptake: higher polyphenol levels increases the proportion of the nitrogen released in organic forms compared with mineral nitrogen (up to 10/1), which are selectively taken up by the pine ectomycorrhiza (Northup *et al.*, 1995). In addition, while mycorrhizas are capable of degrading proteins and transferring amino acids directly to the host plant there is growing evidence that even non-mycorrhizal plants are able to utilize amino acids.

It is possible that some of the plants specializing in the uptake of different forms of nitrogen use this to reduce competition between species. In the tussock tundra of Alaska, for example, McKane *et al.* (2002) found that Labrador tea *Ledum palustre* and dwarf birch *Betula nana* used mainly ammonium while bigelow sedge *Carex bigelowii* used mainly nitrate, and low-bush cranberry (or cowberry) *Vaccinium vitis-idaea* relied mainly on amino acids and ammonium. However, this partitioning of nitrogen types is not universal. In western red cedar–western hemlock (*Thuja plicata–Tsuga heterophylla*) stands in western Canada these two trees and the main understorey plant, salal *Gaultheria shallon*, were found by Bennett and Prescott (2004) to be similar in the types and amounts of nitrogen they used.

Although amino acids allow short-circuiting of the nitrogen cycle, plants and microbes will still be competing for the amino acids and there is a good deal of curiosity as to how plants manage to get any amino acids. In alpine areas there is evidence of temporal partition (plants using most nitrogen in the early summer and microbes in the autumn) but there is little to suggest that this happens in forests. The main mechanism seems to be **differential ability to use different amino acids**; plants can take up glycine faster than heavier amino acids (and it moves by diffusion faster in soils which is to the roots' advantage since they have a lower surface area and less ubiquitous distribution in soils compared with microbes). Moreover, glycine is degraded by microbes more slowly than other amino acids. The rhizosphere and mycorrhizas could also be important in this differential use.

### *8.4.3 Why are nitrogen and phosphorus so often the limiting nutrients in forests?*

As already discussed in Section 8.3.2, nitrogen is usually considered to be the most limiting nutrient in forests although outside of temperate forests phosphorus may be equally or more limiting. Part of the reason for the prominence of these two nutrients rests with the large amounts used by plants. However, it is important to bear in mind that after nitrogen, potassium, magnesium and calcium are often present in higher concentrations in plant tissue than phosphorus. The problem with phosphorus is not so much the amount needed by a plant but its general unavailability in soils. This combination of high needs and availability in soils is reflected in the make-up of common fertilizers based on N, P and K (nitrogen, phosphorus and potassium) – see Section 8.6.2.

The slow recycling of **nitrogen** through the forest makes it more likely to be limiting. There is a stoichiometric difference between plants and animals: plants tend to be carbon rich and nitrogen poor while animals are carbon poor and nitrogen rich. This may well limit the speed at which a herbivore can eat because it has to digest a huge bulk of carbohydrate to extract the small amount of nitrogen. Some animals cheat: aphids sucking sap from trees process huge quantities, stripping out the nitrogen and excreting great quantities of largely unchanged sugars as honeydew. Moreover, since nitrogen is in short supply for herbivores they tend to hoard it, delaying its recycling. The same problem happens with soil microbes, as discussed in Chapter 7; they lock up large amounts of nitrogen, recycling it within the microbe community until the carbon is used up. This hoarding and slowing of the passage of nitrogen is further exacerbated by the use of defensive compounds to reduce herbivory, one of the most important and widespread groups of which are the protein-precipitating phenolic compounds (see Section 7.5.2). These further reduce the availability of nitrogen to herbiovores, and, of course, also slow the decomposition of litter. Lastly, most organic nitrogen is bound to carbon, which makes it difficult and slower to mineralize than other elements (see Vitousek *et al.*, 2002 for more detailed background).

A further reason for nitrogen being in short supply is the way in which it is gained and lost in ecosystems. As explained above, most input comes from biological fixation rather than being available from the underlying rock. The nitrogen cycle is therefore strongly buffered by having a large atmospheric reserve. But nitrogen fixation is an expensive process and may often be limited by the availability of usable carbon. If carbon is locked up in compounds that are difficult to decompose, this in turn will tend to slow nitrogen fixation (again see Vitousek *et al.* (2002) for more detailed reasoning). On top of potentially low input, nitrogen is more readily lost from a forest than most other essential nutrients. Nitrate, being a negatively charged anion, is rapidly leached. More insidious is the **leakage of nitrogen** from terrestrial ecosystems as dissolved organic nitrogen (DON). A large proportion of DON is absorbed onto soil surfaces and so is physically protected in soil aggregates (see Davidson *et al.*, 2003 for details). But while some of the DON is persistent in the soil, it is also readily leached by water. Indeed, DON is often the dominant form of nitrogen exported from many temperate forested watersheds. Hedin *et al.* (1995) found 95% of nitrogen losses in streams flowing from an old-growth temperate forest to be DON, compared with 0.2% as nitrate and 4.8% as ammonia. As an aside, this explains why DON is a very important component of aquatic ecosystems and why many forests with large amounts of organic matter usually have brown streams, coloured by dissolved organic matter (DOM – see Fig. 8.6). The traditional view is that nitrogen losses (as gas or dissolved in water) are controlled by living organisms and so if nitrogen is in short supply, it will be in high demand and losses will be minimal. As Davidson *et al.* (2003) put it, however: 'An N *leak* is fundamentally different from an N *loss* that can be controlled by biological demand [our emphasis]. Leaks occur where biotic systems cannot fully prevent N losses, despite an overall system demand for N'. A gentle leak that organisms cannot stop because they do not control it, can eventually lead to a debilitating loss of nitrogen and may be an important part of the explanation behind why nitrogen is generally limiting in forests.

**Phosphorus** is widely available in bedrock (typically at 1.2 g $kg^{-1}$ of rock), and 99% soil phosphorus occurs as phosphate ($PO_4$). Soils commonly contain between 200–800 mg $kg^{-1}$ although very weathered old soils can have as little as 50 mg $kg^{-1}$. A phosphorus budget has been calculated for a 70-year-old hardwood stand at Hubbard Brook by Yanai (1992) – see Fig. 8.7. Although the amount of phosphorus in the system is high (1756 kg $ha^{-1}$), only 3% (51.6 kg $ha^{-1}$) is stored above ground and 4% (70.9 kg $ha^{-1}$) is stored in the combined above- and below-ground living biomass. Of the vast majority stored in the soil (1685 kg $ha^{-1}$), most is locked up in the mineral soil (95%,

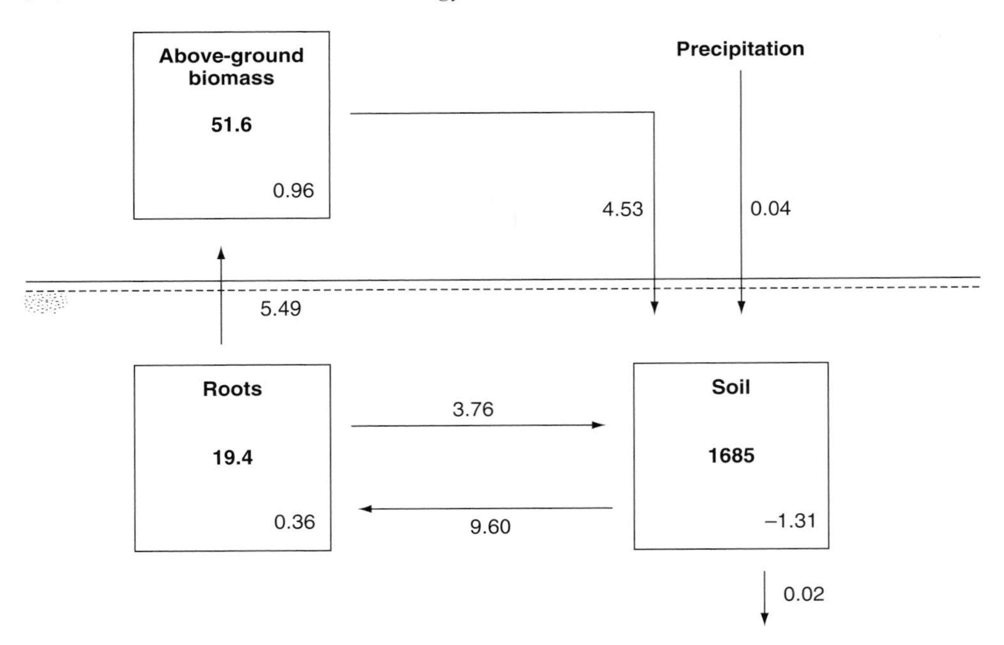


Figure 8.7 Phosphorus budget for a 70-year-old hardwood forest at Hubbard Brook Experimental Forest. Values in bold are the pools of phosphorus (kg $ha^{-1}$); values in the bottom right of the squares are annual rates of changes (kg $ha^{-1}$ $y^{-1}$); arrows show the movement of phosphorus from one pool, to another over the year (kg $ha^{-1}$ $y^{-1}$). (Drawn from data from Yanai, 1992. *Biogeochemistry* 17.)

1600 kg $ha^{-1}$) compared with 5% (85 kg $ha^{-1}$) in the organic matter layer. Turnover of the mineral soil pool is very slow (0.33% per year) compared with 7% per year in the forest floor. The living biomass is accumulating phosphorus at the rate of 1.33 kg $ha^{-1}$ $y^{-1}$ (0.96 above ground and 0.36 in the roots) and the soil is making a net loss of 1.31 kg $ha^{-1}$ $y^{-1}$ to the vegetation. This slow but steady net gain by the vegetation is due to the comparatively young stand growing larger as it matures.

Plants are thought to use primarily inorganic phosphorus (but see above) and unfortunately phosphate, along with other forms of phosphorus, is relatively insoluble so the pool of available inorganic phosphate that plants can utilize is usually very small. Plants can increase the availability of phosphorus through manipulation of the rhizosphere but just how much extra this produces is largely unknown. Much of the remaining soil phosphate is unavailable to plants, either locked up in organic compounds or tightly bound to metal cations in very insoluble complexes. Phosphate is most available to plants around pH 6.5; more alkaline and it is bound progressively to calcium (Ca), more acid and it is bound to aluminium (Al) and iron (Fe). With time, phosphorus is prone to being leached from a soil, and also to becoming less soluble. As calcium is relatively easily leached from a soil, over time, phosphate binds increasingly to aluminium and iron to form compounds that are even less soluble than those formed with calcium. In tropical forests over old and infertile soils, the vegetation has a well-developed **root mat over the top of the soil** to scavenge and recycle phosphorus from decomposing litter directly before it can be absorbed by soil minerals into fixed and insoluble forms. Even in other less exacting forests it is usual for phosphorus to be progressively held in organic forms. This explains why removal of the vegetation on poor soils, such as those in the tropics, can result in very slow recovery.

Soluble phosphorus is remarkably immobile in the soil, moving a matter of centimetres from its place of origin. This is why mycorrhizal fungi are beneficial to the host by foraging in volumes of soil not reached by the plant's roots.

## 8.5 Nutrient dynamics in different forests

Studies of commercial plantations have provided insights into how trees affect soils and how nutrients are cycled through the soil and trees by comparing the before and after picture following tree planting. In the coastal dunes of Argentina, Jobbágy and Jackson (2004) looked at the vertical distribution of potassium (strongly cycled) and sodium (more weakly cycled) after planting with maritime pine *Pinus pinaster*, native to Europe. Fifteen years later, potassium was significantly concentrated in the top 50 cm of the soil while sodium was commonest 1–4 m down. This implies that potassium (a nutrient often in short supply) is much more strongly cycled than sodium, such that the trees are efficient at capturing potassium in the soil and delivering it back to the surface in litter. On the whole, deciduous trees are more active in nutrient cycling than conifers. The difference is particularly marked for calcium which is much more strongly cycled by deciduous trees and is probably responsible for the higher pH brown earths under deciduous trees such as birches, which counteract the leaching process by returning nutrients from deeper horizons to the surface.

Temperate and boreal forest soils tend to be uniformly nitrogen-poor due to a lack of nitrogen fixers, significant losses as DON and, especially in the boreal zone, episodic losses associated with fire and other local disturbances such as wind (see Box 8.3 for a detailed example from Hubbard Brook). McGroddy *et al.* (2004) point out that N/P ratios in foliage vary from $< 5/1$ in coniferous and temperate deciduous forests to $> 30/1$ in tropical forests; that is, there is proportionately more nitrogen in the tropics relative to phosphorus (explained

**Box 8.3 Nutrient dynamics at Hubbard Brook**

The long-term studies revealed that over the last half of the previous century the undisturbed northern hardwood forest showed a long-term net retention (i.e. more came in with precipitation than left in stream water) of hydrogen (H), chlorine (Cl), N and P, and net losses of Ca, Mg, Na, K, sulphur (S), silicon (Si) and Al. Likens (2004) suggests that the net loss of calcium is attributable to acid rain, reducing the input of Ca in precipitation and increasing its leaching from soil. Certainly the amount in stream water has declined over the past 50 years. This has also been associated with a marked decline in concentration and input of Ca in precipitation and similar declines in stream water (Likens *et al.*, 1998). This is discussed further in Chapter 11.

One of the Hubbard Brook watersheds (W2) was clear-felled in 1965–66 and herbicided for 3 years to prevent any regrowth. Stream flow went up (due to reduced evapotranspiration) and decomposition, mineralization and nitrification were greatly accelerated producing high levels of hydrogen ions and nitrate which were rapidly lost in the streams. Net losses of nitrate, Ca and K generally peaked in the second year after felling with each returning to pre-cutting levels at rates unique to each ion. Even decades after clear-cutting, differences in stream water solutes can still be seen, especially in Ca.

An ice storm in January 1998 broke off around 28–30% of the canopy at Hubbard Brook. Following this, levels of nitrate in stream water increased to 7–10 times greater than predisturbance values, peaking over the following winter (Houlton *et al.*, 2003). This was attributed not so much to increased microbial activity (since mineralization, nitrification and denitrification stayed the same) but to a decreased plant uptake of nitrate. It is salutary to note the flux of nitrate was still 2–10 times lower than that expected after complete canopy removal due to forest harvesting.

above by very low levels of P in tropical forests on old infertile soils). Vogt *et al.* (1986) note the same trend in N/P ratios in litter-fall (9–13/1 in boreal and temperate forests, 22–27/1 in the tropics and subtropics) and consequently in the forest floor organic matter (10–11/1 in coniferous and temperate forests and 31–32/1 in tropical forests). Whitmore (1998) points out that such broad-brush pictures can conceal a good deal of variation; for example phosphorus appears limiting in lowland rain forest but in montane rain forests nitrogen seems to be the limiting factor.

In total amounts, subtropical, tropical and Mediterranean forests have a higher concentration of nitrogen in litter-fall than in the forest floor, implying that nutrients are rapidly mineralized and taken back by plants. In temperate and boreal forests, the opposite is true: nitrogen is in higher concentrations in the forest floor (1.13–1.96%) compared with the litter (0.63–1.12%). This implies rapid immobilization and storage in the forest floor (see Vogt *et al.*, 1986 for a worldwide review). So boreal soils may contain a great deal of nitrogen but productivity of the forest is still nitrogen-limited because of slow rates of mineralization (hence the value of the short-circuit outlined in Section 8.4.2). Indeed, boreal forests retain nitrogen for 40–50 times (and sometimes up to 100 times) longer than do temperate deciduous forests (average turnover times are 230 years and 5.5 years, respectively). Phosphorus does not follow the same pattern: in general terms, most forests have similar amounts of phosphorus whether comparing foliage, litter or forest floor except that the colder forest floors of boreal forests have a higher concentration.

Tropical forests deserve looking at in more depth. Whitmore suggests that there are two conventional myths about tropical rain forests and nutrients. Firstly, that most of the nutrients are in above-ground biomass, and secondly that these forests have closed nutrient cycles with little or no leakage. Both Charles Darwin and Alfred Russel Wallace, whose almost simultaneous formulation of the theory of evolution was announced at a meeting of the Linnean Society in 1858, and later elaborated by Darwin (1859), were immensely impressed by the magnificence and variety of species in tropical forests. This is made very clear in Darwin's account of the almost 5-year voyage of *H. M. S. Beagle* first published in 1839 (Darwin, 2003). Terborgh (1992) quotes Wallace at the beginning of his chapter entitled *The Paradox of Tropical Luxuriance*: 'The primeval forests of the equatorial zone are grand and overwhelming by their vastness and by the display of a force of development and vigour of growth rarely or never witnessed in temperate climates.' Yet the soils of these luxuriant rain forests contain very little in the way of mineral nutrients. Figure 2.12 shows that the popular belief that most of the nutrients in a tropical rain forest are in the biomass is seldom true except for certain minerals in certain forests but nevertheless, nutrients are in overall short supply as discussed above. The solution to this paradox of a rich forest over a poor soil is solved by very rapidly recycling when organisms die or parts of them are shed.

Terborgh (1992) demonstrates just how well the trees of particular tropical forests are related to their soils, quoting data concerning the availability of nutrients in the soils of seven forests. Though nutrient availability varies by several orders of magnitude, the above-ground biomass of the least fertile site (San Carlos) is only 20% less than that of the richest (Panama) and the forest of highest biomass (Ivory Coast) is on soil of intermediate fertility. How is this accomplished? The biomass of these forests above ground varies within a factor of 2.3, variation of that below ground is far greater (nearly 12); it is their well-developed and effective root systems that allow the above-ground biomass of tropical forests on extremely nutrient-poor soils to be so high. Nutrient recovery by tree roots is very efficient, aided by mats of roots above the soil surface. In an experiment in an Amazonian forest more than 99% of radioactivity applied as calcium-45 and phosphorus-35 isotopes to the ground surface was retained in the root mat. Termites and fungi, especially decomposer basidiomycetes, destroy dead wood above ground, while root absorption is often assisted by mycorrhizas.

Figure 8.8 illustrates inorganic nutrient cycling in rain forests. Note that in tropical rain forest, 99% of water reaching the ground does so as throughfall and leaches nutrients as it goes. In the New Guinea example shown in the figure, there is great enhancement: K and Mg 9 times, Ca and P 5 times and N 4.6.

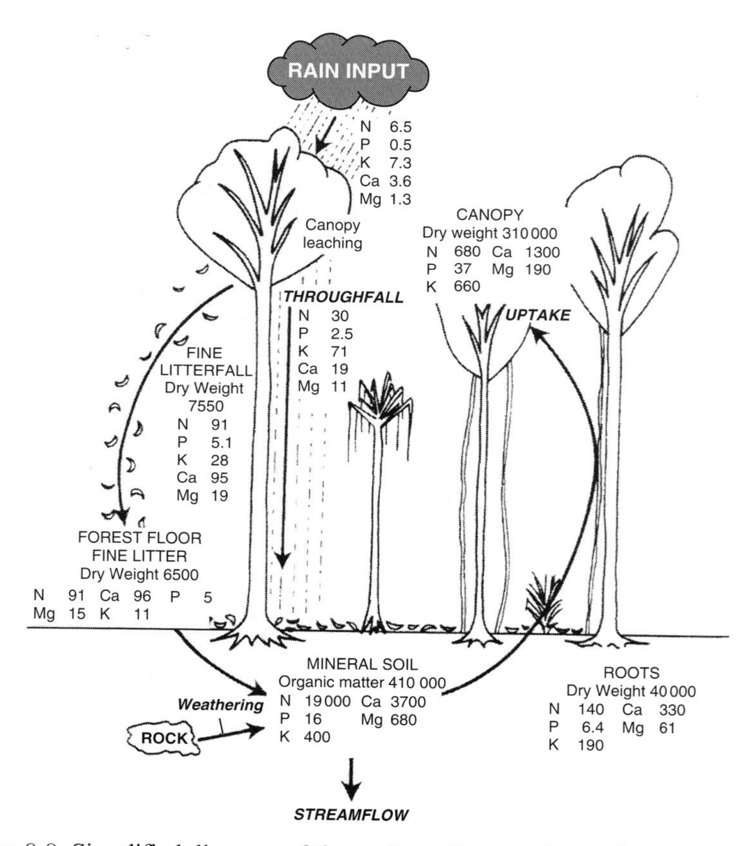


Figure 8.8 Simplified diagram of the cycling of inorganic nutrients in a tropical rain forest. Figures (in kg $ha^{-1}$ or kg $ha^{-1}$ $y^{-1}$) are for the lower montane rain forest at Kerigomna, New Guinea. (After Edwards, 1982. From Whitmore, 1984. *Tropical Rain Forests of the Far East* (2nd edn). Clarendon Press/Oxford University Press.)

The extra nitrogen in the throughfall is unusually high in this forest and is believed to be due to leaching from N-fixing algae epiphytic on the leaves (Whitmore, 1998). Most of the plant roots are in the top 0.1–0.3 m of soil since this is where the bulk of nutrients are released. As already discussed, old forests on well-weathered soils are dependent upon input from outside, particularly for phosphorus, and are less closed (a closed forest shows perfect internal cycling with no nutrient losses or gains). Many rain forests are on hilly terrain where the soil is shallow and continually being rejuvenated by creep and landslip and so unless the parent material is very poor (as are the sedimentary soils in various parts of Sarawak) these ecosystems have a continual input of nutrients. Silver *et al.* (1994) looked at nutrient availability in the tropical montane forest of Puerto Rico and found that cation concentrations and pH increased down the hillside from ridge tops to valley bottoms while soil organic matter and available iron (affected by the pH) decreased. The calcium, potassium and phosphorus content of the vegetation was higher on the ridges and slopes than in the valley bottoms.

## 8.6 Human influences

### *8.6.1 Forest clearance and timber harvesting*

The removal of timber or even whole forests has major implications as far as nutrient distribution is concerned. Original clearing for pasture has in many cases been accompanied by the use of fire, which renders many nutrients available to the plants but also involves considerable losses through leaching. If logs are hauled to a landing site before de-limbing, a 'bird's nest' of branches and needles will accumulate and nutrient concentration in the rest of the forest is lowered.

A number of studies in northern temperate areas and Australia are reported by Attiwill and Weston (2001) and in round terms the nutrient loss associated with whole-tree harvesting (where the whole tree – trunk, branches and sometimes the roots – is taken away rather than just taking the trunk and leaving everything else on site) results in a loss from the total pool of $< 3\%$ of P, Mg and K and $< 8\%$ of N and Ca (rising to 13–19% of Ca loss in oak–hickory forests). In eucalypt forests losses of N and P were similar: 2% in stem-only harvesting and up to 8% in whole-tree harvesting. These losses are relatively small and easily replaced. A detailed study by Yanai (1998) at Hubbard Brook compared potential nutrient loss on W5 that was whole-tree harvested (see Box 8.1) with another watershed that was clearcut, removing just the trunks. Table 8.2 shows that while 70% of the above-ground biomass is held in the trunk, it is poor in nutrients and contains around half or less of the

Table 8.2. *Biomass and nutrient content of above-ground parts of a 70-year-old hardwood forest at Hubbard Brook Experimental Forest (HBEF). The data can be used to work out how much biomass and nutrient loss there will be if forestry operations remove just the trunk in clear-felling or remove all of the tree (except the leaves) in whole-tree harvesting. In reality, some of the material that could be harvested is inevitably left behind because it is too small or inaccessible. The harvest removal ratio shows how much in reality was taken from two watersheds at HBEF when they were actually felled. The ratio shows how much more biomass or nutrient was removed by whole-tree harvesting compared with traditional clear-felling.*

| | Above-ground total kg $ha^{-1}$ | Content of tree parts as a percentage of above-ground total | | | | Harvest removal ratio |
|---|---|---|---|---|---|---|
| | | Trunk | Branch | Twig | Leaf | |
| Biomass | 1 999 000 | 70 | 29 | 0.6 | 2 | 2.7 |
| Phosphorus | 52 | 31 | 56 | 1.0 | 12 | 5.4 |
| Nitrogen | 509 | 40 | 45 | 0.6 | 15 | 3.9 |
| Potassium | 219 | 48 | 36 | 0.5 | 15 | 3.5 |
| Calcium | 584 | 51 | 45 | 0.3 | 4 | 3.5 |
| Magnesium | 54 | 55 | 35 | 0.4 | 1 | 3.0 |
| Sulphur | 59 | 49 | 36 | 0.4 | 10 | 2.8 |

*Source:* Data from Yanai, 1998. *Forest Ecology and Management* 104.

nutrients. By contrast, branches and leaves contain a greater proportion of nutrients per unit of biomass. This is particularly so for phosphorus as the leaves are high in phosphorus but low in biomass while the trunk is low in phosphorus and high in biomass. Consequently the branches hold the greatest reservoir of phosphorus (56% of the tree total).

Other nutrients are distributed in different ways: calcium and magnesium are concentrated in the trunk while for nitrogen and potassium a greater proportion than other nutrients occurs in leaves. From the harvest removal ratios it can be seen that whole-tree harvesting removes disproportionately more phosphorus than other nutrients. Almost three times as much biomass and over five times the amount of phosphorus were removed compared with a clearcut. This whole-tree loss equates to the removal of 50 kg of phosphorus per hectare, equivalent to 71% of the total phosphorus in the living vegetation (70 kg $ha^{-1}$) and 59% of the P in the forest floor (85 kg $ha^{-1}$). Although the amount of phosphorus removed by the harvest was only 3.1% of the total phosphorus in the mineral soil (1600 kg $ha^{-1}$), using the data on phosphorus budgets for the same site given in Fig. 8.7, the whole-tree harvest removed 32% of the most readily available phosphorus pool. The fact that phosphorus is a remarkably immobile nutrient is underlined by the negligible loss to streams (0.2 kg $ha^{-1}$ over 3 years). This is similar to the pattern seen in ammonium after felling in W2, W4 and W6 at Hubbard Brook in an earlier study (Likens *et al.*, 1970). By contrast, nitrate showed a hugely increased concentration in stream water in the same study: it increased from a pre-felling concentration of 0.85–0.94 mg $l^{-1}$ of water to 53 mg $l^{-1}$ (a 60-fold increase) 2 years after cutting, reaching a peak of 82 mg $l^{-1}$ (a 90-fold increase) in W2 with the most drastic felling regime (see Box 8.1). The biggest concern is for calcium depletion (potentially linked to acid rain – see above) which may prove a long-term problem for sustained production.

The different timber-felling treatments on the watersheds of Hubbard Brook (see Box 8.1) provide useful insights. Comparing commercial whole-tree harvest (W5) with felling without any wood removal but with subsequent herbicide treatment for 3 years (W2), streamflow losses of both potassium and calcium were much larger for W2 than W5 during the first 8 years following felling. This is attributed in equal parts to slower decomposition of the smaller amount of logging debris and greater uptake by the regrowing vegetation on W5. The difference between the watersheds was most pronounced in years 1–3 since in year 4 vegetation began to recover following 3 years of herbicide suppression.

Timber removal in the tropics should, according to traditional views (where all nutrients are tied up in the vegetation) be far worse for nutrient depletion. The real problem, however, is not so much where the nutrients are held but that they are in short supply and not rapidly replaced. Removal of a few selected trunks, especially if the bark is left behind, does little since most nutrients are in the branches, twigs and leaves. But commercial logging can have important effects. Table 8.3 shows that removal of the timber makes a significant deficit to nutrients. Moreover, further nutrients are lost by extra leaching. This is caused jointly by: (1) reduced evapotranspiration due to tree removal so more water is available for downward leaching, and (2) there is insufficient vegetation to absorb the flush of soluble nutrients released by decomposition of the remaining debris. There is very little supply of nutrients from the parent rock (the soil is too deep) so rain and dust inputs are the major way of replenishing lost nutrients. It can be seen from Table 8.3 that it will take up to 60 years to restore above-ground nutrients to their former amounts.

### *8.6.2 Fertilizers*

In Northern Europe, major forest fertilization commenced in the early 1950s–60s. Correction of deficiency by addition of fertilizer is a simple and

Table 8.3. *An above-ground nutrient budget estimated for commercially logged lowland dipterocarp rain forest in Bukit Berembun, Malaya. Logging is assumed to remove 60 $m^3$ of timber per hectare, equivalent to 10% of the total biomass of all stems over 5 cm diameter.*

| | Nutrient concentrations (kg $ha^{-1}$) | | | |
|---|---|---|---|---|
| | K | Ca | Mg | Total N |
| Outputs | | | | |
| Removal of timber | 45 | 200 | 20 | 70 |
| Extra leaching | 75 | 30 | 15 | ? |
| Total outputs | 120 | 230 | 35 | 70+ |
| Annual inputs | 5.7 | 3.8 | 0.7 | 11.3 |
| Years to recover | 20 | 60 | 50 | 5–10 |

*Source:* Based on Bruijnzeel, 1992. Data from Whitmore, 1998. *An Introduction to Tropical Rain Forests* (2nd edn). Oxford University Press.

relatively economic business. However, it is used on only a minute fraction of the world's forests because of the perception that nutrient shortage or depletion is rarely a major problem over the long term (compared with large inputs required by the annual crops of agriculture) and is therefore an unnecessary expense. While loss of nutrients is often asserted to lead to loss of productivity (see Section 10.1.3), Attiwill and Weston (2001, p. 176) state that they 'have not found any definitive quantitative evidence in the literature in support of this assertion, nor are fertilizers used regularly in most of the world's forests used for timber production'. They add, however, that fertilizers *are* used regularly and increasingly in *plantations* around the world. This makes economic sense since plantations often use species planted well outside their natural range (e.g. pines planted in Australasia, eucalypts in South America and Europe) on soils marginal for tree growth. Moreover, their rotation length is often shortened, putting more pressure on the soil resources. In such cases nutrient deficiencies and imbalances are usually readily seen and rectified. Potassium appears to be the limiting element in plantation forestry and the cost of applying this from the air is really quite low, as it is with boron, which is required at only 6–8 kg $ha^{-1}$. Agricultural soils require a high pH, but the liming they often require is usually unnecessary on forest soils because most trees, and pines in particular (the staple of temperate forestry), grow well on acid soils. When calculating the amounts of fertilizer required, analyses are made of the pools of both total and available nutrients. There is a dynamic equilibrium between the two pools, of which the second is the one of direct importance to the plants. **Nutrient budgets** are required to estimate long-term balances between inputs and outputs so that productivity can be sustained indefinitely. **Gains** result from atmospheric deposition, leguminous and non-leguminous nitrogen fixation, weathering of rocks, groundwater ingress and fertilizer additions. **Losses** are caused by crop removal, soil erosion, leaching and volatilization including that resulting from burning (Maclaren, 1996, p. 62).

In north-west Europe, Heliövarra and Väisänen (1984) review the effect of fertilization on microbes and invertebrates. With added nutrients, bacteria dominate the microbes and there is a transitory increase in microfauna such as nematodes but large arthropods and above-ground invertebrates show little response. The authors conclude that most of these effects are due to the decrease in acidity rather than fertilization directly, but this is not the whole story since the number of earthworms decrease despite increased pH.

**Soil deterioration**, particularly on very large commercial sites, can result from bad forest practice and the gradual loss of mineral nutrients even though the latter can be replaced without much difficulty. The question of whether trees such as radiata pine *Pinus radiata* are soil degraders or soil improvers is discussed in Sections 2.2.3 and 6.6. With ancient woodlands in Great Britain, the threat is that of **eutrophication** as a result of fertilizer drift (especially NPK) from neighbouring agricultural fields (see Section 11.4.3 as well). This drift influences the margins of major woods but in smaller sites such as Cantryn Wood, Bridgnorth, where much of the woodland clothes the steep margins of a stream, a great deal of the area is subject to its influence. In these areas the increase in stinging nettle *Urtica dioica*, hogweed *Heracleum sphondylium* and sundry other coarse species since a major woodland survey of 1971 is obvious to the casual observer. The decline suffered by 17 ancient woodland species, revealed by a pilot resurvey in 2000 of 14 of the original sites was even worse and in 2003 there was a major resurvey led by English Nature (now called Natural England) and the Woodland Trust. The species concerned included early dog-violet *Viola reichenbachiana*, yellow pimpernel *Lysimachia nemorum*, polypody fern *Polypodium vulgare*, lady fern *Athyrium filix-femina*, three-veined sandwort *Moehringia trinervia* and the nettle-leaved bellflower *Campanula trachelium*. Wood-sorrel *Oxalis acetosella*, wood crane's bill *Geranium sylvaticum* and sweet violet *Viola odorata* were amongst the less uncommon species that had also diminished.

# 9
# Forest change and disturbance

## 9.1 Ecology of past forests

### *9.1.1 The Palaeozoic era and tectonic plate theory*

Woodlands and forests of the past were often very different from those we know today. In this matter, as in so many others, the Greek philosopher Heraclitus was correct in stating that only change is constant. Moreover, the changes known to have occurred in the world's flora and fauna have sometimes been very abrupt (Toghill, 2000), as in the biggest mass extinction of all time which occurred at the end of the Permian period (245 Ma i.e. millions of years ago – see Fig. 1.1). The causes of these mass extinctions may well have been either large asteroids striking the earth or the eruption of **supervolcanoes** such as that now simmering beneath Yellowstone National Park, Wyoming, USA. Each of these possesses a huge magma chamber, the contents of which contain vast amounts of mainly sulphurous gases. Once an eruption begins these previously dissolved gases assume the vapour phase, hurling much of the magma high into the stratosphere where some of it remains as fine ash for such a long period that it entirely disrupts the climate, lowering the temperature to such an extent that few plants and animals survive. Many past changes, however, have been slow and gradual as climate has changed and new types of trees appeared.

If we are to understand why the distributions, as well as the nature, of the forests of the past were often so very different from those of today, we need to understand the basic structure of the planet we live on. At the centre of the Earth is a mainly nickel–iron core (Briggs *et al.*, 1997, p. 31). The crust is the outermost of the five successively less dense layers that surround this core. It consists of a number of abutting **tectonic plates** that form both the continents and the ocean floors (see Section 2.5.2 regarding continental drift). The positions of these plates are not fixed but move large distances over the millennia; our increasing knowledge of the time taken for these movements to occur is gradually enabling us to relate the tectonic history of the Earth to the evolution of forest populations in various parts of the world as Section 9.1.3 illustrates. Plates have in the course of history moved from areas with a warm climate to those which are much colder, as well as vice versa, which explains why forests formerly existed in what is now Antarctica. At destructive plate margins, where one plate is being forced down below another, crustal shortening occurs. The still-ascending Andes of South America exemplify the mountain building that often occurs on the leading edge of an overriding plate, thus leading to the geographical isolation of what were once continuous populations.

Sometimes pressure that has built up over many years between adjoining plates is suddenly released with dramatic results, as when the Burma plate continued to force itself over the Indian plate at the end of 2004. The resulting earthquake, which measured 9.0 on the Richter scale, had an epicentre a little to the west of North Sumatra, and itself caused considerable devastation to the forests of the immediate area. It also triggered off a huge wave that travelled as far as Sri Lanka, India and Kenya reaching 10 metres high and killing many hundreds of thousands as it crashed on the shore. It devastated all in its path, including many coastal forests, whose soils were also rendered saline, especially those of outlying islands. Such a seismic surface wave, which is triggered at sea by seismo-volcanic activity is called a **tsunami**. Relatively low in the open ocean, it rises in elevation on reaching shallow coastal waters.

Severe though this damage was, such waves have caused even greater havoc in the past. Some tsunami are caused by volcanoes. The repeated eruptions of **Krakatoa** in 1883 initiated a series of tsunami and when the entire island blew up in late August – the loudest event ever recorded – the resulting wave was well over twice as high as that of 2004, though its influence was more local and the number of people killed far fewer. This volcanic region is particularly interesting to a forest ecologist. In 1927 there was a violent eruption just below the surface of where the old volcano had been. Since then a large volcanic island has built up and become forested in its cooler areas. We appear to have a cyclic process in which new forested land develops from molten magma welling up from deep within the Earth's crust, only to be destroyed in another vast explosion a few hundred years later.

Vascular plants, which transmit fluids by means of internal plumbing of either tracheids or open-ended vessels, did not arise until the Silurian (about 440–410 Ma) with tree-like forms developing in the Devonian (410–360 Ma), so as Fig. 1.1 shows, the period in which parts of the Earth have been clothed with woodlands and forests is relatively short. The same is true of the Vertebrata, the largest and most important of the four main groups in the phylum Chordata. Fossil traces of invertebrate animals with hard parts, such as belemnites, ammonites and trilobites are found from the beginning of the Cambrian. Traces of typical fish-like vertebrates are found in rocks of Silurian age. These evolved from early chordates not represented in the fossil record, but probably similar to the modern lancelet *Amphioxus*. This animal has an incompressible and elastic rod, the notochord, running down the centre of the body. The earliest vertebrates appear to have been jawless. These early Agnatha included the heavily armoured cephalaspids and pteraspids which were not parasites and scavengers like the modern lampreys and hagfish, some of which still parasitize the fish of woodland streams and rivers. The future, however, lay with the jawed vertebrates (Gnathostomata), which developed in parallel with them, eventually giving rise to the birds and mammals.

The xylem of vascular plants provides a skeleton that supports the body of the plant above the ground, as well as transporting water and mineral salts. Together these plants form the **Tracheophyta** whose four classes, the Psilopsida, Lycopsida (=Lycopodiopsida), Sphenopsida (=Equisetopsida) and Pteropsida, are shown in Fig. 1.1. Members of the **Psilopsida** were abundant in the Silurian (in which they arose) and Devonian periods during which the other three classes evolved from them. Like the modern ferns, plants belonging to this ancient group of vascular plants showed a distinct **alternation of generations** between a small haploid sexual **gametophyte** and a much larger diploid asexual generation (the **sporophyte**) that produces spores that give rise in turn to further short-lived haploid gametophytes (Bell and Hemsley, 2000). The present-day order Psilotales contains two small genera, *Psilotum* and *Tmesipteris*, whose anatomical and reproductive features recall members of the fossil class Psilopsida. There is, however, no continuity in the fossil record, and for various reasons (Bell and Hemsley, 2000) they are now placed in the ferns. The **Lycopsida** and **Sphenopsida** are today in relative decline, while the **Pteropsida** dominate the terrestrial vegetation of the world. This class contains the ferns, conifers, ginkgoes, cycads and the angiosperms, whose most important trees are dicotyledonous. The commonly employed term **Pteridophyta** refers to all tracheophytes apart from the true seed plants i.e. the gymnosperms and angiosperms. Amongst the gymnosperms, the area of the world covered by conifers remains very great.

Two of the most important events which have occurred in the evolution of our woodlands and forests are the rise of the gymnosperms, whose ancestors originated around 370 Ma in the Devonian, and of the angiosperms, whose first fossils date from the Early Cretaceous (130 Ma) in the upper Mesozoic and which include the broadleaved trees so important today. Though the number of species in the gymnosperms is much less now than in the Cretaceous period, they provide a vast amount of softwood timber. The present diversity of the angiosperms is immense, and the dicotyledonous trees covering much of the Earth's surface produce a wide variety of hardwood timbers. The monocotyledonous palms do not yield useful timber, but the dates and coconuts they produce are a valuable food source. Although the plants and animals of many earlier forests were very different from those now existing, general **ecosystem function** including soil relationships was broadly similar.

A brief review of past forests helps us to appreciate just how radical past changes have been. The types of ancient forest which will be briefly described here are the clubmoss (*Lepidodendron*) forests of the Carboniferous period (363–290 Ma), the forests which existed when the dinosaurs roamed, and the more recent interglacial forests which are rather like our own but which were more extensive and possessed far more large animals.

Trees of the **clubmoss forests** (Fig. 9.1), which were up to 40 m tall, are of great economic importance because their fossil remains are the major constituents of coal. The Lepidodendrales (ancestors of our extant, herbaceous clubmosses) had several genera, notably *Lepidodendron* and *Sigillaria*. They possessed a limited degree of secondary thickening and the equivalent of a cork cambium (further description of the biology and evolution of early trees can be found in Ingrouille, 1992 and Thomas, 2000). *Lepidodendron* had spirally arranged leaves and a trunk which reached a metre in diameter. Unlike modern relatives in the Pteridophyta, it produced seeds borne in large cones on the dichotomous branches of the canopy, but was not in the line which subsequently gave rise to the true seed plants of today. *Sigillaria* had a tall columnar stem which was either simple or slightly branched and bore a tuft of simple leaves up to a metre long at the stem apex or apices. The climate of the Carboniferous period was moist and the soils waterlogged; the rhizome-like rhizomorphs of both these trees, which bore slender roots that subsequently broke off, were shallow and repeatedly dichotomously branched. Trees of *Calamites*, a relative of our modern horsetails, were also in the canopy, growing from huge underground rhizomes. Although a variety of invertebrates scavenged and assisted with soil processes, only three of the major vertebrate groups were present in these forests. These were fish, which first appeared in the Silurian period; the amphibia, that arose in the Devonian period (and of which the labyrinthodont shown in Fig. 9.1 was an early representative); and the reptiles, whose earliest known fossils date from the Carboniferous itself.

### *9.1.2 Forests of the Mesozoic era*

The trees, ferns, herbs and lower plants of the world's woodlands have altered almost beyond recognition since Carboniferous times. During the Mesozoic

Figure 9.1 Margin and trees of a clubmoss forest of the Carboniferous Period. The cone-bearing trees in the distance are of *Lepidodendron*, as is the central fallen tree. There is a vertical dead trunk (hulk) of an early cordaitalian member of the gymnosperms on the left; with an unbranched species of *Sigillaria* placed more centrally. The trunk on the right is of a tree fern, some of whose foliage is seen in the top centre of the picture. Grasses and other flowering plants had not yet evolved, but bryophytes, ferns and small horsetails abounded in the undergrowth. The labyrinthodont amphibian shown in the foreground was typical of this early stock that spent much of its life in the water, but had the ability to walk on land. Resting on the fallen trunk is a member of the extinct 'griffenflies' (Protodonata) which included the largest insects that have ever lived and gave rise to the Odonata, the dragonflies and damselflies of today (Grimaldi and Engel, 2005). The biggest of all was *Meganeuropsis permiana* from the early Permian, which had a wingspan of approximately 71 cm. (Drawn by Peter R. Hobson.)

era (215–65 Ma) most important forest trees were gymnosperms, but the angiosperms became increasingly important from their appearance in the early Cretaceous 140 Ma (Fig. 1.1). It is often difficult to identify fossil wood, of which there have been many re-assignations in recent years. The pines (*Pinus* spp.) are now thought to have originated in the early–middle Mesozoic. Palaeozoic fossils formerly ascribed to this genus have been reclassified into extinct pinaceous genera including *Pityostrobus* and *Pseudoaraucaria*. The world's forests changed considerably during the Mesozoic era, when the dinosaurs roamed. The period when almost all the land vertebrates more than a metre long were dinosaurs was initiated by the Permian–Triassic extinction some 250 million years ago when 90% of all animal species, including the trilobites, were eliminated when a meteorite struck what is now Wilkes Land, Antarctica, making a crater 30 miles wide. Within 20 million years the archosaurs, whose descendants the crocodiles and alligators still survive, gave rise to the first primitive dinosaurs. Following another mass extinction 200 Ma, the dinosaurs virtually ruled the land for 135 million years, only to be exterminated when another meteorite struck Mexico at the end of the Cretaceous period, forming a crater six miles across at Chicxulub. Between these two extinctions caused by meteorite impacts, very large land and marine dinosaurs flourished and pterosaurs flew overhead. *Tyrannosaurus rex* and *Baronyx*, another flesh-eater 10 m long discovered as recently as 1986, both lived in the Cretaceous and were amongst the many carnivores preying upon herbivorous dinosaurs and other animals which lived in and around the Mesozoic forests. It is now possible to date the ages of both ancient and modern reptiles by examining the **annual growth rings** in their weight-bearing solid bones. Using this technique it has recently been shown that *T. rex* put on 4 tonnes in a teenage spurt between the ages of 14 and 18. Interestingly, modern elephants do much the same but live to 70, instead of dying at around 30 as did the tyrant lizard king, which also finally attained around 6 tonnes.

Subsequent changes amongst animal populations have been even greater. Today, one of the most characteristic features of woodlands is birdsong, yet the first bird fossil (*Archaeopteryx*), with its toothed beak, huge eyes, expert co-ordination and feathers, arose from an early dinosaur line as late as 147 Ma near the junction between the Jurassic and Cretaceous periods. This animal is now known to have been capable of true flight; fossils of its precursors have yet to be discovered. Ground-dwelling dinosaurs appear originally to have grown feathers for insulation and warmth and only later did the various lines gradually develop the ability to glide or fly. *Microraptor gui* (128–124 Ma), which was about 60 cm long, used its four feathered limbs and tail to glide from tree to tree in much the same way as the modern flying squirrels (all of which are nocturnal) employ the membranes between their front and hind legs. Being warm-blooded (**homoeothermous**) gives both birds and mammals the advantage that they can remain active under cold conditions, their body temperatures being both high and constant. The other groups of vertebrates are cold-blooded (**poikilothermous**), with temperatures only slightly above that of the surrounding water or air, one of the reasons why snakes in cooler countries hibernate in winter and sun themselves in summer.

The most primitive of the three major groups of mammals are the monotremes, which derive their name from the fact that a single tube – the common cloaca – is used for reproduction, urine excretion and the egestion of faeces. The species that persist today are very specialized and unusual, but J. Z. Young's (1950) prediction that fossil evidence regarding the origin of this ancient group would eventually be discovered has been justified by the finding in the Australian Cretaceous of jawbones of species with a dentition similar to that of the juvenile modern monotremes. Modern placental mammals appeared in the Cretaceous, a period which ended 65 Ma with the extinction of 75% of all living species including the mighty dinosaurs. Dominance was now assumed by the mammals which increased rapidly in both diversity and size during the ensuing Tertiary (65–2 Ma). Recent changes have been just as dramatic. The most recent remains of mammoths discovered in western Europe were found in Shropshire in 1986. The animals had been caught in thick mud above permafrost which collapsed on melting into a kettlehole a mere 12 700 years ago.

Knowledge of mammals that could glide from tree to tree in the Late Jurassic or Early Cretaceous forests is very recent (Meng *et al.*, 2006). The discovery in Mongolia of the fossilized remains of *Volaticotherium antiquus* (meaning ancient gliding beast) is one of the most important discoveries of mammals in the Mesozoic era for over a century. This animal, which weighed a mere 70 g, could glide using the membrane stretched between its arms and legs, and is so unusual that a new order of mammals (Volaticotheria) has had to be created in order to accommodate it. It steered itself with a long stiff tail and existed 130 million years ago, whereas the earliest previously known flying mammals were the bats (from 51 Ma) and the flying rodents (from 30 Ma). Its remains are of truly remarkable quality; impressions of the fur and part of the gliding membrane survive in the rock where it was found.

**Isolated faunas** are frequently of very different origins, but their members often show parallel evolution of features found in other groups in distant forest regions. Thus, the three remaining species of monotremes (the duck-billed platypus of Australia and the two spiny echidnas of Australia and New Guinea) like other mammals are warm-blooded and feed their young on milk, but lay eggs (though unlike those of reptiles, these develop in the mother's pouch). The tiny young of marsupials grow rapidly in the pouch of the mother, while young placentals (now dominant in most of the world's forests) are born at a very advanced stage. The island continent of Australia, so long isolated from the rest of the world, developed a most remarkable Pleistocene marsupial population including elephants, lions, large cow-like animals and a huge blunt-faced kangaroo which may have been 3 m tall. All these megafauna are depicted in cave pictures made by the aboriginal population 20 000 years ago in the last Australian ice age (Vandenbeld, 1988). The giant browser *Diprotodon optatum*, nearly 3 m long and 2 m high at the shoulder, was the largest marsupial that ever lived. The demise of this megafauna, as elsewhere, appears to have coincided with the rise of humans.

Though the present separation of New Guinea from northern Australia is relatively recent, their faunas have already diverged. The long-beaked echidna *Zaglossus bruijni*, which feeds on earthworms, is much larger than its Australian equivalent, the burrowing spiny anteater *Tachyglossus auleatus*. Moreover, far more of New Guinea's kangaroos live in the trees, where they move rather clumsily, feeding on the foliage and occupying a niche similar to that of many Old World monkeys.

### 9.1.3 The Cenozoic: Tertiary and Pleistocene forests

In the early Tertiary, which began some 65 Ma, there was a northern forest containing firs, birches, pines and poplars in what is now the Arctic Circle. The more varied temperate angiosperm forest to the south contained maples, buckeyes, horsechestnuts, chestnuts, beech, ash, walnuts, oaks, lime and elms together with evergreen conifers including Douglas fir, firs, hemlocks, neo-cypresses (*Chamaecyparis*) and spruces. At this time the break-up of the ancient supercontinent Pangaea, now understood in terms of tectonic plate theory, was at a relatively early stage and a land bridge connected North America and Asia. The straits between Greenland, Iceland and northern Europe were relatively narrow, and many tree genera were dispersed right across the northern hemisphere. The southward movement of the temperate Tertiary forest (shown in Fig. 9.2), which gave rise to four major forest regions that subsequently developed differently, was initiated by a drop in temperature towards the end of the Tertiary which culminated in the Pleistocene ice age. The upthrust of the Rocky Mountains in North America created the treeless rain shadow of the Great Plains and caused the formation of distinct western and eastern forests. The western forests of North America became dominated by conifers, while angiosperms remained abundant in the east. In Eurasia

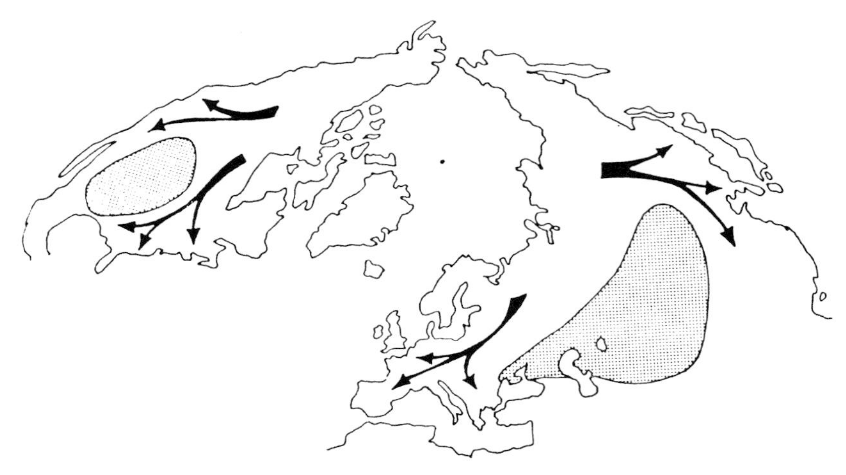

Figure 9.2 Southward movement of the southern Tertiary forest to give the four distinct forest regions now found in the northern hemisphere. Stippling indicates arid areas. (From Gibbs and Wainhouse, 1986. *Forestry*, 59, 141–153, by permission of Oxford University Press.)

the low rainfall of the central area, which was caused by its distance from the sea, separated the eastern and western forest regions.

The present geological period is the Quaternary, which has had a duration of two million years and is divided into two epochs, the Pleistocene and the Flandrian, of which the latter has existed for only 10 000 years. The Quaternary began with the onset of the Pleistocene ice age and has been subject to repeated climatic changes. Toghill (2000, chapter 13) describes the sequence of 17 cycles of cold and temperate climates known to have occurred so far. The end of the successive glaciations of this was marked by a brief, relatively warm period known as the Allerod interstadial, during which the east–west alignment of the Alps and Pyrenees prevented the easy migration of trees and other plants to and from the refuges to the south of them. In eastern Asia, as in North America where the major mountain ranges run north–south, the major elements of the rich Arcto–Tertiary forests retreated south along continental migration routes, rather than being destroyed by ice as were the complex forests of western Eurasia. Even at the end of the Pleistocene glaciations, however, at least one modern species of many tree genera existed in all four major forest regions. These have continued to evolve together with their associated fauna, flora and microbial populations including pests and pathogens. The effects of moving these last two groups from one major region to another can be very serious and are considered in Sections 5.3 and 5.4. Today the magnificence of the Arcto–Tertiary forests is approached only by the mixed mesophytic stands of the southern Appalachians or eastern Asia, which are far richer in trees and understorey species than any other temperate forests. In the early Tertiary the tulip tree *Liriodendron* was widespread and included stands in Iceland and Europe; now it is restricted to eastern China (*L. chinense*) and south-eastern USA (*L. tulipifera*).

Many of the trees, herbs and animals of the Pleistocene interglacial forests were similar or even identical to those which exist now in the current interglacial. Angiosperm trees including birches, aspen, willow, hazel, beeches, elms, oaks and maples were prominent but the firs, pines and spruces continued to dominate many large areas, particularly those that were colder and more exposed. One of the biggest differences was in the animals; the adaptive radiation of the mammals, which at the beginning of the Mesozoic had been only a small group of mammal-like reptiles, was astonishingly rapid during the Tertiary, persisting into the Pleistocene and up to the present day. Pollen diagrams for the last three complete British interglacials, the Cromerian, Hoxnian and Ipswichian (we have yet to reach the next ice age so the current interglacial is as yet incomplete) give us an understanding of the plants, and consequently the climates, which prevailed. At the start of each interglacial there was a gradual rise in temperature which caused the ice sheets to wane and the sea level to rise, reached a peak in the climatic optimum, and then declined. Precipitation increased at the same time, often remaining high for most of the interglacial. Towards the end of the interglacial, heavy leaching took place under cold conditions in which chemical weathering proceeded very slowly. Mineral nutrient levels in the soil dwindled, particularly in upland areas, and did not rise substantially until warming took place in the next interglacial. Plants which escaped the glaciers in elevated areas or survived in refuges to the south and in the case of Britain, to the east, accomplished recolonization of areas devastated by ice.

Interglacial forests contained an astonishing variety of birds and mammals, but some of the latter, particularly the elephants and closely related mammoths, were much larger than those which exist today. The **woolly mammoths** which died out 10 000 years ago were related to **mastodons** which also died out at the end of the last ice age. Male mastodons weighed six tonnes each; recent evidence shows that they fought duels to the death using their curved tusks to crush areas of their opponents' skulls. There was also a gradual impoverishment of the tree flora as glaciation succeeded glaciation. Both Norway maple (*Acer platanoides*) and Norway spruce were native to Britain in the Ipswichian, the last complete interglacial; silver fir (*Abies alba*) and *Pterocarya* (wingnuts) were present in the preceding Hoxnian, and even *Carya* (hickories) occurred in the early Pleistocene. A major climatic warming 10 000 years ago marked both the end of the Pleistocene, and the beginning of the Holocene (or Recent) epoch which still continues. In the last millennium the **Little Ice Age** which began in the mid-1440s, and ran until the mid-1800s, was succeeded by the present period of global warming (see Section 11.3).

### *9.1.4 Holocene changes and the origins of British woodlands*

British woodlands and forests have, like those of almost the entire world, been subject to the increasingly intensive influence of humans for the whole of the Holocene and **ancient semi-natural woodlands** (land continuously wooded since AD 1600) are the closest approach to truly natural forests that remain; they now cover only about 2.3% of the British landscape. They form a cultural landscape that has often been cut over, frequently being used for coppicing (see Section 10.3). **Recent woods** are defined as having been planted or developed naturally on open vegetation at some time in the past 400 years. **Semi-natural woods** are defined as being predominantly composed of trees and shrubs native to the site which have developed from stump regrowth, as in former coppices, or have regenerated naturally from seed rather than plantings; they may be either ancient or recent. The terms **old-growth** and **virgin** forest are frequently used in North America and Europe, respectively, to denote forests which are natural or have been left virtually undisturbed, especially by logging, for very long periods. Subsequent changes to forest vegetation can often be very complex due to human intervention; an example is given in Box 9.1.

**Box 9.1 Influence of humans on the forests of Crete and Cyprus**

The deposits of Knossos, Crete, provide an almost continuous record of the activities of humans since 6000 BC, including Minoan times when wild boar were numerous. Human influence has greatly changed the forests of this Mediterranean island, which is some 160 miles from west to east. Beneath the bare peaks of the limestone mountains, remnants of the once extensive cypress *Cupressus sempervirens* and pine forests clothe the upper ranges. The Calabrian pine *Pinus brutia*, sometimes treated as a subspecies of the extremely drought-resistant Aleppo pine *P. halepensis*, is also associated with the stone or umbrella pine *P. pinea*. Relatively few trees of these three species now remain; two are illustrated in Fig. 9.3a. Old records show that exports of cypress timber to other countries were formerly extensive. Amongst the oaks *Quercus coccifera, Q. macrolepis, Q. cerris* and *Q. brachyphylla*, the last sometimes no more than a large shrub, are still present. The evergreen Holm oak *Q. ilex* and other species of the Mediterranean maquis must also have been much more abundant especially on the lower slopes and the ravines, and other wet places would have had an extensive population of

Figure 9.3 (a) Cypress *Cupressus sempervirens* and stone (= umbrella) pine *Pinus pinea* at Mona Preveli monastery, south-west Crete. Lower branches of the cypress have been trimmed; this tree and the native pines are now of limited distribution on the island. (b) Riverside Cretan palm *Phoenix theophrastii* and tamarisk *Tamarix* forest with giant reed *Arundo donax* near Mona Preveli on the south-west coast of Crete. (Photographs by John R. Packham.)

Figure 9.4 Outlines of Cyprian cedar *Cedrus libani* var. *brevifolia* (= *C. brevifolia*) growing at 1300 m near the tomb of Archbishop Makarios. (Photograph by John R. Packham.)

oriental planes. In contrast, the number of olive trees is said to total 35 million, and there are many fruit trees including almonds, apples and pears. On the positive side many rare orchids and other native herbs characteristic of the very rich Cretan flora are beneath and between the trees in many of these plantations, particularly on the Lesithiou plateau (Sfikas, 2002). Of the many exotic trees Washingtonia palms *Washingtonia robusta* from the New World and eucalypts from Australia are especially prominent, while *Agave americana* is common in more open places. The date palm *Phoenix dactylifera* is now widely planted. In contrast the Cretan palm *P. theophrastii* (Fig. 9.3b), whose fruits are fibrous, blackish and inedible, is now restricted to a few small metapopulations of which the largest is at Vai on the eastern coast. This rare species, here found mainly in shallow damp valleys near the sea, also occurs in south-west Turkey.

Destruction of the native forests has greatly accelerated rates of soil erosion, while extensive agriculture and poorly controlled building in many parts of the island is greatly reducing the area available to natural vegetation. The loss of natural forest together with the introduction of exotic species, many of them weeds, and of disease organisms, are all too commonly part of the globalization being experienced in many parts of the world.

The situation in Cyprus, another large Mediterranean island, is in many ways similar, with extensive areas of Calabrian pine *Pinus brutia*, whose maximum height is around 25 m and which has very narrow paired leaves varying in length from 11–16 cm. The Austrian (or European black) pine *P. nigra* var. *nigra*, that can reach a height of 45 m, is especially prominent in the Troodos Mountains where it was planted many years ago. Strawberry tree *Arbutus unedo* and moderately sized oaks commonly grow beneath it. Cedar of Lebanon *Cedrus libani* var. *brevifolia* is now planted by the Forestry Service in many hilly areas (Fig. 9.4). The wood of acacias, introduced by the British in 1878, is used for firewood as is that of mimosa which grows in the lowlands. Eucalypts are particularly prominent amongst the introduced trees and were planted in a swamp in the Akrotiri Peninsula. The swamp dried out within 30 years and the area is now used for growing pineapples. Few areas have been so extensively worked on as the extensively terraced hills of Cyprus, but some are now falling into disuse and in a few hundred years may be covered by a naturalistic vegetation in much the same way as areas of New England are today (see Section 9.4.2).

### 9.1.5 Presettlement forests of North America

These forests covered an enormous area and few remain in anything like their original condition, but Friedman *et al.* (2001) have used data collected during the General Land Office Survey (GLO) between 1853 and 1917 to throw light on the north-east Minnesota southern or hemiboreal presettlement forest in a landscape which occupied 32 000 km$^2$. The mapping scheme employed dated from as long ago as 1785; it involved the collection of data at intersecting points (marked by trees or cairns) within a nested mapping scheme for the 64 years before the initiation of logging and local settlement. This involved the identification of the tree species present, of which there were 40, and recording individuals in size classes. Black spruce *Picea mariana* was by far the most abundant tree and in some cases was not properly distinguished from the much less abundant white spruce *P. glauca*. White and red pines (*Pinus strobus* and *P. resinosa*) were by far the largest trees and together accounted for 9% of the trees present, with white pine having 20.1% and red pine 7.3% of the total basal area of all trees.

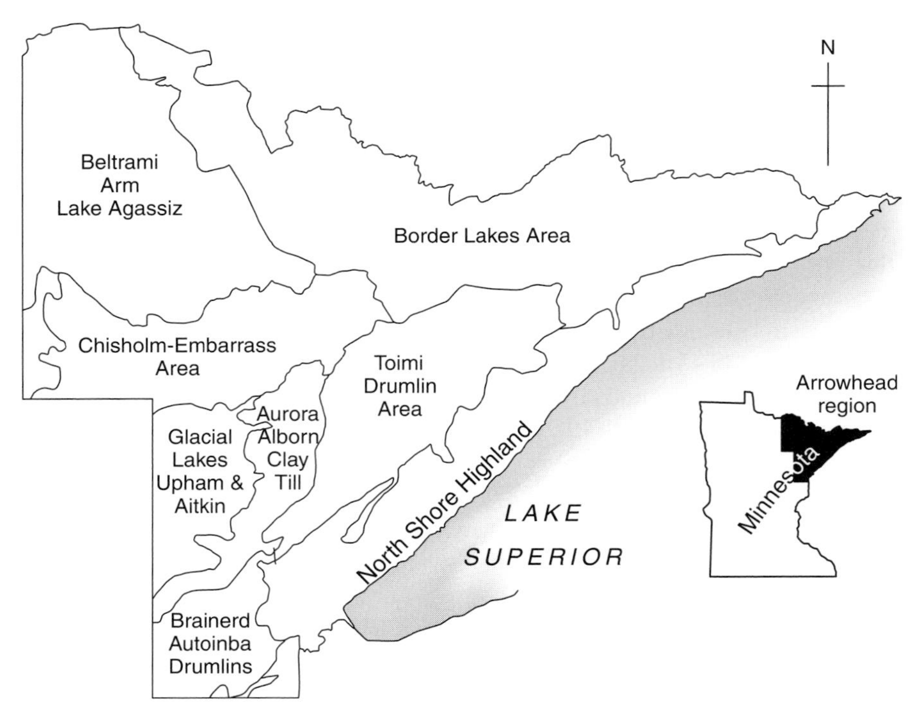


Figure 9.5 Physiographic zones of the Arrowhead Region, north-east Minnesota, USA, used in Fig. 9.6. (Redrawn from Friedman *et al.* 2001. *Journal of Ecology* 89, Blackwell Publishing.)

Figure 9.5 shows the eight physiographic zones present in this heavily glaciated area, while Fig. 9.6 indicates the distributions of the nine most abundant tree species. Most of the Arrowhead region is covered by moraine, but Holocene peats form the dominant geological material in the Glacial Lakes Upham and Aitkin zone, where larch accounted for 43.2% of tree species composition and 31.9% basal area. The Toimi and the Brainerd Autoinba zones are both occupied by **drumlins**, streamlined ridges with their long axes parallel to the flow of the glacial ice that moulded them from soft sediment. The eight zones supported different compositional mixes of the nine tree species which each accounted for more than 1% of the total tree population; the other 31 species accounted for only 6.1% between them. Spruce had a relatively even distribution in most of the Arrowhead region.

Landscape patterns of the tree species were measured at two spatial scales: 1–10 km and 5–50 km. This investigation shows how spatial patterns of forest trees were influenced on local to landscape scales by particular environmental factors, disturbance events and regeneration strategies.

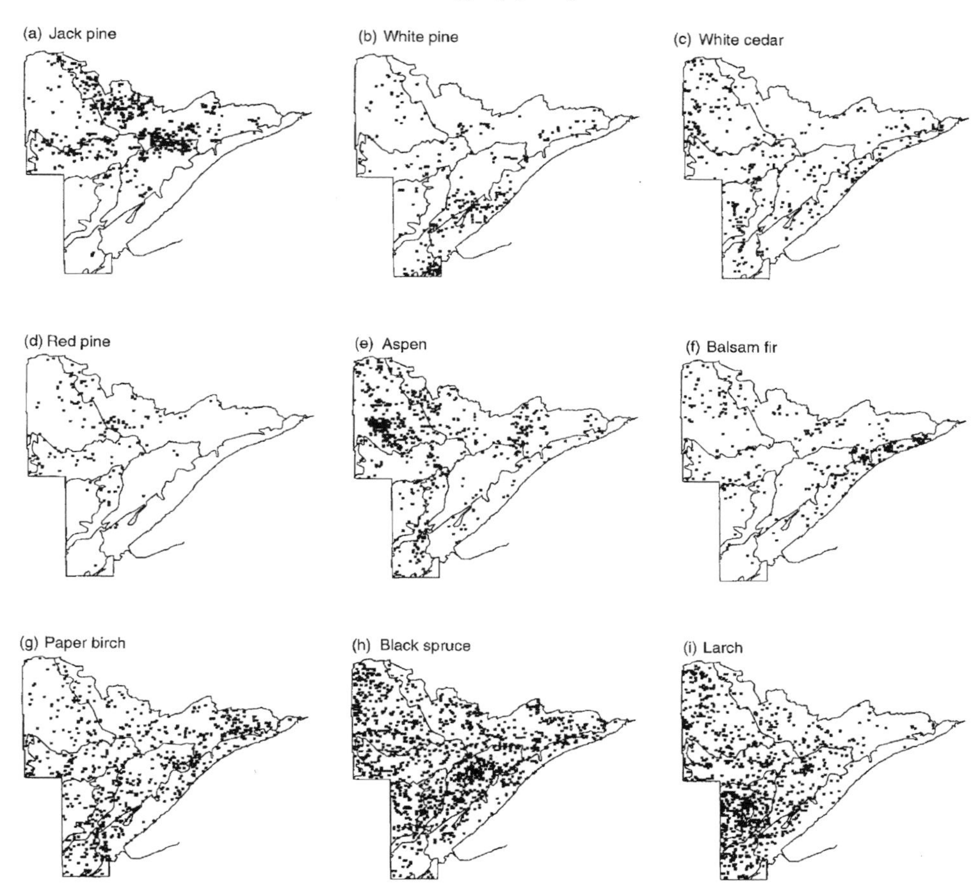


Figure 9.6 Distribution maps of the nine most abundant tree species (during the period 1853 and 1917) in the Arrowhead region, north-east Minnesota. Dots on these maps indicate plots in which three or four individuals of the same species occurred. Colonization patterns of these southern boreal species after stand-killing fires involved different regeneration strategies. Species dependent on seeds from surrounding unburnt trees included (b) white pine *Pinus strobus*, (c) northern white cedar *Thuja occidentalis*, (d) red pine *Pinus resinosa*, (f) balsam fir *Abies balsamea*, (i) larch *Larix laricina* (a late-successional larch) and (not shown) white spruce *Picea glauca*. Trees of (a) jack pine *Pinus banksiana*, (e) aspen *Populus tremuloides* and (h) black spruce *Picea mariana* burn and then regenerate vegetatively or from seeds in canopy-stored serotinous cones (see Section 3.7.1). Paper birch (g) *Betula papyrifera* regenerates from both fire-damaged stumps and blown seed from adjacent stands. (From Friedman *et al.* 2001. *Journal of Ecology* 89, Blackwell Publishing.)

### 9.1.6 Presettlement forests of the Amazon Basin

It should be mentioned here that what appear to be pristine forests might hide a different past. Over recent years the idea has been taking hold that the Amazon basin was not covered in pristine natural forest when the first post-Columbus colonists arrived after 1492. Rather, Amerindians who had been there for thousands of years may have developed a very highly modified landscape consisting of earthworks many kilometres long, major roads, big settlements, agricultural fields associated with *terra preta* soils (Section 2.5.1), and a variety of wetland features such as bridges, causeways, canals and fishing ponds (Heckenberger *et al.*, 2003). In fact, some people believe that in 1491 more people lived in the Americas than in Europe (Crosby, 1972). The reason why we do not see this now, some anthropologists argue, is due to the widespread and catastrophic post-Colombian depopulation from introduced disease, similar to that which happened in North America. Others disagree (e.g. Meggers, 1996) and suggest that the poor soils can never have supported more than small, often shifting populations. There is still no hard and fast consensus but the evidence does seem to be leaning towards the apparently pristine forests really being lush secondary growth following abandonment.

Similar but more recent (and thus documented) depopulation is known to have happened along the eastern seaboard of the USA – see Section 9.4.2 below.

## 9.2 Ecological processes that govern change

### 9.2.1 Succession and community assembly

It is clear from the previous section that major long-term changes in forests have been brought about by changes in climate, geological movements and by the evolution of the plants and animals themselves. In this section, we begin to look at the ecological processes that govern the changes we see in the forests now around us.

Succession is a significant source of change in forests. This starts from either completely bare ground with no organic components (**primary succession**) – e.g. glacial moraine, bare rock (but can also include open water), or from open ground with some living components (**secondary succession**) whose previous vegetation cover has been disturbed, for example by a forest fire or wind storm. **Succession** is then the non-seasonal directional change in the types and numbers of organisms present in a particular habitat over a period of time as it develops from bare ground to a final vegetation type.

**Box 9.2 Does succession always start with plants?**

Traditionally, it has long been assumed that this process of succession is initiated by green plants and that heterotrophs (in effect, eaters of organic matter) move in later. This view has recently been challenged by Hodkinson *et al.* (2002), who have listed many instances worldwide where an **initial heterotrophic stage**, of variable duration, has been described as occurring on fresh substrates devoid of green vegetation. These include newly exposed glacial moraine, shingle beaches, cooled lava flows and ash fields, whose surfaces often teem with moving organisms including active predators and detritivores. Moreover, heterotrophic protozoa and nematodes are often abundant in the interstices between the rock fragments. All these organisms, together with the dead organic matter which provides them with their food, have reached these new habitats before the development of the green plants which will later come to dominate them. It seems that as a general rule community assembly by autotrophs is preceded by this previously little recognized heterotrophic stage in which primary and secondary consumers, including predators and scavengers, form simple communities which conserve mineral nutrients, particularly nitrogen, thus facilitating the later growth of green plants.

Early theories of succession viewed it as a process that was ordered, directional and deterministic (that is, following a set order). Plant communities developed on bare ground under particular climatological, soil and hydrological conditions, passing through a series of stages (**seres**) which eventually resulted in a **climax vegetation** suitable for the environment concerned. This process was largely controlled by competition for light and was very frequently dominated by the largest and tallest plants present. If these were trees, the final stage would be a woodland or forest. More modern views do not envisage such a set order of stages or a fixed endpoint, but simply a probabilistic process largely driven by the availability of propagules and competition between species, interacting with variable environmental conditions. Box 9.2 also indicates that succession may often commence with heterotrophs. This less deterministic process, which has many possible pathways and final vegetation types rather than a set climax, is termed **community assembly**, a phrase now tending to replace the older term succession. In either theory the green plants play a most important role by providing shade, enriching the soil with humus and, together with the associated animals, modifying the often harsh conditions originally present (**facilitation**). In some cases the later-successional (or primary) trees do not invade until the early pioneers create suitable conditions. However, in many developing forests (even in primary successions), all tree species colonize early on (when the ground is bare and there is less competition) and the changing forest is a reflection of changing patterns of dominance (Walker *et al.*, 1986 give a good example). This is driven by a combination of conditions changing to suit different species (e.g. British oaks grow faster with some organic matter in the soil) and differences in growth rates (e.g. oaks grow slower than birches and so take longer to dominate). At the same time, the original pioneer plants tend to be suppressed by later dominants as they are shaded out. An outline of the **tolerance classes** used by foresters, based largely on the ability of tree species to survive shade while growing in the understorey, is given in Box 9.3. Associated communities of insects, birds, mammals and soil fungi also change, particularly when a woodland or forest develops. Recent work concerning the influence of large animals upon the form and species balance of former forests, considered in Section 5.7, has also changed views on the processes operating in truly natural (**old-growth**) forests.

Philosophical approaches to vegetation change are important, influencing both the design of research and the interpretation of results. Pickett and McDonnell (1989), for instance, claim that **community dynamics** 'emphasizes process rather than the end point, accommodates the richness of causes of succession and motivates diverse research approaches'. They also point out that the dynamics which lead to succession are often **allogenic** (controlled by outside factors) rather than **autogenic** (governed by biotic interactions and changes within the community).

### *9.2.2 Cyclic change and gap regeneration*

The variability of succession can be illustrated by examining what happens in gaps created by one or several trees falling over. In deterministic theory, the vegetation that develops in a gap will be a miniature succession, following the same stages that the original forest might have gone through. Drawing on his earlier work, A. S. Watt (1947) gave a classic example for a beech woodland in southern England (Fig. 9.7). Young beech exerts a shade so heavy that there is

**Box 9.3 Relative tolerance classes used to classify the ability of young trees to endure shade**

The light intensity ranges given for each class are the lowest amounts of full sunlight that enable the tree species concerned to survive in the understorey. Trees are shown in three columns, the first two being the gymnosperms and angiosperms of the east. There is no such distinction with the major western trees, which are predominantly gymnosperms. (After Spurr and Barnes, 1980. *Forest Ecology*. Wiley.)

| Eastern North America | | Western North America |
|---|---|---|
| Gymnosperms | Angiosperms | |
| **1. Very tolerant species** can survive when the amount of light is as low as 1–3% of full sunlight. | | |
| Eastern hemlock *Tsuga mertensiana* | Flowering dogwood *Cornus florida* | Western hemlock *Tsuga heterophylla* |
| Red spruce *Picea rubens* | Hop hornbeam *Ostrya virginiana* | Pacific silver fir *Abies amabalis* |
| | Sugar maple *Acer saccharum* | Pacific yew *Taxus brevifolia* |
| | American beech *Fagus grandifolia* | |
| **2. Tolerant species** need 3–10% full sunlight. | | |
| White spruce *Picea glauca* | American lime *Tilia americana* | Spruces *Picea* spp. |
| Black spruce *Picea mariana* | Red maple *Acer rubrum* | Western red cedar *Thuja plicata* |
| | | White fir *Abies concolor* |
| | | Grand fir *Abies grandis* |
| | | Coastal redwood *Sequoia sempervirens* |
| **3. Intermediate species** require 10–30% full sunlight. | | |
| White pine *P. strobus* | Yellow birch *Betula lutea* | Western white pine *P. monticola* |
| Slash pine *Pinus elliottii* | Silver maple *Acer saccharinum* | Sugar pine *Pinus lambertiana* |
| | Most oaks *Quercus* spp. | Douglas fir *Pseudotsuga menziesii* |
| | Hickories *Carya* spp. | Noble fir *Abies nobilis* |
| | White ash *Fraxinus americana* | |
| | Elms *Ulmus* spp. | |
| **4. Intolerant species** must have 30–60% full sunlight. | | |
| Red pine *Pinus resinosa* | Black cherry *Prunus serotina* | Ponderosa pine *Pinus ponderosa* |
| Shortleaf pine *Pinus echinata* | Sweet gum *Liquidambar styraciflua* | Red alder *Alnus rubra* |
| Eastern red cedar *Juniperus virginiana* | American sycamore *Platanus occidentalis* | Madrone *Arbutus menziesii* |
| Loblolly pine *Pinus taeda* | Black walnut *Juglans nigra* | |
| | Scarlet oak *Quercus coccinea* | |
| | Sassafras *Sassfras albidum* | |
| **5. Very intolerant species** need 60% or more full sunlight. | | |
| Jack pine *Pinus banksiana* | Paper birch *Betula papyrifera* | Lodgepole pine *Pinus contorta* |
| Longleaf pine *P. palustris* | Aspens *Populus* spp. | Whitebark pine *P. albicaulis* |
| Virginia pine *P. virginiana* | Black locust *Robinia pseudoacacia* | Grey pine *P. sabiniana* |
| Tamarack *Larix laricina* | Eastern cottonwood *P. deltoides* | Western larch *Larix occidentalis* |
| | Pin cherry *Prunus pensylvanica* | Cottonwoods *Populus* spp. |

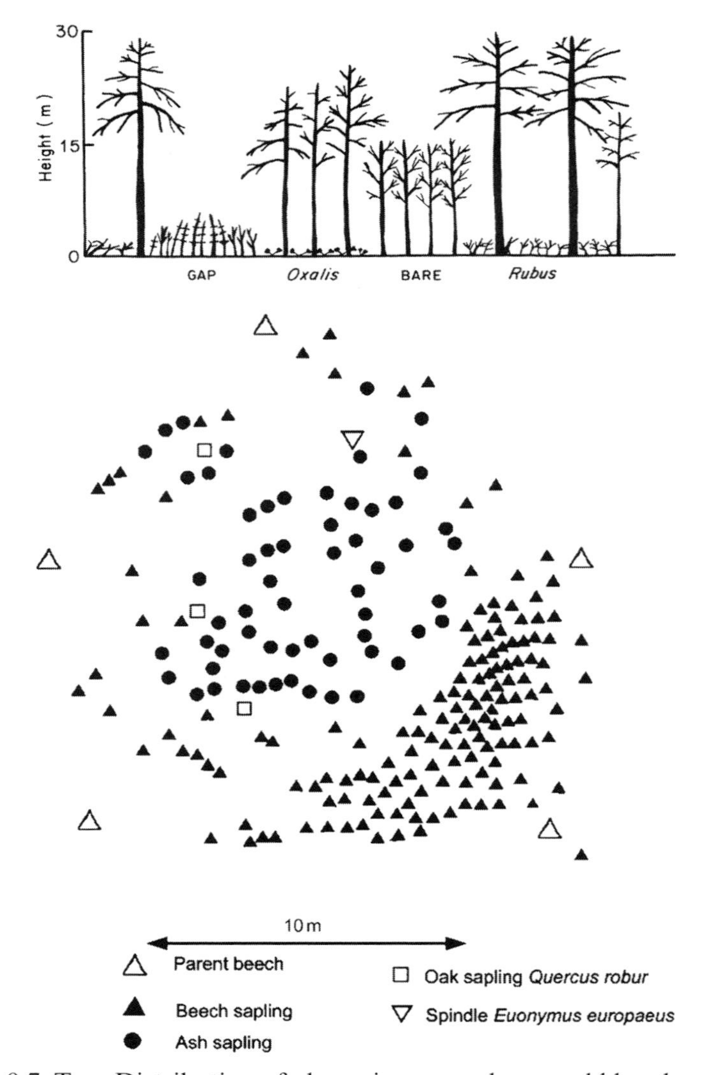


Figure 9.7 Top: Distribution of phases in space when an old beechwood has been left to itself and has trees of all ages. Beech *Fagus sylvatica* is forming a reproduction circle in the gap. Vegetation that subsequently develops in a gap goes through the original succession in minature, with the understorey dominated by bare ground, then wood-sorrel (*Oxalis*) and finally bramble (*Rubus* ). Since gaps appear at different times, all the successional phases can be seen in the woodland. (Redrawn from Watt, 1947. *Journal of Ecology* 35, Blackwell Publishing.) Bottom: Detail of a reproduction circle in a gap which typically has saplings of ash *Fraxinus excelsior* in the centre and those of beech and other trees round the margins. (Redrawn from Watt, 1925. *Journal of Ecology* 13, Blackwell Publishing.)

virtually no field layer beneath it. This bare stage is followed by the *Oxalis* stage when the light reaching the ground beneath more mature beech is sufficient to allow shade-tolerant wood-sorrel *Oxalis acetosella* to establish. Only when very much more light reaches the ground can the bramble of the *Rubus* stage develop. As trees eventually die, the vegetation in the resulting gaps frequently goes through stages similar to those that occurred in the original succession when the forest developed from open ground. This process, in which woodland goes through repeated cycles of similar sequences of development on a small-scale in gaps, is known as **cyclic change** (or **turnover of vegetation**).

In most woodlands and forests successional development does not follow a tight and exact prescribed pattern; modern work emphasizes that there is room for variation in what grows in each gap. In terms of general forest theory Watt himself was well aware of the various influences that could cause the outcome of gap regeneration to vary (see Packham *et al.*, 1992, Section 5.1). With regard to the reproduction circle shown in Fig. 9.7, for example, the size of the gap is important because beech mast formed in high forest drops almost vertically and is largely confined to the margins of wide gaps, the centres of which are often occupied by a dense growth of common ash whose seed is wind-dispersed. The occasional beech seedling has little chance of establishing in the central core of ash until the taller growth and deeper shade of the young peripheral beech, together with the gradual extension of the parent beech canopy, tend to suppress the more light-demanding ash. The outcome varies; in at least some cases the ash in the centre of a gap grows up fast enough to avoid suppression, while in narrow gaps the whole floor is sometimes seeded by beech whose seedlings and saplings may not meet with competition from any other tree.

Numerous studies have now been completed that show that many factors can change the **gap-phase succession** in a forest gap. So much so, that while climate, soil and topography determine the large-scale regional vegetation (e.g. oak forest) it is the small-scale gaps that produce most of the local variation that controls the proportions in which the various species grow.

The starting point is to consider how new plants arise in gaps. Many seeds will arrive as **seed rain** from surrounding trees. In some pines such seed is released from **canopy seed banks** as a result of fire (see Sections 3.7.1, 4.6.2). Pioneer species tend to have light, wind-blown seeds that go further than the heavy seeds of primary species. This is not always clear-cut since a number of trees with heavy seeds use animals to transport the seeds; perhaps the best example is the movement of some pine and oak seeds by corvid birds.

New plants can also arise from a **soil seed bank** (see Section 4.6.2), where seeds become buried under litter and humus, remaining viable for sometimes centuries until stimulated to germinate. It may seem paradoxical at first but it is normal that the most frequent species in the mature vegetation are largely absent from the seed bank and vice versa (Olmsted and Curtis, 1947; Bossuyt *et al.*, 2002). Rather, the soil seed bank is composed largely of early successional species and those of disturbed areas, waiting for another gap to appear. In north-west Europe, common species in woodland soil seed banks are the wild raspberry (*Rubus idaeus*), gorse (*Ulex* spp.) and heather (*Calluna vulgaris*), all representative of early successional stages, with seed densities of up to 1000, 30 000 and almost 70 000 seeds $m^{-2}$, respectively (Thompson *et al.*, 1997). There are some apparent exceptions such as in acid Scots pine woodland in Scotland with an understorey of heather, and a soil seed bank of a mean density of 83 000 seeds $m^{-2}$, 96% of which is heather (Miller and Cummins, 2003). However, this large heather seed bank is probably still a relict of an early woodland stage where heather was much more abundant, aided by the longevity of buried heather seeds (up to 150 years in moist peat).

Most soil seed banks are made up of herbaceous and shrubby species while trees are largely absent. This is primarily because most trees have comparatively large seeds that are readily found and predated on the ground. Supporting this is that trees with small seeds, such as birch, *are* found in seed banks. In Europe, birch has been seen regularly to have hundreds of seeds $m^{-2}$, and exceptionally recorded as up to 12 800 $m^{-2}$ for silver birch (*Betula pendula*). Most viable seeds are near the surface, declining in number with depth and changing in species composition (a facet of the relative longevity of different species). Thus the removal of increasing depth of organic matter by fire or mechanical means will result in a different density and species composition of seedlings.

Existing but damaged trees can also be part of the successional dynamics by producing new growth from buds surviving on branches, trunks and roots – the **bud bank**. Peterken (1996) provides an illustration of vigorous sprouts arising from the horizontal trunk of a wind-thrown small-leaved lime *Tilia cordata*. This species also sprouts vigorously from the bases of old trunks; in some cases thickets of saplings have developed vegetatively from the crowns of fallen trees. Few conifers can do this but it is a common means of regrowth in hardwoods. A final variable is the use of a **seedling bank** by some shade-tolerant species. Here, small, suppressed saplings persist under heavy shade growing very slowly. When a gap appears they have a head start in the race for dominance.

Taking all these variables into account, it is perhaps not surprising that successional pathways may not always be the same. For example, gap size is important. In small gaps created by one tree falling, shade-tolerant trees (the later-successional or primary species) such as beech or firs are more likely to do best and dominate. In larger gaps, early-successional, pioneer trees, such as birch and willow, which invade quickly from light wind-borne seeds and grow quickly, are likely to dominate, giving way later to the shade-tolerant trees. Thus, in small gaps the successional pathway can be truncated, leaving out the earlier stages. In subalpine forests of central Japan, Narukawa and Yamamoto (2001) found that Veitch fir (*Abies veitchii*) and especially Maries' fir (*A. mariesii*) formed seedling banks under the canopy while the Hondo spruce (*Picea jezoensis* var. *hondoensis*) and Japanese hemlock (*Tsuga diversifolia*) were restricted to gaps. In small openings the relatively few fir seedlings would dominate. In larger gaps the spruce and hemlock outnumber the fir seedlings and are more likely to reach the canopy and persist. In very large gaps with high mineral soil exposure, even birch (*Betula ermanii* and *B. corylifolia*) would be able to hold its own (Kohyama, 1984; Yamamoto, 1993).

Larger gaps also vary across their width, with gradients of increasing light and decreasing moisture towards the middle having an effect. The bulk of new seedlings at the edge of big gaps may reflect the high seed rain from surrounding trees, and the seedling bank doing well in partial shade. By contrast, the centre of large gaps will be less dependent upon seed input (except perhaps those from pioneer trees with lighter seed that travel further) and the seedling bank (which may not be able to compete with dense vegetation in light, dry conditions), and more dependent upon soil seed banks and vegetative sprouts from the bud bank. There will also be less below-ground competition away from the root systems of the large trees at the edge. This can be quite crucial; Barberis and Tanner (2005) found that in the seasonal rain forest of Panama, trenching to remove competition from roots of established plants increased tree seedling growth, attributed to increased nutrients in the wet season and increased water and nutrients in the dry season. Other factors may also help establishment; for example, in large gaps the sudden increase in light and temperature may lead to photoinhibition of the understorey vegetation (Houter and Pons, 2005), making it less competitive.

Further variability in seedling establishment is produced by small-scale heterogeneity of the forest floor since seedlings will be successful only if they are in favourable microsites. Kuuluvainen and Juntunen (1998) found in a Scots pine forest in Finland that the pits and mounds of bare mineral soil created by falling trees were important for establishment. Although these bare sites covered just 8.4% of the forest, they held 60% of pine and 91% of birch seedlings and saplings. Bare mineral soil offers less competition and a more constant water supply than the surrounding humus-rich forest floor. Gaps with more bare soil allow birch a better chance to establish and grow quickly enough to persist in the canopy, thus altering the successional pathway. Rotting 'nurse' logs also provide excellent conditions for seedling establishment in moist forests; in this case their main value is in raising the seedlings above the shade of dense understorey vegetation (see Fig. 1.7). Suitable microsites become increasingly important the denser the understorey vegetation and the deeper the litter and humus layer (see Kuuluvainen, 1994 for a review of Finnish boreal forests).

Hytteborn *et al.* (1993) studied small-scale natural disturbance and tree regeneration in two Swedish boreal forests. The canopy of the central boreal forest of Svartnasudden, south of Umea, was relatively sparse and sapling distribution was related to microsite pattern. That of the boreo-nemoral forest of Fiby urskog, near Uppsala, was more pronounced; here sapling distribution was influenced by both microsite pattern and canopy gaps. Both forests are of the *Picea abies–Vaccinium myrtillus* type and have a forest floor covered by a thick layer of moss in which *Pleurozium schreberi, Hylocomium splendens, Ptilium crista-castrensis, Dicranum* spp. and *Polytrichum* spp. are prominent. The nature of the **microsites** where tree regeneration occurs varies with the species concerned. More Norway spruce saplings grow on boulders than on soil, but they are concentrated on highly decomposed logs where they are more successful than on other microsites, whereas goat willow *Salix caprea* and the birches aggregate in the tip-up pits and on the root plates of fallen trees. Aspen *Populus tremula* produces root suckers so its vegetative reproduction is independent of microsites.

## 9.3 Disturbance, patch dynamics and scales of change

Natural disturbance and patch dynamics (Pickett and White, 1985) are involved in the ecology of many woodlands and forests. The various possible forms of disturbance, from a burrowing animal to a forest fire, cause very different degrees and scales of change.

### *9.3.1 Earthquakes and avalanches*

A notable example involves very large (> 0.25 ha) patches in the forests of the Lake District of southern Chile. Many undisturbed forests here have a profusion of lichens, mosses, ferns and vines, yet seedlings and saplings of the huge southern beeches (*Nothofagus*) which can reach over 45 m, have potential lifespans of 400–500 years and tower over the rest of the forest, are virtually absent. Tree species of the lower canopy, in contrast, normally possess individuals of all size classes. Are the southern beeches on their way out? The answer to this riddle appears to lie in the unstable nature of the region, which is amongst the most seismically active in the world.

In 1960, a series of catastrophic earthquakes in this area caused the northern coastline to rise by almost 2 m, with a corresponding lowering in the south. There were thousands of extensive debris avalanches, landslides and mudflows. At the same time volcanoes erupted in the Andes, showering ash throughout the district. When surfaces exposed by the 1960 slides were examined 15 years later, numerous tree saplings with average heights exceeding 3 m were present. In the lower elevation forests Hualle (*Nothofagus obliqua*), *Eucryphia cordifolia* (a eucryphia native to Chile) and *Weinmannia trichospermia* were the most abundant colonizing species, with Coigue (*N. dombeyi*) being the commonest species above 500 m. Nearly all reproductively mature trees of Reuli (*N. alpina*), which is by far the most valuable timber tree in the region, had been felled, but where it was present in more remote areas its saplings were also represented on the exposed slides. The shade-tolerant species common in the adjacent forests, in which their seedlings flourish, were absent or extremely rare on surfaces exposed in 1960. Furthermore, inspection of old-growth forests showed that many had developed on previous landslide scars or debris deposits.

The evidence assembled by Veblen (1985) thus strongly supports the view that the abundance of southern beeches in much of the mid- and low-elevation Andean forests is largely a consequence of the repeated periodic seismic disturbances known to have occurred. Indeed, of the 47 notable earthquakes which occurred in this region of southern Chile between 1520 and 1960, seven were roughly equivalent in magnitude to the main shock of 1960.

### *9.3.2 Wind, cold and wave regeneration*

**Wave regeneration** is a particularly regular form of cyclic change in which mature trees continually die off at the front edge of the wave, which lies behind an opening in the forest canopy and is thus exposed to the prevailing wind. In balsam fir (*Abies balsamea*) forests at over 1000 m in north-eastern USA, young trees spring up after the wave has passed on at a speed of 1–3 m $y^{-1}$ (Fig. 9.8). Though the dead trees often remain standing for some time, they no longer exert sufficient shade to inhibit growth of the young seedlings and saplings growing beneath them. Balsam fir is moderately short-lived and becomes increasingly susceptible to stresses, especially those caused by pathogens, by the age of 50–60 years. Trees bearing the brunt of the prevailing wind

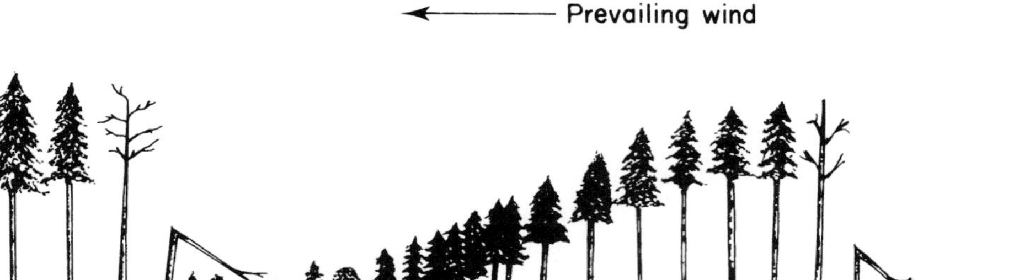


Figure 9.8 Diagrammatic vertical section of a regeneration wave in a balsam fir *Abies balsamea* forest, Whiteface Mountain, New York State, USA. The initially erect dead trees shown are crescentic in plan view; in many cases fusion results in a long multi-crescent wave which progresses up the mountainside. Beyond the ridge top, the waves pass out of the fetch of the prevailing wind, cease to move and eventually disappear. (From Sprugel, 1976. *Journal of Ecology* 64, Blackwell Publishing.)

are very commonly partially senescent, but are finally killed by environmental stresses including damage caused by winter accumulations of rime-ice, winter desiccation and decreased primary productivity which results from summer cooling. Forests with abundant balsam fir in New York State, New Hampshire and Maine all show regeneration waves of a similar type. Balsam fir has a relatively short life-span (80–100 years) compared with the red spruce *Picea rubens* that commonly grows with it which lives up to 300 years and is favoured by long disturbance-free periods. Areas of mixed forest have a high proportion of fir for the first century after cutting or major disturbance, while spruce becomes increasingly dominant as the stand ages and the mature firs die out. It is not surprising, therefore, that firs tend to dominate those areas of the forest where regeneration waves occur. Despite continual change, the overall composition of areas of forest possessing these regeneration waves remains relatively constant and in a **steady state**, provided interventions such as major changes in climate or disease patterns do not occur. Forests of Veitch and Maries' firs (*Abies veitchii* and *A. mariesii*) at 2000–2700 m in Japan show very similar regeneration waves, although these are further apart than those on Whiteface Mountain. This is only to be expected as these species are slower growing and longer lived (to about 100 years).

Crescent-shaped regeneration waves which move straight downwind appear to result from the death of a single tree or small group of trees, thus exposing a downwind arc of living trees to much higher wind speeds. Indeed, wind speeds in the canopy at the exposed edge of the old forest stand may be over 50% higher than those in the rest of the canopy. While a single crescent might

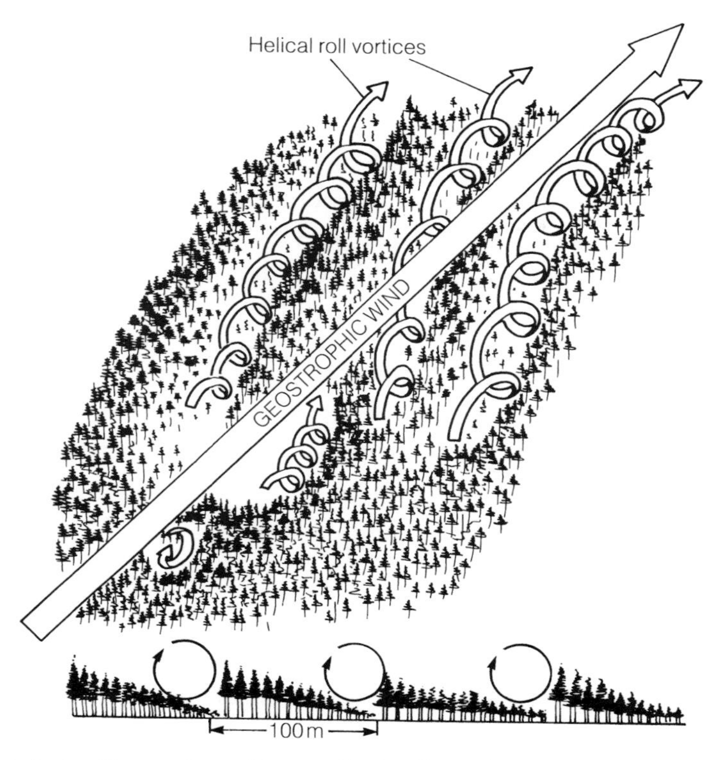


Figure 9.9 An illustration of how helical roll vortices form wave forests. The angle of the major axis of the helical roll vortices in relation to the geostrophic wind direction (generated by atmospheric pressure differences and the Coriolis force) implies the classic Eckman Spiral where surface currents in water flow at an angle to the surface wind. (From Robertson, 1987. *Canadian Journal of Forest Research*, 17.)

possibly move gradually all the way up to the top of the mountain alone, adjacent crescents appear to coalesce in long multi-crescent waves of the type common near the ridgetop. The more complex aerodynamics of the balsam fir forest of Spirity Cove, Newfoundland, the largest wave forest known, were investigated by Robertson (1987), who suggests that the regeneration waves found here are mainly caused by a series of **helical roll vortices** (Goertler vortices). These result in large numbers of sinusoidal dead tree strips, 100–150 m apart and orientated at an acute angle to the direction of the prevailing wind, which move in 55-year cycles (Fig. 9.9). This type of repeated change, literally driven by the wind, is in some respects similar to the **cyclical turnover of vegetation** which Vera (2000) considers to have occurred in the earlier European Holocene forest in response to the activities of large forest animals (see Section 5.9).

### 9.3.3 Wind storms

Constant wind causes damage, deformation and, as described in Section 3.5, even death when accompanied by temperature extremes. Here we consider the major natural disturbances by wind which often result in very extensive felling. This can take the form of 'a mosaic or patchwork which is a state of constant change' of the kind described by Cooper (1913) on the Isle Royale, in the northwest of Lake Superior, where old wind throws regenerated as new ones developed. Fiby urskog, near Uppsala, still shows the **storm-gap structure** developed on unstable boulder and marly moraines described by Sernander (1936), who in 1935 mapped five plots of which Plot 1 is shown in Fig. 9.10a with a superimposed re-map of the same 50 m × 50 m area in 1988/90. The second mapping shows the same basic structure, but the pattern is different because regeneration has occurred in some areas, while in others mature trees have fallen. Sernander allocated a degree of decay to fallen trunks, on which spruce seeds often germinate. His necrotization scale runs from 1–6 (see Box 7.3); in stage 6 the area is usually entirely covered by moss, but a row of young trees may mark the line of the decayed trunk. Plots 1, 2 and 3 were remapped in 1988/90 and the decay states of the logs on the forest floor then

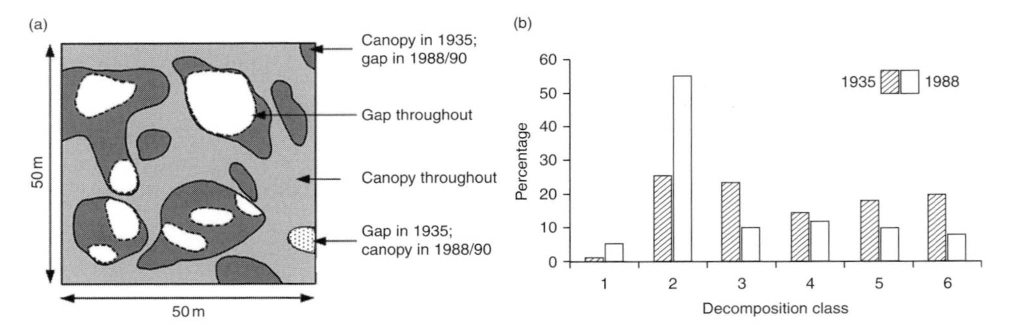


Figure 9.10 Long-term results from the Fiby Forest, near Uppsala, Sweden. (a) shows an area originally mapped by Sernander in 1935 showing storm gaps resulting from windthrow of Norway spruce. Other important trees include Scots pine, birch and aspen. This same Plot 1 was remapped in 1988/90 by Hytteborn *et al.* (1993) and the results obtained in both time periods are shown in the figure. Open areas are gaps on both occasions; dark shaded areas are gaps in 1988/90 but canopy matrix in 1935; the stippled area is a gap in 1935 but canopy in 1988/90; light shading indicates canopy matrix mapped on both occasions. (b) Distribution of fallen logs in the six decomposition classes in Plots 1, 2 and 3 at Fiby urskog in 1935 (n = 169) and 1988/90 (n = 303). Decomposition increases from class 1 to 6; see Box 7.3 for a detailed description of the classes. (Redrawn from Hytteborn *et al.*, 1993. In *Small-scale Natural Disturbance and Tree Regeneration in Boreal Forests* (Liu Qinghong). Ph.D. thesis, University of Uppsala, Sweden.)

and in 1935 are also compared in Fig. 9.10b. There is a much higher proportion of logs in classes 1 and 2 in 1988/90 and the total number of fallen trunks is also larger. Sernander's thesis of 1936 was of great conservation importance in that besides describing the greater susceptibility (in contrast to pine, birch and aspen) of spruce to windthrow and the role of **dwarf trees** in its regeneration, it was also in direct confrontation with the views of Hesselman, who considered the forest of small scientific value. Had Sernander lost the argument the forest, its unique communities and the important epiphytic lichens *Lobaria pulmonaria* and *Nephroma lusitacum* would have been destroyed.

Figures 9.11 and 9.12 also pertain to the small but valuable old-growth forest of Fiby urskog; two of the small trees concerned grew on the granite ridge rather than the moraine, while the large Norway spruce was released by one of the numerous canopy gaps formed by the storm of 1795. This storm was particularly notable but records indicate that all spruce forests in Uppland are hit by major tree-felling storms at least five times every century, indeed Fiby urskog and Granskar lost many spruce in the successive storms of December 1931 and February 1932. There is also considerable interest in the small-scale natural disturbance and tree regeneration in the Norway spruce forests studied by Hytteborn *et al.* (1993) at Fiby urskog and Svartnasudden, northern Sweden (see Section 9.2.2 also). Both these old-growth forests have been free of fire for at least 200 years. Tree death was caused by storm felling, fungal infection and insect attacks, sometimes in combination. Gap sizes in Fiby

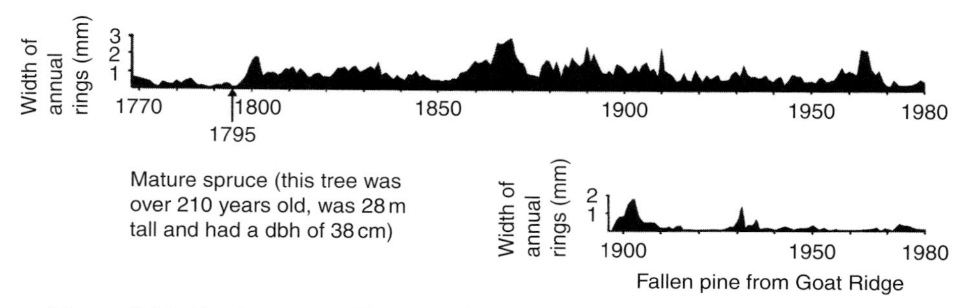


Figure 9.11 Environmentally controlled variation in radial increment in two trees from Fiby urskog, near Uppsala, Sweden. Annual rings of the mature spruce *Picea abies* show a marked increase in width immediately after the great storm of 1795. It appears that the tree was favoured by storm gaps torn in the forest canopy at this time. The fallen Scots pine *Pinus sylvestris* was only 3.9 m high and its roots were superficial, being developed in a mat of humus and vegetation (including lichens) overlying solid granite. Stem diameter where the trunk was bored (10 cm above ground) was only 6.5 cm, yet the increment core had over 80 annual rings, some of them paper-thin. (From Packham and Harding, 1982. *Ecology of Woodland Processes*. Edward Arnold.)

Figure 9.12 Dwarf spruce *Picea abies* which was considerably less than the 1.3 m height from which age increment cores are usually taken. Tree rings taken at ground level showed that this dwarf tree from Fiby urskog, Sweden had an age of at least 43 years. Note the prominent lateral branches and weak leading shoot. Dwarf trees beneath heavy shade retain a wonderful latent vitality. When growing on a suitable substrate they change their form and grow rapidly upwards if a canopy gap appears. (Photograph by Roland Moberg. From Packham *et al.*, 1992. *Functional Ecology of Woodlands and Forests*. Chapman and Hall, fig. 3.3. With kind permission of Springer Science and Business Media.)

urskog ranged from 9–3000 $m^2$ and 98% of the total gap area consisted of gaps less than 250 $m^2$ in area. Most gaps resulted from the fall of a tree or a small group of trees, but the area of such gaps tended to increase with time. Unlike Scots pine and deciduous trees, Norway spruce regenerated in small gaps here. Tree diversity was maintained by gap enlargement and the occasional formation of large gaps.

At the opposite extreme is mass tree mortality. The wind storm that hit northern France on 26–27 December 1999 has the dubious distinction of reaching the Guinness Book of Records as the largest number of trees 'destroyed in a storm' when 270 million trees were felled or broken. Other dramatic storm effects can lead to **mangrove peat collapse** as at Bay Islands, Honduras, after Hurricane Mitch in October 1998 (Cahoon *et al.*, 2003). In some Caribbean mangrove forests on oceanic islands, soil development primarily occurs by the formation of peat derived from mangrove roots, since a continental source of sediment is absent. The continued stability, and even existence, of such forests depends upon the production of thick mangrove peat at a rate considerably greater than loss of elevation due to decomposition of organic matter, for the forest has to maintain itself in the face of local sea-level rise. Powerful storms are known to have caused mass mortality of mangrove forests in the Caribbean and south-west Florida over a prolonged period, but recent study involving detailed measurements of elevation, accretion and root production has actually demonstrated the validity of the concept of mangrove peat collapse put forward in the past.

Low-, medium- and high-wind impact sites were investigated on the two most oceanic of the Bay islands. The impact of Hurricane Mitch was particularly great in Mangrove Bight, Guanaja, causing virtually complete mortality across all tree diameter classes in most places. Trunks of adult red mangroves *Rhizophora mangle* were broken and adult black mangrove *Avicennia germinans* trunks were uprooted. Of the 311 ha of mangrove forests possessed by the island only 3% survived. The mangroves of Roatan, where wind impact was low to medium, were much less severely affected. Here the dominant species were red and black mangrove, with some white mangroves *Laguncularia racemosa* and scattered individuals of buttonwood *Conocarpus erectus*. Most of the mangroves survived the strong winds and tidal inundation caused by Hurricane Mitch, but considerable areas of black mangrove forest on the north shore were killed by defoliation and toppling of the trees. When mangroves are killed peat collapse is likely to continue for some years. Recovery is made more precarious by the fact that successful seedlings cannot develop if the substrate level is at too low an elevation.

### 9.3.4 Exotic animals

**Exotic animals** imported as pets, or for animal collections and zoos, occasionally escape or are deliberately released. Some soon die out but others may not only survive but also cause a major impact on the existing biota, as do exotic plants (see Section 5.6). A number of large cats including leopards and panthers have been set free in the UK; so far there do not seem to have been any human tragedies as a result. Four centuries after being hunted to extinction the wild boar flourishes again in Sussex, and also in Kent, where it occurs along with the marsh frog which is much larger than our native common frog. The alien edible dormouse *Glis glis* (see Section 4.5.1) still flourishes in the Chilterns of southern England, a few animals having been released in 1902 in Tring Park, Hertfordshire, by the first Lord Rothschild, who also introduced the eastern grey squirrel *Sciurus carolinensis*.

### 9.3.5 Drought

**Drought** strongly influences mortality in woodland and forest communities, while young trees often die quite soon after planting if not watered regularly in dry periods. Losses caused by this in plantings associated with the entrance to the English side of the Channel Tunnel were considerable (D.R. Helliwell, pers. comm.). The influence of drought is also often exerted in the most unlikely places, as in the everwet (i.e. normally always very wet) rain forest of the Lambir Hills National Park, Borneo reported by Potts (2003). Here the rainfall between late January and mid-April 1998 fell to less than a fifth of that normally expected. Forest-wide tree mortality rates during the drought period rose to 7.63% per year as compared with 2.40% per year in the period before the drought. Mortality of large rare trees was less than that of large common trees during the drought. This suggests that there may be a compensatory mechanism maintaining the persistence of rare species.

### 9.3.6 Fire

Fire is a naturally occurring ecological force (and interestingly one of the few major forces that humans have any control over, at least in terms of starting fires) found in all continents except Antarctica. Plants and animals in fire-prone forests have often been co-existing with fire for millions of years and have come not just to tolerate fire but need it to ensure survival and reproduction. Most natural fires are caused by the 8 million lightning strikes that hit the Earth each day, although volcanism and earthquakes (causing sparks) can be locally important. Now, of course, humans play their part and in densely populated areas, such as much of Mediterranean Europe, the majority of fires are human-caused.

Fires started by lightning strike to a single tree in damp conditions will produce small gaps similar to a single tree death. However, the main role of fire is in producing large stand-replacing events that may kill trees over hundreds of square kilometres. Most large forest fires are in evergreen, usually coniferous, forests because of the high content of flammable resin in the foliage and the accumulation of masses of small, quick-drying pieces of litter on the forest floor which readily carry fire when dry.

Fire is often treated as a uniform type of disturbance. However, the effect of a fire very much depends on what sort of fire took place. A slow and gentle **surface fire** that trickles through the dry surface litter may kill thin barked trees (such as birch) and herbaceous plants, at least above ground, but will leave thick-barked trees unharmed. Most top-killed plants will resprout from stored

Figure 9.13 Canopy or crown fire in dead and dying balsam fir *Abies balsamea* in eastern Canada. Fire is burning through the whole canopy, producing flames twice the height of the trees. (Photograph courtesy of the Canadian Forest Service.)

buds. Few vertebrates will be killed and the survivors will benefit from the young, nutritious regrowth of vegetation. As surface fires become more intense, they may leap into the tree canopy to add a **canopy or crown fire** above the surface fire. Crown fires can be extremely intense (Fig. 9.13), completely consuming the foliage and small branches, usually killing the tree, and leaving a blackened skeleton. Regeneration is from seed rain from undamaged trees but especially from seed stored in the canopy in fire-proof **serotinous cones** that are stimulated to open by the heat of the fire and shed their seeds a few days after the fire. Serotiny is common in many pines, such as the lodgepole pine *Pinus contorta* and jack pine *P. banksiana* of North America and the bottle-brush trees (*Banksia* spp.) and eucalypts of Australia. Again, relatively few animals, apart from invertebrates, are likely to be killed, and most fire sites attract grazing animals which may have their own effect on the regenerating forest. A third type of fire is the **ground fire** that burns into the organic matter overlying mineral soils, including the living roots of plants. Trees otherwise unmarked can thus be killed when their roots are consumed and the whole tree falls over. Post-fire colonization tends to be from seed rain since stored buds and the soil seed bank are increasingly destroyed as the severity of the ground fire increases.

When an intense surface fire damages the cambium of a tree it leaves a scar. The **fire scar record** of an individual trunk thus indicates actual fire frequencies at its location. In North America native Indians formerly set light to surface litter and a ponderosa pine in the Bitterroot Mountains of western Montana records 21 fires between 1659 and 1915, an average interval of almost 13 years. Such periodic fires created the park-like stand of giant sequoias in the Yosemite National Park shown in a photograph of 1890. Eighty years later thickets of white fir *Abies concolor* occupied much of the ground. If fire were to occur under these conditions the firs would provide a fuel ladder carrying fire up into the canopy. Unlike coastal redwood *Sequoia sempervirens*, the stumps of giant sequoia (or big tree) *Sequoiadendron giganteum* lack the ability to sprout again so the trees would die (Spurr and Barnes, 1980).

A long-term experiment on the influence of fire began in 1929 when three plots were established in an area of the Olokomeji Forest Reserve, Nigeria. Though each was initially burnt, coppiced and cleared of trees, they were all treated differently. Their vegetation was analysed in 1957 (Hopkins, 1965). Grasses were almost absent from the fire-protected plot in which forest vegetation was rapidly re-establishing and a well-developed canopy of trees and shrubs was present. The other two plots were burnt annually, but whereas that burnt at the end of the dry season had become a tree savanna in which fire-scarred trees were widely spaced, that burnt at the start of the dry season, when fires were less intense, contained a fairly mature savanna woodland with some fire-tender forest trees protected inside fire-resistant clumps of closed woodland.

Fires are rarely completely uniform across a landscape, usually forming a mosaic of differently burnt patches and often unburnt islands (Fig. 9.14). This further complicates patterns of seed rain, regrowth from buds or seeds and grazing pressure. It is perhaps not surprising then, that forest fires will often, at least temporarily, increase biodiversity by creating a wide range of niches.

Even a casual interest in the media will have shown that fires have become more common with increasing drought and high temperatures around the world. Conditions previous to the fire are crucial to its subsequent development, as in the case of the lightning strikes that hit drought-stricken forests and created huge fires in Victoria and Canberra, south-east Australia in January 2003. The ability of gum trees *Eucalyptus* spp. to survive the fires which ravage them so rapidly and intensely is described in Section 1.3.1.

The European summer of 2003 was exceptionally hot; the highest temperature ever recorded in England (38.1 °C at Gravesend, Kent) until then occurred in August. Four months of continuous drought preceded the vast **forest fires** in southern France when arson was involved in the virtually simultaneous initiation of some 30 outbreaks in July. Huge areas of forest were destroyed,

Figure 9.14 Part of the Muleshoe burn in Banff National Park, Canadian Rocky Mountains in 1993. The fire has left some areas untouched, some with scorched foliage (e.g. the large tree on the right) which may recover but will cast less shade for a number of years, and other areas where canopy fire (see Fig. 9.13) has killed the trees (e.g. mid foreground and the circles in the distance), leaving an opening in the forest. (Photograph by Peter A. Thomas.)

much damage done and four people killed in conflagrations which jumped from place to place as the winds changed direction. With so much tinder-dry forest even the thousands of fire officers from all over France, together with some from Italy, took a long time to put the blazes out, despite the most modern methods, including water bombing from aircraft, being employed. Huge fires later that year also caused deaths and loss of timber in Mexico, California, Spain, Portugal, Alberta and British Columbia.

### *9.3.7 Combinations of factors*

Damage to forests often results from the combined impact of several factors as in north-west Colorado, where over 10 000 ha of subalpine forest was blown down in 1997 (Kulakowski and Veblen, 2002). The study area, shown in Fig. 9.15, ranges from 2400 to 3400 m (8000–11 000 ft) in altitude, and its forests are dominated by lodgepole pine *Pinus contorta*, quaking aspen *Populus tremuloides*, Engelmann spruce *Picea engelmannii* and subalpine fir *Abies lasiocarpa*. It was concluded that both topographic position and fire history contributed to susceptibility to wind damage. Damage was least in the younger stands, at lower elevations and in areas away from ridges where wind speeds were lower. Stands

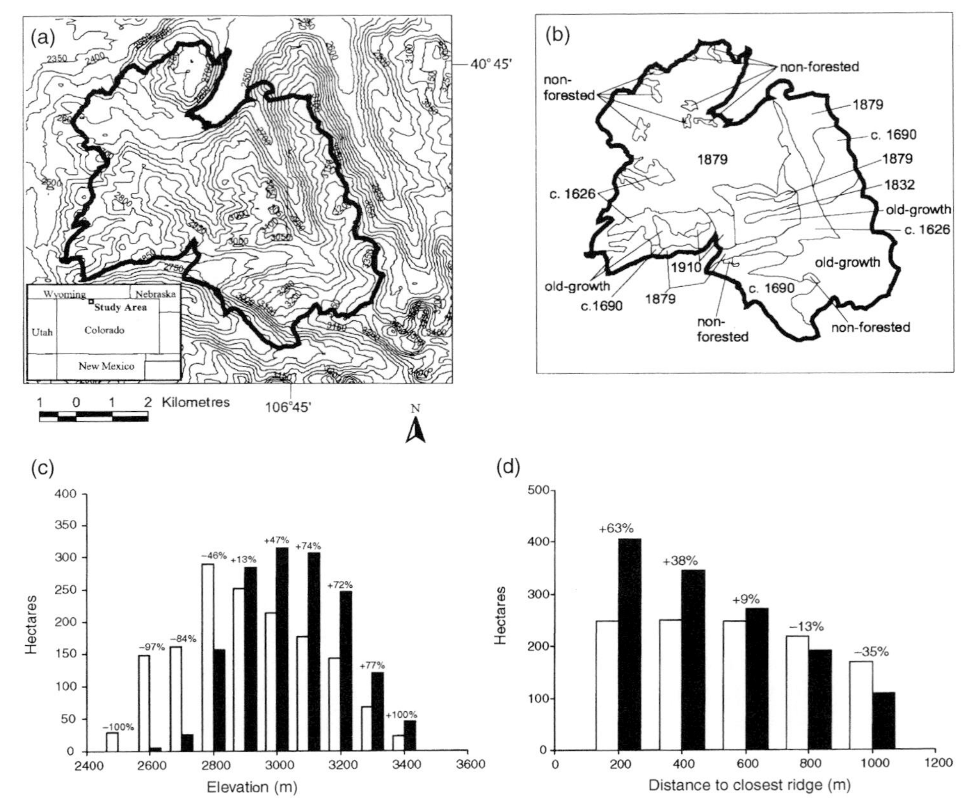


Figure 9.15 (a) Location and (b) fire history classes of a 4400 ha study area of subalpine forest in Colorado, USA. Dates correspond to year of the last stand-replacing fire. (c) Expected (open bars) and observed (filled bars) blowdown of trees at different elevations within the study area. (d) Expected (open bars) and observed (filled bars) blowdown of trees in areas of different distances from the closest ridge. (From Kulakowski and Veblen, 2002. *Journal of Ecology* 90, Blackwell Publishing.)

resulting from recent fires were even less affected because the early pioneer, aspen, is more windfirm than the later pine, and much more so than the eventual spruce-fir forest. When natural disturbances interact, as did fire and blowdown here, forest dynamics are influenced by **synergistic (i.e. combined) effects**, even when disturbance is infrequent.

Cases such as those discussed above, together with the physical and climatic changes experienced to varying degrees by all regions of the world, mean that the existence of vegetational end-points stable in the long term is most unlikely.

## 9.4 Examples of forest change

### *9.4.1 Wistman's Wood, Devon*

This ancient ecosystem (see Box 1.3, p.36) has changed considerably since Hansford Worth recorded a 0.06 ha sample of a typical area in 1921. Further systematic records of the Worth site were made in 1965 (when it was relocated and enlarged to 0.33 ha), 1987 and 1997. The latter two records were of a 0.45 ha near-rectangular area covering much of the southern end of South Wood which included nearly all the 1965 plot. The wood has been much studied by ecologists, so many aspects of its history over the last century are well known. Following the Worth survey came a period when grazing intensity diminished, allowing tree regeneration. This ended soon after the prolonged snow cover of 1962/3, when much of the rowan was debarked and some stems killed by starving sheep. Considerable damage, particularly to old trees, resulted from heavy snowstorms in February 1978. Branches and trunks were snapped, while many boughs were weighed down and subsequently remained prostrate.

Oaks present at the beginning of the twentieth century, that had long been subjected to grazing and severe weather conditions, were low in height and often multi-stemmed. Transplantation to a garden at low altitude demonstrated that the dwarf form did not have a genetic basis. A suggestion that the stunted condition of the oaks was the result of repeated defoliation by winter moth *Operophtera brumata* has not been generally accepted, as attacks by this species were only sporadic. While the defoliation checked growth it did not usually kill entire branches. Some individual trees show remarkable sequences of growth, degeneration and recovery. A tree photographed in 1892 with a low but full and apparently healthy crown, had degenerated substantially by 1921, but is shown with a new regenerated crown in a photograph of 1973.

The oaks now present belong to three generations. The *A generation* consists of the oldest trees; these were present as stunted wind-cut trees in

1900. The *B generation* consists of the erect-trunked individuals which developed after 1900, while the *C generation* is formed by all individuals present as saplings below 1.3 m in height in 1965 and subsequent recruits. The 1997 plot map given by Mountford *et al.* (2001) shows oaks of all three generations, these having been distinguished on the basis of size, vigour and growth form. Past records, including those of trees which have died, enabled rates of increase in canopy height to be calculated for the periods 1965–79, 1979–87 and 1987–97. The *A generation* did well for the middle period, but the total rate of height increase by the *B generation* trees from 1965–97 was much greater. Wistman's Wood has greatly changed in the past century. An oak population which from 1600 until the mid-nineteenth century consisted of 'large bushes', and whose canopy trees had in 1900 an average height of less than 4 m, had individual trees 12 m high and an average canopy tree height of around 8 m a century later. Climate amelioration commencing as early as the mid-nineteenth century appears to be the primary cause of these changes.

### *9.4.2 Forests of New England*

The eastern seaboard of the USA, from Connecticut up to southern Maine, has a number of old-growth forests interspersed with large tracts of what looks like old forest. But in many cases these forests date back less than a century and, as described by Foster (1999) and Foster & Aber (2004), are the latest version of the landscape in a long series of human-induced changes.

Early indigenous people of the eastern seaboard were comparatively few in number but given their skills in manipulating fire, it is highly likely that they had a significant effect on the forest. Certainly, around their habitations, they cleared areas that are recorded as being a few tens of miles in width. It is also highly likely (though not universally agreed) that their influence on the landscape was much wider, resulting in widespread alteration of the landscape, changing the density and species of trees, and creating more clearings over large areas. Archaeological remains of the Iroquois in eastern North America are associated with high charcoal levels and a change from beech *Fagus grandifolia* and sugar maple *Acer saccharum* to oaks and white pine *Pinus strobus*. This created the first 'artificial' landscape of New England.

The arrival of Europeans took this change much further with a rapid and larger-scale conversion of the forest into agricultural land. By the early 1800s more than 60% of the land was in open fields, with the forest remaining as scattered small woodlands. This taming of the landscape was possible (and necessary) due to the small, evenly spread population, having to live off the landscape. So much woodland was removed that by the mid 1850s, wood for everyday use and heating was in short supply and highly sought-after (Foster, 1999).

Beginning in 1830 the industrial revolution arrived, leading to a steady abandonment of the hard-won agricultural land. People moved to the growing urban areas, others migrated westwards to better land or the lure of the gold fields even further west. We have a unique window into this time of abandonment, as Foster (1999) shows, through the prolific journals of Henry Thoreau (1817–1862) who lived in New England as agriculture was on the wane. Land that over two centuries had been made suitable for productive agriculture was now abandoned. This led to an inevitable successional change as weeds, scrub and finally trees invaded the former farmland. White pine, pitch pine *P. rigida*, red cedar *Juniperus virginiana* were the commonest pioneers, with red maple *Acer rubrum* or birches on ploughed land. With time, oak came to dominate these pine forests, and they acquired the look of antiquity, yet old walls, piles of rocks moved from fields, ruined buildings and other artefacts in these forests attest to their former agricultural background (Fig. 9.16).

### *9.4.3 Lowland Polish forest*

Eastern Europe is renowned for its old forests, a number of which may well be primary, that is, they extend back to the last glaciation. Arguably the best-preserved lowland deciduous forest is that of **Białowieża** (pronounced Bee-ow-a-vey-sha) covering 1300 $km^2$ across the borders of Poland and Belarus. Humans have been in the forest since at least the fourteenth century, using it as a royal hunting forest from the fifteenth century but it appears to have been little altered until large-scale timber extraction began after World War I. Fortunately, in 1921, the Białowieża National Park was established in Poland to create a 47 $km^2$ preserve with no further timber extraction or forestry management. What is left is a woodland described by Peterken (1996, p. 73) as 'the largest virgin old-growth stand in lowland Europe' with a full complement of large mammals such as deer, bison, wild boar and wolves, and high species richness (see Packham *et al.*, 1992). This has undoubtedly been there for thousands of years, and is probably as natural a woodland as can be found in lowland Europe. Surely, here is a forest that one might expect to be in a stable condition?

Much is known about the forest, largely due to the work of Janusz Faliński (e.g. Faliński, 1986). Despite its naturalness, two things are notable: firstly, the forest is not static but is undergoing a **natural cycle of change**; and secondly the forest is still recovering from past human intervention.

Figure 9.16 Regrowth area of forest in New England. The ability of woodlands swiftly to recover areas annexed by humans has been remarked by many observers including Rudyard Kipling in his poem *The way through the woods*. The abandonment of land in New England, however, was remarkable both in its extent and in the rapidity of the subsequent successional processes. This view shows an old wall running through what appears to be an old forest at Harvard Forest, Massachusetts. The fields on either side were used for agriculture from the mid 1700s until the mid 1800s when they were abandoned. The forest is thus just 150 years old at most. (Photograph by Peter A. Thomas.)

The forest is noted for the rich **oak–lime–hornbeam** woodland made up mostly of hornbeam *Carpinus betulus* with smaller proportions of small-leaved lime *Tilia cordata*, pedunculate oak *Quercus robur*, Norway maple *Acer platanoides*, Norway spruce *Picea abies* and wych elm *Ulmus glabra*. Hornbeam with varying proportions of Norway maple, lime and wych elm forms the main canopy, 20–25 m above the ground. Above this, a scattering of taller emergent trees of oak and lime reach 30–35 m with even taller Norway spruce reaching over 40 m (Pigott, 1975).

Tree death occurs sporadically through the forests to create gaps, assisted by the wind, at the yearly rate of 200–450 fallen trees per 100 ha (Faliński, 1986). This produces a cyclical change as described in Section 9.2.2 above. Bobiec *et al.* (2000) have described the cycle as starting with a 'young/pole phase' of seedlings and saplings competing and self-thinning over 60 years to produce a dense canopy of trees up to 35 m high and 40 cm dbh (diameter at breast height). This is followed by an 'optimal phase' from 60–200 years which sees the forest opening up, with 4–8 trees per 100 $m^2$, up to 70 cm dbh. Towards the end of this phase, the canopy begins to thin and the 'regeneration phase' is entered with seedlings appearing below the canopy, increasingly so as gaps form in the canopy. Bobiec *et al.* (2000) showed that in the parts of the National Park they sampled, the three phases (young/pole, optimal and regeneration) occupy about a third of the area each, and they considered the forest to be in some sort of cyclical balance.

On a smaller scale, earlier work by Pigott (1975) showed that this smooth cycle was not always following the same path. He found that the new groups of even-aged saplings in the young phase tended to be made up of a single species. For example, of 20 groups surveyed, 19 of them were either all hornbeam or lime, with the other one being pure elm. He also found similar groups of pure Norway maple and ash. Several gaps dominated by lime can be seen in the forest cross-section shown in Fig. 9.17. You will notice the odd elm and Norway spruce at the edge of these gaps growing in slightly more shaded conditions.

The exact reasons for these single species groups are still unclear but some partial answers are possible. Groups of young lime usually appear under the partial shade of hornbeam (as with the right-hand group in Figure 9.17), regeneration from seed being aided in places by suckering from roots and rooting of branches (Peterken, 1996). As the hornbeam canopy undergoes self-thinning with age (i.e. the shortest trees are eventually outcompeted and die) more light reaches the ground beneath (*c.* 200–300 kJ $m^{-2}$ $day^{-1}$) without an open gap appearing in the canopy. As this happens, seedlings of lime, elm and maple are able to establish in the shaded conditions. So, why not mixed groups? Some parts of the answer are straightforward. Each species tends not to establish under its own canopy (the exact reasons for this are still unknown). Also, where nettles are most vigorous it is only the elm (and a few lime which are soon lost) that can establish and grow. Interestingly, Pigott points out that while wych elm of English origin is less shade tolerant than small-leaved lime, elm may prosper because it is the least susceptible to fungal disease and invertebrate herbivores harboured by the nettles. Wild boar (*Sus scrofa*) may also play a part since seedlings of lime and hornbeam were found to

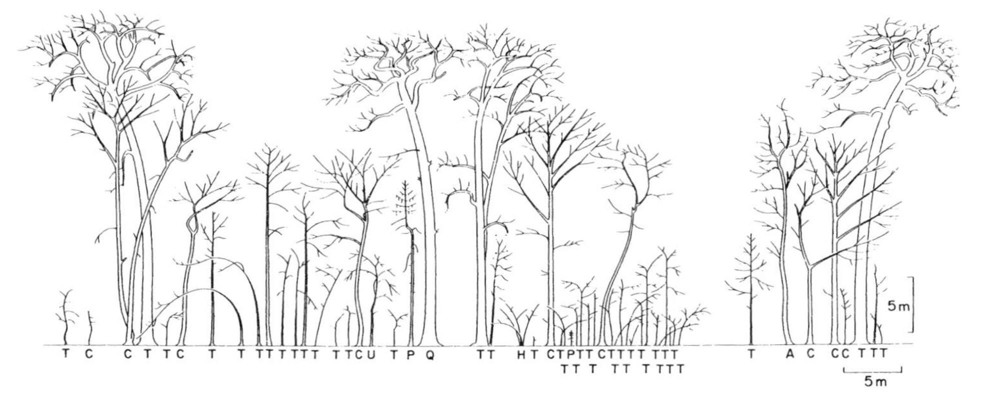


Figure 9.17 Vertical section through part of the Białowieża Forest, Poland, showing a strip 2-m wide, passing through two main groups of saplings. T, small-leaved lime *Tilia cordata*; C, hornbeam *Carpinus betulus*; U, wych elm *Ulmus glabra*; P, Norway spruce *Picea abies*; A, Norway maple *Acer platanoides*; Q, pedunculate oak *Quercus robur*; H, hazel *Corylus avellana*. (From Pigott, 1975. Natural regeneration of *Tilia cordata* in relation to forest structure in the forest of Białowieża, Poland. *Philosophical Transactions of the Royal Society of London. Series B* 270, fig. 2, facing page 162. Courtesy of the Royal Society.)

coincide exactly with areas disturbed by boar rooting (possibly due to temporary removal of root competition). It seems, however, that an important cause of single species clumps in these developing gaps is that small variations and combinations of conditions of light, moisture and other factors favour one species over others. It is certainly evident that most groups are single species from the start rather than jostling for dominance once established. There may also be a fair degree of chance in the process such as the proximity of a parent tree and optimal conditions in a gap coinciding with a large crop of fruit. Supporting this idea of the importance of critical combinations of conditions is that in large open gaps there is a greater tendency for groups of seedlings to contain several species; '... conditions are generally favourable and there is no selection for slight differences between the sensitivities of species' (Pigott, 1975, p. 176).

Human influence has played a major role in causing change in the Białowieża Forest, where small-leaved lime is more abundant in the central region, which includes the National Park. Pigott highlights earlier work of Paczoski who showed that before 1928 there was a long period with little regeneration of lime; trees between 0.05 and 0.4 m diameter were absent. These trees below 0.05 m appear to be the saplings that Pigott found in the 1970s. Paczoski attributed this to the sensitivity of the lime to exploitation, the central area having been the least heavily managed. Much of the forest away from the central area had the same herbaceous vegetation, differing only in the much smaller proportion of small-leaved lime and a corresponding increase in pedunculate oak and Norway maple. The gap in the regeneration of *Tilia cordata* extends backwards from about 1923 to before 1870, a period when the area was maintained as a hunting reserve by the Russian Czars. During this period the number of carnivores was kept low and many deer and European bison were kept for hunting. If this helped cause the failure of regeneration of small-leaved lime it did not prevent that of other trees, and the subsequent reduction in numbers of large mammals, such as wild boar, was influential. Cattle grazing, artificial increases and decreases in deer numbers (and hence grazing pressure), and raking and collection of leaf-litter as compost may also have influenced the situation.

Vigorous regeneration since 1923 has caused the proportion of young trees in the small-leaved lime population to become very high, so the forest still has features initiated by earlier management. In the oak–lime–hornbeam forest small-leaved lime appears to be the potential dominant. The mature trees are long-lived and very tall, and the species is very shade-tolerant when young. Hornbeam *Carpinus betulus* is shorter-lived and forms a lower canopy; it is likely to become displaced, at least for a time, from its present predominant role. Pedunculate oak *Quercus robur* is found most commonly in the National Park; it is much less frequent in the whole forest than formerly, having been heavily exploited in the twentieth century, like Scots pine *Pinus sylvestris*. Norway spruce *Picea abies*, which is favoured by high numbers of game animals, now controlled, has diminished in numbers since 1950 and is also subject to the influence of acid rain. Climate allowing, the present situation makes it appear that much of the forest will undergo a slow but inevitable change over the next few centuries leading to a dominance of lime with a subcanopy of hornbeam and Norway maple (Peterken, 1996).

Humans are still having an effect on the forest used for timber extraction around the National Park. Bobiec *et al.* (2000) show that the composition of the forest has been altered; only 19% is now in the optimal (mature) phase while 61–68% is in the young/pole phase.

### 9.4.4 Southern English yew woodlands

Woodlands formed by yew (*Taxus baccata*) are primarily found on the thin, dry, calcareous chalk soils in south-east England (Fig. 9.18) that are otherwise dominated by grazed grassland. Despite being poisonous, yew seedlings are prone to being grazed, so yew only usually develops beneath the spiny cover of hawthorn, *Crataegus monogyna*, and especially juniper, *Juniperus communis* bushes (Thomas and Polwart, 2003), where it is protected from herbivores.

Figure 9.18 Yew woodlands on chalk soils of Box Hill, southern England. The dense shade cast by the yews results in a forest floor bare except for litter, dead wood and the dead remains of nurse shrubs that aided yew establishment. (Photograph by Peter A. Thomas.)

Owing to its extreme tolerance of shade and the dense shade it creates, yew eventually outcompetes its nurse shrubs and any other trees that establish in the same way, and develops progressively into pure yew woodland, often with the dead woody remains of the former scrub beneath. The occasional holly *Ilex aquifolium*, or whitebeam *Sorbus aria*, may initially survive by outgrowing the yew in earlier years but as they die they are not replaced and the canopy of the yews fuses into a virtually continuous layer.

Though yew is extremely shade tolerant it cannot regenerate under its own canopy so its woodlands persist in a cycle of growth and destruction. As old trees die, a new growth of juniper or hawthorn in the gap creates the protected conditions for new yew seedlings to establish – this is cyclic succession as described above. However, since these gaps are often still too shaded for good shrub growth, most yew regeneration happens around the edges of existing woods, and the gaps left by dying trees revert back to grassland. So, most yew woods are single generation stands 'moving' across the landscape by regeneration at the edge, leaving individual relict trees behind until eventually even they die. As shrubs invade the now open grassland, the whole cycle can start again.

This helps to explain why Tittensor (1980) and others point out that most yew woods in England are quite young, largely resulting from the abandonment of land during the last two centuries as a result of the Napoleonic wars, agricultural neglect in the 1920s and myxomatosis killing rabbits.

## 9.5 Stability and diversity

When looking at forest change, it is the implicit assumption that the forest will recover after a disturbance. The exact pathway by which this happens may be different each time but it will return to something similar to what was there before the disturbance. This can be defined as **stability**, the ability of a forest, or indeed any other ecosystem, to return to equilibrium following a disturbance. To put it another way, stability is the ability to resist permanent change. (There is some debate about precisely how stability should be defined; Johnson *et al.* (1996) and McCann (2000) give good reviews.)

The role that diversity plays in stability is one of the most controversial concepts in ecology. Beginning with a seminal paper by MacArthur in 1955, a school of thought developed that believed that a more diverse community was more stable. This can be thought of as a spider's web, with the threads representing trophic or feeding links. A web with more links should be more robust and less liable to collapse if some of the links are broken. In the same way, a forest with a higher diversity of species should be less affected with the loss of a species from the food web than a forest with fewer species.

Building on this, it has been claimed by some that communities have **functional redundancy**; some species can be lost with no effect on overall efficiency because their function in the food web is taken over by other species. In other words, a community is remarkably robust when species are lost. But care is needed, as pointed out by Schulze and Mooney (1993). They liken an ecosystem to a moving car. We can see that many components are essential for the car to keep moving (such as the spark plugs, dynamo and the radiator), while others can be lost without major apparent effect (such as the ash tray). But caution is needed since some components, such as the bumpers/fenders and air-bags, may appear dispensable until there is a low- or high-speed collision, respectively; i.e. some parts function only occasionally. In the same way, some species in a forest may appear to be redundant but in a time of disturbance, for example, may play a crucial role in the early stages of recovery. It can be argued, with justification, that in the real forest, all species play a crucial function at some time in some way.

What practical evidence is there for the idea that diversity leads to greater stability? The relationship between diversity and stability is still unclear and it is difficult to generalize. In Ecotron (artificially enclosed ecosystem) experiments by Naeem *et al.* (1994) it was shown that high-diversity communities fixed more carbon and had higher cover. Moreover, there have been a number of studies that have shown less year-to-year variation in biomass production in species-rich grasslands compared with species-poor ones (e.g. Dodd *et al.*, 1994). But this may be specific to these types of communities and the evidence for forests is less clear.

When this whole idea was first raised, it was considered that the high stability of tropical forests (at least those that have remained largely unchanged for thousands of years) was due to their high diversity. However, the true picture is probably exactly the opposite way round: it is the stability of the tropical environment that has led to the high species diversity by allowing each species to occupy a narrower niche safely, as Wright (2002) argues. A bird, for example, can feed on one species of flower in the tropics but in a northern conifer forest where the climate is less predictable, and certain plants may flower poorly in some years, or the forest may burn, each species must have a wide niche in order to be able to survive. There is therefore less room for a large number of species in the conifer forest.

The discussion on the link between diversity and stability continues!

# 10
# Working forests

## 10.1 Forest resources and products

### *10.1.1 Timber production and trade: loss of natural habitat*

**Forestry** not only comprises the art and science of initiating, regenerating and cultivating woodlands and forests (silviculture, see Section 10.3.1), but also all the practical procedures involved including road construction and maintenance, felling and selling timber, control of deer and other animal populations. The main product of commercial forestry is wood, which is employed as a fuel, a most important sustainable raw material, the basis of the paper industry, and for at least 10 000 other uses (Sutton, 1999). The economics of plantation forestry in particular are complex; huge amounts of capital are involved in the ownership of the ground, and the costs of establishing a particular crop are not recovered for many years. Whereas a bakery will sell its products within a day or two, it is only in the case of the very fast-growing introduced tree species that a forester has a chance of cropping the trees he plants within his working life. The impact of the compound interest of establishment and maintenance costs on the industry is very great indeed, while the value of particular timbers varies widely with time. In the past huge areas of **natural forest** have been virtually plundered; exploited and cut down without regard for the long-term consequences, a process often called 'mining', that is treating forests as a non-renewable resource. Half of the world's forest growing 8000 years ago has already been lost. On top of this, however, there has been a loss of quality of much of what remains. Over half of what we have left is badly degraded and fragmented and only 20% of the original forest cover remains intact and in large tracts (Bryant *et al.*, 1997). The area of truly natural (= old-growth) forest in the world is now substantially less than that; indeed virtually no entirely natural woodland survives in many industrialized countries. That which remains is highly valued; it represents one of the world's most important ecosystems, has great biodiversity, while its stored carbon may slow the development of global warming (Section 11.3). By far the greatest remaining natural forest area is that of Amazonia where 80% of logging is at present illegal and the total area of the forest is, after a period of reasonable stability, again shrinking at a considerable rate with much of the land being used for rearing cattle or growing crops. These losses are a tragedy in that the plundered areas are unlikely to recover in the foreseeable future, whereas the 20% of legal logging, in which selected mature trees are immediately replaced by appropriate young plantings, allows the forest to maintain itself. The area lost in 2003 was over 2.3 million ha (see Fig. 11.1); much of it was subsequently used to grow non-GM soya beans for the European market.

In some areas natural forests are simply plundered for their finest and tallest trees, often taking out a very small percentage of the trees present, but leaving an impoverished ecosystem that may never fully recover. Studies of the effects of such **selective logging** in Amazonia, Africa and South-East Asia, have compared animal populations in undisturbed primary forest with those in which only the valuable species have been removed. Despite differences in the flora and fauna across such vast distances, results of these studies are remarkably similar (Terborgh, 1992). Disruption of the forest canopy caused a decline in numbers of **frugivorous** (fruit-eating) birds and primates, particularly the larger ones. This probably results from the loss of important food resources. Numbers of leaf-eaters (**foliovores**) are often unaffected and may even increase when the regenerating vegetation provides more young leaves. On occasions when commercially valueless fig trees were poisoned an essential part of the diet of many rain-forest birds and mammals was lost.

When a particular West Malaysian dipterocarp forest was harvested only 3% of the trees were extracted, yet Johns (1988) found that 51% of the trees were destroyed as 5% were pushed aside in road building and 43% injured or killed during felling and timber dragging. Canopy height was lowered and much debris remained. While the harvested dipterocarps were not important in primate nutrition, loss of other species reduced fruit production although young foliage continued to be abundant. In other cases the harvested trees have previously provided food for primates.

Former long-distance trade in timber was much greater than is often realized. Recent dendrochronological investigations of the roof timbers used in the construction of Salisbury cathedral in southern England, whose foundation stone was laid in 1220, show that many beams came from trees which were more than 300 years old when felled in the area around Dublin, Ireland. Timbers were used green and some still have their bark on, so that both the year and season of felling can be determined; the earliest date from the spring of 1222. The cathedral was built with extraordinary speed, being completed in 1258, and when the spire was added it was at 404 feet the tallest medieval structure in the world.

At the present time wood is often transported from one side of the world to the other and much of the trade is now sustainable (Section 10.6). The **profitability of any timber crop** is determined by five factors, of which two of the most important are average price per unit volume and the length of rotation required to reach a given volume. The three others are the quantity which can be sold; the costs of production including tree stocks, planting, thinning, pruning and logging; and the risks which involve management, marketing, disease and climate (Maclaren, 1996). Length of rotation is particularly important when paying compound interest on money borrowed to pay for planting and preparation costs. Nevertheless slow-growing trees, for example, in northern Sweden can still be profitable providing the value of timber is not too low.

### *10.1.2 Timber quality and seasoning*

**Timber quality** varies with speed of growth and the conditions under which trees are grown; its nature varies in the different plant groups. Timbers are usually classified as **softwoods** coming from conifers (gymnosperms) and **hardwoods** from broadleaved trees such as oak and beech (angiosperms). It should be noted that these are commercial timber expressions since 'softwoods' range in density from 380 kg $m^{-3}$ in western red cedar *Thuja plicata* to 672 kg $m^{-3}$ in yew *Taxus baccata*, and in 'hardwoods' from 140 kg $m^{-3}$ in balsa *Ochroma pyramidale* to 1280–1370 kg $m^{-3}$ in lignum vitae *Guaiacum officinale*. (Water has a density of 1000 kg $m^{-3}$ or a specific gravity of 1.) The strength of timbers depends upon their structure and the speed of their growth. As noted in Chapter 1 most timbers contain concentric rings which in seasonal climates outside the tropics are produced annually. These annual rings are composed of **earlywood** grown in the spring and **latewood** grown in the summer. In the softwoods water conduction up through the wood is by means of thin-walled tracheids, hollow cells with valves called pits. The earlywood tends to be low density (wide tracheids with thin walls) and the latewood denser (narrow tubes with thick walls). Conduction in the hardwoods of angiosperm trees is through pipe-like collections of hollow vessels. These vessels are organized within the annual rings in two patterns. In **ring-porous** hardwoods such as that of the oak there is every year a conspicuous line of very large earlywood vessels which contrast with the numerous small vessels found in the latewood zone. **Diffuse-porous** hardwoods such as the beech have numerous small vessels distributed relatively evenly. Rapidly grown softwoods such as pine have a proportionately wider low-density earlywood so to get strong timber from pine it must

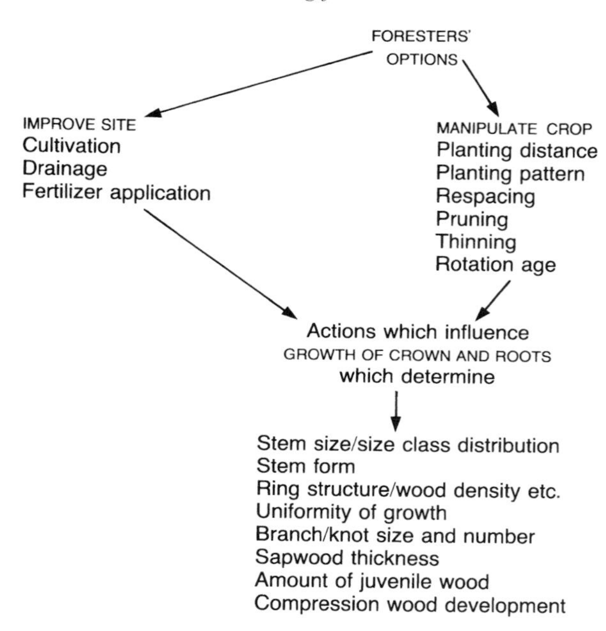


Figure 10.1 Ways in which foresters can influence tree growth and timber development. (From Brazier, 1979. Information Paper IP 12/79. Building Research Establishment. Reproduced by permission of BRE.)

be grown slowly. Conversely, and counter-intuitively, rapidly grown ring-porous timber such as oak has proportionately wider dense latewood and so provides strong hard timber. The major ways in which foresters can influence the growth of trees and the timber they contain are illustrated in Fig. 10.1.

Turning trees into well-seasoned strong and stable timber is a complex and painstaking process well described by Sherrill (2003), who points out the potential dangers faced by anyone who is trimming or felling trees, particularly if these are old and diseased as is not uncommon in urban parklands. Once the trunk and any major branches are converted into logs of convenient length, the wood is transported to the mill and converted into planks, beams or even veneer. Milling for maximum usable volume is a science in itself, while Fig. 10.2 demonstrates how the distortion during shrinkage associated with seasoning is related to the fundamental structure of the tree. In the sequence C–B–A the end-grain of the boards becomes increasing perpendicular to their faces with the result that the board is less likely to cup. Board C cups because it is more flat sawn on the bark side than on the centre side. Board A is perfectly quarter sawn, the grain is perpendicular and it should remain flat throughout a very long life. Though E is mainly quarter sawn, it contains the heart and may well warp and split right through the pith in the centre. G is a rift-sawn beam

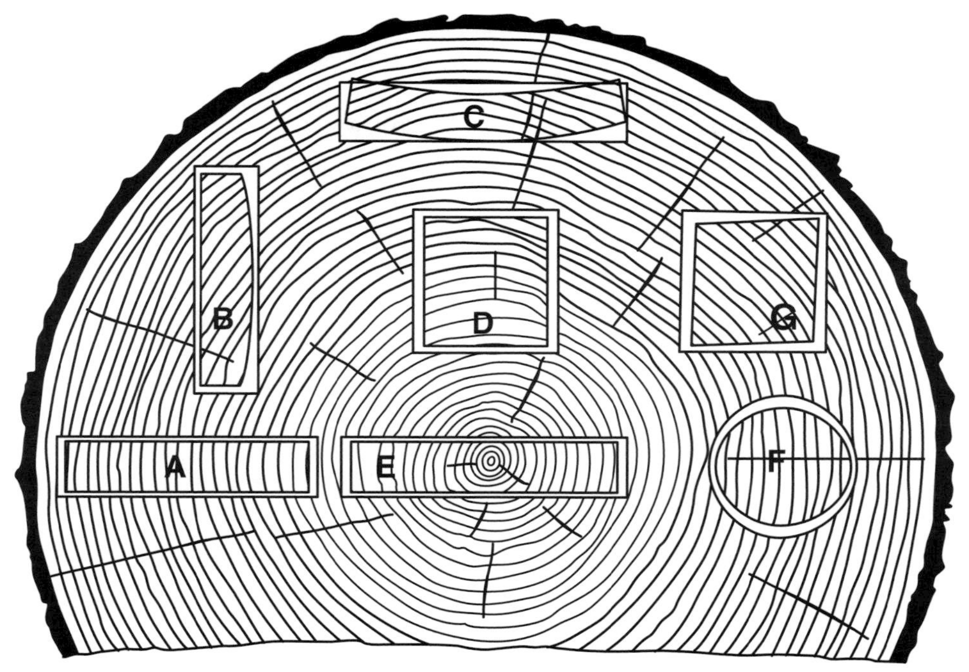


Figure 10.2 Cross-section of a log showing the cuts that can be made, and the likely changes in shape following drying out. Distortion continues until the boards come to EMC (equilibrium moisture content). This does not remain completely constant but varies with the moisture content of the surrounding environment being higher in rainy and humid periods. (From Forest Products Laboratory, 1999. *Wood Handbook, Wood as an Engineering Material*, General Technical Report, FPL-GTR-113, US Department of Agriculture, Forest Service, Forest Products Laboratory, Madison, WI, pp. 3–8.)

which takes on a slight diamond shape, while F shows how a spindle or dowel changes shape as it dries.

Trees, logs and sawn timber are all influenced by stain fungi. **Sap stain** is also known as **blue stain**, though its colour may vary from blue to brown or black depending on the species of wood-inhabiting fungus and the type of tree in which it is growing. Sap stain is generally caused by three groups of fungi (Breuil *et al.*, 2005). These are ophiostomatoids, mainly *Ceratocystis, Ophiostoma* and *Leptographium*, black yeasts such as *Hormonema dematiodes* and *Phialodophora* spp. and dark moulds including *Alternaria alternata* and *Cladosporium* spp. Wood products stained by them are substantially devalued because of their cosmetic appearance. Customers may also assume that such timber is also infected with moulds, white or brown rot. More seriously, several sap stain fungi are pathogenic and if dispersed by bark beetles or wood-borers contribute to the death of standing trees by disrupting the flow of water to the tree crown. The dead trees may then be attacked by decay fungi or form the initial fuel of forest fires. The colour patterning caused by these fungi, known as **spalting**, is sometimes shown to advantage by wood turners and furniture makers.

An enormous number of tree species around the world are used for timber. Their uses are prodigious and often highly specialized. For example, some 200 tree species, of which more than 70 are endangered, are used in making musical instruments. This special case of wood use is so important that a special group, known as **SoundWood**, an international conservation programme of Fauna & Flora International (FFI) has been established to ensure the conservation of the trees involved (May, 2003; plus see www.soundwood.org/). The timber of many of these endangered trees is very valuable, that of the African Blackwood *Dalbergia melanoxylon* sells for up to $20 000 a cubic metre. Seedlings of this tree, called mpingo in Swahili, are now being cultivated and existing adults safeguarded wherever possible. The mahoganies, the larger timber-yielding ebonies, rosewoods and the Brazilian pernambuco *Guilandina echinata* used to make violin bows are all at risk. In North America tight-grained, and thus usually old-growth and slow-grown, wood of Sitka and Engelman spruce (*Picea sitchensis* and *P. engelmanii*) is used for the soundboards of instruments from guitars to pianos. Recent investigations of the violins made by Stradivarius (1644–1737) at Cremona, northern Italy, are of interest in this respect. The wood he used probably came from the Alpine region north of the city, in an area still known as 'The Forest of Violins'. The very dense wood of his 'Messiah' violin was recently found to have about 220 annual rings within a zone of Norway spruce *Picea abies* 15 cm wide, compared with 90–150 rings in a modern violin. Annual rings formed during the Little Ice Age (see Section 9.1.3) were closer than those of more recent years, and Stradivarius lived during its very coldest period, known as the Maunder Minimum after E. W. Maunder who recorded a drop in solar activity at this point.

### 10.1.3 Assessment of production, use of the yield class system and the importance of spacing in commercial forestry

Detailed assessment of timber and of tree growth is accomplished using forest management tables such as those employed by the British Forestry Commission (Hamilton and Christie, 1971; Edwards, 1983). As trees grow they increase in fresh weight, dry weight, height and volume, of which foresters usually measure the last two. Measurable volume is normally taken as that of stemwood exceeding 7 cm diameter overbark, and the pattern of growth in an even-aged stand is described in terms of annual volume increment. **Current**

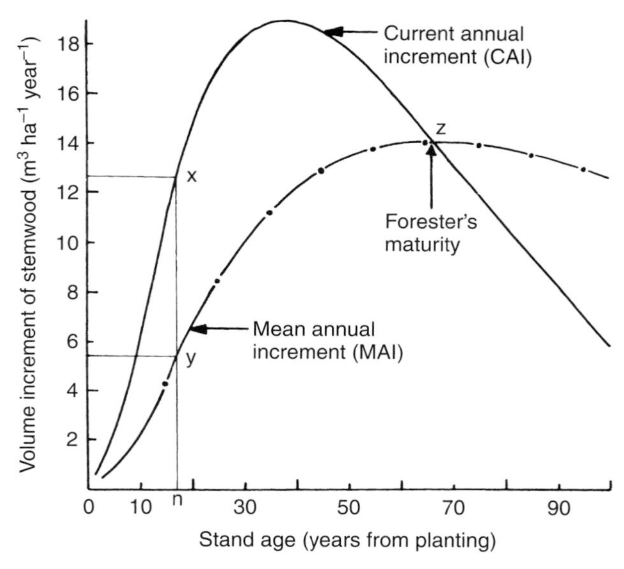


Figure 10.3 Patterns of volume increment in an even-aged stand of Scots pine *Pinus sylvestris* belonging to yield class 14. For definitions see text. (Drawn from the data of Hamilton and Christie, 1971. From Packham and Harding, 1982. *Ecology of Woodland Processes*. Edward Arnold.)

**annual increment (CAI)**, the volume of wood added in a year, increases for some years after planting and subsequently declines. **Forester's maturity** is the time at which the forest stand has the maximum average rate of volume increment that can be achieved by a particular species on the site concerned. It is the point where the downward curve for CAI on a graph crosses that for **mean annual increment (MAI)** – the average rate of CAI from planting to a given point in time (see Fig. 10.3). In theory the maximum average rate of volume production could be permanently maintained by repeatedly felling the stand at forester's maturity and replanting with the same species. In practice yields sometimes fall gradually over a number of rotations as a result of soil deterioration, the build-up of pests and pathogens and other factors.

**Yield classes** are based on maximum MAI, i.e. the maximum average rate of volume production per year reached by a particular species on a given site, irrespective of when this occurs. Under British conditions this may be as low as $4\,m^3\,ha^{-1}$ for many hardwood, larch and pines, but exceed $30\,m^3\,ha^{-1}$ for grand fir (Fig. 10.4). As yield classes represent the number of cubic metres (to the nearest even number) of timber produced per hectare at maximum MAI, yield class 12 includes any stand whose maximum MAI lies between $11\,m^3\,ha^{-1}$ and $13\,m^3\,ha^{-1}$. It takes longer to reach maximal MAI values in slower-growing stands as the yield class curves for oak demonstrate (Fig. 10.4).

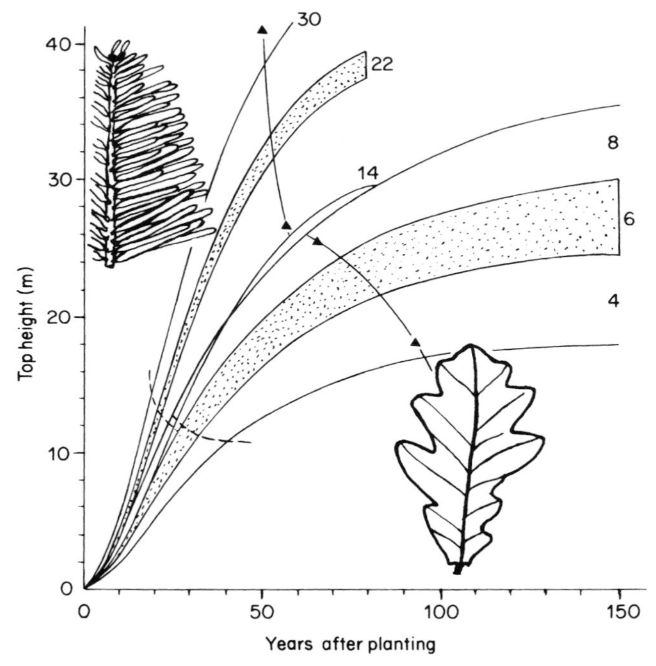


Figure 10.4 General yield class curves for grand fir *Abies grandis* and native British oak (*Quercus robur* and *Q. petraea*) under English conditions. The stippled areas represent the median yield classes for each species; the other figures indicate the lowest and highest general yield classes normally encountered. - - - -, time of first thinning; ▲–▲, age of maximum annual increment (MAI). (Redrawn from Hamilton and Christie, 1971. From Packham and Harding, 1982. *Ecology of Woodland Processes*. Edward Arnold.)

Though the maximum mean annual increments of different species may be of the same magnitude, they are often reached at quite different ages in the life of the tree, e.g. 35–40 years for poplar *Populus* spp., 60 for Douglas fir *Pseudotsuga menziesii* and 80 for Norway spruce *Picea abies* (all trees in yield class 12 under British conditions).

Direct measurement of tree volume in a stand is time consuming, but there is a fairly close relationship between top height and the total cumulative volume production of a stand. The **top height** of a stand is the mean height of the 100 trees of the largest **diameter at breast height (dbh** –1.3 m above ground) per hectare. The use of this measure makes sense because it is concerned with those trees that are going to form the major part of the crop. The average height of all the trees present is not a sensible measure, since the shortest individuals in a closely spaced stand are likely to be suppressed very soon. **General yield class curves** are plotted on graphs whose vertical axis shows the top height and the horizontal axis gives the number of years since planting. Practical foresters

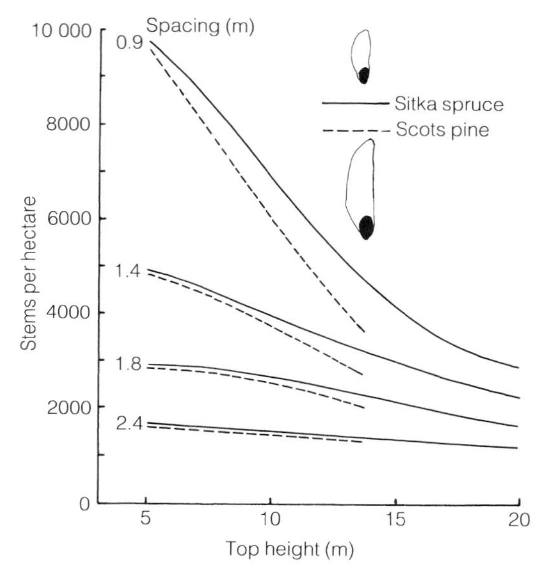


Figure 10.5 Surviving stem numbers in unthinned Sitka spruce *Picea sitchensis* and Scots pine *Pinus sylvestris* for four initial spacings, together with scale outlines of the winged seeds. Survivorship trends of Norway and Sitka spruce are very similar. (Redrawn from Hamilton and Christie, 1974. From Packham *et al.*, 1992. *Functional Ecology of Woodlands and Forests*. Chapman and Hall, Fig. 10.7. With kind permission of Springer Science and Business Media.)

obtain the yield class of a stand by entering its top height in the appropriate year after planting. For example, in Fig. 10.4, oaks that have a top height of 20 m at age 60 would have a yield class of 6, that is, they have been adding an average of 6 cubic metres of wood per hectare each year.

**Tree spacing** exerts a very important influence on the planting costs, weed control and the volume and quality of wood obtained from a forest plantation. Many of the experiments conducted by the British Forestry Commission with Scots pine, Norway spruce *Picea abies*, Sitka spruce, European larch *Larix decidua*, Japanese larch *L. kaempferi* and Douglas fir *Pseudotsuga menziesii* to investigate this involved initial spacings of 0.9, 1.4, 1.8 or 2.4 m (3, 4.5, 6 or 8 ft), both between and along the rows. The numbers of seedlings planted per hectare were 11 960, 5315, 2990 and 1682, respectively. Closer spacing caused competition to begin earlier and for mortality prior to the thinning stage to be much greater. Species in the trials varied slightly, but some 90% of the seedlings in the 2.4 m spacing plots survived to the 10 m top height stage, whereas almost half of those in the 0.9 m plots were dead when survivors reached this height (Fig. 10.5). Weeds, however, were suppressed earlier, as the canopy closes much more rapidly when the trees are planted close together.

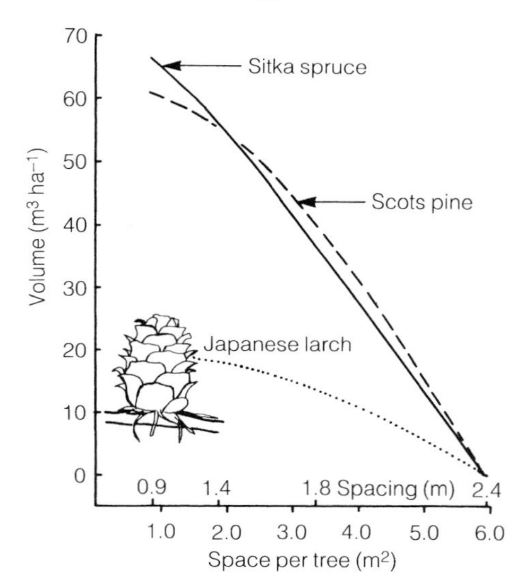


Figure 10.6 The **additional volumes** of timber obtained by growing trees of three different species at spacings closer than 2.4 m. The figure shows differences in total volume production at the different spacing relative to that of 2.4 m at the age of maximal mean annual increment. The curve for Norway spruce *Picea abies* (not shown) is very similar to those of Sitka spruce *Picea sitchensis* and Scots pine *Pinus sylvestris*. Note the reflexed cone scales of the Japanese larch *Larix kaempferi*. (Redrawn from Hamilton and Christie, 1974. From Packham *et al.*, 1992. *Functional Ecology of Woodlands and Forests*. Chapman and Hall, Fig. 10.9. With kind permission of Springer Science and Business Media.)

Although the trunks of trees planted at wider spacing grow to a greater diameter than those planted close together, the taper rate in the lower parts of their stems is much greater leading to more waste when cutting timber from the logs. Greater volumes of wood are formed in a stand when trees are planted close together (Fig. 10.6), and this also reduces branching from the lower parts of the trunks which accordingly yield timber with fewer knots. The amounts of additional timber obtained by growing trees at closer spacings than 2.4 m, are not so great if a 7 cm dbh limit is applied to exclude stems that are not of 'merchantable quality', i.e. not suitable for saw mills. This material, however, can sometimes be sold for rustic work in gardens so may itself be quite valuable. The **basal area** of a tree is the overbark cross-sectional area of the trunk 1.3 m above ground. Like total volume, basal areas per hectare are considerably greater when trees are closely spaced.

**Forest mensuration**, the measurement of standing and felled timber, is very important when timber is being bought and sold (Hamilton, 1985). The sizes and shapes of trunks and major limbs strongly influence the value of a given volume of timber. Wide trunks with low taper produce the largest area of usable planks per unit volume of uncut wood. Isolated trees frequently have a form unsuited to economic conversion to sawn timber. They are likely to be highly tapered and, as in European oaks, can have major limbs that spread out widely from less than three metres above the ground. The poet Longfellow who sat his village blacksmith 'under a spreading chestnut tree' knew what he was talking about. Species such as western red cedar *Thuja plicata* sometimes have major limbs bowing down and rooting in the soil.

Commercial foresters are making every effort to improve the efficiency of their industry. Improved tree genotypes, sometimes involving polyploids (having multiple sets of chromosomes) and hybrids, are constantly coming into general use, and every effort is made to reduce volume losses caused by pests and disease.

**Fertilizer additions** can be important in improving productivity in plantations (see Section 8.6.2). On a world basis nitrogen and phosphorus are most commonly supplied by foresters, though potassium additions have proved useful in Europe. In North America nitrogen is far more important in limiting tree growth than any other element (partly due to lower rates of nitrogen pollution – see Section 11.4.3), but the requirements of different species vary greatly. Basswood (American lime) *Tilia americana*, white ash *Fraxinus americana* and the tulip tree *Liriodendron tulipifera* in eastern North America are at an advantage on soils with abundant nitrogen, whereas species such as red and white oak (*Quercus rubra* and *Q. alba*), red maple *Acer rubrum* and aspen *Populus tremuloides*, compete well on sites with low nitrogen status. Engelmann spruce *Picea engelmannii* responds vigorously to ammonium nitrogen but not to nitrate. In the UK and much of continental Europe phosphate additions are often helpful; here much soil nitrogen originates through the fixation of atmospheric nitrogen.

Fertilizer additions influence the forest ecosystem as a whole, often causing understorey plants and 'weed trees' to grow so much more vigorously that the application of herbicides or mechanical weeding become necessary. Patterns of disease in forest trees, the development of mycorrhizas and insect populations may also be affected in complex ways. In young stands the resultant increase in leaf biomass, and thus of water transpiration, may be great enough to reduce run-off into streams significantly. The term **eutrophication** is often used to describe the enrichment of both the soil and of the aquatic systems associated with the forest, in whose waters primary production may be increased.

### *10.1.4 Ecosystem values of plantations and natural forests*

Recent studies in many parts of the world have demonstrated the importance of maintaining ancient forests if woodlands of the highest conservation value, known as **woodland key habitats**, are to be both identified and retained (see Sections 6.4.1–2). However, as demand for timber grows, it is inevitable that plantations will increase in area. More of our commercial timber already comes from plantations (34% – despite their comparatively small area; Table 1.1) and managed secondary growth forests – i.e. felled and regrowing (22%) – than from less managed forests (34%) (Sedjo and Botkin, 1997). Such plantations often lack the biodiversity of unmanaged forests. In temperate and northern forests, the field and ground layer vegetation are key components in maintaining biodiversity. Fungi, lichens, herbs and shrub species differ considerably between plantation forests grown on former pasture and the sites of felled older woodlands, though all are subject to shade as the woodland matures, so plants characteristic of open sites are gradually lost, often with some of the associated animals. Most, but not all, comparisons of unmanaged forests and plantations show an impoverishment of plants and animals in the plantations. For example, in Britain, lowland Norway spruce *Picea abies* plantations contain less than 10 species of lichen while lowland oakwood can contain over 40 species (Humphrey *et al.*, 2002) while in montane rain forest 81 species of vascular epiphyte were found in untouched forest compared with 13 species in a hardwood plantation, made up of fewer ferns and orchids but more bromeliads (Barthlott *et al.*, 2001). Similar examples can be given for animals: in tropical forests, Carlson (1986) found forest-interior birds (i.e. living away from forest edges) to be 2–3 times as abundant in intact tropical forest as in radiata pine *Pinus radiata* plantations, and Vallan (2002) observed more than twice the number of amphibian species in intact Madagascan forests than in eucalypt plantations. Foresters sometimes argue, with reason, that plantations should be viewed in the same way as an agricultural crop (to which they compare very favourably) and not as a replacement for natural forests. Plantations usually include non-native species, are even-aged and often monocultures and so are very distinct from natural forests. But as plantations become a larger part of many countries it is inevitable that they become more vital as reservoirs of forest species.

It is important to appreciate that like any other forest type, plantations vary from even-aged exotic monocultures to much more structurally diverse stands of native species, and so some are better substitutes for natural forest than others. A study in Britain by Ferris *et al.* (2000) showed that lowland conifer plantations were least like natural forests but mature and over-mature (see Section 10.1.3) pine and spruce stands in the uplands closely matched native pine and oak woodlands. Older, more open stands allow more light to reach the ground, encouraging the vital field and ground vegetation. There is plenty of evidence that the management of any sort of plantation can be altered so that it better conserves biodiversity and is more aesthetically appealing without losing much profitability. This is the idea behind multi-use forests discussed below.

Plantations have a distinct value not only in what species they hold but in their position in the landscape. Plantations can be used as a buffer between natural forests and an agricultural landscape and may be useful in connecting existing forest patches, allowing species to migrate. In England and Wales 83% of ancient woodlands (amounting to 31% of ancient woodland total area) are less than 20 ha in size and there is a great need to facilitate movement of species to prevent local extinctions and inbreeding. Moreover, plantations planted against older woodlands will more speedily acquire species. This is all the more important now that we know that the rates at which many of the key species migrate from ancient sites to more recent woodland are often exceedingly slow (see Section 6.4).

## 10.2 Single- and multi-use forests

Human populations have risen very rapidly in the last few centuries, but in New Zealand and a few other areas they remain so low that very large areas are devoted to little besides timber production. The area of indigenous tall forest, mainly dominated by various southern beeches *Nothofagus* spp., is so well protected as to remain almost constant at 23% of the total land surface of New Zealand. Exotic plantation forest on the other hand, which had an area of 1.5 million ha in 1996, is increasing rapidly. Though some is planted on scrubland, the majority is on improved or unimproved pasture, most of which was originally developed on areas of felled and often burnt-over native forest. Though some has a degree of recreational value (including picnicking, motorbike riding, orienteering and pig-hunting) much of this plantation forest is used solely for timber production. Most of this forest is not only single-use, it is also dominated by a single tree species and large areas of it are even-aged.

**Even-aged single-species stands** also occur naturally but they are the exception. Examples of species that do so in North America include Douglas fir, lodgepole pine and radiata (Monterey) pine; in California the last rapidly colonizes after fire but can also form the climax vegetation. The southern beeches of New Zealand also form natural **monocultures**. Four species of *Nothofagus* occupy 2.87 million ha (47%) of the indigenous forest of New Zealand and large areas of this have just one tree species.

In most parts of the world, especially where forests are small and near population centres, they are increasingly required to fulfil several functions. **Multi-use forests** all possess the same broad objectives as those set forth in the

Multiple-Use-Sustained-Yield Act passed by the US Congress in 1960. This stated that the 'most judicious use of land' in respect of the renewable resources of the national forests should specifically concern outdoor recreation, soil, timber, watershed, wildlife and timber. This act also made provision for the establishment and maintenance of wilderness areas, a concept now also realized in other countries where there is interest in the operation of truly natural ecosystems. Individual multi-use forests differ considerably in both the trees and other species present and in their major functions. Some are mainly scientific study areas, while in others the major emphasis is on recreational use. This is considered further under conservation in the multi-purpose landscape (Section 10.7.3).

## 10.3 Silviculture and the replacement of trees

### *10.3.1 Silvicultural systems*

**Silviculture** treats forest trees as crops which are established, tended, harvested and then replaced by others. Essentially it is that part of forestry which involves an understanding of how trees modify, and are influenced by, the ecosystems in which they live. A **silvicultural system** encompasses (a) the regeneration of trees, (b) the form of the crop produced and (c) the orderly arrangement of the crops over the forest as a whole. The main groups of forest system described below have many variants; all produce woodlands of distinctive character. Choice of silvicultural system is particularly critical where soil and avalanche protection are involved. Well-established forest industries, particularly in Europe, are usually based on the concept of the **normal forest**, from which the same quantities of timber and other forest products are taken every year or period of years. The complete succession of age classes in such forests is so balanced that as each class matures and is harvested, it provides a yield similar to that of the classes which preceded it. Regular and sustained, rather than intermittent or spasmodic yields are highly desirable in theory, but problems can arise when the market is saturated with cheap timber from abroad or that salvaged after great storms. There are also pressures to fell more than usual if the value of timber is especially high.

Very many variants of the main silvicultural systems have developed in response to the needs of particular markets and local forestry conditions; these should be carefully considered before new forests are established or old ones radically altered. Matthews (1989) gives a detailed assessment of how they have been employed in both temperate and tropical areas, but does not deal with continuous cover forestry as such (Section 10.3.3). An unusual, but very effective, system used only in South-East Asia, is the **Malaysian Uniform**

**System** (MUS). It takes advantage of the mast fruiting of the Dipterocarpaceae (Sections 4.4.3–4), a family containing many of the most valuable Asiatic timber trees. Devised before World War II, its basis is the felling of the **over-storey** (harvest of mature trees) at a time when there are many dipterocarp seedlings and saplings present on the forest floor. These will have resulted from one of the irregularly produced very heavy mast years found in this family. Once released by the increased light resulting from felling, they grow away rapidly and give rise to the type of forest desired despite the presence of many other less desirable tree species.

### *10.3.2 Traditional management: coppicing and pollarding*

Many hardwood trees and a few conifers develop from shoots which arise at the base of a cut stem; indeed **coppicing** and its variants are the most ancient form of silvicultural system. Using the methods shown in Fig. 10.7, workmen in much of Europe maintained stands for several centuries, often without the use of seedlings. In contrast, natural woodlands contain mainly **maiden trees**, which have arisen directly from seed. In the UK hazel coppice and tall oaks formed the most characteristic **coppice-with-standards**. Distinction was made between the *wood* produced by the coppice and the *timber* produced by the standard trees; the peasants often had rights to the wood while the timber belonged to the landowner. **Storing** is the process by which coppice is returned to high forest by removing all but one pole from each coppice stool. New trees of aspen *Populus tremula*, American beech *Fagus grandifolia* and most species

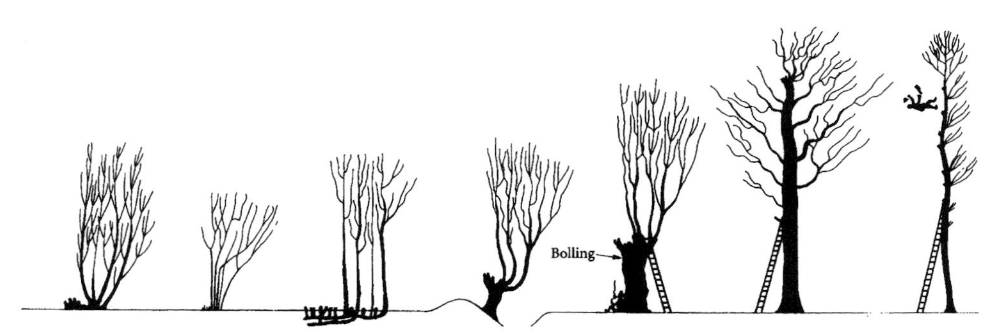


Figure 10.7 Production of wood from trees. Left to right: coppice stool above ground (e.g. ash); coppice stool below ground (e.g. hazel); clone of suckers (e.g. elm); stub on boundary bank; pollard; high pollard; shredded tree. The left-hand side of each has just been cut; the right-hand half is fully regenerated and is about to be cut again. Note that the area of a coppice system cut in any one year is called a **cant, hagg** or **fall**. (From Rackham, 2003. *Ancient Woodland; its History, Vegetation and Uses in England.* Castlepoint Press.)

Figure 10.8 Above: Stump of common oak *Quercus robur* showing early coppice growth at Chaddesley National Nature Reserve near Kidderminster, UK. This is one of the few reserves run by Natural England (the English government body responsible for nature conservation) on semi-commercial lines. Right: Stand of American beech *Fagus grandifolia* which has arisen from suckers, Massachusetts, USA. (Photograph above by John R. Packham, and right, by Peter A. Thomas.)

of elm *Ulmus* spp. arise from genetically identical **suckers** produced by the roots (Fig. 10.8); the stumps normally die.

Many trees are effectively rejuvenated by coppicing. Ash, alder, maples, wych elm, oaks, sweet chestnut, bass and lime (*Tilia* spp.) and the tulip tree *Liriodendron tulipifera* are coppiced in temperate parts of the world, as are eucalypts and teak *Tectona grandis* in the tropics. Few conifers coppice; coastal redwood *Sequoia sempervirens*, canary pine *Pinus canariensis* and the monkey

Figure 10.8 (cont.)

puzzle tree *Araucaria araucana* being important exceptions. Coppicing of eucalypts by people, although comparatively recent, is an important development. Rackham (2003) describes an area in New South Wales where eucalypts, mainly silvertop ash *E. sieberiana*, had recently been cut for the third time. These three rotations of felling and regrowth had converted eucalyptus forest into an ecosystem very similar to a coppice-wood with its own native coppicing flora. Bluebell (Fig. 10.9) is a characteristic member of such a flora in the coppiced oakwoods of southern England, forming sheets of vivid blue in spring.

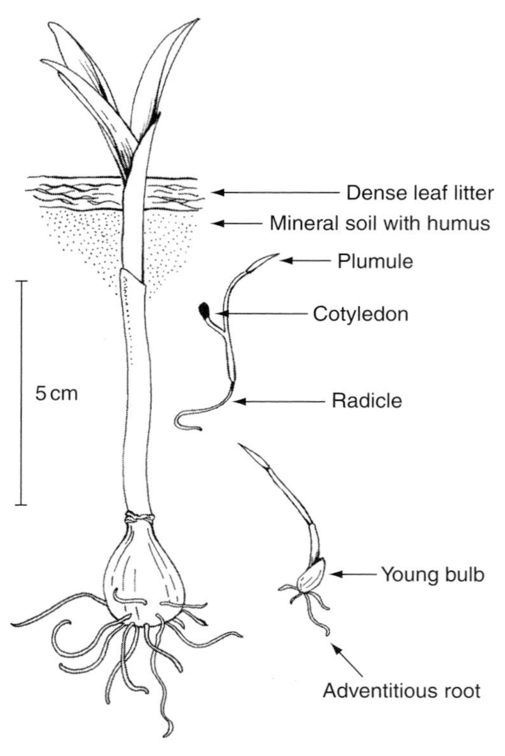


Figure 10.9 Bluebell *Hyacinthoides non-scripta*, an important member of the shade flora of coppiced woodlands. This shade-evading bulbous geophyte (main growing point below ground) has a pointed shoot that can pierce thick litter and expand rapidly on reaching the light. The black seeds germinate at the surface; as the bulbs develop they form contractile roots that draw the bulbs downwards. The rarity of bluebell on the chalk scarps of south-east England appears (Blackman and Rutter, 1954) to be due to the rapid drying out of the thin surface layer of soil and the physical barrier of the underlying chalk (which prevents the bulbs being drawn down into a deeper and moister zone), rather than to the high pH and calcium content of the soil. (From Packham and Harding, 1982. *Ecology of Woodland Processes*. Edward Arnold.)

**Pollards** are trees whose trunks have been cut between 1.8 and 4.5 m above the ground. Old trackways in ancient woods are often marked by lines of them in the UK. They were also used where deer and cattle could browse young shoots from freshly coppiced trees, but were unable to reach shoots at the top of the remaining trunk (**bolling** – rhyming with rolling) of a pollard. **Stubs** are very short pollards 1.2–1.8 m high often found at land boundaries; margins of medieval woods were often marked by banks with stubs and can sometimes be traced long after the woods have gone. **Shredding** is a peculiar practice is which the side branches are repeatedly cut off leaving a tall trunk with a tuft at the top. Some trees managed in this way remain in France, Italy and Africa, though not in Britain.

### *10.3.3 High forest management*

During the twentieth century large-scale high forest management normally produced crops of seedling origin, almost invariably employing variants of clear cutting, successive regeneration fellings, shelterwood or selection systems. Tree seedlings were often produced in enormous numbers in forest nurseries. In the last two decades, however, there has, in Europe at least, been a considerable movement towards continuous cover forestry where the aim is a much more natural, and often more species-rich forest, in which harvesting is gradual, natural regeneration is important, and the amount of bare ground at any one time relatively low.

With **clear cutting** the whole crop is cleared by a single felling and is even-aged. Stands within a forest may be felled at different times creating large open areas within the forest. Systems of **successive regeneration fellings** involve the crop being cleared by two or more successive fellings on the same area; in this case the crop may be somewhat uneven-aged. *Uniform systems* of this type involve an even opening of the canopy by removing a percentage of the trees. In *group systems* the canopy is opened by scattered gaps. Fellings may be confined to particular strips (*strip systems*) or wedges (*wedge systems*). Where the opening of the canopy is irregular and gradual the young crop in the resulting **shelterwood system** is somewhat uneven-aged. Two accessory systems are sometimes employed. These are **two-storied high forest** where a fresh young crop is planted beneath an existing immature one, and **high forest with reserves** in which a proportion of the old crop is retained after regeneration is complete. The rapidly developing practice of agroforestry is considered in Section 11.6.1.

When **shelterwood systems** are established the existing trees are harvested on two or more occasions, allowing seedlings which will form the future forest to develop before the canopy above recloses completely. The first stage is to remove inferior trees so that seeds falling on the forest floor are from good quality parents. If all goes well and some good seed years ensue, the resulting seedlings make rapid progress and many of the parent trees can be felled, thus providing an economic return while opening up the stand. The last of the mature trees are removed when the now transformed forest is well established. If natural regeneration is patchy, commercial seedlings may have to be employed. This is particularly the case in the extreme variation known as the **seed tree system**, where very few parent trees are left. These retained trees offer very little protection to the young seedlings that may be rapidly overgrown by weeds, and are themselves vulnerable to windthrow.

**Selection systems** come into their own where scenic beauty is important, or when the forest is on steep slopes since it helps prevent soil erosion or avalanches. They involve continuous felling and regeneration throughout the area so the crop is irregular i.e. wholly uneven-aged. The aim is to retain a balance between trees of different sizes, and often of different species. Having reached an appropriate balance the amount of timber felled each year should ideally be equal to the net annual increment. Forest structure remains broadly similar for long periods; unthrifty trees are felled along with others of various sizes. Gaps in the canopy of protection forest using this system close quickly, while the soil is never liable to severe erosion. Trees in such forests are also wind-firm, and it is best if their seedlings are shade-tolerant (as are those of hornbeam *Carpinus betulinus*, beech *Fagus sylvatica* and limes *Tilia* spp. in the UK) if they are to develop under such conditions. Shade-intolerant species (such as the British oaks *Quercus petraea* and *Q. robur* and Scots pine *Pinus sylvestris*) are best grown in shelterwood systems with cuts up to 5 ha in size. There are situations, however, when shade-intolerant species can be grown in selection systems, but then the trees must be particularly well suited to the site and the canopy more open. Useful though selection systems are, they demand a great deal of skill on behalf of both management and workers and the scattered logs are difficult to collect even with mechanized skidders which drag logs to the roadside.

The transformation of even-aged woodland to a **continuous cover system**, a modern concept which has much in common with a selection system, is not a simple matter; the aim is to reduce the influence of shade and root competition in some parts of the stand to a point where regeneration can commence (Helliwell, 2002). This system, defined by Mason *et al.* (1999) as the maintenance of the forest canopy without clear felling areas greater than 0.25 ha, involves a great deal of judgement on the part of the forester who marks the trees to be cut down. If felling is done at 5-yearly intervals, regenerating 12.5% of the total area on each occasion would in theory make it fully uneven-aged in 35 years. Some early attempts at such transformations in Sweden were based on the creation of regularly distributed gaps of fairly uniform size; these clearings attracted elk and other deer with the result that due to selective browsing most of the new trees were birch or willow rather than the pines or spruce which were needed. Even if new seedlings arise naturally or are planted, after the first 10 or 15 years the older parts of the forest gradually become fragmented so finding space for new gaps becomes increasingly difficult. It becomes virtually impossible to proceed further on an area basis, so the stand has to be treated as if it were fully uneven-aged. In practice it is usually best simply to commence reduction of the standing volume more heavily and less regularly than normally occurs during the thinning of an even-aged forest.

When this is done, as it has been with much greater success elsewhere in Sweden, the range of tree size classes is kept as large as possible and the forest becomes more open in some places without deliberately clearing gaps of predetermined size. It is sensible to keep a check on the developing situation by plotting the number of stems by diameter class although the curve obtained will be very uneven in the early stages of the transformation. Whilst it is highly desirable to maintain the forest by natural regeneration, this may not always be possible particularly as in many trees seed production is irregular (see Fig. 4.9), and it may be necessary to plant seedlings in unfavourable sites or years. The end result of a transformation to uneven-aged forest from a clear cutting or shelterwood system is effectively a form of selection forest.

### *10.3.4 Thinning and crown classification*

Thinning is an important aspect of commercial forestry as it promotes the growth of the remaining trees, gives intermediate financial returns and may increase the total yield of usable timber over the life of the stand. There are three major factors in a thinning regime. **Thinning intensity** is the rate at which timber is removed per year; e.g. $8\,m^3\,ha^{-1}\,y^{-1}$. **Thinning yield**, on the other hand, is the actual timber volume removed in any one thinning. For a **thinning cycle** (time between one thinning and the next) of 5 years, the thinning yield would be $40\,m^3\,ha^{-1}$ in this case. If thinning intensity is low, stands of normal initial spacing become so overstocked that cumulative production of usable timber is reduced and suppressed trees die before harvesting. If thinning intensity is high it enables the remaining trees to increase in diameter more rapidly than if the stand is left unthinned, but they do not use the extra growing space fully in cases of excessive thinning. So either under- or over-thinning reduces timber production. It turns out that in practice cumulative volume production remains the same over a wide range of thinning intensities. In general foresters prefer to thin as much as possible because it brings in more money earlier and the resulting fewer but large trees can be more valuable per cubic metre. **Marginal thinning intensity** is the maximum thinning that can be used without incurring loss in this total volume. Once the stand is sufficiently well developed for thinning to commence this intensity is about 70% of maximum mean annual volume increment per year until the stand reaches maximum MAI (Rollinson, 1985). Thus for a stand with a yield class of 10 the marginal thinning intensity is $7\,m^3\,ha^{-1}\,y^{-1}$, and this would give a thinning yield of $42\,m^3\,ha^{-1}$ with a 6-year thinning cycle.

Trees in even-aged forests, such as that shown in Fig. 10.10, may be the same age but they differ in both size and form. This is of considerable importance

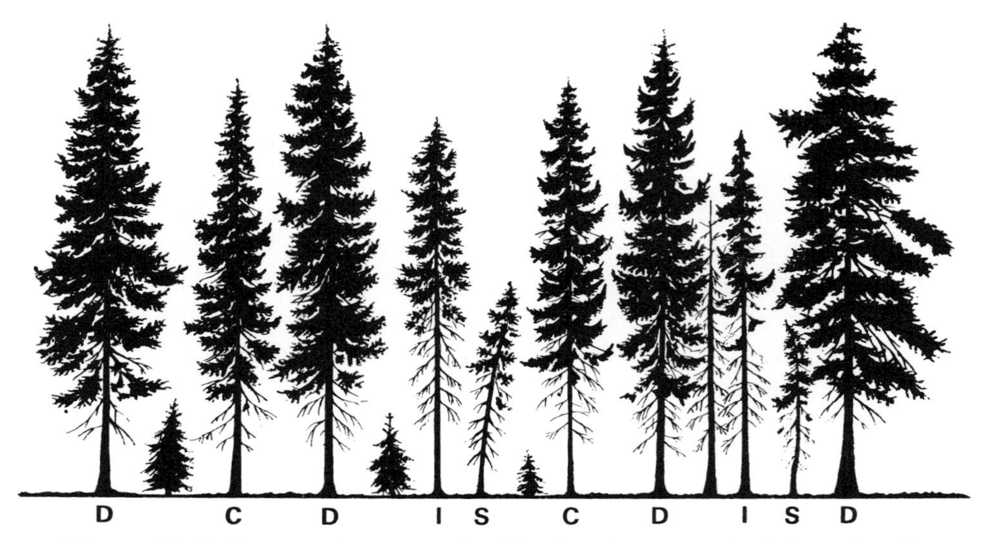


Figure 10.10 Even-aged crown classification in a stand of Douglas fir *Pseudotsuga menziesii* in the Wyre Forest, England. Naturally regenerating saplings are unlabelled. The tree fourth from the right is dead, while that on the extreme right, adjacent to a glade, is as close to a wolf tree (see text) as occurs in this area of Worcestershire. D, dominant; C, co-dominant; I, intermediate; S, suppressed. (Drawn by Peter R. Hobson. From Packham *et al.*, 1992. *Functional Ecology of Woodlands and Forests*. Chapman and Hall, Fig. 1.7. With kind permission of Springer Science and Business Media.)

as the crop ages and has to be thinned. In some cases thinning is done on a systematic basis, such as the removal of alternate rows, but usually judgements are made about individual trees which fall into the five classes illustrated below. Crowns of **dominant trees** receive full light from above and, to some extent, the side. The slightly shorter **co-dominant trees** receive full light from above, but are obscured laterally by the dominants. These two classes of tree form the main canopy and are the most thrifty.

The lower crowns of **intermediate trees** get some direct light through holes in the canopy, but are subject to severe lateral competition from their larger neighbours. **Suppressed trees** are in far worse state, being strongly over-topped by the previous three classes and dependent on sun flecks and light filtering through the canopy. Slow-growing and usually weak, these are often doomed to fall into the last category of the **dead trees** commonly found in unthinned plantations. The largest dominant trees may grow into **wolf trees** which are coarse, heavy-limbed individuals which lack effective lateral competition and have crowns so broad that they inhibit growth of more thrifty neighbours. Such trees should be removed from timber crops, although they may be of conservation interest. They are more common in stands that are not entirely even-aged.

Other classifications have been devised for stands of uneven age where the main criterion is degree of vigour rather than size. Vigorous trees are considered important because they are less susceptible to both disease and insect attack; it is important to identify vigorous young individuals which will respond positively to release if older and larger trees around them are felled.

### *10.3.5 Seed collection and preparation*

Despite advances in natural regeneration, world forestry still relies heavily on seedlings raised in a nursery. Seed collection, treatment, storage and germination were formerly undertaken by local craftsmen to great effect. Due to the large scale now involved, much greater efficiency and consistently higher standards are required; bulk handling of tree seed is highly scientific and can only satisfactorily be undertaken by organizations with the appropriate machinery and well-equipped cold stores. The aim is both to reduce needless losses and to improve the quality of the seedlings produced. Particular care has to be taken with the seeds of forest broadleaves if losses are to be avoided; much is known about the maturation and harvest of fruits and seeds, transportation of seeds, temporary storage, cleaning and grading (Suska *et al.*, 1996). Upon collection, seeds have to be dried and stored under conditions that reduce losses caused by excessive drying and by disease. These need particular consideration in such species as silver maple *Acer saccharinum*, sycamore *Acer pseudoplatanus* and the oaks, the viability of whose **recalcitrant seeds** is damaged if dried below a certain moisture level (40–48% of fresh weight in the case of acorns). **Orthodox seeds** by contrast are relatively long-lived and able to withstand dehydration to very low moisture contents without losing viability. The seeds of beech *Fagus sylvatica*, which crops irregularly, are well understood; they can now be stored for long periods and sown after the elimination of dormancy. Figure 10.11 demonstrates the critical importance of moisture content upon germination capacity of seeds stored for very long periods. Above a moisture content of 9% germination capacity decreases progressively, none at 12.9% germinating after 6 years. When moisture contents are reduced to 5.7% or lower, germination rates can drop dramatically shortly after drying.

Hardwoods such as ash and beech possess hardy seeds which can be stored for considerable periods; others give more difficulty. In Britain, trees of oaks (*Quercus robur* and *Q. petraea*), sweet chestnut *Castanea sativa*, horse chestnut *Aesculus hippocastanum* and sycamore *Acer pseudoplatanus* are very frequently grown from seed. All have extremely perishable fruits that are shed at high

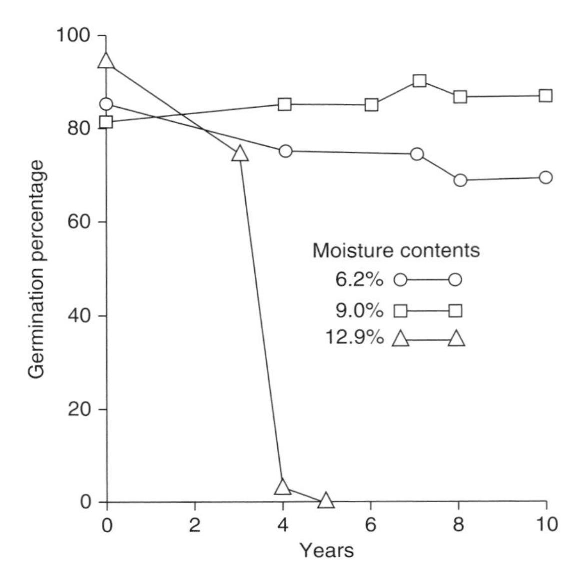


Figure 10.11 Influence of seed moisture content on percentage germination of beech *Fagus sylvatica* during long-term storage at −5 °C. After 10 years germination was highest (87%) in seeds stored at 9.0% moisture and lower (69%) at 6.2% moisture. Seeds remained viable for 5 years maximum when stored at 12.9% moisture. (Redrawn from Suska *et al.*, 1996. *Seeds of Forest Broadleaves: from Harvest to Sowing*. INRA.)

moisture contents, are frequently infected by fungi, and are killed by very little drying. They cannot be stored in sealed containers as they need sufficient oxygen for respiration, and die at temperatures below −3 °C. Their germination rates drop sharply during the winter; only half the acorns planted in the spring will germinate, despite having a germination rate of 90% when collected in the autumn.

## 10.4 Improving the forest: choice of species and provenance

Forest productivity can be improved in various ways, but is most easily done using plantation trees. **Suitable choices of species and origin** are absolutely essential in this process. The terms provenance and origin have very precise meanings in forestry. A **provenance** is the place where any group of trees, whether native or not, is growing. **Origin**, on the other hand, is the place in which a stand of native trees is growing, or that from which a non-native stand was originally introduced.

There are many examples of trees that have been introduced to new areas with great benefits to the local population. The **neem tree** *Azadirachta indica*, which may well have originated in Assam and Myanmar (Burma), and is important in India where it is considered sacred, is amongst the species likely to have much greater significance in the future (Shultz *et al.*, 1992). Numerous uses are made of its products. Neem oil extracted from its seeds has long been used as a hair tonic, to treat skin diseases, and a source of soap, while an extract from the bark is used in preparations that reduce dental caries and inflammation of the mouth. A wide variety of complex organic compounds can be extracted from this tree and many are proving of value as medicines, insecticides, nematicides (killing nematodes) and fungicides. Its timber is also valuable and this substantial evergreen tree of the mahogany family (Meliaceae) grows rapidly in the lowland tropics, reaching a height of 30 m and a dbh of 80 cm. The tree is often established by planting seeds or seedlings, but saplings, root suckers and tissue culture have also been employed. It is now grown in many parts of the world including much of Asia, Saudi Arabia, South America, parts of the USA and many African countries, being used even in dry infertile sites on the southern margin of the Sahara. Its deep and extensive root system is very effective in preventing erosion and it survives well in droughts. In contrast the tree cannot resist frost and dies quickly if the soil becomes waterlogged. Neem pollards are the source of much firewood in Africa.

Plantation forests consisting of non-native monocultures are often extremely successful as timber crops as in the case of the Monterey or **radiata pine** *Pinus radiata*. This is a Mediterranean coastal species native to California and Baja California where it can exceed 40 m in height, which flourishes as an alien in many countries and particularly New Zealand. The foresters here have spent a great deal of time selecting and improving particular strains and clones of this fast-growing tree. A 25-year-old plantation at Rotorua, North Island which had been pruned and grown for saw logs was in 1999 stocked at 400 stems with a mean diameter of 50 cm and a total volume of 940 $m^3$ of timber $ha^{-1}$. This is equivalent to a yield class of nearly 38 (see Fig. 10.4). Seed from this plantation still has the origin of California, but the provenance is Rotorua. Some plantations, incidentally, now have even greater growth rates. This is not at major expense to soil nutrients; studies of these plantations in New Zealand have shown that their annual uptake of major nutrients is only a thirtieth to a hundredth of that of major agricultural crops.

Topography has a major influence on the forest industry. New Zealand is always thought of as a major forestry centre and most of its wood production is exported, yet half the country slopes at an angle greater than 28 degrees and two-thirds is over 12 degrees. Three-quarters has an altitude greater than 200 m. Shallow soils, long wind runs, frosts during the growing season, and browsing by the tree-climbing marsupial possum *Trichosurus vulpecula* introduced from Australia, all limit the possibilities for commercial forestry in many areas. Yet with the choice of the right provenance of the right species forestry is very successful.

Forests can also be improved by **tree breeding**, crossing trees that have useful characteristics to try and get all the good traits into one tree. Seeds collected from these breeding trials would be grown and the best trees, usually referred to as **plus** or **superior trees** are identified. Seed from these superior trees would then be used for the next crop. In this way it is possible to select for desirable growth characteristics (fast growth, straight trunk, low taper, etc.) and for other desirable traits such as disease resistance. This sort of breeding programme has been used in agriculture for millennia and is certainly not a new technique. The major drawback with such programmes is that the seed collected from superior trees is genetically mixed (each seed has two parents) so not all the seeds will perform as well as the superior parent. Various solutions to this problem are possible. Forest geneticists have used **seed orchards** where superior trees are grafted onto established root stocks. Since the resulting trees are close together, they will most likely exchange pollen increasing the probability of a greater number of the seeds being superior. This is an expensive process and tends to be used for valuable fruit or nut trees rather than for timber trees. **Genetic engineering techniques** can be used to transfer desirable genes directly from one individual to another but this requires a knowledge of which genes are responsible for the desired trait. Gene transfer is currently being used to address a range of issues (Giri *et al.*, 2004) including:

- reducing the time until onset of seed production;
- modifying tree hormones to manipulate tree shape and growth;
- modifying wood to allow easier separation of cellulose and lignin during papermaking.

Another way to reduce genetic variability is to take cuttings from a superior tree, root them and use them as seedlings for the next plantation. Since all the new trees are genetically identical they are clones and this has given rise to **clonal forestry**. Again, this is not a new process in the wider world. The vast cultivated olive *Olea europaea* ssp. *europaea* plantations of the Mediterranean region are based upon the repeated careful selection and subsequent cloning by cuttings of favoured genotypes of the wild tree *O. europaea* ssp. *sylvestris*, most populations of which have wider leaves, thornier branches and smaller fruits than the commercial varieties. The ultimate result is the production of large numbers of genetically identical and highly productive trees. Commercial cloning of timber trees has been carried out for a number of decades now, involving poplars *Populus* spp., Norway spruce *Picea abies* and many other trees in temperate areas, and eucalypts throughout the tropics.

Foresters initiating plantations of clonal trees often use several clones in order to reduce the risk of disease devastating the whole plantation. Cloning works well for a number of generations (usually around five) but thereafter does not work indefinitely. Eventually there is a build-up of **topophysis**, the long-term persistence of age or position, such that new trees are physiologically old, and cuttings from the side of a tree continue to grow sideways when rooted. This and the high costs have dogged the long promised and much awaited clonal forestry programme in Europe using Norway spruce, and its use has been more or less abandoned in northern Europe (Anon., 2002). A solution to topophysis, and a way of increasing the number and cost effectiveness of producing clones, is to use **somatic embryogenesis**. Tissue from selected trees is cultured under laboratory conditions (tissue culture) with hormones and nutrients, stimulating them to quickly produce a number of bare embryos. These can be separated and grown into whole young clonal trees. The stock of tissue from a superior tree can be stored in liquid nitrogen which prevents it from physiologically ageing. This technique is still being developed but Giri *et al.* (2004) list more than 50 trees that are under investigation.

Trees can often be improved by using **polyploids** (having extra sets of chromosomes). Polyploids are rare amongst conifers but common amongst hardwoods (angiosperms). While many of these are of no great commercial value, others such as the polyploid forms of white birch *Betula papyrifera* and Japanese alder *Alnus japonica* grow larger and faster than their normal diploid cousins (which have the normal two sets of chromosomes in each cell). **Hybrid** offspring from the crossing of two separate species frequently show more rapid growth than either parent. Besides this **hybrid vigour**, they often also bring together useful features of the parent species into one new plant. For example, the hybrid larch *Larix* × *eurolepis* (the × preceding the species name denotes a hybrid) brings together the good shape and fast growth of the European larch *L. decidua* and the disease resistance of the Japanese larch *L. kaempferi*.

**Global warming** will undoubtedly cause major changes in tree planting practice worldwide. Just how much change partly depends upon the sensitivity of trees to climate. To test how different clones of common European species would cope with climate change, Kramer (1995) relocated clones of European larch *Larix decidua*, Downy birch *Betula pubescens*, small-leaved lime *Tilia cordata*, grey poplar *Populus canescens*, English oak *Quercus robur*, beech *Fagus sylvatica* and Norway spruce *Picea abies* over a large latitudinal range in Europe. He compared the timing of spring events such as bud opening and leaf opening in these transplanted clones with trees of the same species already growing in the areas which were assumed to have adapted to their local climates. The response of the transplants was very similar to the indigenous individuals showing that these species possess considerable plasticity in their response to climate and should be able to cope with changes in their local climate. However, with long-lived trees, it is difficult to predict exactly what might happen to a valuable crop over decades or centuries and economic prudence will require caution. Beech *Fagus sylvatica* has grown splendidly for millennia in the south-east counties of England and in Sussex particularly. How will this native English tree fare in the increasingly dry and droughty conditions brought about by climate change? If left to nature beech woods might well be increasingly invaded by oak and ash that can withstand drought far more effectively. Perhaps the answer is to encourage more planting of beech in northern England and Scotland. The effect of climate change on forests is considered further in Section 11.3.1.

## 10.5 Forest practices

### *10.5.1 Initial species choices and quarantine*

Forest practices are most important in making the best use of the land, preserving and creating biodiversity, and in limiting disease and pest problems. The major initial decision is whether to grow a monoculture or to have a multi-species forest as is common in Europe and in natural rain forests (though the rubber tree *Hevea brasiliensis* has been grown as a monoculture in the tropics for well over a century). The widespread practice of planting monocultures of radiata pine and other species produces even-aged single species coniferous forests of a type which occurs naturally in many parts of the world, including British Columbia and Alberta where lodgepole pine dominates large areas. Tree mixtures are very useful in some countries and trees such as eucalypts, which produce naturally durable timber can be very valuable, but in New Zealand, and often elsewhere, chemical treatment is used to prevent rot when timber lacking natural resistance is in contact with the ground. If clear felling is adopted the size of the logging **coupe** (an area in a forest to be clear felled) must be appropriate to considerations of erosion, flooding, water quality and landscape values. Site preparation must be suitable; burning is easy but often has detrimental effects and windrowing (which heaps woody slash and litter in piles or rows) often depletes soil productivity for decades. Timber extraction (logging) and road construction cause major environmental problems in terms of erosion, loss of nutrients and soil compaction (Li, 2004; McIntosh and Laffan, 2005; Mieth and Bork, 2005).

Codes of forestry practice used in New Zealand and most other countries are intended to educate and act as guidelines, rather than being mandatory. The area demanding the utmost vigilance is that of **quarantine**: preventing a further worldwide spread of the pests and diseases which have already caused a great deal of damage, as Chapter 5 indicates. Carter (1989) reported that 212 insects and 92 fungi known or suspected to harm radiata pine had not yet entered New Zealand. A major advantage of planting exotic trees is that they usually do better than in their places of origin because they have left their pests and diseases behind them. In some ways New Zealand is very vulnerable to introduced pests. The possum already carries bovine TB in many parts of the country, and if wood-consuming termites ever became established they could cause havoc to the many wooden houses, apart from anything else.

### *10.5.2 Plantation forests and the strategy of risk*

In countries with huge areas of plantation forest such as New Zealand, there is a choice between planting a number of tree species or just one. With several tree species significant losses are likely to occur annually because there will be a greater variety of damage-causing pests and pathogens; with a single species at present without many pest and pathogens the annual risk is low but there is the possibility of a catastrophe should they arrive. As Maclaren (1996) puts it, the forest owner has to balance the risks of a high-probability but low-impact damage against a low-probability but high-impact damage. In choosing the second option and planting radiata pine in New Zealand, foresters have lowered the costs of research and protection; the real concern is that these forests will no longer be so productive when, as is almost inevitable, pathogens and pests from other parts of the world finally invade. Similar arguments apply to other species and other places.

The greatest difficulties arise when genetic variability in a crop is low (as in clonal forestry) since a disease or pest overcoming **host resistance** can easily sweep through the entire population. The opposite problem is equally damaging where a pest or pathogen has evolved the ability to overcome a wide range of host resistances and can invade populations even with high genetic variability. This is the case with honey fungus *Armillaria* spp. and the gypsy moth *Lymantria dispar*. Honey fungus is one of the world's most important and voracious root diseases, made up of some 40 species worldwide (not all of which are pathogens) to which very few woody plants are immune (see Section 5.4.2). The gypsy moth is one of North America's most devastating forest pests, introduced from Europe in 1868–9, and able to feed on the foliage of literally hundreds of species of plant. If this moth becomes established in New Zealand it will attack a wide range of hosts including hardwoods, just as *Armillaria mellea* does already.

The ultimate example of an introduced tree lacking genetic variability is that of the English elm *Ulmus procera*, all of whose individuals are now believed to have arisen clonally from a single individual brought to Britain by the Romans. Its origin has been traced to the Atinian elm, a variety from central Italy that reproduces asexually. It was widely employed to support and train vines in wine production, having been recommended for this purpose by Columella in his AD 50 classic *De Re Rustica* (Of Rural Affairs). Gil *et al.* (2004) compared DNA from the English elm to four lineages drawn from across Europe in efforts to trace its origin. The closest match was with lineage C, that also occurs in the Iberian peninsula, and appears to originate in the region around Rome. It seems, therefore that the English elm reached Britain via Spain – which was also under Roman control in the first century AD. The rapid and devastating spread of Dutch elm disease in the 1970s (Section 5.4.5), when more than 25 million trees were killed, was promoted by the fact that this susceptible variety had been preserved genetically unaltered as the core of the English elm population for 2000 years.

### *10.5.3 Reclaiming disturbed land for forestry*

On a world basis, land is increasingly in short supply so there is every reason to reclaim so-called brown-field sites which have been disturbed by mineral workings or other industrial processes, and are now unused. This is particularly the case when attempts are being made to improve the countryside near population centres. Besides being unproductive, derelict land is often dangerous with old buildings liable to collapse, crumbling quarry faces or exposure to chemical contamination. During a 12-year period centred on the 1980s over 157 000 ha of English land was reclaimed for beneficial use, including forestry, at a cost of almost £750 million. The British Forestry Commission devoted considerable resources to assessing and solving the many problems involved when the ultimate use is forestry (Moffat and McNeill, 1994).

On relatively infertile sites, including many former colliery spoil tips, conifers prove to be both appropriate and economically productive. In other places broadleaves (such as willows *Salix* spp., alders *Alnus* spp. and birches *Betula* spp.) can yield valuable timber and are also desirable for wildlife and aesthetic reasons. In both cases sewage sludge greatly improves growth, being cheaper and promoting a greater increase in growth than mineral fertilizers. Foliar analysis provides a good indication of nutrient absorption by young trees. Different substrates pose different problems, for example fresh pulverized fuel ash (pfa) produced by the burning of coal in thermal power stations commonly has a pH of 11 or 12. Though this drops to pH 8 or 9 after being stored in a lagoon (lagooning), it is still too alkaline for sensitive species such as beech, ash or sycamore. Sitka spruce and alder can be planted directly in it, though admixture with a surface layer of soil or colliery shale improves growth. Drainage is a key factor and on many sites deep discing or ripping of the substrate is essential. Ridge and furrow methods with young trees being planted and firmed up in notches on the ridges help soil aeration.

Restoration plans and planning applications need particular care and it is essential to devise and execute a long-term programme of **aftercare** to ensure that drainage, soil nutrient levels, weed control and tree density are properly maintained. In the case of new mineral extractions careful soil stripping and storage will make subsequent restoration of the site very much easier. It is vital that those who may be responsible for the site in later years be made aware of problems that may arise if deeply buried contaminants such as iron pyrites are disturbed.

## 10.6 Sustainable forest management

Since the early 1980s, foresters and others interested in forests have been debating and defining **sustainable forest management**. Many definitions have been produced with a similar theme, exemplified by the **forest principles** set out by the UNCED conference (United Nations Conference on Environment and Development, also called the Earth Summit or Rio Summit) held in Rio de Janeiro, Brazil in 1992:

> Forest resources and forest lands should be sustainably managed to meet the social, economic, ecological, cultural, and spiritual needs of present and future generations.

It is easy to see the value of sustainability as encapsulated by such a definition but it is notoriously difficult to achieve. This is partly because different groups have different priorities within this definition; some are concerned mostly with water quality, others the conservation of biodiversity, yet others with maintaining wood production, and the list could go on. In theory these are all compatible but in practice different objectives often require different management. Moreover there are problems in defining and predicting the needs of future generations. For example, if forests are managed to produce timber with the cost that two species of insect become extinct, how do we judge whether this is important to future generations? If we decide that all species must be preserved just in case and this reduces timber output in poor tropical countries to the detriment of the present population, is this acceptable? Using the above definition, the answer is clearly that the needs of the present human population are paramount, so where does the compromise lie? Even when objectives have been set it can be equally difficult to define exactly what these mean in practical terms that can be measured. This outline of the problems is in no way meant to undermine the need for sustainability but to highlight the difficulty in making sustainable forest management a reality. Discussion is still going on within the forestry community (e.g. Floyd *et al.*, 2001). Regardless of the ongoing debate, a number of bodies are putting sustainable forestry into practice. The **Forest Stewardship Council (FSC)** based in Germany (see www.fsc.org/en/) has established a worldwide certification scheme based on the UNCED forest principles. Now, many major retailers in Europe, North America, South America and Asia ask for FSC certification when buying timber or forest products.

Before sustainable forest management as described above became an issue, foresters were primarily concerned with **sustainable yield**, the amount of wood that can be continuously produced at a given intensity of management. This is comparatively straightforward to measure since it involves the healthy production of at least as much timber per successive rotation (of the same duration) over a prolonged period. This aim has undoubtedly been achieved in many European forests that have been carefully tended for several centuries. There is concern about the rate at which infection by honey fungus *Armillaria mellea* has built up in radiata pine stands in North Island, New Zealand; frequently 20% of second rotation trees are infected whereas the figure on ex-pasture sites is only 2%. If more severe problems occur, the possibility of a crop rotation system involving clear felling, the use of the land for agricultural crops or pasture, and the complete decay of woody material before more tree seedlings are planted should be considered.

Problems of tree health frequently involve pests as well as pathogens and here effective control involves a careful consideration of the biology of both the pest and its host. This has led to the concept of **integrated pest management (IPM)**, a very effective approach considered in relation to tropical trees by Speight and Wylie (2001). A classic example is that of grasshopper attack on eucalypts in Paraguay. This can cause losses of 70–80% due to ring-barking followed by tree fall during the first 3 months after establishment. The tougher grey bark of older trees seems relatively immune to attack so the solution to the problem is the use of older transplants (planted seedlings) in less risk-prone sites. Attempts to employ IPM with *Hypsipyla* shoot borers have been less successful; these Lepidopteran pests virtually prevent the plantation cultivation of mahoganies (*Swietania, Khaya*), cedars (*Cedrella, Toona*) and other valuable members of the Meliaceae in South-East Asia and Australia. Attempts to achieve classical biocontrol by introducing parasitoids and predators of these shoot borers have not eliminated the problem in their native country or in introduced populations elsewhere. Cases where susceptible species of Meliaceae have been grown with minimal shoot-borer damage often involve rearing the desirable tree in mixtures with other tree species; there are many opinions as to exactly why this procedure is sometimes successful. As with pathogens, some genetic strains of vulnerable tree species are less at risk than others, and in both cases attempts are being made to improve the situation by the breeding and selection of more resistant genotypes.

## 10.7 Landscape ecology and forests

### *10.7.1 Metapopulations and diversity*

There is a need for a conceptual framework to facilitate the description of ecological processes occurring on a landscape scale (see Section 10.7.3) above the level of the local population. This is provided by the concept of the **metapopulation**, an assemblage of partially isolated populations of the same species existing in a balance between extinction and colonization. Animal and plant species are seldom distributed continuously in space; they are usually organized into local populations (each of which may be increasing or decreasing) connected to a greater or lesser extent through dispersal. This spatial structure implies that the demography and genetics of populations will be the product of both local environmental conditions and processes operating on a regional scale. The metapopulation concept is still vigorously discussed, here we will define some of the most important variables involved and discuss a relevant woodland example.

Whether regional populations form a linked metapopulation or a series of discrete populations depends on the proportion of suitable habitat potentially available to the species concerned. Where suitable (S) habitats cover a much larger area than unsuitable (U) habitats, i.e. $S \gg U$, plants and animals can be described as developing **spatially extended populations**. Where the reverse applies, i.e. $U \gg S$, species form **regional ensembles** in which migration between patches of suitable habitat is effectively prevented by isolation, and re-colonization does not occur after local extinctions. If, however, suitable habitats cover an intermediate area relative to those which are unsuitable, it is considered by Freckleton and Watkinson (2002, 2003) that metapopulations are particularly likely to exist. In such cases, the regional dynamics (movement between populations) are determined by the relationship between local birth and death processes (B and D) and regional immigration and emigration rates (I and E).

Whilst it is generally accepted that the relative size of I and E versus B and D describes the dynamics within the local population units, it is also essential to realize that variation in any or all of I, E, B or D can be crucial for regional dynamics. In practice very few plant metapopulations have been described, and there are good reasons why plants may not develop such populations in the same way as short-lived mobile animals (Ehrlen and Eriksson, 2003). The existence of long-lived seeds or vegetative ramets enables local populations to exist for a long time even though the habitat patch has become unsuitable. Additionally, successful dispersal and recruitment may be very sporadic so re-colonization may be unlikely after local extinction, while dispersal over long distances can be governed by chance events.

Consideration of metapopulations can be a useful tool for identifying constraints on populations. In the central hardwoods region of Indiana in central USA, the North American red squirrel (*Tamiasciurus hudsonicus*) has gradually colonized all but the most heavily forested areas since the early 1990s. This range expansion by the red squirrel has coincided with a reduction in numbers of the grey squirrel (*Sciurus carolinensis*). It may appear at first sight that the red squirrel is detrimentally interacting with or outcompeting the grey in the same way that the same grey squirrel introduced into the UK is detrimentally affecting the native red squirrel (*Sciurus vulgaris*). The reason for the changes in Indiana, however, is more to do with metapopulations. Goheen *et al.* (2003) investigated the problem in an area of Indiana where forest patches (covering 10% of the land) are embedded within an agricultural matrix and joined by wooded fencerows (hedgerows). The grey squirrel is more sensitive to forest fragmentation than the red and the former are found only in the largest most continuous forested sites. As importantly, they found that the red squirrel is more able to move between local populations along the fencerows. Overall survival of grey squirrels moved into fencerows was 71% compared with 95% in red squirrels, presumably due to higher predation risks for the grey. The grey squirrel is thus found in fewer, more isolated populations in a regional ensemble rather than a metapopulation, and has a low ability to move between them. Thus the grey squirrel is constrained both by individual area requirements (further fragmentation will reduce their suitable habitat) and poor dispersal ability (local extinction is hard to replace by movement). By contrast the red squirrel can live in smaller fragments but can also move between local populations much more easily to maintain viable numbers if, for example, local disturbance reduces population size in one area. Such movement makes the future survival of the red squirrel metapopulation much more probable.

### *10.7.2 Fragmentation and connectivity*

The Earth's surface now supports a relatively minute proportion of truly natural landscape. As we have seen in the previous chapter, this is particularly true of woodlands and forests. One of the most insidious effects of these changes is the fragmentation of forests into smaller parcels. Even large countries with relatively unpopulated areas such as the USA, where forest covers 31.4% of the land, are suffering from extensive forest fragmentation. Riitters *et al.* (2002) determined that 44% of the forest area within the contiguous US states is within 90 m of a forest edge and 62% is within 150 m. People can drive to within a kilometre of 82% of all land in the contiguous US states (Riitters and Wickham, 2003). Nevertheless the forest still tends to be in fairly large areas such that at least 73% of all forest is in landscapes that are at least 60% forested. Consequently, around half of the forest edge found by Riitters *et al.* is created by small **perforations** in forest of less than 7.29 ha. Thus, the forest is large with holes cut within it. In contrast, the British landscape is just the opposite. England has no virgin forests and until very recently it seemed that even its last reserves of **ancient woodland** (land continuously wooded since AD 1600) faced a very uncertain future. Only 11.8% of the UK is covered by forest and much of this is introduced conifers. Not only is the amount of ancient woodland very low; it is also highly fragmented, being scattered in very many relatively small sites which are often far apart. In contrast to the USA, the agricultural landscape makes up the broad **matrix** (the most extensive and connected landscape element), perforated by small areas of woodland. In England and Wales, 83% of ancient woodlands (amounting to 31% of ancient woodland total area) is less than 20 ha in size, and more than 40% is smaller than 5 ha (Thomas *et al.*, 1997).

The contrast between England, with its high human population density and fragmentary ancient forest, and New Zealand, with its very large areas of well-protected and largely natural forest, is extreme. Before the relatively recent arrival of Polynesians and then Europeans, the only mammals present were either marine or bats. It is suggested that globally 4% of the boreal forest has been removed or fragmented compared to >50% of temperate broadleaf forest and nearly 25% of tropical rain forest (Wade *et al.*, 2003).

Fragmentation can have a large effect on the living biota in a forest. As noted above under metapopulations, the resulting lack of **connectivity** between forest areas makes it much more difficult for native species to migrate and expand within this important habitat. In recent years, however, vigorous action has been taken in many parts of the world (see Box 10.1) and especially the UK. Thomas *et al.* (1997) present an encouraging picture of firm action being taken in the UK to prevent further loss of ancient semi-natural woodland and to de-coniferize and reclaim areas where woods long dominated by broadleaved trees, especially the English oaks, had been planted over. New

### Box 10.1 Connectivity in a fragmented landscape

A good example of woodland fragmentation and connectivity is provided by communities of the maritime juniper *Juniperus oxycedrus* ssp. *macrocarpa*, a Mediterranean species tolerant of saline spray along the south-west coast of Spain, which are both vulnerable and ecologically important. Though now greatly reduced in area, it seems that these formerly extended from the El Rompido cliff in Huelva to the west all the way to Gibraltar in the east. Studies by Muñoz-Reinoso (2004) are designed to form the basis for future research and restoration policy in an area where these communities have been gravely diminished by both the planting of Mediterranean stone pines *Pinus pinea*, and the increasing urbanization of the coast. Here the tendency is for the sizes of previously large communities of maritime juniper to be greatly reduced. There are now many small sites where regeneration is often poor, indeed of a total population estimated at *c.* 24 500 individuals, 93.6% were present at just three locations. High proportions of young plants were found in large and protected populations such as that at the Doñana National Park. Adult individuals dominate the smaller communities under the extensive plantations of *Pinus pinea*. These communities have been subjected to fragmentation and are adversely affected by the deposition of pine needles, deficient pollination and invasion by inland plant species.

Besides the landscape value of its woodlands, which harbour several endangered plants including the labiates *Sideritis arborescens* and *Thymus carnosus*, maritime juniper is itself listed as endangered in this area of Spain. Moreover, fruiting individuals are an important food source for vertebrates such as badgers, foxes, wild boar and rabbits. If typical communities of maritime juniper are to be restored, it is important that they be large enough to support adequate populations of associated plants. The communities themselves show considerable variation; climate and soil texture largely control the amount of water available to plants and thus separate xeric from mesic communities. On a smaller scale, coastal physiography (dunes and cliffs), together with such factors as soil calcium carbonate content and sand mobility, also result in differences in floristic composition.

woodlands of native hardwoods are also being planted to provide stepping stones to promote connectivity between small fragments of ancient woodland.

In addition to reducing connectivity, fragmentation also leads directly to species loss. Island biogeography theory (Section 11.4.1) predicts that smaller forest fragments will hold fewer species, and more isolated fragments (i.e. more distant from other areas of forest) will also have fewer species. The effect of small fragments is made worse by the number of **forest interior** species that require large tracts of unfragmented forest to persist and which can be sensitive

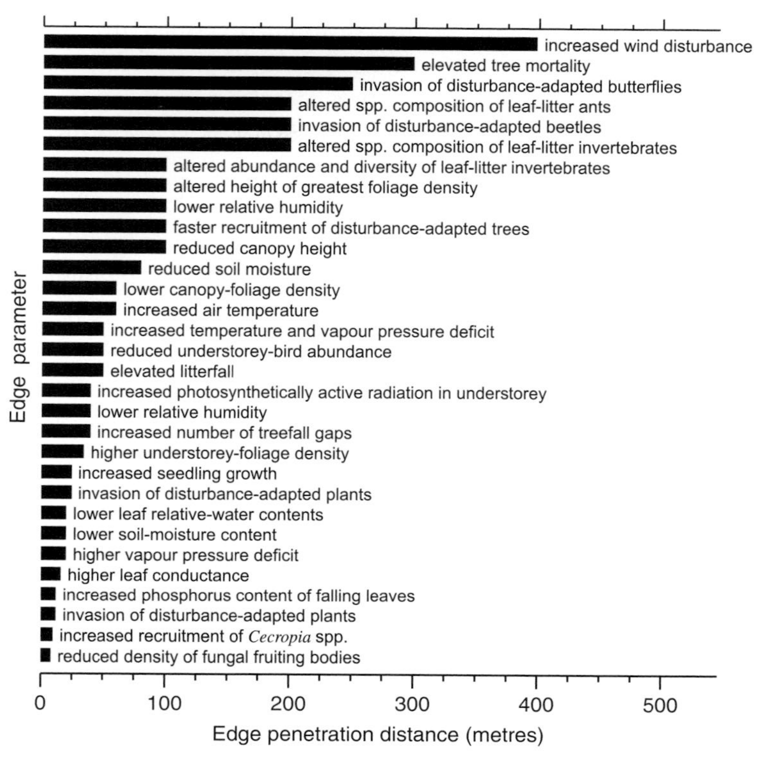


Figure 10.12 Penetration distances of different edge effects into forest fragments in central Amazonia. These data are part of the Biological Dynamics of Forest Fragments Project. Started in 1979 and covering 1000 km$^2$, this is the longest running and largest study on habitat fragmentation. (Redrawn from: Laurance *et al.*, 2002. *Conservation Biology* 16.)

to a road dissecting an otherwise large pristine forest. Most primates fall into this category. The **edge effect**, the change in conditions which permeate into a forest from the edge, makes the apparent forest fragments even smaller. Figure 10.12 illustrates some of the biological and physical changes that occur at a tropical forest edge where it is usually lighter, less humid and with higher extremes of temperature than in the forest interior. The figure shows that some of these effects are felt 300–400 m into the forest. In conifer forests of Wyoming, Reed *et al.* (1996) found that the landscape area affected by clear-cuts and roads was 2.5–3.5 times their actual area.

Fragmentation into smaller units aided by the edge effect can have a profound influence upon forest species. For example, taking birds, Burke and Nol (2000) working in the mixed forests of southern Ontario found that most bird species bred successfully in large forest fragments (mean size 121 ha) but less well so (below replacement levels) in small fragments (mean of 7.8 ha).

Jones *et al.* (2000), using results of the National Breeding Bird Survey for the mid-Atlantic region of the USA in a GIS (geographic information system) model, found that of all landscape features affecting species richness of birds, forest edge was the most important. Generally, forest specialist birds were negatively correlated with forest edge and generalist birds were positively associated. In addition to differences in environmental factors and vegetation, there is evidence that bird differences between interiors and edges are influenced by greater nest predation and particularly greater incidence of brood parasitism by the brown-headed cowbird *Molothrus ater* (Chalfoun *et al.*, 2002). Like the European cuckoo *Cuculus canorus*, the cowbird lays its eggs in other birds' nests and the chick thrives at the expense of the hosts. This parasitism is greatest near forest edges since it is closer to the open land favoured by the cowbird. However, not all forest edges are detrimental. A study of the birds in hardwood and mixed forests of Vermont by Ortega and Capen (2002) found that four of the 18 forest interior species had lower abundance adjacent to roads but two species had higher abundance and 12 were unaffected. Of the six edge species, four were more abundant near roads and two were unaffected. A number of likely reasons could be put forward based on the characteristics of this particular forest or the ecological requirements of the birds, but the important point is that forest systems are so complex that conclusions that apply to one forest may not apply to others.

Fragmentation in tropical forests may have severe consequences since many tropical species are rare across their entire range so each forest fragment is less likely to hold a viable population. Since tropical forests are dense by nature, large openings can act as significant barriers. Even unpaved roads only 30–40 m wide in Amazonian forests act to inhibit the movement of many species. But even here edges are not completely bad news; many species that can utilize more open conditions thrive in the forest edge including termites, light-loving butterflies, hummingbirds and fruit-eating bats (Laurance *et al.*, 2002). In fact some animals such as frogs and small mammals have a greater species richness in forest fragments than in intact forest (Gascon and Lovejoy, 1998).

Forest edges can differ from interior areas not just in their fauna but also in their plant species. In the hardwood forests in the Piedmont area of North Carolina, USA there are noticeable differences in tree species composition up to 5 m into the forest. This may seem a small distance but comparison with the third bar up from the bottom in Fig. 10.12 shows that invasive plants establish best near to an edge. The edge zone is more open and so readily invaded in the USA by non-native trees such as the tree-of-heaven *Ailanthus altissima*. It seems likely that once established these trees can provide a large seed source with potential to spread into gaps further into the forest (Rob McDonald, pers. comm.). In the central Amazonian rain forests, lianas are more important in forest edges than in interior forests. Along edges (defined as a 100-m wide strip) Laurance *et al.* (2001) found that though there is not a greater biomass of lianas along edges, they are more numerous with a higher proportion of small lianas (2–3 cm diameter) and relatively few bigger ones (4 or more cm diameter). More trees are infested and tree growth along the edge is reduced due to competition above and below ground. Laurance *et al.* suggest that since lianas are very prone to damage by fire and wind, and because they inhibit the invasion of trees, they create a positive feedback loop by promoting and exacerbating subsequent disturbance, making the gaps larger. On the positive side, these abundant lianas with some new growth of pioneer species can partially 'seal' the edge of the forest within 5 years, reducing the extent of the edge effect until a disturbance opens it again.

Roads can produce particularly large edge effects. In the Rocky Mountains in Wyoming, Reed *et al.* (1996) found that the edge effect of roads penetrated 1.54–1.98 times further into the conifer forest than along the edge of a forestry clearcut. Forman (2000) estimates that 15–22% of total land area of the contiguous US states is ecologically affected by roads, based on an edge effect of 200 m from secondary roads, 305–365 m from primary roads in forests and grassland (with 10 000 vehicles per day) and 810 m for busy roads in urban areas (50 000 vehicles per day). In addition to chemical pollution and difficulties in crossing roads, birds can suffer from acoustic masking of their songs, reducing their ability to attract mates. Birds with high-pitched song with frequencies well above those of traffic noise are less susceptible to noise pollution. In tropical forests, presence of roads is one of the main factors leading to increased deforestation and further fragmentation (Laurance *et al.*, 2002). Reviews on the ecological effects of roads have been produced by Forman and Alexander (1998) and Spellerberg (1998).

### *10.7.3 Conservation and biodiversity in a multi-purpose landscape*

Although in some instances a particular species is so rare that its needs are paramount, a major objective of conservation is normally the promotion or preservation of biodiversity. As we have already seen, biodiversity can be high in exotic woodlands, but these lack the integrity and uniqueness of ancient communities, whose species and surroundings can tell us much about the past, as Rackham (2003) demonstrates.

Throughout this chapter we have indicated that the philosophy of forest management in developed countries has been changing from concentrating purely on timber production towards multi-use forests managed sustainably.

We have also seen how fragmentation of forests and the need for connectivity to allow movement within metapopulations demonstrates the need to view forests within the wider landscape. This is the domain of **landscape ecology** which involves the study of patterns within the landscape, how patches (e.g. areas of forest) interact within a landscape mosaic, and how these patterns and interactions change over time.

The use of forests for timber production and biodiversity conservation has been considered under multi-use forests (Section 10.2) but a large number of battles have been, and are being, fought on the use of different forests in the landscape. This is especially true when pristine forests, which are increasingly seen as strongholds of biological diversity, are threatened with logging. This is exemplified by the bitter conflict over plans to fell old-growth Douglas fir forests *Pseudotsuga menziesii* in north-western North America which are the home of the northern spotted owl *Strix occidentalis caurina*. At its most polarized, loggers feel that valuable large timber in these stands should not be wasted while conservationists uphold the rights of the owl to exist unmolested. Compromises are suggested – harvesting some of the timber in certain areas and leaving others untouched – but it is a truism that good compromises leave everyone unhappy. Much of the argument over this and many similar cases in temperate forests revolve around the long-standing view that biodiversity is best preserved in reserves from which logging is excluded.

Landscape ecology approaches are showing, however, that this is often not sufficient. Reserves are without doubt very important, but in most types of temperate forest these typically make up less than 10–15% of the land area so much of the biodiversity remains outside the reserve. Even large reserves are likely to be dependent upon populations outside the reserve because the species concerned are migratory or, as in the case of carnivores, they have large ranges. In other instances there is a need to supplement small populations; here the need to conserve forest species in the wider landscape, albeit at lower levels, also becomes important. This is where multi-use forests are valuable in providing some biodiversity conservation. As noted above, in a very hostile landscape (such as forest fragments in a matrix of intensive-use agricultural land), wooded corridors, such as provided by fencerows/hedgerows and wooded stepping stones (which reduce the distance organisms need to move in one go to colonize new areas) are increasingly valuable. These need to be placed in the landscape, if there is a choice, to maximize their use; much research is currently directed at this.

Conservation in the wider landscape has been taken to heart in the design of multi-use forests, particularly in plantations. Plantation design has changed considerably over recent decades. Emphasis is often put on **visual landscaping**, based on the principle that people prefer forests that look natural and mature. Fortunately, many of the resulting features are also useful for biodiversity conservation (see the section above). Lindenmayer and Franklin (1997), Peterken (1999) and Hartley (2002) suggest a number of measures to increase the natural features that should be built into forest design and management.

(1) Species composition. Use native species and mixed-species stands if possible. Where this is not feasible, native species should be left as mature **retention trees** in openings in the stands and new trees should be encouraged around the perimeters, in wet areas and as an open understorey below the crop.
(2) Site preparation. Retain as many components of the original forest as possible to maintain biodiversity – acting as a **lifeboat**: the soil should be disturbed as little as possible; standing dead snags and the maximum amount of dead wood should be left (see Section 7.7 for a discussion on the value of dead wood).
(3) Open space. Leave up to 15% of the plantation as **open space**, which should be concentrated along roads and in areas intrinsically unfavourable to tree growth such as damp hollows. Retain 10–15% cover within the cut areas to enhance connectivity.
(4) Treeline. Leave a more natural irregular treeline along upper margins of plantations in hilly areas, possibly planting scrub above this.
(5) Tending. Fell trees in large irregularly shaped areas, thin some areas heavily to encourage the understorey, leave others unthinned to create dense thickets, or better still move to group selection, shelterwood or continuous cover forestry systems (see Section 10.3.3). Divide the forest into short- and long-rotation areas to help simulate a more varied age structure with many young trees but also areas of older trees. Leave unfelled or native vegetation buffer zones around the cut areas to reduce the edge effect.

The Wyre Forest on the Shropshire/Worcestershire border, central England is an excellent example of a multi-use forest. It resembles many others, including tropical forests – as we now realize – in having been managed for its timber for an exceedingly long time, in this case for over 900 years at the very least. The ecology, management and history of its 2430 ha are described in a symposium volume (Packham and Harding, 1995) which clearly illustrates the large number of factors responsible for its present biodiversity in terms of its vascular and lower plant communities, invertebrates, amphibians, reptiles, birds and mammals. The way in which ancient forest records and present evidence help us to unravel the patterns of the past is of particular interest (Hobson, 1995a, 1997). Figure 1.4 illustrates the way in which many of the oaks have the basal curve typical of trees which have regrown from coppice stools, a relict of the times when coppicing by the charcoal burner took place regularly. Baskets, brooms, rustic garden furniture and fencing were all made

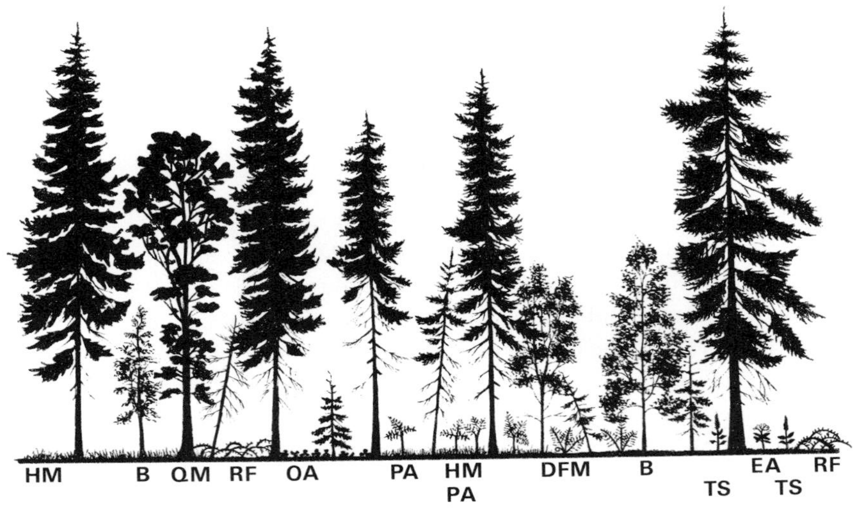


Figure 10.13 Mixed conifer-broadleaved area of the Wyre Forest, England. Douglas fir *Pseudotsuga menziesii* (unlabelled) planted by the Forestry Commission is regenerating naturally from seed, though suffering losses from windthrow in some places. Sessile oak (QM, *Quercus petraea*) has regrown from a single coppice stool, and birch (B, *Betula* spp.) established from windblown seed in regeneration gaps is here outcompeting young Douglas fir. HM, creeping soft-grass *Holcus mollis* and other low-growing herbs; RF, bramble *Rubus fruticosus*; OA, wood-sorrel *Oxalis acetosella*; PA, bracken *Pteridium aquilinum*; DFM, male-fern *Dryopteris filix-mas*; TS, wood sage *Teucrium scorodonia*; EA, wood spurge *Euphorbia amygdaloides*. (Drawn by Peter R. Hobson. From Packham *et al.*, 1992. *Functional Ecology of Woodlands and Forests*. Chapman and Hall, Fig. 10.2. With kind permission of Springer Science and Business Media.)

using forest produce, while peeled bark was used for tanning (Hobson, 1995b). The forest is also interesting in that part of it has been managed by private enterprise, while that depicted in Figs 10.10 and 10.13 is in areas taken over by the Forestry Commission whose first purchases here were made in 1925/6.

The area of the Wyre Forest coincides closely with an outlier of Middle Coal Measure of Upper Carboniferous age and has had a complex tectonic history with much folding and faulting. Rocks exposed at the surface vary from flaggy sandstones to clays, silts and shales; the soils to which they have given rise show major differences in nutrient content, pH and ability to retain moisture. All these features greatly influence the biodiversity of the forest, some of whose rarest plants were associated with flushes or acidic bogs. Until the 1930s bog asphodel *Narthecium ossifragum*, the insectivorous round-leaved sundew *Drosera rotundifolia*, marsh lousewort *Pedicularis palustris* and bogbean *Menyanthes trifoliata* occurred in their hundreds in a bog which is now part

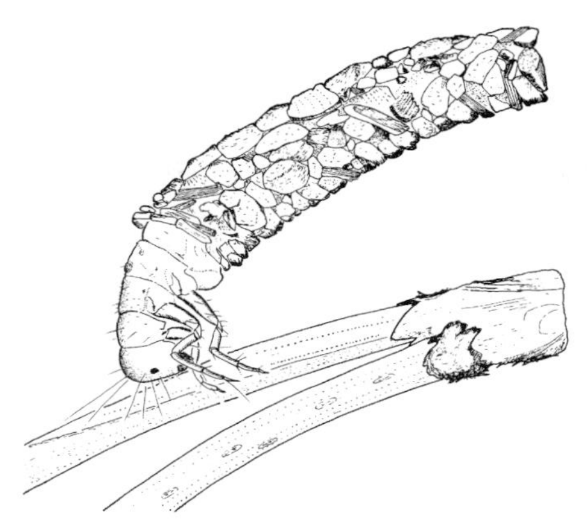

Figure 10.14 Larva of terrestrial caddis *Enoicyla pusilla* on pine needles in Chaddesley Woods, Worcestershire, UK. (Drawn by David J. L. Harding. From Packham *et al.*, 1992. *Functional Ecology of Woodlands and Forests*, page 268. Chapman and Hall. With kind permission of Springer Science and Business Media.)

of a golf course and of which only a mown fragment remains. Losses of vascular plants in particular can be traced through botanical records kept over the centuries, but although much has gone very much remains and the forest and its orchids are still carefully conserved. The forest contains many fascinating insects including the terrestrial caddis *Enoicyla pusilla*, which has also been studied (Harding, 1995) nearby at Chaddesley Woods National Nature Reserve near Kidderminster and Shrawley Wood south of Stourport (Fig. 10.14).

Fruiting bodies of fungi abound in autumn when some of the edible species including the parasol *Macrolepiota procera*, shaggy parasol *M. rhacodes*, lawyer's wig *Coprinus comatus*, wood blewit *Lepista nuda*, wood hedgehog *Hydnum repandum* and penny bun *Boletus edulis* are frequently collected and cooked, particularly by those with East European connections. As mentioned previously, under the right conditions almost all woody plants appear able to exist in mutually advantageous symbiotic relationships with a suitable species of fungus. Larches associate in this way with both *Suillus grevillei* and *S. luteus*. Such mycorrhizas, which are very important to the trees of the forest, are often more sensitive to local factors including pH than the trees. In consequence the same species of tree may be associated with different species of mycorrhizal fungi in different habitats. The forest trees also have a much less helpful relationship with parasitic fungi; although the forest contained relatively few elms virtually every one was affected by the epidemics of Dutch elm disease. Honey fungus *Armillaria mellea* is present throughout Wyre Forest, but as in so many mixed UK woodlands, it appears to be primarily saprotrophic. This species causes greater damage in simplified situations with much smaller populations of other fungi (see Section 5.4.2).

One of the most encouraging aspects of recent management developments in Wyre has been the way the many different interests have been reconciled so that, as Ian Hickman put it at the symposium, cross-purpose has become multi-purpose. The Forestry Act 1967 allowed the Forestry Commission to build car parks, forest walks and visitor centres on its land and this bore fruit with the creation of the Callow Hill Visitor Centre with its forest walks at the southern boundary of Wyre Forest.

Stages in the preparation of a charcoal burn. (Drawn by Peter R. Hobson.)

# 11

# The future – how will our forests change?

## 11.1 Threats to forests and the increasing demand for timber

### *11.1.1 Global demands*

The world's woodlands and forests have always changed in response to alterations in climate and the impact of animals, especially the grazers and disease organisms. The major difference in the present Flandrian interglacial is the overwhelming impact of humans, whose world population rose from under two billion in 1900 to over six billion in 2000, after having been comparatively negligible in Neolithic times. Accurate estimation of annual global wood consumption is difficult but it was at least 3.36 billion $m^3$ in 1996, of which some 55% was used as fuel, largely in developing countries. The other 45% went for industrial use, largely as saw logs but also as chemical raw material. Sugar produced by trees in photosynthesis can be converted into a host of complex chemicals, then combined into a cellular structure with a high strength to weight ratio. Lignin, cellulose, resins, turpentine, wood alcohol and many other products are all derived from trees.

Twenty per cent of the world population lives in **developed economies** where consumption is high and the fertility rate (average number of children per woman) low; the remaining 80% exist in the so-called **developing economies** where consumption is low and the fertility rate may be as high as seven. Average use of wood per person per year in 1990 was around $2.3\,m^3$ in the USA, nearly four times the 1996 global average of $0.58\,m^3$. The world's resources could not cope with an unlimited continuation of the population increase seen in the twentieth century. Instead, future factors are expected to result in an annual demand for timber which will be stable, but much higher than it is now. Recent predictions (Sutton, 1999, 2000) are that a large rise in consumption levels in the developing economies will be accompanied by a fall in population increase as occurred in the existing developed economies. This is expected to result in a stabilization of the world population at around ten billion by 2050. Total demand for timber will thus enlarge; it will also increase on a per capita basis as timber is substituted for ever-diminishing fossil fuels. A small but significant proportion could well come from the efficient harvesting and use of urban timber, a process now well underway in parts of the USA, where many people became disgusted by the way in which urban trees were used as landfill or just burnt in a convenient spot (Sherrill, 2003).

If this scenario becomes reality, timber production will have to increase on a massive scale. Total world forest area will have to be greatly increased, as will productivity per unit area. Sutton (2000) predicts that by the year 3000 the total annual wood harvest will have increased to nearly 28 billion $m^3$, mainly from plantation forests. His views are interesting and although even short-term prediction is difficult, what he puts forward seems to be within the realm of technical possibility; one hopes that the human population will in the future become sufficiently disciplined to plan accordingly.

### *11.1.2 Rain-forest distribution and vulnerability*

There is widespread concern about the decline of forests around the world, especially at lower latitudes. Tropical forests are causing particular concern given that they contain more than 50% of all the species on Earth, and have a large economic value. Even more critically, more than 50 million indigenous people live in tropical forests; Amazonian forests alone hold 400 indigenous groups composed of 1 million people (Bryant *et al.*, 1997). Costanza *et al.* (2003) calculate that the current economic value of tropical forests collectively is $\$3.8 \times 10^{12}$ (\$3.8 million million) per year, the most valuable of all terrestrial ecosystems. This is over four times the value of all temperate and boreal forests together even though they are 1.5 times the area of tropical forests. Of all the tropical forests it is the rain forests that are considered most at risk. The plight of tropical rain forests caused particular concern as early as 1850, when the British Association for the Advancement of Science appointed a committee to consider the probable economic and physical effects of the destruction of tropical forests. This concern continues; the strong pressure that has for many years been exerted by both individuals and conservation groups, is recorded in Goldsmith (1998). It is difficult to get an accurate picture of the current rate of loss. As Watson *et al.* (2000) have pointed out 'for tropical countries, deforestation estimates are very uncertain and could be in error by as much as 50%'. FAO (2005) give an average loss of forest in Brazil of 3.1 M ha per year between 2000 and 2005. Data for just the Brazilian Amazonia has been put forward by Laurance *et al.* (2004) – see Fig. 11.1 – which underlines the general

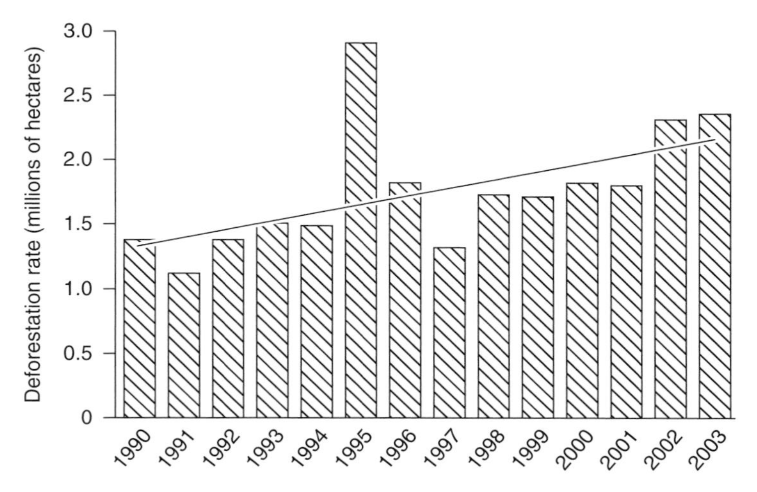


Figure 11.1 Deforestation in Brazilian Amazonia since 1990. The data are from Brazil's National Space Agency. The fitted regression line is the best fit to the data and shows a significant upward increase in annual deforestation. (Redrawn from data from Laurance *et al.*, 2004. *Science* 304.)

feeling that globally deforestation is still increasing. This is especially tragic since Amazonia holds over half the Earth's remaining tropical rain forests. The primary underlying reasons are the need for cropland, increasing population density and the need to earn revenue to pay off external debt (Han *et al.*, 1997). A more insidious problem is not just the loss of rain forest but its increasing fragmentation. Estimates suggest that 4% of the boreal forest has been fragmented or removed by humans, compared with over half of temperate broadleaf and nearly one-quarter of the tropical rain forest (Wade *et al.*, 2003). In their pleading letter to the journal *Science*, Laurance *et al.* (2004) point out that the Brazilian Government's plans to expand dramatically highways in the Amazon basin will not just lead to further development and forest loss, but will also lead to 'fragmentation of surviving forests on an unprecedented spatial scale'. In 1993, Skole and Tucker published a study that showed that the area of Amazonian forest within 1 km of the numerous edges caused by felling was 1.5 times larger than that of the cleared forest. The effects of fragmentation are considered in Section 10.7.2.

When large areas of primary forest are cut, the successional pathways which occur if the land is subsequently abandoned are of major interest; Mesquita *et al.* (2001) report studies of such patterns in Amazonia. The pioneer genus which dominated clearcut sites after 6–10 years differed with subsequent land use, being *Cecropia* (Moraceae) if the land was abandoned directly after clearcut and *Vismia* (Clusiaceae) if it was subsequently used as pasture before abandonment.

In the *Cecropia* stands 58 plant families and 300 species were identified, of which 77 species were also found in the 147 (belonging to 43 families) beneath the *Vismia* canopy, where regeneration was dominated by small *Vismia* individuals. Regeneration under *Cecropia* was more diverse and did not include a single small *Cecropia*. Not surprisingly the number of regenerating plants and plant diversity decreased sharply with distance from primary forest in both cases. This work clearly has important implications for **rain forest conservation**; if primary forest is to be felled this should be done in relatively narrow strips allowing adequate seed dispersal from the unfelled trees, while the clearcut areas themselves should be left to regenerate directly after felling.

The world distribution, climates and classification of tropical rain forests are described by Whitmore (1998), who provides maps showing outlines of the areas still occupied by them in central and south America, central Africa and Madagascar, and areas running from the west coast of peninsular India through to the Malayan and Australian rain forests. Other, non-tropical, rain forests formerly covered very large areas and some still do. Temperate rain forests occur in five major areas of the world: The Valdivian of Argentina and Chile, the Pacific Northwest of North America, the western Black Sea, New Zealand and Tasmania. All face pressure of felling.

Although small in area, the **coastal temperate rain forests of Scotland** represent communities whose extent was much greater before the influence of humans. Their relevance globally is discussed by Rhind (2003), who provides illustrations of a boulder-littered oak woodland on the island of Mull, west Scotland, and of rowan trees in Glen Affric, Scotland covered with bryophytes and lichens, including *Lobaria pulmonaria*. This lichen is well known in Sweden as a plant characteristic of primary beech woodland (see Section 6.4.1). The amount of coastal temperate rain forest remaining in north-west Scotland is estimated at 6896 $km^2$ out of 302 227 $km^2$ for the whole earth. These woodlands are typically dominated by sessile oak *Quercus petraea* and downy birch *Betula pubescens*; many fungus species abound as well as an abundance of moisture-loving invertebrates. Oceanic bryophytes present include *Dicranum scottianum, Frullania teneriffae, Leptoscyphys cuneifolius, Lophocolea fragrans, Metzgeria leptoneura, Radula aquilegia* and *Sematophyllum micans*. 'Old growth' Lobarion lichens often include *Lobaria amplissima, L. scrobiculata, Pseudocyphellareia crocata* and *Sticta canariensis*.

### *11.1.3 Changing patterns of animal distribution*

While data concerning the numbers and world distributions of many species of birds are often freely available, less is known concerning other groups of vertebrates. Information about the status of British mammals collected by members of the Mammal Society shows just what is needed not only in Britain, but in many other places, if we are to make informed judgements about forest policy in relation to the animals present. An increase in badgers *Meles meles* in Britain to a quarter of a million has been accompanied by many deaths through road accidents. Increases in otter *Lutra lutra* numbers have been accompanied by losses of the alien mink, though not necessarily caused by them. Even so the range and numbers of otter in the UK are considerably less than 50 years ago. The very large increase in the rabbit *Oryctolagus cuniculus* population to some 40 million is bad news for both farmers and foresters due to their incessant grazing and browsing. The rise in polecat *Mustela putorius* numbers since World War II is attributed to the loss of gamekeepers, who formerly kept its numbers down. Numbers of feral ferrets, the tamed version of the polecat, are increasing but still relatively small. Due to competition with alien grey squirrels, the still declining red squirrel is extinct in most of England and Wales (Section 5.7.2).

Perhaps even more serious is the decline of the dormouse *Muscardinus avellanarius* in England, and of the wild cat *Felis silvestris* in Scotland. Harvest mouse *Micromys minutus* numbers have also declined by around 75% due to habitat loss, while those of water vole *Arvicola amphibius* are even worse with a decline of over 90% attributed to mink predation and habitat changes. Deer on the other hand, particularly muntjac, fallow and roe, are increasing markedly and impacting heavily on the shrubs, ground flora and tree regeneration of British woodlands (Section 5.7.1). Besides our interest in their distribution, there is also much to learn about the behaviour patterns of mammals and the use of remote-controlled concealed cameras has recently proved most useful in studying that of the giant panda.

Over the longer term there is great interest in the changing forms of the animals that have inhabited the Earth. Attenborough (1980, 1985, 2002) gives well-illustrated accounts of the animals that inhabit our forests and tree canopies (Fig. 11.2), as well as describing those of the past, including the bizarre grazing mammals which evolved two million years ago when South America was an island. How will evolution go, and how greatly will the human population be affected by global warming or another ice age?

## 11.2 Desertification

Desertification first became a topical issue in the 1930s with the dust bowl of the Great Plains of North America. Since then desertification has become a global problem, particularly with the devastating droughts of 1968–73 in parts

Figure 11.2 Mounted head of European bison *Bison bonasus*, which was amongst the big-game animals ruthlessly hunted by *big-game* hunters more interested in the thrill of the chase than conservation. This particular species, however, is no longer in danger of extinction (see also Section 5.8.2). (Drawn by Peter R. Hobson.)

of the **Sahel** where 100 000 people and 12 million cattle died. The Sahel region is a narrow band of West Africa sandwiched between the Sahara desert to the north and the savannas and tropical forest to the south. The United Nations Conference on Desertification (UNCOD) defined desertification as land degradation in arid, semi-arid and dry subhumid areas resulting mainly from adverse human impact. Despite the definition, the underlying complex causes of desertification also involve climate patterns and long-term climate change. Le Houérou (1997) points out that the Sahara has expanded and shrunk a number of times over the past 1.7 million years. Moreover, the drought in the Sahel (a 20–40% reduction in rainfall over three decades in parts of the Sahel) is similar to a dry episode that occurred during most of the first half of the 1800s (Nicholson, 2001; see Section 11.3.1 also). However, it is readily apparent that human impact on the landscape has aided the decline of forests and other vegetation during these droughts through soil compaction by domestic animals and by gradual clearing of trees for firewood.

As mentioned in Section 3.3, forests, in repeatedly recycling the amount of water passing from the oceans to the land, increase rainfall on a continental scale. This is a major aspect of the reasoning behind present attempts to re-afforest the Sahel region. Mixed plantings of eucalypts (particularly *Eucalyptus camaldulensis*) and nitrogen-fixing acacias are proving most useful in stabilizing soils and thus helping to bring land back into production, although there is suggestion that the eucalypts can lead to soil acidification. *Acacia senegal* and *A. albida* are particularly useful as crops.

## 11.3 Climate change

### *11.3.1 Global warming*

The climate of the Earth has changed naturally many times since the origin of life at the beginning of the Cambrian. The present global warming (the temperature of the Earth has increased during the twentieth century by 0.76 °C) differs from those of the past in that it has almost certainly been initiated by the activities of humans. During the twentieth century the atmospheric concentration of carbon dioxide ($CO_2$) rose by 35% from 280 parts per million (ppm) to 379 ppm in 2005. The concentrations of other greenhouse gases have also risen since 1750, the beginning of industrialization, notably methane (151% increase) and nitrous oxide (18% increase). While there is general agreement that climate is changing, the amount of change we can expect is not as certain. Links between changes in the Earth's atmosphere and global warming have been investigated by mathematical modelling. Although the final results are difficult to predict with accuracy, in 2007 the Intergovernmental Panel on Climate Change estimated that the long-term increase in surface air temperatures would by 2100 be between 1.8 and 4.0 °C with a best estimate of 2.5 °C. Houghton *et al.* (2001) contains more detail on predicted changes. The expected changes will not be felt uniformly over the whole globe. Northern latitudes are expected to show greatest temperature changes and in the UK, south-east England will show the greatest change in rainfall, with 20–30% less in summer and 10–15% more in winter than at present. Good news for the tourist industry but the increase in summer droughts is likely to be bad news for forests since we know that the droughts of 1976 and 1995 in the UK led to earlier defoliation of trees and shrubs as well as the death of drought-intolerant trees.

Climate change is not only a fact; its onset is rapid, and as the world becomes warmer the sea level is rising faster than had been expected. Sea level rose 0.1–0.2 m during the twentieth century and is projected to rise somewhere between 0.18 and 0.59 m by 2100 as the Arctic and Antarctic ice caps melt. The Thames barrier, built to protect London after the loss of 300 lives due to flooding in 1953, is the largest structure of its kind in the world. It had allowances for sea-level rise built into it, but the increasingly stormy weather and greater than expected sea-level rise caused it to be used many more times than was expected by 2003, when plans to raise its height by more than a metre had to be commissioned. Sea-level rise and the possibility of **tsunami**, huge and unusual ocean waves sweeping round the world, will have a considerable influence on both low-lying coasts and coastal forests.

When considering the influence of global warming on forests, it is important to bear in mind that since trees are long-lived, many forests reflect past climate changes over the last millennium. Gillson and Willis (2004) point out that trees that are more than 150 years old established during the Little Ice Age (1590–1850) and reflect a climate long gone. So if the old trees are failing to regenerate, or the forest is changing, it may be a reflection of changes in past climate rather than a present-day 'problem'. Due to this, some of the current changes we are seeing in forests can be hard to interpret. Nevertheless, it does seem that forests are already responding to global warming, especially in the northern hemisphere where the impact of climate change is largest. Many examples could be selected from the published literature to illustrate current and predicted changes, and the underlying mechanisms. The overall picture is complex given that the responses of forests and individual species occur at different spatial and time scales.

Some changes are readily seen on the small scale. On the basis of an investigation of a late-successional hemlock–northern hardwood forest, Woods (2004) suggests that climate change may induce changes in forest ecosystems by changing disturbance patterns, with many more instances of **intermediate disturbance** due to storms. An intermediate disturbance is one that has a marked influence on an ecosystem without destroying all the individuals present, as occurs when fire kills all the trees in a particular area. Tree mortality in the Dukes Research Natural Area in Michigan, before and

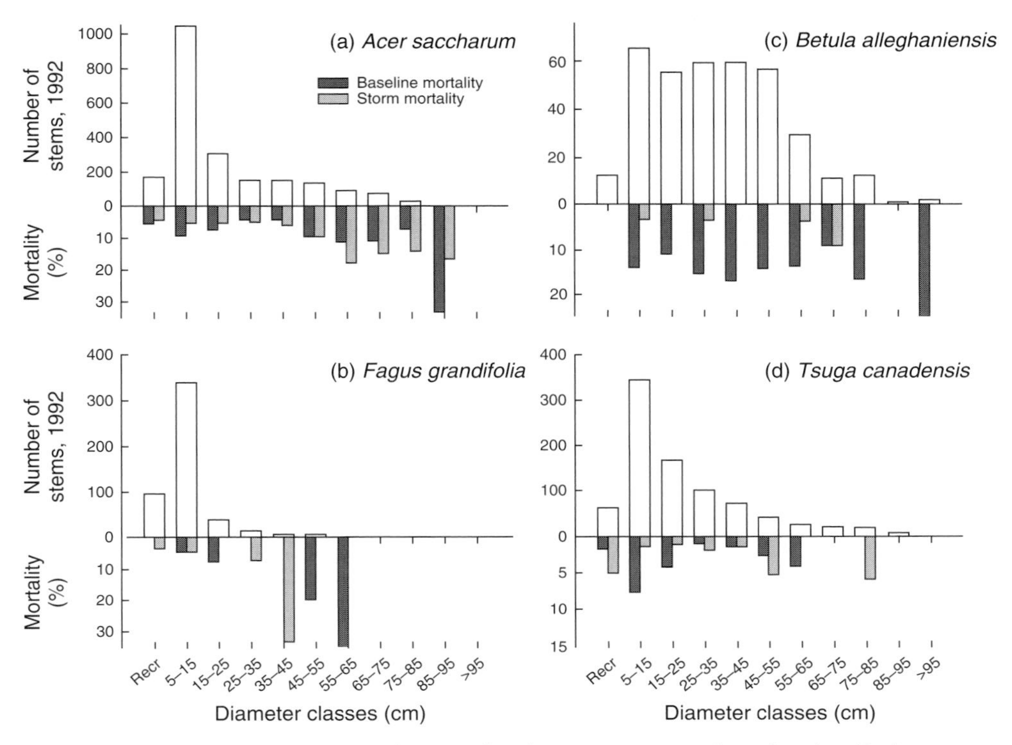


Figure 11.3 Size structure by species in permanent plots in the Dukes Research Natural Area, an old growth mesic hemlock–northern hardwood forest in northern Michigan, USA. Ascending bars show distribution of live trees in 10 cm dbh (diameter at breast height) classes in 1992. Descending bars show mortality as percentage of size class; lighter shading indicates baseline (1992–2002) mortality, darker shading indicates mortality due to a storm in July 2002. Open-ended bars extend below the graphed range. The 'Recr' category includes stems reaching 5 cm between 1992 and 2002. (From Woods, 2004. *Journal of Ecology* 92, Blackwell Publishing.)

after a severe storm in July 2002 showed that it did not influence all tree species and all sizes of tree similarly (Fig. 11.3). Yellow birch *Betula alleghaniensis* was little affected by the storm but had a high baseline mortality (normal losses during the period 1992–2002). Losses of both types were low in eastern hemlock *Tsuga canadensis*, while the shade-tolerant maples, especially sugar maple *Acer saccharum*, and American beech *Fagus grandifolia* had similar patterns of baseline and storm mortality with storm mortality being higher for the larger stems.

The storm of 2002 thus reduced the dominance of the larger shade-tolerant trees, caused changes in canopy structure and distinctively affected the understorey and ground flora by letting in more light. This work demonstrates that the consequences of rare, intermediate disturbances can differ markedly from what would be predicted by simply scaling up patterns resulting from frequent, less intense events. A feature of most climate-change scenarios is that the rare events will become more common. When these are added to the likelihood that severe disturbances such as fire and insect attack are also likely to become more common (see Dale *et al.*, 2001 and Logan *et al.*, 2003 for reviews), and changes in vegetation may exacerbate temperature increases (for instance, vegetation growing further north would absorb more heat – see Foley *et al.* 2003 for a review) the predicted large-scale effects of climate change become easier to believe.

In other cases, changes to the forest may be more extreme, involving the loss of key species or even whole forests. This is likely where the forest or species are closely tied to specific climatic conditions, show great sensitivity to climate or are currently living near the edge of their environmental tolerance. For example, in southern Germany predicted changes will result in hot dry summers but with a wetter spring and severe flooding. Beech *Fagus sylvatica*, an important forest tree species in this region, is drought and flooding sensitive and is already showing reduced growth and competitive ability especially as seedlings on extreme sites. Beech is likely to further suffer since it has limited capacity to respond to increased $CO_2$ levels compared with other forest trees. It is predicted that if climate-change scenarios are correct then beech will disappear from southern Germany (Rennenberg *et al.*, 2004). Tropical cloud forests, growing in altitudinal bands of cloud formation on mountains, rely on regular immersion in low-lying clouds for their existence and are a classic case of forests at high risk from climate change. Under current scenarios, there will be less cloud and it is likely to shift higher up the mountains. This will upset the delicate equilibrium of these very high biodiversity forests and will require species to move vertically (that is, if they have not already reached the top of the mountain) and between mountain peaks. Foster (2001) points out that while this has certainly happened in the past, mature cloud forest may take 200–300 years to develop and the suitable climate envelope may have moved on by then given the fast rate of climatic change predicted. Moreover, other climate change threats such as the incidence of typhoons may increase which would be particularly damaging to cloud forests since gaps in the canopy are very slow to regenerate.

To add spice to the arguments, recent discoveries imply that the climate of Europe could quite suddenly become much colder, as it has done in numerous past ice ages. The cause lies in the increasing dilution of the waters of the North Atlantic as the Greenland ice sheets continue to melt at an ever-increasing rate and more fresh water pours into the sea from Siberian rivers fed by higher

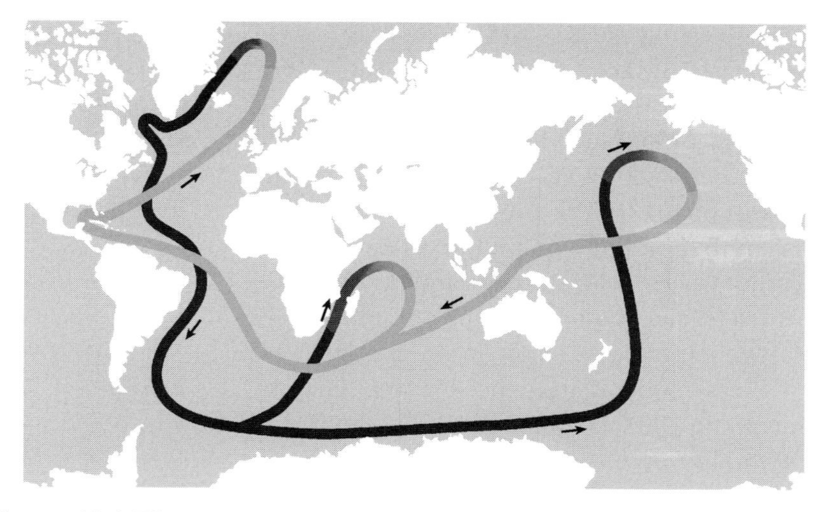

Figure 11.4 The great ocean conveyor current that joins warm surface currents (light grey) to cold ocean floor currents (dark grey). The flowing of this giant conveyor current is dependent upon warm water cooling by Greenland and sinking, but if too much melting ice mixes with the seawater it may become too buoyant to sink, causing the conveyor to halt irreversibly, with dire consequences to the climate of western Europe and beyond. Arrows show direction of flow.

rainfall. The present mild climate of Britain and western Europe is mediated by the continual warming provided by the Gulf Stream and its continuation, the North Atlantic Drift. This water eventually cools, sinks to a great depth and is returned to the tropics by the so-called **conveyor current** (Fig. 11.4). The problem is that if this salt water is diluted it fails to sink as effectively and the conveyor current, which has slowed appreciably and faltered over the past few decades, could eventually halt as we now know it has in the past. The measured salinity of this water is falling and there is a real risk that the conveyor counter current, and consequently the Gulf Stream, will halt long-term, although recent predictions suggest this is unlikely to happen within the modelling time frame between now and 2100 (Houghton *et al.*, 2001). If and when it does halt, however, Europe may again have a climate like that of Alaska, which is at similar latitudes, with sea ice off the coast of England. This would not be a local change: the climate of the whole world would be affected and the connected loss of monsoon rains would cause tropical forests to change into grasslands.

It has become accepted that atmospheric pollution is responsible for not just global warming but also another major effect, that of cooling. **Global dimming**, coined by Stanhill and Cohen (2001), causes less sunlight, between

9 and 30% in various parts of the world, to reach the surface of the Earth than would be the case if the atmosphere was not suffering from particulate pollution resulting from the emissions of power stations, surface vehicles and aircraft, as well as others such as forest fires. After the tragedy of 11 September 2001 there were virtually no flights by commercial aircraft in the USA. This diminution in pollution alone was enough to increase the local daily temperature range in that period by an unprecedented 1 °C.

Meteorologists have for many years, in some cases for as long as a century, kept a record of **pan evaporation**, the amount of water that has to be added to bring the open surface of water in an exposed vessel to the same level each day. Pan evaporation is influenced particularly by the amount of sunlight as the incoming photons virtually kick water molecules into the air, with relative humidity and wind force as subsidiary factors. Temperature has a still smaller effect. All over the world pan evaporation diminished during the twentieth century (by 22% between the 1950s and the 1990s in Israel) and a few scientists eventually realized that the cause was that atmospheric pollution changes cloud structure. Normal clouds consist of relatively few large droplets of water whereas polluted clouds consist of roughly ten times as many much smaller droplets centred around particles of ash, soot, sulphates, nitrates and other particulate impurities. While sunlight passes through normal clouds, these polluted clouds act as mirrors reflecting it back into space.

Global dimming is much stronger in the northern than the southern hemisphere and it has been responsible for a cooling effect so strong that it resulted in a further Sahel famine in 1984 (see Section 11.2 also), following failure of the annual monsoon for several years. If the Asian monsoon were to fail the effects on the world population would be even worse. Europe has made a big effort to cut down atmospheric pollution in recent years, causing a reduction in global dimming in areas downwind of it (see Roderick, 2006). Global dimming has the potential to cause huge crop losses, yet without it the effects of global warming would be far greater than they already are. Unless the countries of the world realize this and take effective action, the outlook for the forests of the world is bleak indeed.

### *11.3.2 Phenological changes: springs are getting earlier*

**Phenology** is the study of natural phenomena recurring each year, especially in relation to climatic conditions (Section 4.6.1). Records of this type go back a very long way, indeed in Japan and China some of the dates of flowering of cherry and peach trees are known from the eighth century. The earliest known

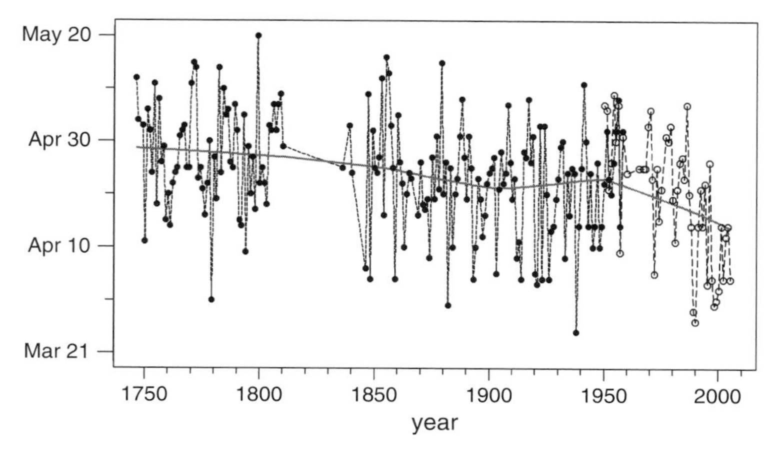


Figure 11.5 First leafing time series for oak in Norfolk and Surrey, provided, respectively, by the Marsham and Combes records. A smoothed (LOWESS) line has been superimposed. (Courtesy of Tim Sparks, Centre for Ecology and Hydrology, Monks Wood.)

surviving series in the UK was commenced by Robert Marsham FRS in 1736. After his death in 1798 records of the first leafing of the trees on his estate in Norfolk were continued by successive generations of his family. Figure 11.5 shows the Marsham first leafing dates for oak, the final records for the twentieth century being provided by Jean Combes, who has kept a summary of the leafing of oak, ash, lime and horse chestnut since 1947. Phenology is becoming increasingly important in an era of global climate change, and it is realized that phenological events, especially those in spring, can be extremely sensitive to climate fluctuations (attributable to more pronounced changes in winter and spring temperatures). A Royal Meteorological Society network kept phenological records for very many years (1875–1947). The UK Phenology Network, organized in 1998 with the support of the Woodland Trust, has effectively replaced it. The new network had more than 2000 registered recorders in 2005 and more than one million records in its database.

The season of spring, occurring as it does between winter and summer, technically begins on the day when the sun is over the equator (21 March in the northern hemisphere, 21 September in the southern). In practice, and particularly in the public mind, the start of the season is associated with particular biological events. In central and northern Europe it coincides with the arrival of the first skylark *Alauda arvensis*. This bird remains a resident in mild British winters, so in Britain it is the appearance of the first flowers, such as the snowdrop *Galanthus nivalis* and primrose *Primula vulgaris*, or the first budburst of various trees and shrubs, that are often used as the biological indicators of this important event. Data presented by Sparks (2000) and by Sparks and Smithers (2002) demonstrate quite clearly that, over very long periods, higher temperatures early in the year consistently led to earlier flowering in woodland herbs such as wood anenome *Anemone nemorosa* and earlier budburst in trees such as oak *Quercus* spp., hawthorn *Crataegus monogyna* and blackthorn *Prunus spinosa*.

There is thus no doubt that higher temperature in the early part of the year (or possibly a climatic variable correlated with temperature) causes phenological events, including first observations of particular insects such as the brimstone butterfly *Gonepteryx rhamni*, to occur earlier than they otherwise would. Observations since 1950 in Northumberland, and since the early sixties in Norfolk, show that first flowering dates for snowdrop have become increasingly early. When a smoothed line is applied to the data, which oscillate somewhat from year to year, it can be seen that whereas snowdrops in Northumberland flowered around 24 February in 1950, by 2000 they were doing so on 20 January. In warmer Norfolk this had advanced to the end of the first week of the year. Oak, which in Surrey was leafing around 2 May in 1950, was by 2000 doing so in the first week of April. The average date of first flowering of 385 British plant species was found by Fitter and Fitter (2002) to have advanced by 4.5 days during the past decade compared with the previous four decades. Of these, 16% of species showed an average advancement of 15 days in a decade, while 3% of species flowered significantly later in the 1990s than previously. There are fewer data for autumn events but these suggest an elongation of the growing season at that end of the year as well.

Anecdotal evidence suggests that the variation between years is also becoming more pronounced. The weather in the winter of 2002/3 in the English Midlands was quite extraordinary in this respect. Following an autumn so mild that the leaves of many trees were retained for much longer than usual, February was unusually cold with 21 frosty nights. The summer of 2003 on the other hand, was so very hot that Batsford Arboretum in Gloucestershire arranged tours demonstrating autumn leaf colours a fortnight earlier than usual. Moreover, the European winter of 2005/6 was very severe, greatly delaying foliage development of herbs, shrubs and trees. Climate warming is indeed associated with erratic short-term weather events, including wind storms.

The above data give clear evidence that warming is undoubtedly occurring and virtually all natural events, including the times at which birds nest, butterflies emerge, amphibians breed and birds migrate, are taking place earlier in the year. That much is certain, but it leads to the further question of how global warming will affect the distribution of species.

### 11.3.3 How will global warming influence the distribution of forest plants and animals?

As global temperatures increase, there will be a **shift of species ranges** towards the poles and towards higher altitudes based on species' **climatic thresholds**. In England acceptance of this factor has already led to change in the official attitude towards the planting of beech *Fagus sylvatica* (Section 10.4) with more being planted towards the north end of its range in northern England and Scotland. The USA Forestry Service, concerned about future changes in natural tree distribution, has devised a computer matrix simulation model called SHIFT whose function is to estimate potential migration of five major tree species due to global warming in the next 100 years (Iverson *et al.*, 2004). The species concerned are persimmon *Diospyros virginiana*, sweetgum *Liquidambar styraciflua*, sourwood *Oxydendrum arboreum*, loblolly pine *Pinus taeda* and southern red oak *Quercus falcata* var. *falcata*. The model allows very long-distance events in which colonization resulting from seed dispersal by wind or animals could occur up to 500 km beyond the present distribution boundary. The abundance of these species near their range boundaries proved to have more influence than percentage forest cover in influencing migration rates. Though migration patterns for the five species differed, model outputs predicted migration of a generally limited nature for all of them over the next century. In this period there was a high probability of migration into zones 10–20 km distant from present boundaries, but very little of a greater migration attributable to climate change, though some remote outliers of particular species might result from rare long-distance dispersal of tree seeds.

Each species will respond to its own environmental needs and tolerances, so biomes and forest types as we know them may alter considerably as some species are left behind or lost and others are acquired (Walther, 2003 gives a more detailed review).

### 11.3.4 Carbon sequestration by forests

One strategy for reducing the effects of climate change is to remove $CO_2$ from the atmosphere by locking up (sequestering) carbon in vegetation. Since forests cover around one-third of the Earth's land area and store more than 80% of terrestrial carbon, they are the prime candidate for absorbing extra carbon. Pastures on the other hand not only have a lower carbon density, they contribute both nitrous oxide and methane (from their livestock), which make a considerable contribution to global warming. The startling discovery by Keppler *et al.* (2006) that vegetation, especially tropical forests and grasslands,

Table 11.1. *The amount of carbon (Gt) stored in the three main types of forest and the amounts found above ground in the vegetation and below ground in the soil. Pg or petagrams, $10^{15}$ g are the same as Gt, gigatonnes, $10^{9}$ t or a thousand million tonnes. Some carbon is also found in other forest types, such as tropical savannas, which has been ignored here for simplicity.*

| | Vegetation | | Soil | | Total | |
|---|---|---|---|---|---|---|
| Forest type | Gt | % | Gt | % | Gt | % |
| Tropical | 212 | 59 | 216 | 27 | 428 | 37 |
| Temperate | 59 | 16 | 100 | 13 | 159 | 14 |
| Boreal | 88 | 25 | 471 | 60 | 559 | 49 |
| Total in forests | 359 | 100 | 787 | 100 | 1146 | 100 |
| Total in all terrestrial vegetation | 466 | | 2011 | | 2477 | |
| Per cent in forests | 77% | | 40% | | 46% | |

*Source:* (Data from Houghton *et al.*, 2001. *Climate Change 2001: The Scientific Basis.* Cambridge University Press.)

appear to produce around 149 000 Gt (gigatonnes) of methane annually (see Table 11.1 for a comparison of units), a potent greenhouse gas, led to speculation that forests may not be as valuable in fighting climate change as was thought. However, Frank Keppler and his colleagues have pointed out that methane production would reduce the value of carbon sequestration of new forests by less then 4%.

In Chapter 8 the concept of ecosystem productivity was explored. This shows that once a forest reaches maturity as much carbon is lost in respiration in a year as is captured by the growing vegetation. Thus, in theory, some carbon is withdrawn from the atmosphere as a forest grows, but once it is mature it becomes carbon neutral. Empirical evidence shows, however, that on a global scale temperate and northern forests are a net sink of carbon, i.e. they are sequestering extra carbon. The cause of this increase has been extensively debated and has been put down to two main possibilities: **regrowth or enhanced growth**. The initial idea was to attribute the extra sink to **carbon dioxide fertilization**, an increase or enhancement in forest growth stimulated by the extra carbon dioxide in the atmosphere, making the trees bigger. This makes intuitive good sense since $CO_2$ is a key ingredient in photosynthesis and it has long been known that growth can be stimulated in glasshouses by pumping in extra $CO_2$. Many experiments were carried out under field conditions to see if the same thing happened in the wild. It is now clear that while some trees and forests show increased growth this tends to be a short-term increase and within a few years, growth rates have fallen back to previous levels. The poor fertilization effect is usually because growth is limited by other environmental factors such as increasing temperatures and drought causing stress, and low nitrogen availability. As discussed in Chapter 8, nitrogen is often in short supply and, at least in temperate forests, is the main factor limiting growth. Carbon dioxide fertilization may initially lock up extra carbon in new wood and in the soil, but it will also lock up more nitrogen causing a further shortage and reduced growth. In the large areas of forests where nitrogen is an abundant pollutant (see Section 11.4.3 below) this problem should be removed but studies in eastern North America and Europe have shown that, despite extra growth, $<10\%$ extra carbon is sequestered in woody tissues (Nadelhoffer *et al.*, 1999). Since leaves, twigs and fine roots are short-lived and then decompose, it is the carbon in woody tissue with a longer turnover time that is of value. Moreover, excessive nitrogen can lead to decline in forest growth and even wholesale death (see Section 11.4.3).

In conclusion, it seems that global forests are sequestering carbon not so much because of **enhanced growth** but due to **regrowth** (although the debate is not completely settled – e.g. Houghton, 2003). The net carbon sink (sequestration) by forests appear to be primarily due to afforestation (planting trees on new ground), reforestation (replanting former forests) and, most importantly, the recovery of existing forests from historical land uses (i.e. the forests are still growing back to maturity and so have a positive net ecosystem productivity). In 1996 the Intergovernmental Panel on Climate Change (IPCC) suggested that 700 million hectares of forestland might be available for sequestration globally, due to slowed deforestation (138 Mha), regeneration of tropical forests (217 Mha) and plantations and agroforestry (345 Mha). They suggest that over the period until 2050, sequestration rates could reach a maximum of 2.2 gigatonnes (Gt), compared to the net increase in atmospheric carbon of 3.3 Gt per year during the 1990s. Beedlow *et al.* (2004) provide a good readable overview of this subject and underlines that we must protect existing forests if we are to maintain the forest carbon pools and perhaps even sequester more carbon. Gains can also be made by using woody material, particularly forest thinnings, as **biofuel** to replace fossil fuels. **Short-rotation coppice** (SRC) based on poplar or willow varieties cut every 1–3 years could also be employed though its productivity is not that high. In Europe it is estimated that the total carbon pool (above and below ground) in a mixed hardwood forest after 150 years is in the order of 324 t $ha^{-1}$ $y^{-1}$ compared with 162 t $ha^{-1}$ $y^{-1}$ in a SRC. However, when carbon savings from not using fossil fuels are taken into account, the effective annual sequestration of carbon in SRC is 24–29 tonnes of $CO_2$ per hectare compared to 6–7 t $CO_2$ $ha^{-1}$ in mixed forest (Deckmyn *et al.*, 2004). Growing enough biofuel to be

### Box 11.1 Chronic Nitrogen Amendment Study at Harvard Forest, USA

Starting in spring 1988, a series of long-term experiments was set up as part of the Harvard Forest Long-Term Ecological Research (LTER) project. Two adjacent stands were used: an even-aged red pine *Pinus resinosa* stand planted in 1926 and a 50-year-old mixed hardwood stand. The hardwood stand was dominated by black oak *Quercus velutina* and red oak *Q. rubra* with black birch *Betula lenta*, red maple *Acer rubrum* and American beech *Fagus grandifolia*. Within each, three main plots were established: control (with no extra nitrogen added), low N (50 kg of N added $ha^{-1}$ $y^{-1}$) and high N (150 kg N $ha^{-1}$ $y^{-1}$). An additional plot of low nitrogen with added sulphur was used for a few years but proved similar to just low nitrogen so the sulphur treatment was abandoned. The background rate of nitrogen deposition is 7–8 kg $ha^{-1}$ $y^{-1}$ (moderate by north-eastern standards) so the additions of ammonium nitrate fertilizer ($NH_4$ $NO_3$) correspond to approximately 6 and 18 times the current background rate. Bear in mind that the natural deposition level, without any pollution, is around 1 kg $ha^{-1}$ $y^{-1}$.

The two stands reacted to the nitrogen in different ways. After 14 years, the mortality of the pine trees (calculated as the fraction of biomass in dead trees) rose with nitrogen levels (control 12%; low N 23%; high N 56%) and was associated with reductions in foliage and wood production, leaf area (see photographs accompanying this box) and photosynthetic capacity. Complete mortality in the high N stand is likely in the near future. In the hardwood stands, however, mortality was lowest on low N treatment (control 27%; low N 19%; high N 49%) and growth of the surviving trees on the high N plots was 30% higher than on the control over 14 years and on the low N plots was 11% higher. Polyamine production (particularly putrescine) is known to increase in cells subjected to abiotic stress (such as low pH, nutrient stress, low temperatures). In the pine plots putrescine level in the foliage was higher in both nitrogen treatments as compared with the control plot. In the hardwood stand, it was higher only in the high N treatment (Minocha *et al.*, 2000), giving additional evidence that the pine stand was under more stress. However, the high N input is having an effect in the hardwood stands because all the white birch, 73% of the red maple and 58% of the black birch had died by 2002, suggesting environmental stress. This strongly suggests that while any addition of nitrogen has been harmful to the pines, the hardwoods benefited from low N addition because the initial site conditions were limiting to growth. This was mirrored in above-ground production. In the pine stands the mean annual wood production was highest in the control and decreased 31% and 54% relative to the control for the low N and high N plots, respectively.

Losses of inorganic forms of nitrogen were high in the high N plots (higher in pines than hardwoods) but dissolved *organic* nitrogen (DON – see Section 8.4.2) showed no changes over time. This may well be due to inorganic nitrogen being

Box 11.1 Chronic Nitrogen Amendment Study at Harvard Forest, USA. Chronic nitrogen plots in red pine (top to bottom): high nitrogen, low nitrogen, control. (Photographs taken December 2004 by Peter A. Thomas.)

**Box 11.1 (cont.)**

converted to DON by the process described in the ferrous wheel hypothesis (see Section 8.4.2). Although nitrogen was lost, it was not as much as was expected: 85% was retained over the first 6 years. The cause is still being explored.

Active fungal biomass (i.e. biomass that was functional) was reduced in both stands but more so in the pine stands: biomass was 27–62% (low – high N plots, respectively) lower than the control in the hardwood stand and 42–69% lower than the control in the pine stand (Frey *et al.*, 2004).

Further background information can be obtained from Magill *et al.* (2004). This and 11 other papers make up a special section in *Forest Ecology and Management* (2004, **196**, 1, pp. 1–186) entitled: 'The Harvard Forest (USA) nitrogen saturation experiment: results from the first 15 years.'

An earlier series of 29 papers on nitrogen enrichment in Europe can be found in *Forest Ecology and Management* (1998, **101**, 1–3, pp. 1–363) entitled: 'The whole ecosystem experiments in the NITREX and EXMAN projects', including Boxman *et al.* (1998). (**NITREX** = Nitrogen saturation experiments; **EXMAN** = Experimental manipulation of forest ecosystems in Europe.)

There is also a series of 5 papers published in *BioScience* (2003, **53**, 4, pp. 341–420) on nitrogen pollution problems in the USA (including Galloway *et al.*, 2003).

self-sufficient in energy would require prohibitive amounts of space but using energy from biomass could reduce carbon emissions in EU countries by 2–30% (Schwaiger and Schlamadinger, 1998).

About two-thirds of the global pool of forest carbon is underground in roots and soils, amounting to 787 Gt (Table 11.1) and thus much research has been aimed at understanding the dynamics of soil carbon, in particular the carbon found in the passive carbon pool (Section 7.6.1) with its very slow turnover. Tropical forests form 42% of the area of all forests but hold only 27% of the soil carbon (Table 11.1). Rather it is the northern forests making up 33% of all forests that hold 60% of the soil carbon. But they are not necessarily sponges ready to soak up extra carbon. In fact both tropical and northern soils are vulnerable to carbon loss because northern forests contain so much carbon (and warming is predicted to be greatest nearer the poles) and in tropical forests small changes in temperature may be enough to degrade greatly the forest and lead to carbon loss (Attiwill and Weston, 2001). Moreover, climate warming is thought to reduce soil carbon by increasing decomposition in the soil. Experiments using heating cables in a northern hardwood forest indicate that decomposition is increased exponentially with increasing soil temperature (McHale *et al.*, 1998). However, it is unclear whether this would continue once the more readily decomposed carbon has gone.

Recent evidence from nitrogen fertilization studies in temperate forests indicate that soils rather than plants are the dominant long-term sink of applied nitrogen. But sequestration of nitrogen in soils promotes less carbon storage than would plant uptake and so nitrogen pollution may diminish the ability of forest soils to sequester and hold carbon (Nadelhoffer *et al.*, 1999).

## 11.4 Other causes of forest decline

### *11.4.1 Effects of plant strategies on forest stability and sustainability*

Woodlands and forests, like all ecosystems, are subject to climatic and biotic impacts that threaten their viability and survival. We read almost daily of areas of forest suffering from drought, frost, fires, floods or windstorms on the one hand, or over-grazing, disease, cultivation, pollution or urbanization on the other. There are two key aspects to the response of an individual plant or stand of vegetation to such an extreme event, for which the terms resistance and resilience were coined by Westman (1978). **Resistance** is the ability to resist displacement from its current state. **Resilience** is the speed and completeness of the return to the original state (sometimes referred to as elasticity). The two terms are included in the overarching term of **stability**, defined as the ability of a forest to resist permanent change (the relationship between stability and diversity is discussed in Section 9.5). As Grime (2001) emphasizes, when ecosystems of any sort are exposed to an increasing severity or frequency of extreme events due to such things as climatic change or rising human population density, attention switches from an interest in how the ecosystem reacts to a concern with its sustainability. There may come a point where a forest is so repeatedly disturbed or altered that it cannot return to its original state and may even be permanently changed. A very severe storm or severe fire could, for example, alter the species composition of a forest ( despite high resistance) but the forest might return to its original condition within a short time (high resilience). However, frequent storms or fires may give insufficient time for recovery before the next disturbance so the forest never gets beyond an early stage of recovery or the trees may fail altogether and a different habitat develops such as savanna or grassland. Such a change is readily seen with heavy grazing which prevents tree regeneration, eventually leading to a grassland habitat once the existing trees die. Even if the forest does not completely disappear, at what point is the essential character of a woodland or forest ecosystem so impaired that its value is irreparably degraded?

The opportunities for analysis and prediction of ecosystem resistance, resilience and sustainability that arise from recognition of CSR plant strategies by Grime

(2001) – see Section 4.1 for more detail – makes fascinating reading; here there is space only for discussion regarding wooded ecosystems. The four major points made here by Grime apply particularly to these very systems:

(1) Remnants of older landscape such as ancient woodland or old-growth forests, which are usually highly diverse and of lower productivity than their modern counterparts, will continue to diminish in size and become progressively more isolated from similar landscape fragments. The classical island biogeographical theory (MacArthur and Wilson, 1967) then applies. Floras and faunas of such relict systems will lose diversity as species extinctions fail to be compensated by immigrations from remote 'nearest neighbour' populations.
(2) In ecosystems dominated by slow-growing, long-lived plants such as forests, resistance to change is considerable. Once resistance is broken and population reductions and extinctions occur, however, the low reproductive output, limited dispersal range and long juvenile phase of many stress-tolerant species reduce the prospect of successful recovery. Many of the plants and some of the animals that exploit the deteriorating fragments of isolated forests are able to do so because of their life histories and functional characteristics that allow long-distance dispersal and rapid invasion.
(3) Conditions in forests enriched by humans or intensive exploitation are very different from those in fragments of old forests of low productivity. In richer forests there is a more dynamic flux, in space and time, of plant and animal populations with 'weedy' (i.e. R or C) traits. Vegetation processes often proceed rapidly and tend to be constantly returned to an early successional stage by eutrophication (nutrient enrichment) and disturbance in both industrial and agricultural areas.
(4) Species capable of rapid colonization are often already present in a forest at low frequency or in nearby areas and consequently are in a position to cause quick and unpredictable changes in the flora and fauna once a landscape is modernized. Rhododendron *Rhododendron ponticum* in the UK is a classic example of an alien woodland invader, but the course of invasions by such species remains difficult to predict and there is much to learn regarding the invisibility of plant invaders and the conditions that allow a low-frequency invader to explosively take over.

### *11.4.2 Pollution, acid rain and forest decline*

Acid rain became a problem when combustion of fossil fuels increased after the industrial revolution, releasing increasing quantities of sulphur dioxide ($SO_2$), nitrogen oxides ($NO_x$ – see next section for a definition) and other acidifying particles into the atmosphere to join what is naturally produced. The sulphur dioxide and nitrous oxides dissolve in atmospheric moisture to form sulphuric ($H_2SO_4$) and nitric ($HNO_3$) acids which are brought to ground as acid precipitation. Unpolluted rain has a pH of 5.6 (on the acidic side of chemical neutrality due to $CO_2$ dissolving in it to form weak carbonic acid) while acidic rain usually has a pH between 4.0–4.5 with extreme cases of pH 2.0 (more acidic than lemon juice) occasionally reported. Acid rain effects became apparent in the 1970s, and by the 1980s acid rain was widespread around the globe and particularly bad in eastern North America, north-west Europe, eastern Asia and parts of the southern hemisphere such as Brazil. Although the acidic components are generated mainly by industrial centres, long-distance transport has resulted in widespread problems.

Sulphur dioxide emissions, the main cause of acid rain, are reducing in many countries due to legislative initiatives and the improvement of technology in industry. Nitrous oxide emissions, however, are on the whole less regulated and are increasing globally. They are thus likely to make up the main component of acid rain in the future, supplanting sulphuric acid (see Likens (2004) for a discussion on legislative changes on emissions in the USA). In the early days of taking acid rain seriously, it was hotly debated whether sulphur dioxide was ultimately responsible for acid rain. The long-term studies at the Hubbard Brook Experimental Forest (HBEF – see Box 8.1) have shown categorically that concentrations of sulphur dioxide in precipitation and stream water are correlated with emissions of sulphur dioxide upwind of HBEF (Likens *et al.*, 2002). Similarly, atmospheric depositions of nitrate ($NO_3$) at HBEF are correlated with nitrous oxide emissions.

Acid rain acidifies those terrestrial and aquatic ecosystems that are not sufficiently buffered causing loss of species and accelerated leaching of nutrients particularly calcium and magnesium, but also potassium and sodium. At HBEF between 1940 and 1955 precipitation supplied 29% of the calcium required to balance that taken up in biomass and lost in streamflow (the rest was made up from weathering and from the soil storage complex). By 1976–1993 precipitation was supplying only 12%. Around a half of the pool of available calcium in the soils at HBEF has been depleted during the past 50 years by acid rain. This loss leads to a lack of buffering capacity to absorb further acidity, so forest and aquatic ecosystems become more sensitive to further damage by acid rain. The effects of acid rain on such things as fish and trees are well-known but the effects are more widespread. For example, acidification of soil leads to a loss of soil fauna with a consequent reduction in the rates of organic matter decomposition. Larger animals are not immune: calcium shortage can limit reproduction of passerine birds (small perching birds living near the ground). Tilgar *et al.* (2002) found that great tits *Parus major* in coniferous and hardwood stands in Estonia laid more eggs and raised more and bigger fledglings when calcium was abundant.

By the 1980s there were widespread fears that acid rain was leading to wholesale **forest decline**, partly driven by the severe die-back of forests seen in parts of

Germany (termed 'Waldsterben', meaning forest death). By the late 1980s most European countries had more than 15% of trees with severe defoliation (defined as trees with crown densities less than 25% that of healthy trees). The causes of this decline, while so apparent to the media, were in reality not so straightforward. There are certainly direct effects of the acidity leading to calcium and magnesium loss and aluminium toxicity. In addition, the abundant sulphur and nitrogen causes imbalance (stoichiometric problems) that results in chlorosis and potentially death. The problem may also be exacerbated by high nitrogen input (discussed below) which increases growth, puts greater demands on magnesium and produces softer growth more at risk from insects, pathogens, frost and drought. High ozone levels are also known to directly damage leaf structure. Finally, the hot dry summers and cold winters of the 1970s and 1980s may also have taken their toll. The frustrating part is that direct cause and effect are difficult to pinpoint on already acidic soils.

Fortunately the effects of acid rain, in terrestrial ecosystems at least, have been less serious than originally feared, and by the 1990s recovery seemed evident in many forests as sulphur emissions declined. Bear in mind that at the height of concern, severe decline covered 8000 $km^2$ of Europe which amounts to less than 0.5% of the forest area. Acid rain sceptics also point out that some forests were showing improved growth during this period. Between 1971 and 1990 European forests increased their growing stock (total volume of timber) and growth rates (wood produced per year) by 30% (Kauppi *et al.*, 1992). Other forests are not faring as well; forest biomass increase at HBEF has declined to a small rate since 1987 (Likens *et al.*, 1996). Calcium is undoubtedly being lost with potentially dire consequences (Fig. 11.6) but there is evidence that young hardwood stands and conifer stands are able in some way to access calcium from pools not traditionally considered available (Hamburg *et al.*, 2003; Jandl *et al.*, 2004), so the potential for acid rain to cause depletion problems may be less than feared. Levels of soil phosphorus are often a limiting factor in natural ecosystems, as discussed in Section 8.4.3. Unlike nitrogen, this element cannot be supplied by fixation from the atmosphere, and thus limited supply is a quite fundamental feature of many woodlands, particularly as soil phosphate in ancient woodland soils is normally less than that in secondary woodland.

Despite the fact that acid rain has ceased to be front page news, it is still a long-term concern in as much that it is one more pressure on forests. Moreover, much of the emphasis has been on reduced productivity and mortality of the trees. Acid rain and other pollution such as nitrogen enrichment (see below) at sublethal levels may have more far-reaching effects on species composition leading to large changes in native vegetation.

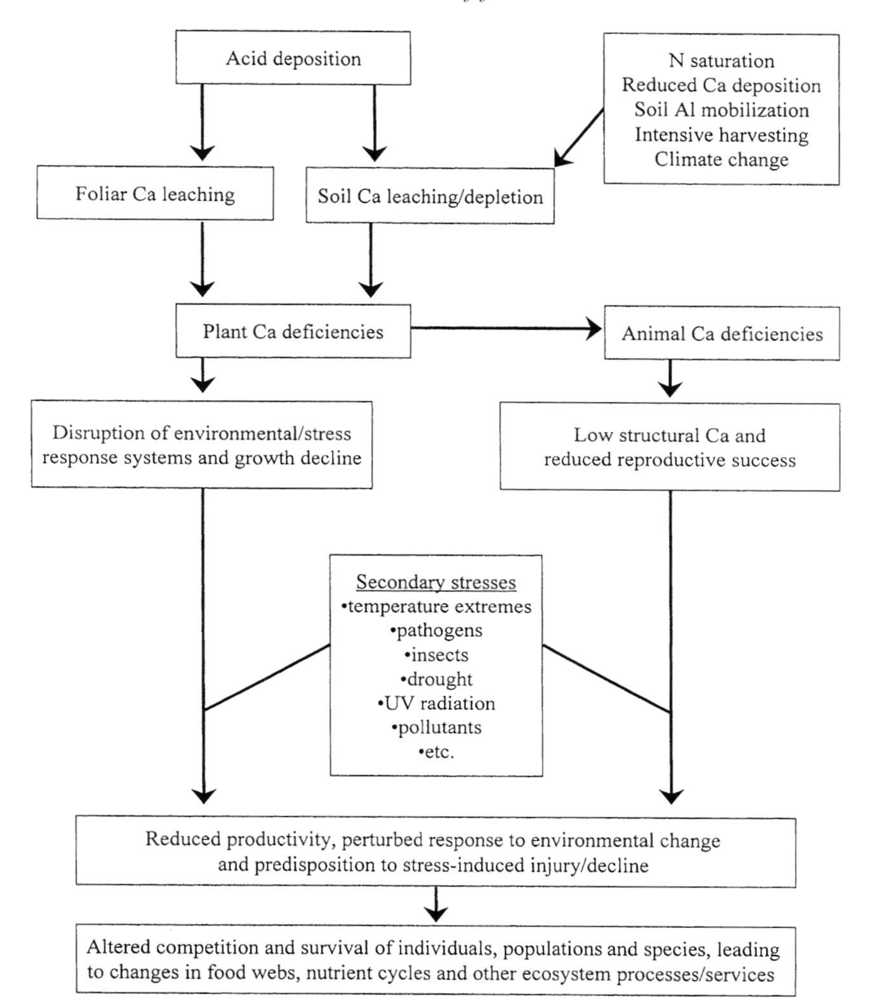


Figure 11.6 Pathways by which depletion of calcium (Ca) could lead to declines of forest ecosystem health. A reduction in Ca in plants (through low uptake or excessive loss) could lead to reduced ability to detect or react to environmental stress. Low calcium levels in animals may impede reproduction. The net result could make forest ecosystems prone to other stresses, threatening long-term health, function and stability of the ecosystem. Concern about soil phosphate levels (see Section 8.4.3) must also be remembered. (Modified from Schaberg *et al.*, 2001.*Ecosystem Health* 7, Blackwell Publishing.)

### *11.4.3 Nitrogen pollution*

As discussed in Chapter 8, nitrogen is in short supply in many forests – particularly in temperate areas – due to it being locked up in various unavailable forms in the soil and due to ready leaching of any temporary excess into groundwater and streams. It might then be expected that a little extra nitrogen in the form of pollution would be a good thing, and in small quantities this is true. The current problems, particularly in temperate areas, of **long-term nitrogen enrichment** stem from too much of a good thing (see Nosengo, 2003 for a short, readable review). Although the atmosphere is 78% nitrogen this is in a form unusable by most organisms (see Section 8.4.1); nitrogen pollution comes from excess quantities of **ammonia** ($NH_3$) and **nitrogen oxides** ($NO_x$). (In this context, $NO_x$ refers primarily to nitric oxide (NO), nitrogen dioxide ($NO_2$) and nitrous oxide ($N_2O$).) The largest source of ammonia is from agriculture; from crop fertilizers (especially those based on urea) and emissions from livestock and their manure (from the breakdown of urea). In the UK, 80% of the total annual atmospheric ammonia production of 320 000 tonnes comes from agriculture: 47% from cattle, 14% from poultry, 9% from pigs and 11% from fertilizers. Non-agricultural sources of nitrogen include sewage sludge spread on land, landfill sites and vehicles with catalytic converters. Nitrogen oxides are primarily from industrial and vehicular combustion of fossil fuels. Anthropogenic production of nitrogen has increased from around 15 million tonnes (Mt) year $^{-1}$ in 1860 to around 165 Mt year$^{-1}$ in 2000. Background deposition of nitrogen is rarely more than 1 kg ha$^{-1}$ year$^{-1}$ but since the 1980s, nitrogen enrichment has increased this by 10–40 times or even higher. As with acid rain, this nitrogen enrichment is mainly a northern hemisphere problem, especially in Europe, North America and Japan, a reflection of population density and intensity of land use.

Part of the problem with nitrogen is that the same atom of nitrogen can have many effects in the atmosphere, terrestrial, freshwater and marine ecosystems, and on human health, one after the other. Nitrogen oxides, for example, while in the atmosphere can reduce atmospheric visibility, are involved in harmful levels of low-level ozone production and contribute to acid rain. Once on the ground they cause problems for forest vegetation as discussed below, and then are flushed into streams and the oceans causing eutrophication. Galloway *et al.* (2003) refer to this extending list as the **nitrogen cascade**.

Box 11.1 (page 458) shows that forests can become **nitrogen saturated**, i.e. have an availability of ammonium and nitrate in excess of the total combined demand of plants and microbes (Aber *et al.*, 1998). Figure 11.7 gives four hypothetical stages that a temperate forest goes through as nitrogen enrichment increases. Stage 0 represents a highly nitrogen-limited forest unaffected by external nitrogen. In stage 1 there is increased nitrogen deposition which relieves some of the nitrogen limitation and nitrogen mineralization is increased; the plants contain more nitrogen (leaf N) and growth as measured by net primary productivity (NPP) goes up: the extra nitrogen is beneficial. By

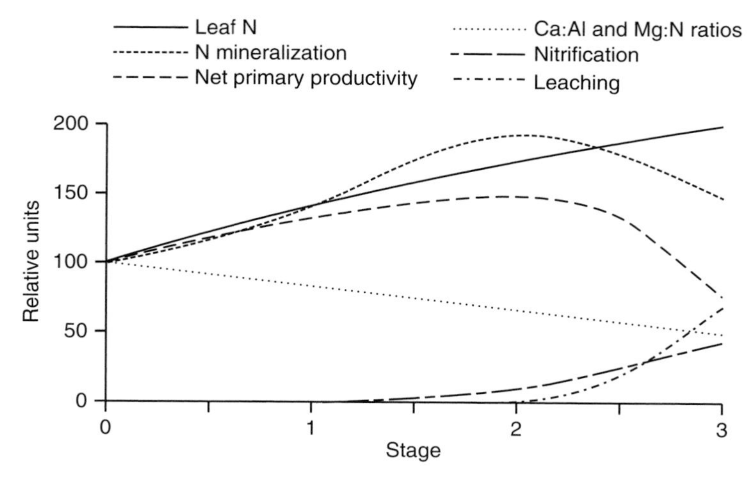


Figure 11.7 Hypothetical responses of a temperate forest to long-term chronic nitrogen additions at various stages through time. (Redrawn from Aber *et al.*, 1998. *BioScience* 48.)

stage 2, nitrification (the conversion of ammonium to nitrate) starts and since nitrate is readily leached, some nitrogen starts being washed out of the soil. Growth may not be as high as expected since much of the extra nitrogen is either leached away or locked up in the soil. Beyond this, in stage 3, the amount of nitrogen is so high that it is no longer limiting to growth but growth (NPP) will decline and may lead to death. A variety of reasons for the decline in growth can be put forward, mostly covered under 'forest decline' above, but an important issue is the increasing imbalance of nutrients and the pressure put on dwindling magnesium and calcium supplies due to acidic conditions (nitrogen compounds can bind with these two cations forming compounds that can be readily leached). In this final stage the leaching of nitrate is also high, increasing water pollution – part of the nitrogen cascade.

The key point in this hypothesis is that growth is stimulated at low enrichment but decreases once a critical threshold is passed. The pine plots at Harvard Forest discussed in Box 11.1 are clearly entering stage 3 while the hardwood plots are nearer stage 1. This shows that different forests move between stages at different rates, have different starting points (degrees of nitrogen-limitation), and different critical thresholds. Certainly not all forests are in decline from nitrogen enrichment. In Sweden nitrogen deposition is fairly modest even at its highest (27 kg $ha^{-1}$ $y^{-1}$). Binkley and Högberg (1997) observed that in forests around Sweden the amount of nitrogen being leached is less than is arriving, suggesting that the forest is not nitrogen saturated. Adding other nutrients gave no extra growth so nitrogen still seems to be limiting. In fact Swedish forests increased in growth by 30% (measured as an increase in basal area) between 1953 and 1992. The sensitivity of forests to nitrogen enrichment depends to some degree on the soil nutrient status and the buffering capacity of soils to extra nitrogen (the ability to absorb and lock-up nitrogen), sensitivity to increased acidity of soil (as described above, nitrogen compounds contribute to acid rain) and the rate and duration of nitrogen deposition.

As with acid rain, stand decline and death are not the only concerns. At fairly low enrichment levels species diversity may increase as more **nitrophilic species** (nitrogen loving) and acid-resistant species invade, but diversity will decline in the long-term as native species are progressively lost (see Bobbink *et al.*, 1998 for more details). They are lost not just because of unsuitable nutrient conditions but due to changes in competitive relations. Secondary stresses from pathogens, frost and drought may also contribute to species loss and decline of vigour.

Fortunately, the effects of nitrogen-enrichment look to be reversible. The **NITREX project** in Europe (see the end of Box 11.1) have investigated the effects of removing nitrogen by using transparent roofs above plots watered with a mixture approximating pre-nitrogen-enrichment rain. Boxman *et al.* (1998) used these in a Scots pine stand in the Netherlands. Within their plots, the reduction in nitrogen led almost immediately to reduced leaching of nitrate and nitrogen levels in the soil remained high. Nevertheless within 3–4 years the nitrogen content of the pine needles had declined and magnesium and potassium levels had increased, helping to correct the previous nutrient imbalance. This was matched by an increase in diameter growth of the dominant trees. There was also evidence that the understorey vegetation was recovering, with a decline in nitrophilous species such as brambles *Rubus* spp. and the broad-buckler fern *Dryopteris dilatata* and an increase in the fruiting bodies of mycorrhizal fungi. Boxman and colleagues suggest that the effects of reducing nitrogen follow a reversed nitrogen cascade, with a reaction first of all in the soil water, then the soil and then the vegetation.

Nitrogen pollution is also having an impact in indirect ways. **Volatile organic compounds** (VOCs) produced by such things as solvent-rich paints and varnishes (**anthropogenic VOCs**), have declined due to legislation but in eastern USA, particularly the south (and undoubtedly elsewhere in the world), VOCs produced by forest trees (**biogenic VOCs**) have increased (Purves *et al.*, 2004). During heat-waves (when more VOCs are emitted, particularly isoprenes and monoterpenes) biogenic VOCs may greatly outweigh anthropogenic VOCs. Woody plants are the main source and pioneer species such as aspen, poplars and sweetgum *Liquidambar styraciflua* emit high levels of

VOCs while late successional species such as beech and maples tend to emit none (no VOCs have been detected from any native USA maple). Moreover, common plantation species such as poplars, pines and eucalypts are high emitters. The increase in biotic VOCs in the USA is primarily due to changes in land use, both the doubling in forest cover in north-eastern USA over the last 100 years due to farm abandonment (involving pioneer species), and an increase in plantations. Where does nitrogen come into this? Volatile organic compounds interact with chemicals such as carbon monoxide and methane, potent greenhouse gases, with consequent impacts on climate change. Photochemical oxidation of VOCs in the presence of nitrogen oxides also produces ground-surface **ozone** ($O_3$), harmful to human health and agriculture. Damaging ozone levels (defined as $> 60$ ppb – parts per billion) occur over 24% of the world's forest area and are predicted to affect 50% of forest area by 2100 (Fowler *et al.*, 1999).

**Nitrous oxide** ($N_2O$) is a potent greenhouse gas that has been increasing in the atmosphere over recent decades with about one third of the increase directly due to human causes. A large source of this increase is from tropical soils due to forest clearing, agricultural fertilization and forest regrowth dominated by legumes. As such, any disturbance is likely to increase nitrous oxide production. Hurricane Georges crossing Puerto Rico in September 1998 caused extensive defoliation, and emissions of nitrous oxide were five times above normal over the next 7 months and still twice normal after 2 years (Erickson and Ayala, 2004). Assuming hurricanes come through once every 50 years on average, they calculated that each hurricane would contribute only 18% of the production of nitrous oxides over a 50-year period but this would obviously be more significant if hurricane frequency increased.

### 11.4.4 Heavy metal pollution

Acid rain and nitrogen enrichment are frequently accompanied by heavy metal pollution. This can come from the mobilization of heavy metals already in soils due to increasing acidity, possibly supplemented by pollution associated with the industrial processes causing acid rain. Bedrock can contain a suite of up to 38 heavy metals (defined as those with a density greater than 5 g $cm^{-3}$) which are potentially toxic to life, usually by damaging proteins and enzymes. The commonest found in polluted soils are: cadmium, arsenic, chromium, mercury (very toxic); lead, nickel (moderately toxic); boron, copper and zinc (least toxic). Some of these are, of course, essential micronutrients in low concentrations.

Heavy metals can cause direct toxicity to roots and they can disrupt nutrient uptake causing deficiencies. A large number of trees are sensitive to metal contamination; in Europe these include silver birch *Betula pendula*, ash *Fraxinus excelsior*, rowan *Sorbus aucuparia*, small-leaved lime *Tilia cordata* and apple *Malus domestica* (Sawidis *et al.*, 1995). Others are more tolerant and take up some metals without apparent ill-effect. These are useful as **biomonitors** of metal contamination. European examples include white willow *Salix alba*, silver lime *Tilia tomentosa*, elder *Sambucus nigra*, pedunculate oak *Quercus robur*, beech *Fagus sylvatica* and Italian poplar *Populus nigra* ssp. *italica*. Trees can also act as biomonitors by intercepting and holding particulate pollution on the outside of the leaf. In Greece, Sawidis *et al.* (1995) studied a selection of tree species as biomonitors of zinc and copper. The investigators discovered that the strongest metal accumulators had rougher leaf surfaces which more effectively trapped and held the particles.

It is well known that some herbaceous plants, including grasses have **metal-tolerant ecotypes**, varieties within the species that can tolerate high metal contamination, although the tolerance is almost always metal-specific. These appear to be less common in trees, although there are examples of zinc-tolerant birch and clones of the North American aspen *Populus tremuloides* that tolerate metal-rich air pollution. However, care is needed since Turner and Dickinson (1993) found that sycamore *Acer pseudoplatanus* growing on contaminated sites in the UK could grow abundant roots in uncontaminated pockets of soil and so just look tolerant. Moreover, seedlings on completely contaminated sites could grow for at least 3 years even though they were doomed to die. In some species, a careful pre-exposure to non-toxic levels of heavy metals can lead to more tolerance of what are normally toxic levels. Punshon and Dickinson (1997) exposed various clones of willow to subtoxic concentrations of single metals (0.15 mg of copper $l^{-1}$, 0.15 mg cadmium $l^{-1}$ or 2.5 mg zinc $l^{-1}$) with concentrations gradually raised 10-fold over 128 days, and found reduced phytotoxicity and increased resistance, most notably to cadmium. But increased tolerance is not found in all species. Wisniewski and Dickinson (2003) found pre-exposure of pedunculate oak *Quercus robur* to copper did not produce any acclimation. They concluded that survival of seedlings on copper-rich soil was due to soil or rhizosphere processes alleviating metal toxicity. Mycorrhizas also play a role: various pines such as jack pine *Pinus banksiana* and Scots pine *Pinus sylvestris* inoculated with fungi of the *Suillus* genus can tolerate increasing concentrations of lead, nickel and zinc until the metals damage the fungus itself. The fungus helps by sequestering the metals in the fungal mycelium so that tree roots experience lower concentrations of the metals.

Trees that will grow on heavy metal-enriched soil can be used for **phyto-remediation**, the use of green plants to clean-up contaminated soils or sediments (phyto – Greek for plant, remedium – Latin word meaning to correct or remove an evil). Suresh and Ravishankar (2004) highlight that this is brought about in two possible ways. The first is that the tree can facilitate microbial degradation or fixing of the metals in the rhizosphere (or even further out into the soil) – see Section 5.5.1. Secondly, trees can extract and store metals in their woody skeleton and leaves. This works especially well for metals like nickel, zinc, copper, lead, chromium and cadmium. Poplars have proved useful in that they can accumulate relatively high levels of metals, especially cadmium, zinc and aluminium. Laureysens *et al.* (2004) found lowest concentrations of the metals in the wood and highest concentrations usually in the leaves being shed in the autumn. Thus the wood could be used as biomass for energy production without putting too many pollutants back into circulation, and the fallen leaves could be collected to effectively remove the metals from the site.

Phytoremediation is a simple and inexpensive means of extracting contaminants from subsurface soils and water that is 2–5 times less expensive than traditional capping, sealing in the contamination with an impervious layer. The main limitations are that the contaminants below the rooting depth cannot be extracted by the plant root system and plants will not grow on the most polluted sites.

## 11.5 Problems in urban forests – the social interface

Humans have been influencing other animals and their environment for a very long time, as the recent discovery of a 400 000-year-old skeleton of the ancient elephant *Palaeoloxodon antiquus* at Ebbsfleet, Kent, demonstrated yet again. It was surrounded by crude flint tools used for stripping flesh, and appears to have been consumed raw by modern human's ancestor *Homo heidelbergensis*, after being driven into a bog or becoming stuck at the edge of an ancient lake accidentally. Today few forests are more strongly influenced by humans than those in urban areas. A large number of woodlands are affected, not always for the best, by nearby centres of population. However, in far-sighted countries, there is a growing realization of the value of **urban forests**, most often defined as forests near urban centres used for the cultivation and management of trees for their contribution to the psychological, sociological and economic well-being of the urban society. These are distinct from cultivated green spaces with trees such as gardens and parks, distinguished by urban forests being used for timber production and their uncultivated understorey vegetation.

Unfortunately many urban woodlands, as well as being used as adventure playgrounds by local children who frequently damage the trees, are also often visited by undesirables including burglars and drug dealers. Nevertheless there can be an abundance of wildlife amidst a great deal of waste, which includes beer cans and bottles as well as larger debris such as burnt-out cars and three-piece suites. These woodlands are frequently used for the illegal grazing of horses, especially in cases where the council has fenced off the woodland. There are many other places where urban woodlands are better treated, and there is no doubt that in virtually all instances they do give town and city dwellers an opportunity to see something of local wildlife.

In an area centred on Sheffield, Gilbert (1989) deals with the ecology of the whole gamut of urban habitats. His sections on private gardens and cemeteries have much of value on trees, while that on woodlands considers a number of individual cases in detail. These include the archaeology of a small urban wood, the ecology of an ancient semi-natural woodland, plantations on the sites of ancient woodland, the use of 'naturalistic' plantations and the bird life of urban woods. In many ways it is sensible to consider urban wildlife as a whole, bearing in mind the fact that many animals move into and out of the local woodlands and that many successional wasteland sites are on their way to becoming new woodland.

Many of the plants and animals – including slugs – encountered in towns and cities are introduced species, while many of the native taxa live in communities quite different from those they are found in elsewhere. Some animals adapt less well than the birds, although several rodent species do so all too well. The life expectancy of town foxes *Vulpes vulpes* is short, averaging 14 months in London and 18 months in Bristol, while large family groups fail to develop. As with hedgehogs *Erinaceus europaeus*, many of these animals become road casualties. Gilbert advises us to accept and enjoy town wildlife for what it is, and is not over keen on attempts to create ecological parks with small areas of particular habitat types in the middle of cities.

An exemplary attitude to urban forests is found in Sweden. In the 1960s and 1970s new housing developments were supplied with cultivated green areas since they were believed 'to be superior to "virgin and raw" nature, and nature was not trusted to "stand the wear" of the urban citizen's recreational use' (Rydberg and Falck, 2000, p. 3). But in the 1980s, an interest arose in combining the green spaces and traditional forestry areas into multiple-use areas. Rydberg and Falck (2000) point out in their review of Swedish urban forestry the numerous benefits of these forests in terms of amenity, aesthetics, health (mental and physical) and recreation. In 1985 Swedes made more than 200 million forest visits per year, 55% of which were to urban forests.

Similar statistics are found elsewhere; for example, there were 222 million woodland leisure trips in England in 2002/3 (Anon., 2004) and 67% of adults in the UK made a leisure visit to a woodland in recent years, 50% of which were to woodlands in and around towns (Anon., 2003). Trees in urban forests are also beneficial in that, being large, they are excellent at filtering out particulate pollution (Beckett *et al.*, 1998). Despite the benefits of urban woodlands, they are not immune to the effects of visitors. Littlemore and Barker (2003) have shown that oak woodland in the English Midlands can be remarkably sensitive to trampling. Bluebells *Hyacinthoides non-scripta* were particularly sensitive; the equivalent of 500 passes of a person on foot over one summer prevented the trampled plants from producing any seeds even 2 years later. Brambles *Rubus fruticosus*, with their tough, woody and flexible stems were very much more resistant.

In England, two major initiatives have been built around the need for urban forests. The National Forest (started in 1995) is a central government initiative to link two old forests (Needwood and Charnwood Forests) in central England, aiming to increase woodland cover from 6% to 33%, planting on farmland, derelict land and conurbations over an area of 50 200 ha (200 square miles). Emphasis is very much on recreation: 74% of the area will have public access. The other initiative was to set up Community Forests, starting in 1989. The aim of these is to improve degraded landscapes and provide for recreation and education in the countryside close to 12 major English towns. While this discussion has been on urban forests we should not ignore cultivated parks. Frederick Law Olmstead, landscape architect of New York City's Central Park, is said to have proclaimed that a city park was to 'provide a natural verdant and sylvan scenery for the refreshment of town-strained, men, women, and children'.

Despite the desire to visit woodland, many people live in places where access to natural woodlands is difficult, often due to ownership constraints. This is driven near major population centres by concerns about the likelihood of damage to the understorey vegetation from trampling. How are these beautiful natural ecosystems to be made available and how are they to be protected? A splendid example is provided by the seven miles of Wenlock Edge owned by the National Trust (NT) in the West Midlands of England. Access to this geologically interesting wooded escarpment is via a footpath leading from an NT car park in the village of Much Wenlock, Shropshire. The walk takes you over a series of Silurian rocks dipping towards the south-east (Toghill, 1990). A lane leads up over the Aymestry Limestone, down the Lower Ludlow Shales and up the dip slope of the Wenlock Limestones of the major escarpment. The two limestones have both been extensively quarried in the past; the soils to which they give rise are species-rich with some notable orchids in the woods and grasslands of the reserve. Well-situated notices draw attention to the geology, biota and the various routes available through the area. One of the most interesting woods here is Blakemore Coppice, whose upkeep is supervised by Forest Enterprise and in which repeated visits in spring are rewarded by a fine illustration of an aspect society, with huge vistas – and indeed the local air – dominated by ramsons *Allium ursinum* in late May. The paths are metalled with local stone and so many people can walk through the woods without damaging the delicate ground flora. This provision is expensive but vital for the continued biodiversity of a wood visited by people from Birmingham and further afield.

## 11.6 Agroforestry and new forests

### *11.6.1 Growing trees with other crops*

The World Agroforestry Centre (formerly called ICRAF, the International Council for Research in Agroforestry) in Nairobi, Kenya is the main international agroforestry body. They define **agroforestry** as a collective term for all land-use practices in which woody perennials (trees and shrubs) are grown on the same land as pastured animals (**sylvo-pastoralism**) or crops (**agri-silviculture**). This is most commonly practised in the tropics but can also be found in temperate areas north and south of the equator, including such things as the production of maple syrup from trees between fields in New England. Agroforestry systems produce major improvements in both production and soil protection and there is no doubt that more and more areas will be involved worldwide. In many ways, agroforestry provides the best of both worlds. It produces benefits from the trees that can include fuel, fodder, food, fertilizer, fibre and fruits (the six Fs of conservation forestry) and possibly medicines and insecticides (such as from the neem tree *Azadirachta indica*, native to South Asia and widely planted in drier parts of the subtropics and tropics). The trees can also help to rehabilitate degraded land by building soil organic matter, act as biological pumps to remove excess irrigated water and so prevent salt build-up on the surface, and protect against erosion (and consequent loss of soil organic matter and nutrients) by the litter and roots.

Rather than a uniform layer of trees over a crop or grass, agroforestry usually works best with some segregation. The usual system is alleys of crops between rows or strips of trees. The rows are typically 4–8 m apart and run from east to west to minimize shading of the alleys between them. This intercropping can result in higher biomass production than either the trees or crop alone, partly because of differing moisture needs, taken from different horizons in the soil or at different times, the extended growing season of crops from the ameliorated microclimate and the use of nitrogen-fixing trees. Care is obviously needed in choosing compatible crops and trees. A whole field of cereal grains or pulses may be more valuable than the mixed agroforestry produce in the short-term; the advantage of the latter is its long-term sustainability and long-term profitability.

Although agroforestry is most commonly practised in the tropics, forms of it are appearing elsewhere in the world. For example, alternating plantations of pines with sheep pasture or agricultural crops may be a means of lessening problems with tree pests and diseases as well as conserving the soil from the worst effects of long-term agriculture. Although financial returns from the trees and crops are separated in time this is a potentially valuable form of agroforestry. Farm forestry has rapidly expanded in the UK since the production of livestock has become less profitable. Areas used for agriculture usually have much higher phosphate levels than woodland soils and indeed many woodlands planted a century ago have approximately twice as much phosphate as ancient woodlands. Not surprisingly these woodlands have shown good growth. However, as modern agricultural soils have still higher fertilizer contents, weeds, particularly stinging nettle *Urtica dioica*, are abundant in farm woodlands whose soils verge on the edge of eutrophication, limiting the development of a more natural understorey vegetation.

### *11.6.2 Habitat creation and conservation*

In recent years the creation of attractive herb communities, for both meadows and woodland floors, has become increasingly common in areas used by the public (Buckley, 1989). Helliwell (1996), who has worked on habitat transfers involving grassland, marshland and woodland, provides a valuable summary of the general principles involved, paying particular attention to soils and the need to appreciate the considerable differences between the various layers of strongly stratified soils when this material is moved to a receptor site. This section, however, is largely based on experience of woodland habitat creation in Telford New Town and the Wolverhampton area in central England (Packham *et al.*, 1995; Cohn *et al.*, 2000) in which many of the species involved were ancient woodland vascular plant indicators **AWVPs** (see Fig. 3.9 and Sections 4.1 and 5.9). The first experiment was made in the (11.2 $m^2$) plot of pedunculate oak woodland planted in 1981 and used for defoliation experiments (Fig. 5.2). Introductions to this, the Old Compton site, were made in the autumn of 1989 by means of plantings of wood-sorrel *Oxalis acetosella*, bluebell *Hyacinthoides non-scripta* and yellow archangel *Lamiastrum galeobdolon*; this had the advantage that the precise positions where plants of these species started were clearly demarcated. The entire experimental area was also uniformly sown

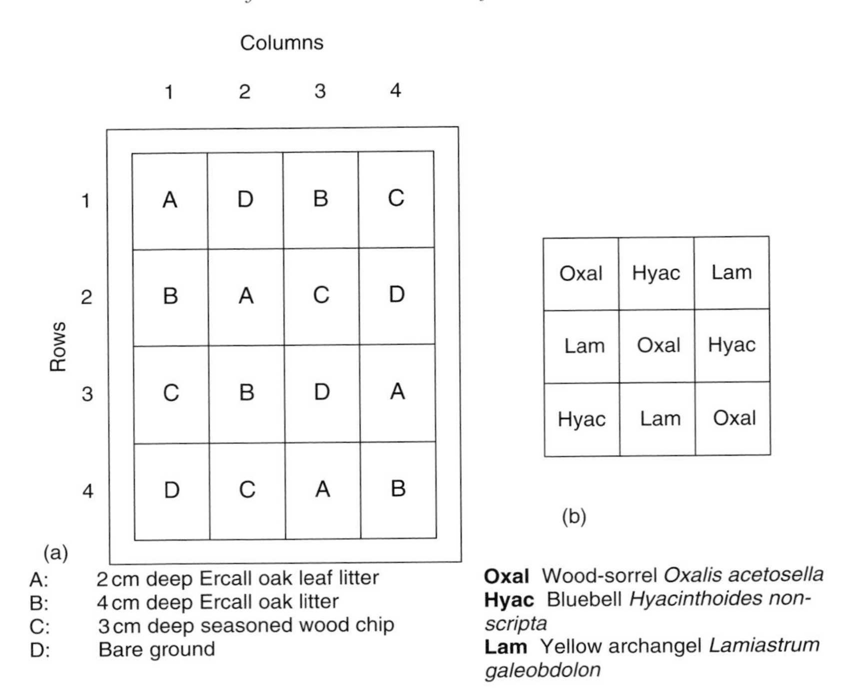


Figure 11.8 Layout of the Old Compton experiment in central England to investigate the influence of litter on the establishment of woodland forbs. (a) A buffer zone surrounds sixteen 2 × 2.3 m plots; four to each litter treatment. (b) Plan of each 2 × 2.3 m plot. (Modified from Cohn and Packham, 1993. *Arboricultural Journal* 17.)

with seeds of red campion *Silene dioica* and foxglove *Digitalis purpurea* at this time, using a rate of 100 seeds $m^{-2}$ for both species. The aim of the experiment was firstly to compare the effects of litter and mulch treatments on spontaneous vegetation and on the establishment of introduced woodland herbs with differing regeneration strategies, and secondly to observe community development and plant dynamics over a prolonged period. The sown portion of the site measured 8 × 9.2 m, divided into 16 plots each measuring 2 × 2.3 m and subdivided into 9 subplots each of 0.5 $m^2$, thus providing a shaded buffer zone a metre or more wide round the experiment. The four treatments employed are listed in Fig. 11.8 and were replicated four times in a Latin Square (Fig. 11.8).

Statistically significant differences for total annual weed cover in May 1990 were in the order $D \gg A \gg B \gg C$ ($\gg$, very much greater than; $>$ greater than); for perennial weeds the order was $D > A \gg C > B$. Foxglove had most young plants on the bare ground and least on the woodchip and deeper oak leaf litter in January 1991; at this time the results for red campion were $A + D \gg B > C$ (+, same as). The numbers of rooted nodes of yellow archangel per treatment were C > B > A ≫ D in January 1991. Two years later seeds of enchanter's nightshade *Circaea lutetiana*, sweet woodruff *Galium odoratum* and greater stitchwort *Stellaria holostea* were also sown right across the experimental area, each at a rate of 25 seeds $m^{-2}$.

Sixteen years after the establishment of the plot the subsequent fates of the introduced species were seen to be closely related to their C–S–R strategies (strategies of Competition, Stress-tolerant, Ruderal, which are discussed in Chapter 4). Wood-sorrel (S to C–S–R) is now present at some level over the entire site, growing particularly well along those margins heavily shaded by the oaks. Yellow archangel, another species noted for its tolerance of shade, is also doing well and even growing out from the woodland margin. Foxglove (C–R to C–S–R) became increasingly less successful as occupation of the woodland floor became more complete. This experiment still continues to give interesting results, although the influence of the original four surface treatments diminished fairly quickly. A considerable number of ash (and Norway maple *Acer platanoides*) seedlings have established; oak is represented only by the maturing trees. The herbs introduced into the Old Compton plot were all **forbs** (herbaceous plants other than grasses) but graminoids are also important; grasses such as wood millet *Milium effusum* and wood false-brome *Brachypodium sylvaticum* are amongst the many other species which have been introduced successfully elsewhere.

Developments in this and numerous other habitat-creation plantings since then have clearly demonstrated the very slow natural rates of spread associated with at least some of the ancient woodland vascular plant indicators mentioned previously (Section 6.4.1). Bluebell, a noted shade-evader, has very heavy seeds for a woodland herb and its first crop of seedlings normally occurs at a distance which virtually corresponds to the height of the flowering stalk, which finally topples over taking the seeds with it. Primrose *Primula vulgaris* is another plant with an inefficient seed-dispersal mechanism but when mature plants are stolen from public urban woodlands by the local inhabitants, a narrow ring of young primroses often develops around the position formerly occupied by the parent plant. Wood-sorrel on the other hand has no difficulty in spreading at least a metre from the parent plant as its seeds are released by an explosive mechanism which fires them in an arc whose height must exceed 50 cm at apogee. Yellow archangel has quite complex germination requirements and some workers have consequently recorded very low germination rates. Despite this the plant reproduces effectively from seed in nature, and in woodlands of the English Midlands and the south large populations of seedlings can be seen year after year. The seeds are distributed by ants, particularly *Formica rufa* and *Lasius niger*, which, attracted by the oil in the elaiosome (an oil-rich appendage to the seed), carry the seed away before discarding it once the oils are eaten. Yellow archangel also spreads very effectively by means of its stolons (above-ground horizontal stems), which can result in annual radial extension from the parent plant of 0.5–1.5 m.

Herbs may be introduced to a site by planting, often in soil plugs, or seed. On a large scale the use of seed is in the long term cheaper and more effective. Apparently small environmental differences may lead to the success or failure of particular species within the complex mosaic of the woodland floor, with all its variations in shading, fertilizer content, soil texture and competition from other plants. To use expensive plantings, many of which will not be in suitable positions, is unwise, especially as it is considerably cheaper to sow multi-species seed mixes in the autumn. There are exceptions: yellow archangel, the dormancy of whose stored seed is difficult to break, is best introduced by planting in the spring, using suitable transplants or seedlings developed from ripe seed sown directly after collection and left outside so that they are subject to repeated temperature changes over the winter. It is essential to collect or purchase the seed of the various species individually and make up the seed mix oneself. Some seed merchants are tempted to supply old seed and their seed mixes often contain a high proportion of useless non-woodland grasses, ostensibly to provide a protective canopy. The seed should be subjected to germination tests and of known provenance, the more local the better. It is, of course, essential to obtain the landowner's permission when collecting seed. Even distribution of seed is much easier when sowing if the seed mix is first diluted with silver sand.

Effort spent on site preparation is usually well rewarded, it should be borne in mind that many unwanted existing species can persist vegetatively at light levels that are sufficiently low that they would find it difficult or even impossible to establish from seed. They thus form a residual population able to take advantage of any opportunity. Nutrient levels in natural woodland soils are usually low, as has been emphasized earlier. Soil nutrient levels at the Old Compton site (Fig. 11.8) were moderate, considerably lower than at the Telford woodland, Shropshire, where, as is so often the case in such eutrophic sites, stinging nettles are both abundant and vigorous. It is virtually impossible to eliminate weeds from such sites, but they can be visually improved by planting more attractive species which flower earlier in the year.

Recent investigations on the relative importance of light and soil fertility upon field layer development in urban woodlands demonstrated that in most situations light was the more important factor. Although soil fertility in urban woodlands is usually far above that of more natural woodlands, vigorous growth of many weed species can be constrained by restricting the amount of light available. It also helps if a large buffer zone is maintained around the site ensuring adequate distance from sources of weed seeds. High soil fertility presents a much greater obstacle to the development of diversity in grassland restoration schemes than in woodlands (Bryant, 2003).

Attempts have been made to create new woodland by translocating either loose woodland soil or blocks cut from the woodland floor complete with as many of the plants growing as possible. One of the largest attempts is that of a 5 ha remnant of a once larger ancient woodland known as Biggins Wood at Folkestone, Kent, southern England. This site was part of the area which had to be disturbed when the Channel Tunnel was constructed. One hectare of the wood had already been disturbed by clay extraction for brick making. A total of 11 000 $m^2$ of topsoil was moved to a prepared receptor site nearby. The site was then planted with nursery-grown trees and shrubs with localized weed control to help them establish. In practice only a narrow band along alternate rows was treated, the others being left. After 5 years, 16 of the original 99 woodland species had not been re-recorded, but 93 additional (mainly ruderal) species had been found.

The three woodland types present in the wood were (A) **invasive elm woodland** dominated by suckering or small-leaved elm *Ulmus carpinifolia* (= *U. minor* ssp. *minor*), with ash, field maple and hazel; (B) **oak–ash–maple woodland** (*Quercus robur*, *Fraxinus excelsior*, *Acer campestre*) with ground ivy *Glechoma hederacea*, enchanter's-nightshade *Circaea lutetiana*, bugle *Ajuga reptans*, bramble *Rubus* spp. and pendulous sedge *Carex pendula* abundant in the field layer; and (C) **disturbed ash–oak woodland** in which the most interesting field layer species were moschatel *Adoxa moscchatellina*, wood speedwell *Veronica montana*, primrose *Primula vulgaris* and red campion *Silene dioica*.These three woodlands showed a broad affinity to the **ash–field maple–dog's mercury (W8 woodland)** of the National Vegetation Classification (Rodwell, 1991), but due to the absence of dog's mercury *Mercurialis perennis* caused by poor drainage most of the wood fell into the tufted hair-grass *Deschampsia* subcommunity. Localized flushes had abundant grey willow *Salix cinerea*, pendulous sedge, angelica *Angelica sylvestris*, meadowsweet *Filipendula ulmaria*, and creeping buttercup *Ranunculus repens*. Soil moisture within Biggins Wood thus varied quite markedly as Fig. 11.9 demonstrates, but all the soils employed had a pH of around 7.2. The uppermost 20 cm of soil was transferred to the receptor site between 5 and 13 September 1988. No attempt was made to segregate the different soil layers or indeed the vegetation itself. In practice roughly 20% of the understorey plants were either at the surface or close enough to it to resume growth. A smaller proportion arose from buried corms, bulbs, rhizomes and other storage organs. Some came from plants arising from the seed bank.

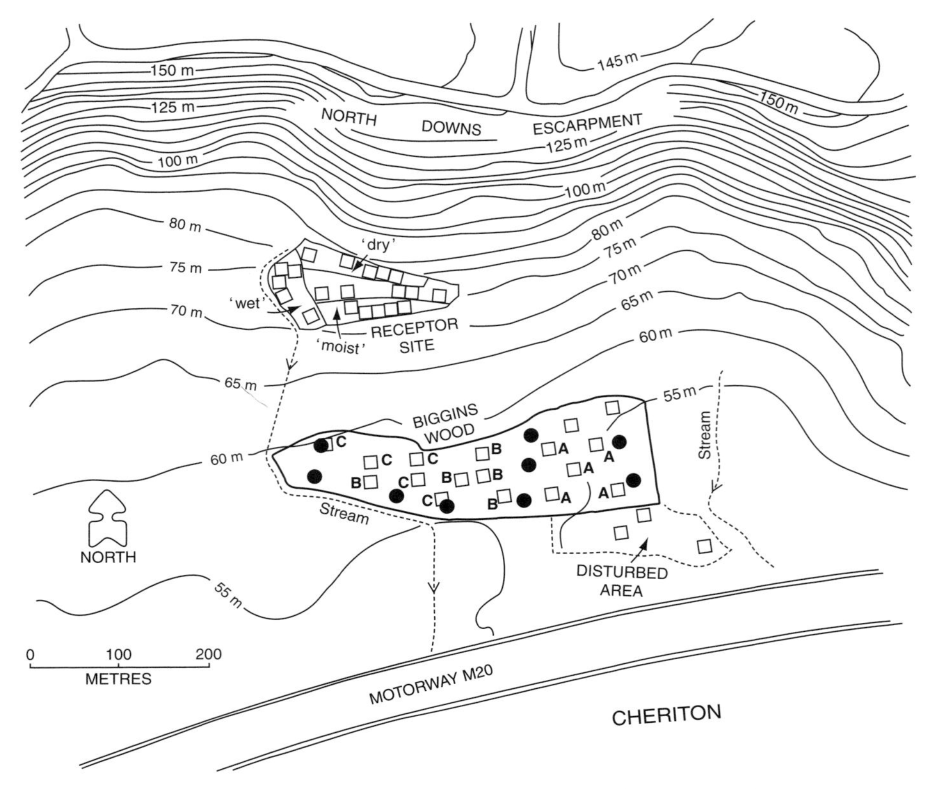


Figure 11.9 Biggins Wood, Kent, southern England, in 1988, showing areas of 'dry', 'moist' and 'wet' vegetation types suitable for transfer to the receptor site some 20 m to the north above it on the North Downs. The woodland types A, B and C are described in the text. (From Helliwell *et al.*, 1996. Vegetation succession on a relocated ancient woodland soil. *Forestry*, 69, 57–74, by permission of Oxford University Press.)

The site was planted with nursery-grown trees and shrubs at 1.5 m spacing in February/March 1989. The ten species involved were field maple *Acer campestre* (8%), alder *Alnus glutinosa* (2%), silver birch *Betula pendula* (18%), hazel *Corylus avellana* (16%), spindle *Euonymus europaeus* (8%), ash *Fraxinus excelsior* (20%), crab apple *Malus sylvestris* (4%), wild cherry *Prunus avium* (4%), pedunculate oak (12%) and guelder-rose *Viburnum opulus* (8%).

The New Biggins Wood thus had a very promising start so it is sad that its subsequent progress was marred by lack of maintenance. The consultants advising on the project were very conscious of the fact that bramble *Rubus fruticosus* agg. tends to become rampant in large canopy gaps in southern English woodlands. It was thus essential to keep it in check until the trees were large enough to provide effective shade. This was not done and the consequent growth of bramble was so overwhelming that for several years it was not possible to make a survey to establish which transferred species had survived and which had not. By 2004, however, shade from the developing trees had become so much greater that the growth of bramble and other ruderal species had diminished, so Helliwell was able to survey the site again and ascertain the fate of the transferred woodland species. Vernal species such as moschatel, ransoms, lords-and-ladies, celandine, common dog violet and primrose had survived, but many of the other species, including most of the sedges and many of the 'woodland edge' species had not. The loss of moisture-loving species was not unexpected as the receptor site is, in the main, more freely drained than the original site, but a larger range of other species could probably have been retained if there had been suitable management in the 10 years prior to 2004.

Habitat creation is here to stay, especially in areas where the natural vegetation has been extensively modified. But besides the practical issues it poses **ethical and philosophical problems**, particularly when it involves the conversion of valuable ecosystems of other kinds in order to create initially second-rate woodland. It enhances the countryside and helps redress a severe decline in woodland species, but the full integrity of old woodland – once destroyed – can never be fully restored in the short time-frame we humans work in. Woodland creation works best where a greatly reduced woodland is now being expanded again. If seed, plants and even bryophytes from the remaining woodland can be employed in gradual new plantings, it may well be possible to produce a community with many features of the original. Above all, it is most important that any attempt at habitat creation be notified to a competent authority. One of the most encouraging trends in Great Britain in recent years is the planting of high-quality forests intended to encourage wildlife. The Woodland Trust Scotland acquired both Glen Quey and its neighbour Glen Sherup in Scotland at the turn of the millennium, creating a combined area of over 1000 ha. This area of the Ochils hills was planted with around a million trees between 2000–2003, and transformed from two grassy hills overgrazed by sheep to a diverse dynamic mosaic of habitats. The main aim is to recreate an upland birchwood where declining species such as pearl-bordered fritillary *Boloria euphrosyne*, spotted flycatcher *Muscicapa striata*, song thrush *Turdus philomelos*, redstart *Phoenicurus phoenicurus* and wood warbler *Phylloscopus sibilatrix* will thrive. Various birds of prey such as short-eared owl *Asio flammeus*, sparrowhawk *Accipiter nisus* and kestrel *Falco tinnunculus* have already started to appear more frequently.

## 11.7 The final challenge

The future of the world's woodlands and forests is one of the greatest problems we face. During the last 10 000 years humans have greatly changed the world's vegetation. Many distinct plant and animal communities have been lost and others greatly modified, while the majority of the really large marsupial and placental mammals have been eliminated. When the Polynesian ancestors of the Maoris first came to New Zealand, the last major forested area to be reached by humans, around AD 1200 they found a landscape dominated by birds. Many were ground-living and some, the giant moas, extremely large. The top carnivore was the fast-flying and deadly Haast eagle which, with a wingspan of 3 m, struck fear into the Polynesians who imported rats and dogs as well as a number of food plants which failed to flourish in the much colder climate. Within a relatively short period many of the native birds perished. The giant moas were eaten by the Maoris, so the Haast eagle lost its prey, while rats consumed the eggs and young of other species. Irreversible changes like this have occurred throughout the world; the challenge is to maintain at least some of the remaining forests and other vegetation in as natural a state as possible.

Is it possible to facilitate the development of more natural forests even in highly populated regions like western Europe? It might seem very far-fetched, but the Dutch are considering the creation of **mega-fauna corridors** whose forested landscape would enable bison to move west from Poland to Holland as they could do earlier in the Holocene.

There are almost certainly other unusual species still awaiting discovery, particularly on unexplored and densely forested islands. At a more fundamental level, as Fitter (2005) has demonstrated so clearly, there is still much to learn about the nature and population processes of the micro-organisms that inhabit the soils of woodlands, forests and other ecosystems, and which exert a profound influence on the plants rooted in them. This would be difficult enough if climatic and other environmental conditions were to stay relatively stable, which will certainly not be the case. The major economies of the world seem to be ignoring both the implications of global warming and the importance of dwindling oil, gas and coal reserves as raw materials rather than energy sources. Not only does the human population of the world now expect a much higher standard of living, it is also increasing at an alarming rate. Under these circumstances the maintenance of at least some major examples of all the natural ecosystems, particularly of woodlands and forests, though not impossible, will demand a great deal of determination.

# References


Aber, A., McDowell, W., Nadelhoffer, K. *et al.* (1998) Nitrogen saturation in temperate forest ecosystems. *BioScience*, **48**, 921–34.

Aerts, R. (1996) Nutrient resorption from senescing leaves of perennials: are there general patterns? *Journal of Ecology*, **84**, 597–608.

Ågren, G. I., Bosatta, E. and Magill, A. H. (2001) Combining theory and experiment to understand effects of inorganic nitrogen on litter decomposition. *Oecologia*, **128**, 94–8.

Ainsworth, A. M. (2004) *Developing Tools for Assessing Fungal Interest in Habitats. 1. Beech Woodland Saprotrophs*. English Nature Research Report 597. Peterborough: English Nature

Ainsworth, A. M. (2005) Identifying important sites for beech deadwood fungi. *Field Mycology*, **6**, 41–61.

Anderson, J. M. (1971) Observations on the vertical distribution of Oribatei (Acarina) in two woodland soils. In *IV. Colloquium Pedobiologiae*. Paris: INRA, pp. 257–272.

Anderson, J. M. (1975) Succession, diversity and trophic relationships of some soil animals in decomposing leaf litter. *Journal of Animal Ecology*, **44**, 475–495.

Anderson, J. M. and Swift, M. J. (1983) Decomposition in tropical forests. In *Tropical Rain Forest: Ecology and Management*, Ed. Sutton, S. L., Whitmore, T. C. and Chadwick, A. C. Oxford: Blackwell Scientific Publications, pp. 287–309.

Anderson, R. L., Foster, D. R. and Motzkin, G. (2003) Integrating lateral expansion into models of peatland development in temperate New England. *Journal of Ecology*, **91**, 68–76.

Anderson, W. B. and Eickmeier, W. G. (1998) Physiological and morphological responses to shade and nutrient additions of *Claytonia virginica* (Portulacaceae): implications for the "vernal dam" hypothesis. *Canadian Journal of Botany*, **76**, 1340–9.

Anderson, W. B. and Eickmeier, W. G. (2000) Nutrient resorption in *Claytonia virginica* L.: implications for deciduous forest nutrient cycling. *Canadian Journal of Botany*, **78**, 832–9.

Andersson, L. I. and Hytteborn, H. (1991) Bryophytes and decaying wood – a comparison between managed and natural forest. *Holarctic Ecology*, **14**, 121–30.

Anon. (2002) Clonal forestry – who are you kidding? *Scandinavian Journal of Forest Research*, **17**, 485.

Anon. (2003) *UK Public Opinion of Forestry 2003*. Edinburgh: Forestry Commission.

Anon. (2004) *Report of the 2002–03 Great Britain Day Vists Survey*. A survey undertaken by the Countryside Agency, Countryside Council for Wales, British Waterways, Department for Culture, Media and Sport, Environment Agency, Forestry Commission, Scottish Natural Heritage, VisitBritian, VisitScotland and Wales Tourist Board (see: www.countryside.gov.uk).

Archibold, O. W. (1995) *Ecology of World Vegetation*. London: Chapman & Hall.

Arno, S. F. and Hammerley, R. P. (1984) *Timberline: Mountain and Arctic Forest Frontiers*. Seattle: The Mountaineers.

Arnolds, E. and de Vries, B. (1993) Conservation of fungi in Europe. In *Fungi of Europe: Investigation, Recording and Conservation*, Ed. Pegler, D. N., Boddy, L., Ing, B. and Kirk, P. M. Richmond, UK: The Royal Botanic Gardens, Kew, pp. 211–30.

Ashman, M. R. and Puri, G. (2002) *Essential Soil Science*. Oxford: Blackwell Science.

Ashton, P. S. (1964) Ecological studies in the mixed dipterocarp forests of Brunei State. *Oxford Forestry Memoirs*, **25**, 1–75.

Ashton, P. S., Givnish, T. J. and Appanah, S. (1988) Staggered flowering in the Dipterocarpaceae: new insights into floral induction and the evolution of mast fruiting in the aseasonal tropics. *American Naturalist*, **132**, 44–66.

Attenborough, D. (1980) *Life on Earth*. London: Reader's Digest/Collins/BBC.

Attenborough, D. (1985) *The Living Planet*. London: Reader's Digest/Collins/BBC.

Attenborough, D. (2002) *The Life of Mammals*. London: BBC Books.

Attiwill, P. M. and Adams, M. A. (1993) Nutrient cycling in forests. *New Phytologist*, **124**, 561–82.

Attiwill, P. M. and Weston, C. J. (2001) Forest soils. In *The Forests Handbook*. Vol 1. *An Overview of Forest Science*, Ed. Evans, J. Oxford: Blackwell Science, pp. 157–87.

Avissar, R. and Werth, D. (2005) Global hydroclimatological teleconnections resulting from tropical deforestation. *Journal of Hydrometeorology*, **6**, 134–45.

Baillie, I. C. (1996) Soils of the humid tropics. In *The Tropical Rain Forest*, Ed. Richards, P. W. Cambridge: Cambridge University Press, pp. 256–86.

Baines, D., Sage, R. B. and Baines, M. M. (1994) The implications of red deer grazing to ground vegetation and invertebrate community structure of Scottish native pinewoods. *Journal of Applied Ecology*, **31**, 776–83.

Barberis, I. M. and Tanner, E. V. J. (2005) Gaps and root trenching increase tree seedling growth in Panamanian semi-evergreen forest. *Ecology*, **86**, 667–74.

Bardgett, R. D. (2005) *Biology of Soil: a Community and Ecosystem Approach*. Oxford: Oxford University Press.

Bardgett, R. D., Usher, M. B. and Hopkins, D. W. (2005) *Biological Diversity and Function in Soil*. Cambridge: Cambridge University Press.

Barthlott, W., Schmit-Neuerburg, V., Nieder, J. and Engwald, S. (2001) Diversity and abundance of vascular epiphytes: a comparison of secondary vegetation and primary montane rain forest in the Venezuelan Andes. *Plant Ecology*, **152**, 145–56.

Bates, C. G. and Roeser Jr, J. (1928) Light intensities required for growth of coniferous seedlings. *American Journal of Botany*, **15**, 185–94.

Beckett, K. P., Freer-Smith, P. H. and Taylor, G. (1998) Urban woodlands: their role in reducing the effects of particulate pollution. *Environmental Pollution*, **99**, 347–60.

Beedlow, P. A., Tingey, D. T., Phillips, D. L., Hogsett, W. E. and Olszyk, D. M. (2004) Rising atmospheric $CO_2$ and carbon sequestration in forests. *Frontiers in Ecology and the Environment*, **2**, 315–22.

Bell, P. R. and Hemsley, A. R. (2000) *Green Plants: Their Origin and Diversity*. (Second Edition). Cambridge: Cambridge University Press.

Bending, G. D. (2003) Litter decomposition, ectomycorrhizal roots and the 'Gadgil' effect. *New Phytologist*, **158**, 228–9.

Bennett, J. N. and Prescott, C. E. (2004) Organic and inorganic nitrogen nutrition of western red cedar, western hemlock and salal in mineral N-limited cedar-hemlock forests. *Oecologia*, **141**, 468–76.

Benton, M. J. (1991) *The Rise of the Mammals*. London: Quantum Publishing.

Berg, A., Ehnstrom, B., Gustafsson, L. *et al.* (1994) Threatened plant, animal, and fungus species in Swedish forests: distribution and habitat associations. *Conservation Biology*, **8**, 718–31.

Berg, H. and Redbo-Torstensson, P. (1998) Cleistogamy as a bet-hedging strategy in *Oxalis acetosella*, a perennial herb. *Journal of Ecology*, **86**, 491–500.

Berryman, A. A. (1996) What causes population cycles of forest Lepidoptera? *Trends in Ecology and Evolution*, **11**, 28–32.

Bills, G. F. and Polishook, J. D. (1994) Abundance and diversity of microfungi in leaf litter of a lowland rain forest in Costa Rica. *Mycologia*, **86**, 187–98.

Binkley, D. and Högberg, P. (1997) Does atmospheric deposition of nitrogen threaten Swedish forests? *Forest Ecology and Management*, **92**, 119–52.

Blackman, G. E. and Rutter, A. J. (1954) *Endymion non-scriptus* (L.) Garcke. Biological flora of the British Isles. *Journal of Ecology*, **42**, 629–38.

Bloor, J. M. G. and Grubb, P. J. (2003) Growth and mortality in high and low light: trends among 15 shade-tolerant tropical rain forest tree species. *Journal of Ecology*, **91**, 77–85.

Blundell, A. G. and Peart, D. R. (2001) Growth strategies of a shade-intolerant tropical tree: the interactive effects of canopy gaps and simulated herbivory. *Journal of Ecology*, **89**, 608–15.

Boardman, N. K. (1977) Comparative photosynthesis of sun and shade plants. *Annual Review of Plant Physiology*, **28**, 355–77.

Bobbink, R., Hornung, M. and Roelofs, J. G. M. (1998) The effects of air-borne nitrogen pollutants on species diversity in natural and semi-natural European vegetation. *Journal of Ecology*, **86**, 717–38.

Bobiec, A. (2002) Living stands and dead wood in the Białowieża forest: suggestions for restoration management. *Forest Ecology and Management*, **165**, 125–40.

Bobiec, A., van der Burgt, H., Meijer, K. *et al.* (2000) Rich deciduous forests in Białowieża as a dynamic mosaic of developmental phases: premises for nature conservation and restoration management. *Forest Ecology and Management*, **130**, 159–75.

Bohlen, P. J., Scheu, S., Hale, C. M. *et al.* (2004) Non-native invasive earthworms as agents of change in northern temperate forests. *Frontiers in Ecology and the Environment*, **2**, 427–35.

Bond, W. J. (1989) The tortoise and the hare: ecology of angiosperm dominance and gymnosperm persistence. *Biological Journal of the Linnean Society*, **36**, 227–49.

Booth, R. E. and Grime, J. P. (2003) Effects of genetic impoverishment on plant community diversity. *Journal of Ecology*, **91**, 721–30.

Bopp, W. (2005) Australia; a world of astonishing plants. *The Garden*, **130**, 14–17.

Boring, L. R. and Swank, W. T. (1984) The role of black locust (*Robinia pseudoacacia*) in forest succession. *Journal of Ecology*, **72**, 749–66.

Bormann, F. H. and Likens, G. E. (1979) *Pattern and Process in a Forested Ecosystem*. New York: Springer-Verlag.

Bosch, J. M. and Hewlett, J. D. (1982) A review of catchment experiments to determine the effect of vegetation changes on water yield and evapotranspiration. *Journal of Hydrology*, **55**, 3–23.

Bossuyt, B., Heyn, M. and Hermy, M. (2002) Seed bank and vegetation composition of forest stands of varying age in central Belgium: consequences for regeneration of ancient forest vegetation. *Plant Ecology*, **162**, 33–48.

Bouget, C. and Duelli, P. (2004) The effects of windthrow on forest insect communities: a literature review. *Biological Conservation*, **118**, 281–99.

Bowman, D. M. J. S. (2000) *Australian Rainforests: Islands of Green in a Land of Fire*. Cambridge: Cambridge University Press.

Boxman, A. W., van der Ven, P. J. M. and Roelofs, J. G. M. (1998) Ecosystem recovery after a decrease in nitrogen input to a Scots pine stand at Ysselsteyn, the Netherlands. *Forest Ecology and Management*, **101**, 155–63.

Brazier, J. D. (1979) Never mind the trees, what about the wood? *Information Paper 12/79*. Princes Risborough, UK: Building Research Establishment, Princes Risborough Laboratory.

Breuil, C., Fleet, C. and Loppnau, P. (2005) Sap stain in trees, logs and lumber: fungi, pigment and pigment biosynthetic pathways. In *Forest Pathology: From Genes to Landscapes*, Ed. Lundquist, J. E. and Hamelin, R. C. St. Paul, MN: APS Press, pp. 69–77.

Briggs, D., Smithson, P., Addison, K. and Atkinson, K. (1997) *Fundamentals of the Physical Environment*. (Second Edition). London: Routledge.

Briggs, P. and Morris, P. (2002) Putting dormice back on the map. *British Wildlife*, **14**, 91–100.

Broadmeadow, S. and Nisbet, T. R. (2004) The effects of riparian forest management on the freshwater environment: a literature review of best management practice. *Hydrology and Earth System Sciences*, **8**, 286–305.

Brook, B. W. and Bowman, D. M. J. S. (2005) One equation fits overkill: why allometry underpins both prehistoric and modern body size-biased extinctions. *Population Ecology*, **47**, 137–41.

Brooks, R. T. (2004) Early regeneration following the presalvage cutting of hemlock from hemlock-dominated stands. *Northern Journal of Applied Forestry*, **21**, 12–18.

Brown, A. E., Zhang, L., McMahon, T. A., Western, A. W. and Vertessy, R. A. (2005) A review of paired catchment studies for determining changes in water yield resulting from alterations in vegetation. *Journal of Hydrology*, **310**, 28–61.

Bruijnzeel, L. A. (1992) Managing tropical watersheds for production: where contradictory theory and practice co-exist. In *Wise Management of Tropical Forests*, Ed. Miller, F. R. and Adam, K. I. Oxford: Oxford Forestry Institute, pp. 37–75.

Brunet, J. and von Oheimb, G. (1998) Migration of vascular plants to secondary woodlands in southern Sweden. *Journal of Ecology*, **86**, 429–38.

Bryant, D., Nielsen, D. and Tangley, L. (1997) *The Last Frontier Forests: Ecosystems and Economies on the Edge*. Washington, DC: World Resources Institute.

Bryant, M. J. (2003) *The influence of soil fertility and light intensity on field layer development in urban secondary woodlands*. Ph.D. thesis, University of Wolverhampton, UK.

Brys, R., Jacquemyn, H., Endels, P. *et al.* (2004) Reduced reproductive success in small populations of the self-incompatible *Primula vulgaris*. *Journal of Ecology*, **92**, 5–14.

Bryson, B. (2004) *A Short History of Nearly Everything*. London: Black Swan.
Buckley, G. P. (1989) *Biological Habitat Reconstruction*. London: Belhaven Press.
Burke, D. M. and Nol, E. (2000) Landscape and fragment size effects on reproductive success of forest-breeding birds in Ontario. *Ecological Applications*, **10**, 1749–61.
Burnham, C. P. (1970) The regional pattern of soil formation in Great Britain. *Scottish Geographical Magazine*, **89**, 25–34.
Burnham, C. P. and Mackney, D. F. (1964) Soils of Shropshire. *Field Studies*, **2**, 83–113.
Burrows, C. J. (1990) *Processes of Vegetation Change*. London: Unwin Hyman.
Burton, T. M. and Likens, G. E. (1975) Salamander populations and biomass in the Hubbard Brook Experimental Forest, New Hampshire. *Copeia*, **1975**, 541–6.
Bury, R. B. (2004) Wildfire, fuel reduction, and the herpetofaunas across diverse landscape mosaics in northwestern forests. *Conservation Biology*, **18**, 968–75.
Cahoon, D. R., Hensel, P., Rybczk, J. *et al.* (2003) Mass tree mortality leads to mangrove peat collapse at Bay Islands, Honduras after Hurricane Mitch. *Journal of Ecology*, **91**, 1093–105.
Cain, M. L., Damman, H. and Muir, A. (1998) Seed dispersal and the Holocene migration of woodland herbs. *Ecological Monographs*, **68**, 325–47.
Calow, P. (1998) *The Encyclopedia of Ecology and Environmental Management*. Oxford: Blackwell Science.
Cameron-Smith, B. (1991) *Australian Rainforests*. Sydney: Royal Botanic Gardens.
Campbell, F. T. (2002) Invasive species: a growing concern. In *Symposium on the Hemlock Woolly Adelgid in Eastern North America*, February 5–7, 2002, Ed. Onken, B., Reardon, R. and Lashomb, J. New Jersey: Rutgers University, pp. 1–8.
Carline, K. A., Jones, H. E. and Bardgett, R. D. (2005) Large herbivores affect the stoichiometry of nutrients in a regenerating woodland ecosystem. *Oikos*, **110**, 453–60.
Carlson, A. (1986) A comparison of birds inhabiting pine plantation and indigenous forest patches in a tropical mountain area. *Biological Conservation*, **35**, 195–204.
Carter, P. C. S. (1989) Risk assessment and pest detection surveys for exotic pests and diseases which threaten commercial forests in New Zealand. *New Zealand Journal of Forestry Science*, **19**, 353–74.
Cederlund, G. and Bergström, R. (1996) Trends in the moose–forest system in Fennoscandia, with special reference to Sweden. In *Conservation of Faunal Diversity in Forested Landscapes*, Ed. DeGraaf, R. M. and Miller, R. I. London: Chapman & Hall, pp. 265–81.
Chadwick, O. A., Derry, L. A., Vitousek, P. M., Huebert, B. J. and Hedin, L. O. (1999) Changing sources of nutrients during four million years of ecosystem development. *Nature*, **397**, 491–7.
Chalfoun, A. D., Ratnaswamy, M. J. and Thompson III, F. R. (2002) Songbird nest predators in forest–pasture edge and forest interior in a fragmented landscape. *Ecological Applications*, **12**, 858–67.
Chazdon, R. L. and Whitmore, T. C. (2002) *Foundations of Tropical Forest Biology*. Chicago: Chicago University Press.
Cherubini, P., Fontana, G., Rigling, D. *et al.* (2002) Tree-life history prior to death: two fungal root pathogens affect tree-ring growth differently. *Journal of Ecology*, **90**, 839–50.

Christenson, L. M., Lovett, G. M., Mitchell, M. J. and Groffman, P. M. (2002) The fate of nitrogen in gypsy moth frass deposited to an oak forest floor. *Oecologia*, **131**, 444–52.

Clark, F. E. (1967) Bacteria in soil. In *Soil Biology*, Ed. Burges, A. and Raw, F. London: Academic Press, pp. 15–49.

Clarke, P. J., Kerrigan, R. A., and Westphal, C. J. (2001) Dispersal potential and early growth in 14 tropical mangroves: do early life history traits correlate with patterns of adult distribution? *Journal of Ecology*, **89**, 648–59.

Clements, F. C. and Shelford, V. E. (1939) *Bioecology*. New York: Wiley.

Cline, S. P., Berg, A. B. and Wight, H. M. (1980) Snag characteristics and dynamics in Douglas-fir forests, western Oregon. *Journal of Wildlife Management*, **44**, 773–86.

Cohn, E. V. J. and Packham, J. R. (1993) The introduction and manipulation of woodland field layers: seeds, plants, timing and economics. *Arboricultural Journal*, **17**, 69–83.

Cohn, E. V. J., Trueman, I. C. and Packham, J. R. (2000) More than just trees. *Aspects of Applied Biology*, **58**, 93–100.

Colquhoun, M. K. and Morley, A. (1943) Vertical zonation in woodland bird communities. *Journal of Animal Ecology*, **12**, 75–81.

Connell, J. H. (1971) On the role of natural enemies in preventing competitive exclusion in some marine animals and in rain forest trees. In *Dynamics of Populations*, Ed. Den Boer, P. J. and Gradwell, G. R. Wageningen, The Netherlands: PUDOC, pp. 298–312.

Coomes, D. A., Allen, R. B., Bentley, W. A. *et al.* (2005) The hare, the tortoise and the crocodile: the ecology of angiosperm dominance, conifer persistence and fern filtering. *Journal of Ecology*, **93**, 918–35.

Cooper, W. S. (1913) The climax forest of the Isle Royale, Lake Superior, and its development. I. *Botanical Gazette*, **55**, 1–44.

Corner, E. J. H. (1964) *The Life of Plants*. London: Weidenfield and Nicolson.

Cosgrove, P., Amphlett, A., Elliott, A. *et al.* (2005) Aspen: Britain's missing link with the boreal forest. *British Wildlife*, **17**, 107–15.

Costa Neves, H., Valente, A. V., Favila, B. *et al.* (1996). *Laurissilva da Madeira. Caracterização Quantitativa e Qualitativa*. Funchal: Secretaria Regional de Agricultura, Florestas e Pescas, Parque Natural da Madeira.

Costanza, R., d'Arge, R., de Groot, R. *et al.* (2003) The value of the world's ecosystem services and natural capital. *Nature*, **387**, 253–60.

Côté, S. D., Rooney, T. P., Tremblay, J.-P., Dussault, C. and Waller, D. M. (2004) Ecological impacts of deer overabundance. *Annual Review of Ecology, Evolution and Systematics*, **35**, 113–47.

Covington, W. W. (1981) Changes in the forest floor organic matter and nutrient content following clear cutting in northern hardwoods. *Ecology*, **62**, 41–8.

Crawford, R. M. M. (1990) Studies in plant survival. In *Studies in Ecology*, **11**, Ed. Anderson, D. J. Greig-Smith, P. and Pitelka, F. A. Oxford: Blackwell Scientific, p. 113.

Crawley, M. J. and Akhteruzzaman, M. (1988). Individual variation in the phenology of oak trees and its consequences for herbivorous insects. *Functional Ecology*, **2**, 409–15.

Cronin, L. (1987) *Key Guide to Australian Wild Flowers*. Chatsworth, New South Wales: Reed Books.

Cronin, L. (1989) *Key Guide to Australian Palms, Ferns and Allies*. Chatsworth, New South Wales: Reed Books.
Crosby, A. W. (1972) *The Columbian Exchange; Biological and Cultural Consequences of 1492*. Westport CT: Greenwood.
Cunnington, J. H. and Pascoe, I. G. (2003) Post entry quarantine interception of chestnut blight in Victoria. *Australasian Plant Pathology*, **32**, 569–70.
Dale, V. H., Joyce, L. A., McNulty, S. *et al.* (2001) Climate change and forest disturbance. *BioScience*, **51**, 723–34.
Danielsen, F. (1997) Stable environments and fragile communities: does history determine the resilience of avian rain-forest communities in habitat degradation? *Biodiversity and Conservation*, **6**, 423–33.
Darwin, C. (1859) *Origin of Species*. London: John Murray.
Darwin, C. (2003) *The Voyage of H. M. S. Beagle*. Folio edition. First published as single volume in 1839.
Davidson, E. A., Chorover, J. and Dail, D. B. (2003) A mechanism of abiotic immobilization of nitrate in forest ecosystems: the ferrous wheel hypothesis. *Global Change Biology*, **9**, 228–36.
Davie, T. and Fahey, B. (2005) Forestry and water yield – current knowledge and further work. *New Zealand Journal of Forestry*, **49**, 3–8.
Deacon, J. (2006) *Fungal Biology*. (Fourth Edition). Oxford: Blackwell Publishing.
Deckmyn, G., Muys, B., Quijano, J. C. and Ceulemans, R. (2004) Carbon sequestration following afforestation of agricultural soils: comparing oak/beech forest to short-rotation coppice combining a process and a carbon accounting model. *Global Change Biology*, **10**, 1482–91.
Deevey, E. S. (1947) Life tables for natural populations of animals. *Quarterly Review of Biology*, **22**, 283–314.
Dickie, I. A. and Reich, P. B. (2005) Ectomycorrhizal communities at forest edges. *Journal of Ecology*, **93**, 244–55.
Dickinson, C. H. and Preece, T. F. (1976) *Microbiology of Aerial Plant Surfaces*. London: Academic Press.
Dodd, M. E., Silvertown, J., McConway, K., Potts, J. and Crawley, M. (1994) Stability in the plant communities of the Park Grass Experiment: the relationships between species richness, soil pH, and biomass variability. *Philosophical Transactions of the Royal Society, London B*, **346**, 185–93.
Domec, J.-C., Warren, J. M., Meinzer, F. C., Brooks, J. R. and Coulombe, R. (2004) Native root xylem embolism and stomatal closure in stands of Douglas-fir and ponderosa pine: mitigation by hydraulic redistribution. *Oecologia*, **141**, 7–16.
Duvigneaud, P. (1971) Concepts sur la productivité primaire des écosystèmes forestiers. In *Symposium on the Productivity of Forest Ecosystems*, Ed. Duvigneaud, P. Paris: UNESCO, pp. 111–140.
Duvigneaud, P. and Denaeyer-De Smet, S. (1970) Biological cycling of minerals in temperate deciduous forests. In *Analysis of Temperate Forest Ecosystems*, Ed. Reichle, D. E. Berlin: Springer, pp. 199–225.
Dzwonko, Z. (2001) Effect of proximity to ancient deciduous woodland on restoration of the field layer vegetation in a pine plantation. *Ecography*, **24**, 198–204.
Edwards, P. J. (1982) Studies of mineral cycling in a montane rain forest in New Guinea. *Journal of Ecology*, **70**, 807–27.
Edwards, P. N. (1983) *Timber Measurement: a Field Guide*. Forestry Commission Booklet No. 49. London: HMSO.

Ehrlen, J. and Eriksson, O. (2003) Large-scale spatial dynamics of plants: a response to Freckleton & Watkinson. *Journal of Ecology*, **91**, 316–20.

Ellenberg, H. (1988) *Vegetation Ecology of Central Europe*. (Fourth Edition). Cambridge: Cambridge University Press.

Elton, C. S. (1958) *The Ecology of Invasions by Animals and Plants*. London: Methuen.

Emerman, S. H. and Dawson, T. E. (1996) Hydraulic lift and its influence on the water content of the rhizosphere: An example from sugar maple, *Acer saccharum*. *Oecologia*, **108**, 273–8.

Erickson, C. (2003) Historical ecology and future explorations. In *Amazonian Dark Earths: Origin, Properties, Management*, Ed. Lehmann, J., Kern, D. C., Glaser B. and Woods, W. I. Dordrecht: Kluwer, pp. 455–500.

Erickson, H. E. and Ayala, G. (2004) Hurricane-induced nitrous oxide fluxes from a wet tropical forest. *Global Change Biology*, **10**, 1155–62.

Ernst, W. H. O. (1979) Population biology of *Allium ursinum* in northern Germany. *Journal of Ecology*, **67**, 347–62.

Evans, G. C. (1956) An area survey method of investigating the distribution of light intensity in woodlands, with particular reference to sunflecks, including an analysis of data from rain forest in Southern Nigeria. *Journal of Ecology*, **44**, 391–428.

Evans, J. (1984) *Silviculture of Broadleaved Woodland*. Forestry Commission Bulletin No. 64. London: HMSO.

Evans, J. (1999) *Sustainability of Forestry Plantations – the Evidence*. Report Commissioned by the Department of International Development, London.

Fahey, T. J. and Hughes, J. W. (1994) Fine root dynamics in a northern hardwood forest ecosystem, Hubbard Brook Experimental Forest, NH. *Journal of Ecology*, **82**, 533–48.

Falik, O., Reides, P., Gersani, M. and Novoplansky, A. (2003) Self/non-self discrimination in roots. *Journal of Ecology*, **91**, 525–31.

Faliński, J. B. (1986) *Vegetation Dynamics in Temperate Lowland Primeval Forests: Ecological Studies in Białowieża Forest*. Lancaster: Junk.

FAO (2005) *Global Forest Resources Assessment 2005*. Rome: Food and Agriculture Organization of the United Nations.

Fay, N. (2002) Environmental arboriculture, tree ecology and veteran tree management. *Arboricultural Journal*, **26**, 213–38.

Federer, C. A. (1984) Organic matter and nitrogen content of the forest floor in even-aged northern hardwoods. *Canadian Journal of Forest Research*, **14**, 763–7.

Ferris, R., Peace, A. J., Humphrey, J. W. and Broome, A. C. (2000) Relationships between vegetation, site type and stand structure in coniferous plantations in Britain. *Forest Ecology and Management*, **136**, 35–51.

Filella, I. and Penuelas, J. (2003–4) Indications of hydraulic lift by *Pinus halepensis* and its effects on the water relations of neighbour shrubs. *Biologia Plantarum*, **47**, 209–14.

Finzi, C. A. and Canham, C. D. (2000) Sapling growth in response to light and nitrogen availability in a southern New England forest. *Forest Ecology and Management*, **131**, 153–65.

Fitter, A. H. (2005) Darkness visible: reflections on underground ecology. *Journal of Ecology*, **93**, 231–43.

Fitter, A. H. and Fitter, R. S. R. (2002) Rapid changes in flowering time in British plants. *Science*, **296** (5573), 1689–91.

Fjeldså, J., Erlich, D., Lambin, E. and Prins, E. (1997) Are biodiversity 'hotspots' related with ecoclimate stability? A pilot study using the NOAA-AVHRR remote sensing data. *Biodiversity and Conservation*, **6**, 401–22.

Fjeldså, J. and Lovett, J. C. (1997) Geographical patterns of old and young species in African forest biota: the significance of specific montane areas as evolutionary centres. *Biodiversity and Conservation*, **6**, 325–46.

Flowerdew, J. R. and Ellwood, S. A. (2001) Impacts of woodland deer on small mammal ecology. *Forestry*, **74**, 277–87.

Floyd, D. W., Vonhof, S. L. and Seyfang, H. E. (2001) Forest sustainability: a discussion guide for professional resource managers. *Journal of Forestry*, **99**, 8–28.

Fogg, K. (1988) The effect of added nitrogen on the rate of decomposition of organic matter. *Biological Review*, **63**, 433–62.

Foley, J. A., Costa, M. H., Delire, C., Ramankutty, N. and Snyder, P. (2003) Green surprise? How terrestrial ecosystems could affect earth's climate. *Frontiers in Ecology and the Environment*, **1**, 38–44.

Forest Products Laboratory (1999) *Wood Handbook: Wood as an Engineering Material*. General Technical Report, FPL-GTR-113. Madison, WI: US Department of Agriculture, Forest Service, Forest Products Laboratory.

Forman, R. T. T. (2000) Estimate of the area affected ecologically by the road system in the United States. *Conservation Biology*, **14**, 31–5.

Forman, R. T. T. and Alexander, L. E. (1998) Roads and their major ecological effects. *Annual Review of Ecology and Systematics*, **29**, 207–31.

Foster, D. R. (1999) *Thoreau's Country Journey through a Transformed Landscape*. Cambridge, MA: Harvard University Press.

Foster, D. R. and Aber, J. D. (2004) *Forests in Time: the Environmental Consequences of 1000 Years of Change in New England*. Yale: Yale University Press.

Foster, P. (2001) The potential negative impact of global climate change on tropical montane cloud forests. *Earth-Science Reviews*, **55**, 73–106.

Fowler, D., Cape, J. N., Coyle, M. *et al.* (1999) The global exposure of forests to air pollutants. *Water, Air, and Soil Pollution*, **116**, 5–32.

Franklin, J. F., Cromack, K., Jr, McKee, A. *et al.* (1981) *Ecological Characteristics of Old-Growth Douglas-Fir Forests*. General Technical Report PNW-118. Portland, OR: US Department of Agriculture, Forest Service, Pacific Northwest Forest and Range Experiment Station.

Freckleton, R. P. and Watkinson, A. R. (2002) Large-scale spatial dynamics of plants, metapopulations, regional ensembles and patchy populations. *Journal of Ecology*, **90**, 419–34.

Freckleton, R. P. and Watkinson, A. R. (2003) Are all plant populations metapopulations? *Journal of Ecology*, **91**, 321–4.

Frederickson, M. E., Greene, M. J. and Gordon, D. M. (2005) 'Devil's gardens' bedevilled by ants. *Nature*, **437**, 495–6.

Frey, S. D., Knorr, M., Parrent, J. L. and Simpson, R. T. (2004) Chronic nitrogen enrichment affects the structure and function of the soil microbial community in temperate hardwood and pine forests. *Forest Ecology and Management*, **196**, 159–71.

Friedman, S. K., Reich, P. B. and Frelich, L. E. (2001) Multiple scale composition and spatial distribution patterns of the north-eastern Minnesota presettlement forest. *Journal of Ecology*, **89**, 538–54.

Fujita, T., Itaya, A., Miura, M., Manabe, T. and Yamamoto, S.-I. (2003) Long-term canopy dynamics analysed by aerial photographs in a temperate old-growth evergreen broad-leaved forest. *Journal of Ecology*, **91**, 686–93.

Fuller, R. J. (1995) *Bird Life in Woodland and Forest*. Cambridge: Cambridge University Press.

Gadgil, R. L. and Gadgil, P. D. (1975) Suppression of litter decomposition by mycorrhizal roots of *Pinus radiata*. *New Zealand Journal of Forest Science*, **5**, 33–41.

Galloway, J. N., Aber, J. D., Erisman, J. W. *et al*. (2003) The nitrogen cascade. *BioScience*, **53**, 341–56.

Ganjegunte, G. K., Condron, L. M., Clinton, P. W., Davis, M. R. and Mahieu, N. (2004) Decomposition and nutrient release from radiata pine (*Pinus radiata*) coarse woody debris. *Forest Ecology and Management*, **187**, 197–211.

Garbaye, J. (1994) Helper bacteria: a new dimension to the mycorrhizal symbiosis. *New Phytologist*, **128**, 197–210.

Gardner, G. (1977) The reproductive capacity of *Fraxinus excelsior* on the Derbyshire limestone. *Journal of Ecology*, **65**, 107–18.

Gartner, T. B. and Cordon, Z. G. (2004) Decomposition dynamics in mixed-species leaf litter. *Oikos*, **104**, 230–46.

Gascon, C. and Lovejoy, T. E. (1998) Ecological impacts of forest fragmentation in central Amazonia. *Zoology – Analysis of Complex Systems*, **101**, 273–80.

Gerhardt, F. (1993) *Physiographic and historical influences of forest composition in central New England, USA*. M.F.S. thesis, Harvard University, MA, USA.

Gersani, M., Brown, J. S., O'Brien, E. E., Maina, G. M. and Abramsky, Z. (2001) Tragedy of the commons as a result of root competition. *Journal of Ecology*, **89**, 660–9.

Gibbs, J. N. and Wainhouse, D. (1986) Spread of forest pests and pathogens in the northern hemisphere. *Forestry*, **59**, 141–53.

Gibbs, J. N., Brasier, C. M. and Webber, J. F. (1994) Dutch elm disease in Britain. *Research Note 252*. Forestry Commission. London: HMSO.

Gil, L., Fuentes-Utrilla, P., Soto, Á., Cervera, M. T. and Collada, C. (2004) English elm is a 2,000-year-old Roman clone. *Nature*, **431**, 1053.

Gilbert, O. L. (1989) *The Ecology of Urban Habitats*. London: Chapman & Hall.

Gillson, L. and Willis, K. J. (2004) 'As Earth's testimonies tell': wilderness conservation in a changing world. *Ecology Letters*, **7**, 990–8.

Gimingham, C. H. and Birse, E. M. (1957) Ecological studies on growth-form in bryophytes. I. Correlations between growth-form and habitat. *Journal of Ecology*, **45**, 533–45.

Giri, C. C., Shyamkumar, B. and Anjaneyulu, C. (2004) Progress in tissue culture, genetic transformation and applications of biotechnology to trees: an overview. *Trees*, **18**, 115–35.

Girling, M. A. and Greig, J. (1985) A first fossil record for *Scolytus scolytus* (F.) (elm bark beetle): its occurrence in elm decline deposits from London and the implications for Neolithic elm disease. *Journal of Archaeological Science*, **12**, 347–51.

Goheen, J. R., Swihart, R. K., Gehring, T. M. and Miller, M. S. (2003) Forces structuring tree squirrel communities in landscapes fragmented by agriculture: species differences in perceptions of forest connectivity and carrying capacity. *Oikos*, **102**, 95–103.

Goheen, J. R., Keesing, F., Allan, B. F., Ogada, D. and Ostfeld, R. S. (2004) Net effects of large mammals on *Acacia* seedling survival in an African savanna. *Ecology*, **85**, 1555–61.

Golding, D. L. (1970) The effects of forests on precipitation. *Forestry Chronicle*, **46**, 397–402.

Goldsmith, F. B. (1998) *Tropical Rain Forest: a Wider Perspective*. London: Chapman & Hall.

Gonthier, P., Warner, R., Nicolotti, G., Mazzaglia, A. and Garbelotto, M. M. (2004) Pathogen introduction as a collateral effect of military activity. *Mycological Research*, **108**, 468–70.

Gosz, J. R., Holmes, R. T., Likens, G. E. and Bormann, F. H. (1978) The flow of energy in a forest ecosystem. *Scientific American*, **238**, 92–102.

Grabham, P. N. and Packham, J. R. (1983) A comparative study of the bluebell *Hyacinthoides non-scripta* (L.) Chouard, in two woodland situations in the West Midlands, England. *Biological Conservation*, **26**, 105–26.

Green, T. (2001) Should ancient trees be designated as Sites of Special Scientific Interest? *British Wildlife*, **12**, 164–6.

Greenup, M. (1998) Managing *Chamaecyparis lawsoniana* (Port-Orford-Cedar) to control the root disease caused by *Phytophthora lateralis* in the Pacific Northwest, USA. In *Coastally Restricted Forests*, Ed. Laderman, A. D. Oxford: Oxford University Press, pp. 93–100.

Gregg, J. W., Jones, C. G. and Dawson, T. E. (2003) Urbanization effects on tree growth in the vicinity of New York City. *Nature*, **424**, 183–7.

Grier, C. C., Vogt, K. A., Keyes, M. R. and Edmonds, R. L. (1981) Biomass distribution and above- and below-ground production in young and mature *Abies amabilis* zone ecosystems of the Washington Cascades. *Canadian Journal of Forest Research*, **11**, 155–67.

Griffin, G. J., Robbins, N., Hogan, E. P. and Farias-Santopietro, G. (2004) Nucleotide sequence identification of Cryphonectria hypovirus 1 infecting *Cryphonectria parasitica* on grafted American chestnut trees 12–18 years after inoculation with a hypovirulent strain mixture. *Forest Pathology*, **34**, 33–46.

Grimaldi, D. and Engel, M. S. (2005) *Evolution of the Insects*. Cambridge: Cambridge University Press.

Grime, J. P. (1974) Vegetation classification by reference to strategies. *Nature*, **250**, 26–31.

Grime, J. P. (1979) *Plant Strategies and Vegetation Processes*. Chichester: John Wiley & Sons.

Grime, J. P. (2001, reprinted 2002) *Plant Strategies, Vegetation Processes and Ecosystem Properties*. (Second Edition). Chichester: John Wiley & Sons.

Grime, J. P., Hodgson, J. G. and Hunt, R. (1988) *Comparative Plant Ecology*. London: Unwin Hyman.

Grosse, W., Buchel, H. B. and Lattermann, S. (1998) Root aeration in wetland trees and its eco-physiological significance. In *Coastally Restricted Forests*, Ed. Laderman, A. D. Oxford: Oxford University Press, pp. 293–305.

Grubb, P. J. (1992) A positive distrust of simplicity – lessons from plant defences and from competition among plants and among animals. *Journal of Ecology*, **80**, 585–610.

Grundy, J. H. (1981) *Arthropods of Medical Importance*. Chilbolton, UK: Noble Books.

Gurnell, J., Wauters, L. A., Lurz, P. P. W. and Tosi, G. (2004) Alien species and interspecific competition: effects of introduced eastern grey squirrels on red squirrel population dynamics. *Journal of Animal Ecology*, **73**, 26–35.

Hamann, H. (2004) Flowering and fruiting phenology of a Philippine submontane rain forest: climatic factors as proximate and ultimate causes. *Journal of Ecology*, **92**, 24–31.

Hamburg, S. P., Yanai, R. D., Arthur, M. A., Blum, J. D. and Siccama, T. G. (2003) Biotic control of calcium cycling in northern hardwood forests: acid rain and aging forests. *Ecosystems*, **6**, 399–406.

Hamilton, G. J. (1985) *Forest Mensuration Handbook*. London: HMSO.

Hamilton, G. J. and Christie, J. M. (1971) *Forest Management Tables (Metric)*. Forestry Commission Booklet, No. 34. London: HMSO.

Hamilton, G. J. and Christie, J. M. (1974) *Influence of Spacing on Crop Characteristics and Yield*. Forestry Commission Booklet, No. 52. London: HMSO.

Han, W., Lindsay, S. M., Dlakic, M. and Harrington, R. E. (1997) Socioeconomic factors and tropical deforestation. *Nature*, **386**, 562–3.

Hansen, R. A. (1999) Red oak promotes a microarthropod functional group that accelerates its decomposition. *Plant and Soil*, **209**, 37–45.

Hansen, R. A. and Coleman, D. C. (1998) Litter complexity and composition are determinants of the diversity and species composition of oribatid mites (Acari: Oribatida) in litterbags. *Journal of Applied Soil Ecology*, **9**, 17–23.

Harding, D. J. L. (1992) Defoliation patterns amongst Chaddesley oaks. In *Ecology of Woodland Processes*, Ed. Packham, J. R. and Harding, D. J. L. London: Edward Arnold, pp. 69–79.

Harding, D. J. L. (1995) The terrestrial caddis *Enoicyla pusilla*. In *Ecology, Management and History of the Wyre Forest*, Ed. Packham, J. R. and Harding, D. J. L. University of Wolverhampton and the British Ecological Society, pp. 108–113.

Harding, D. J. L. (1996) *Geometrids foretell the Millennium*. Manuscript of paper delivered at British Ecological Society Winter Meeting at Durham, December 1996.

Harding, D. J. L. (2000) Two decades of data on oak defoliation in a Worcestershire woodland NNR. In *Long-term Studies in British Woodland*, Ed. Kirby, K. J. and Morecroft, M. D. English Nature Science No. 34. Peterborough: English Nature, pp. 87–97.

Harding, D. J. L. (2002) Where have all the caterpillars gone? *Quarterly Journal of Forestry*, **96**, 278–83.

Harding, D. J. L. and Stuttard, R. A. (1974) Microarthropods. In *Biology of Plant Litter Decomposition*, Vols I and II, Ed. Dickinson, C. H. and Pugh, G. J. F. London: Academic Press, pp. 489–532.

Harmon, M. E., Franklin, J. F., Swanson, F. J. *et al.* (1986) Ecology of coarse woody debris in temperate ecosystems. *Advances in Ecological Research*, **15**, 133–302.

Harper, J. L. (1977) *Population Biology of Plants*. London: Academic Press.

Hartley, M. J. (2002) Rationale and methods for conserving biodiversity in plantation forests. *Forest Ecology and Management*, **155**, 81–95.

Hättenschwiler, S. and Vitousek, P. M. (2000) The role of polyphenols in terrestrial ecosystem nutrient cycling. *Trends in Ecology and Evolution*, **15**, 238–43.

Hawksworth, D. L. (1991) The fungal dimension of biodiversity: magnitude, significance, and conservation. *Mycological Research*, **95**, 641–55.

Heckenberger, M. J., Kuikuro, A., Kuikuro, U. T. *et al.* (2003) Amazonia 1492: pristine forest or cultural parkland? *Science*, **301**, 1710–14.

Hedin, L. O., Armesto, J. J. and Johnson, A. H. (1995) Patterns of nutrient loss from unpolluted, old-growth temperate forests: evaluation of biogeochemical theory. *Ecology*, **76**, 493–509.

Heilmann-Clausen, J. and Christensen, M. (2000) Svampe pa bogestammer – indikatorer for vaerdifulde lovskovslokaliteter. *Svampe*, **42**, 35–47.

Heilmann-Clausen, J. and Christensen, M. (2003) Fungal diversity on decaying beech logs – implications for sustainable forestry. *Biodiversity and Conservation*, **12**, 953–73.

Heliövarra, K. and Väisänen, R. (1984) Effects of modern forestry on northwestern European forest invertebrates: a synthesis. *Acta Forestalia Fennica*, **189**, 1–32.

Heliövarra, K., Väisänen, R. and Simon, C. (1994) Evolutionary study of periodical insects. *Trends in Ecological Evolution*, **9**, 475–80.

Helliwell, D. R. (1996) Habitat transference. *Aspects of Applied Biology*, **44**, 401–5.

Helliwell, D. R. (2002) *Continuous Cover Forestry*. (Second Edition). Privately Published Wirksworth, Derbyshire.

Helliwell, D. R., Buckley, G. P., Fordham, S. J. and Paul, T. A. (1996) Vegetation succession on a relocated ancient woodland soil. *Forestry*, **69**, 57–74.

Helms, J. A. (2002) Forest, forestry and forester: what do these terms mean? *Journal of Forestry*, **100**, 15–19.

Hennon, P. E., Shaw, C. G. III and Hansen, E. M. (1998) Reproduction and forest decline of *Chamaecyparis nootkatensis* (yellow-cedar) in southeast Alaska, USA. In *Coastally Restricted Forests*, Ed. Laderman, A. D. Oxford: Oxford University Press, pp. 54–69.

Herrera, C. M., Jorrdano, P., Guitan, J. and Travaset, A. (1998) Annual variability in seed production by woody plants and the masting concept: reassessment of principles and relationship to pollination and seed dispersal. *The American Naturalist*, **152**, 576–94.

Hill, M. O. (1979a) *DECORANA – A FORTRAN Program for Detrended Correspondence Analysis and Reciprocal Averaging*. Ithaca, NY: Ecology and Systematics, Cornell University.

Hill, M. O. (1979b) *TWINSPAN – A FORTRAN Program for arranging Multivariate Data in an Ordered Two-way Table by Classification of the Individuals and Attributes*. Ithaca, NY: Ecology and Systematics, Cornell University.

Hill, M. O., Bunce, R. G. H. and Shaw, M. W. (1975) Indicator species analysis: a divisive polythetic method of classification and its application to a survey of native pinewoods in Scotland. *Journal of Ecology*, **63**, 597–613.

Hilton, G. M. and Packham, J. R. (2003) Variation in the masting of common beech (*Fagus sylvatica* L.) in northern Europe over two centuries (1800–2001). *Forestry*, **76**, 319–28.

Hilton, G. M., Packham, J. R. and Willis, A. J. (1987) Effects of experimental defoliation on a population of pedunculate oak (*Quercus robur* L.). *New Phytologist*, **107**, 603–12.

Hobson, P. R. (1995a) Patterns of the past: ancient records and present evidence. In *Ecology, Management & History of the Wyre Forest*, Ed. Packham, J. R. and Harding, D. J. L. University of Wolverhampton and the British Ecological Society, pp. 6–24.

Hobson, P. R. (1995b) Charcoal burning and other forest industries in Wyre Forest. In *Ecology, Management & History of the Wyre Forest*, Ed. Packham, J. R. and Harding, D. J. L. University of Wolverhampton and the British Ecological Society, pp. 126–132.

Hobson, P. R. (1997) *Ecology, history, management and conservation of the multipurpose Forest of Wyre*. Ph.D. thesis, University of Wolverhampton, UK.

Hodder, K. H., Bullock, J. M., Buckland, P. C. and Kirby, K. J. (2005) *Large Herbivores in the Wildwood and Modern Naturalistic Grazing Systems*. English Nature Research Reports, No. 648. Peterborough: English Nature.

Hodkinson, I. D., Webb, N. R. and Coulson, S. J. (2002) Primary community assembly on land – the missing stages: why are heterotrophic organisms always there first? *Journal of Ecology*, **90**, 569–77.

Hole, F. D. (1981) Effects of animals on soil. *Geoderma*, **25**, 75–112.

Hopkin, S. (2004) Millipedes. *British Wildlife*, **16**, 77–84.

Hopkins, B. (1965) *Forest and Savanna*. London: Heinemann.

Hora, F. B. (1981) *The Oxford Encyclopedia of Trees of the World*. Oxford: Oxford University Press.

Hornbeck, J. W., Martin, C. W. and Eagar, C. (1997) Summary of water yield experiments at Hubbard Brook Experimental Forest, New Hampshire. *Canadian Journal of Forest Research*, **27**, 2043–52.

Horton, T. R., Bruns, T. D. and Parker, V. T. (1999) Ectomycorrhizal fungi associated with *Arctostaphylos* contribute to *Pseudotsuga menziesii* establishment. *Canadian Journal of Botany*, **77**, 93–102.

Houghton, J. T., Ding, Y., Griggs, D. J. *et al.* (2001) *Climate Change 2001: The Scientific Basis*. Cambridge: Cambridge University Press.

Houghton, R. A. (2003) Why are estimates of the terrestrial carbon balance so different? *Global Change Biology*, **9**, 500–9.

Houlton, B. Z., Driscoll, C. T., Fahey, T. J. *et al.* (2003) Nitrogen dynamics in ice storm-damaged forest ecosystems: implications for nitrogen limitation theory. *Ecosystems*, **6**, 431–43.

Houter, N. C. and Pons, T. L. (2005) Gap size effects on photoinhibition in understorey saplings in tropical rainforest. *Plant Ecology*, **179**, 43–51.

Hubbell, S. P. (2001) *Unified Neutral Theory of Biodiversity and Biogeography*. Princeton, NJ: Princeton University Press.

Hughes, A. P. (1959) Effects of the environment on leaf development in *Impatiens parviflora* D.C. *Journal of the Linnean Society (Botany)*, **56**, 161–5.

Hultengren, S. (1999) The project "The epiphytic lichens of southwestern Sweden"– a short presentation. *Symbolae Botanicae Upsalienses*, **32** (2), 181–93.

Humphrey, J. W. (2005) Benefits to biodiversity from developing old-growth conditions in British upland spruce plantations: a review and recommendations. *Forestry*, **78**, 33–53.

Humphrey, J. W., Davey, S., Peace, A. J., Ferris, R. and Harding, K. (2002) Lichens and bryophyte communities of planted and semi-natural forests in Britain: the influence of site type, stand structure and deadwood. *Biological Conservation*, **107**, 165–80.

Hutchings, M. J. and Slade, A. J. (1988) Foraging for resources and the structure of plants. *Plants Today*, **1**, 28–33.

Hyatt, L. A., Rosenberg, M. S., Howard, T. G. *et al.* (2003) The distance dependence prediction of the Janzen–Connell hypothesis: a meta-analysis. *Oikos*, **103**, 590–602.

Hytteborn, H. and Packham, J. R. (1985) Left to nature: forest structure and regeneration in Fiby urskog, central Sweden. *Arboricultural Journal*, **9**, 1–11.

Hytteborn, H. and Packham, J. R. (1987) Decay rate of *Picea abies* logs and the storm gap theory: a re-examination of Sernander Plot III Fiby urskog, central Sweden. *Arboricultural Journal*, **11**, 299–311.

Hytteborn, H., Packham, J. R. and Verwijst, T. (1987) Tree population dynamics, stand structure and species composition in the montane virgin forest of Vallibacken, northern Sweden. *Vegetatio*, **72**, 3–19.

Hytteborn, H., Liu Qinghong and Verwijst, T. (1993) Small-scale disturbance and stand structure dynamics in a primeval *Picea abies* forest over 54 years, central Sweden. In *Small-scale natural disturbance and tree regeneration in boreal forests* (Liu Qinghong). Ph.D. thesis, University of Uppsala, Sweden.

Hytteborn, H., Maslov, A. A., Nazimova, D. I. and Rysin, L. P. (2005) Boreal forests of Eurasia. In *Coniferous Forests. Ecosystems of the World 6*, Ed. Andersson, F. A. Amsterdam: Elsevier, pp. 23–99.

Ingrouille, M. (1992) *Diversity and Evolution of Land Plants*. London: Chapman & Hall.

Iverson, L. R., Schwarts, M. W. and Prasad, A. M. (2004) How fast and how far might tree species migrate in the eastern United States due to climate change? *Global Ecology and Biogeography*, **13**, 209–19.

Jabin, M., Mohr, D., Kappes, H. and Topp, W. (2004) Influence of deadwood on density of soil macro-arthropods in a managed oak–beech forest. *Forest Ecology and Management*, **194**, 61–9.

Jackson, R. B., Canadell, J., Ehleringer, J. R. *et al.* (1996) A global analysis of root distributions for terrestrial biomes. *Oecologia*, **108**, 389–411.

Jacobs, D. F. and Severeid, L. R. (2004) Dominance of interplanted American chestnut (*Castanea dentata*) in southwestern Wisconsin, USA. *Forest Ecology and Management*, **191**, 111–20.

Jandl, R., Alewell, C. and Prietzel, J. (2004) Calcium loss in Central European forest soils. *Soil Science Society of America Journal*, **68**, 588–95.

Janzen, D. H. (1970) Herbivores and the number of tree species in tropical forests. *American Naturalist*, **104**, 501–28.

Janzen, D. H. (1971) Seed predation by animals. *Annual Review of Ecological Systematics*, **2**, 465–92.

Janzen, D. H. (1978) Seeding patterns of tropical trees. In *Tropical Trees as Living Systems*, Ed. Tomlinson, P. B. and Zimmerman, M. H. Cambridge: Cambridge University Press, pp. 83–128.

Jenik, J. (1979) *Pictorial Encyclopedia of Forests*. London: Hamlyn.

Jobbágy, E. G. and Jackson, R. B. (2004) The uplift of soil nutrients by plants: biogeochemical consequences across scales. *Ecology*, **85**, 2380–9.

Johns, A. D. (1988) Effects of selective timber extraction on rain-forest structure and composition and some consequences for frugivores and folivores. *Biotropica*, **20**, 31–7.

Johnson, K. H., Vogt, K. A., Clark, H. J., Schmitz, O. J. and Vogt, D. J. (1996) Biodiversity and the productivity and stability of ecosystems. *Trends in Ecology and Evolution*, **11**, 372–7.

Jones, C. G., Ostfeld, R. S., Richard, M. P., Schauber, E. M. and Wolff, J. O. (1998) Chain reactions linking acorns to gypsy moth outbreaks and Lyme disease risk. *Science*, **279**, 1023–6.

Jones, D. L. and Elliot, W. R. (1986) *Pests, Diseases and Ailments of Australian Plants*. Melbourne: Lothian.

Jones, E. W. (1959) *Quercus* L. Biological flora of the British Isles. *Journal of Ecology*, **47**, 169–222.

Jones, K. B., Neale, A. C., Nash, M. S. *et al.* (2000) Landscape correlates of breeding bird richness across the United States mid-Atlantic region. *Environmental Monitoring and Assessment*, **63**, 159–74.

Jones, W., Hill, K. and Allen, J. (1995) *Wollemia nobilis*, a new living Australian genus and species in the Arucariaceae. *Telopea*, September 1995, 173–6.

Kauppi, P. E., Mielikainen, K. and Kuusela, K. (1992) Biomass and carbon budget of European forests, 1971 to 1990. *Science*, **256**, 70–4.

Kelly, D. and Sork, V. L. (2002) Mast seeding in perennial plants; why, how, where? *Annual Review of Evolutionary Systematics*, **33**, 427–47.

Kennedy, P. G., Izzo, A. D. and Bruns, T. D. (2003) There is a high potential for the formation of common mycorrhizal networks between understorey and canopy trees in a mixed evergreen forest. *Journal of Ecology*, **91**, 1071–80.

Keppler, F., Hamilton, J. T. G., Braß, M. and Röckmann, T. (2006) Methane emissions from terrestrial plants under aerobic conditions. *Nature*, **439**, 187–91.

Khomentovsky, P. A. (1998) *Pinus pumila* (Siberian Dwarf Pine) on the Khamchatka Peninsula, Northeast Asia: ecology of seed production. In *Coastally Restricted Forests*, Ed. Laderman, A. D. Oxford: Oxford University Press, pp. 199–220.

Killingbeck, K. T. (1996) Nutrients in senesced leaves: keys to the search for potential resorption and resorption proficiency. *Ecology*, **77**, 1716–27.

Kimmins, H. (1992) *Balancing Anti-environmental Issues in Forestry*. Vancouver: UBC Press.

Kimmins, J. P. (1997) *Forest Ecology*. New Jersey: Prentice Hall.

Kira, T. (1975) Primary production of forests. In *Photosynthesis and Productivity in Different Environments*, Ed. Cooper, J. P. Cambridge: Cambridge University Press, pp. 5–40.

Kirby, K. J. (2003) *What Might a British Forest-Landscape Driven by Large Carnivores Look Like?* English Nature Research Reports, No. 530. Peterborough: English Nature.

Kirby, K. J. and Drake, M. (1993) *Dead Wood Matters*. English Nature Science No 7. Peterborough: English Nature.

Kitajima, K. and Augspurger, C. K. (1989) Seed and seedling ecology of a monocarpic tropical tree, *Tachigali versicolor*. *Ecology*, **70**, 1102–14.

Kitayama, K., Suzuki, S., Hori, M. *et al.* (2004) On the relationships between leaf-litter lignin and net primary productivity in tropical rain forests. *Oecologia*, **140**, 335–9.

Kizlinski, M. L., Orwig, D. A., Cobb, R. C. and Foster, D. R. (2002) Direct and indirect ecosystem consequences of an invasive pest on forests dominated by eastern hemlock. *Journal of Biogeography*, **29**, 1489–504.

Klironomos, J. N. and Hart, M. M. (2001) Animal nitrogen swap for plant carbon. *Nature*, **410**, 651–2.

Knops, J. M. H., Nash, III T. H. and Schlesinger, W. H. (1996) The influence of epiphytic lichens on the nutrient cycling of an oak woodland. *Ecological Monographs*, **66**, 159–79.

Knops, J. M. H., Bradley, K. L. and Wedin, D. A. (2002) Mechanisms of plant species impacts on ecosystem nitrogen cycling. *Ecology Letters*, **5**, 454–66.

Koenig, W. D. and Knops, J. M. H. (1998) Scale of mast-seeding and tree-ring growth. *Nature*, **396**, 225–6.

Koenig, W. D. and Knops, J. M. H. (2000) Patterns of annual seed production by northern hemisphere trees: a global perspective. *The American Naturalist*, **155**, 59–69.

Koenig, W. D. and Liebhold, A. M. (2003) Regional impacts of periodical cicadas on oak radial increment. *Canadian Journal of Forest Research*, **33**, 1084–9.

Köhl, M., Traub, B. and Päivinen, R. (2000) Harmonisation and standardisation in multi-national environmental statistics – mission impossible? *Environmental Monitoring and Assessment*, **63**, 361–80.

Kohyama, T. (1984) Regeneration and coexistence of two *Abies* species dominating subalpine forests in central Japan. *Oecologia*, **62**, 156–61.
Komonen, A. (2003) Hotspots of insect diversity in boreal forests. *Conservation Biology*, **17**, 976–91.
Konrad, H., Kirisits, T., Riegler, M., Halmschlager, E. and Stauffer, C. (2002) Genetic evidence for natural hybridization between the Dutch elm disease pathogens *Ophiostoma novo-ulmi* ssp. *novo-ulmi* and *O. novo-ulmi* ssp. *americana*. *Plant Pathology*, **51**, 78–84.
Kramer, K. (1995) Phenotypic plasticity of the phenology of 7 European tree species in relation to climatic warming. *Plant Cell and Environment*, **18**, 93–104.
Kraus, T. E. C., Dahlgren, R. A. and Zasoski, R. J. (2003) Tannins in nutrient dynamics of forest ecosystems – a review. *Plant and Soil*, **256**, 41–66.
Kulakowski, D. and Veblen, T. T. (2002) Influences of fire history and topography on the pattern of a severe wind blowdown in a Colorado subalpine forest. *Journal of Ecology*, **90**, 806–19.
Kuuluvainen, T. (1994) Gap disturbance, ground microtopography, and the regeneration dynamics of boreal coniferous forests in Finland: a review. *Annales Zoologici Fennici*, **31**, 35–51.
Kuuluvainen, T. and Juntunen, P. (1998) Seedling establishment in relation to microhabitat variation in a windthrow gap in a boreal *Pinus sylvestris* forest. *Journal of Vegetation Science*, **9**, 551–62.
Laderman, A. D. (1998) *Coastally Restricted Forests*. Oxford: Oxford University Press.
Laiho, R. and Prescott, C. E. (2004) Decay and nutrient dynamics of coarse woody debris in northern coniferous forests: a synthesis. *Canadian Journal of Forest Research*, **34**, 763–77.
Lanner, R. M. (1996) *Made For Each Other: a Symbiosis of Birds and Pines*. New York: Oxford University Press.
Larcher, W. (1975) *Physiological Plant Ecology*. Berlin: Springer.
Laurance, W. F., Pérez-Salicrup, D., Delamônica, P. *et al.* (2001) Rain forest fragmentation and the structure of Amazonian liana communities. *Ecology*, **82**, 105–16.
Laurance, W. F., Lovejoy, T. E., Vasconcelos, H. L. *et al.* (2002) Ecosystem decay of Amazonian forest fragments: a 22-year investigation. *Conservation Biology*, **16**, 605–18.
Laurance, W. F., Albernaz, A. K. M., Fearnside, P. M., Vasconcelos, H. L. and Ferreira, L. V. (2004) Deforestation in Amazonia. *Science*, **304**, 1109.
Laureysens, I., Blust, R., De Temmerman, L., Lemmens, C. and Ceulemans, R. (2004) Clonal variation in heavy metal accumulation and biomass production in a poplar coppice culture: I. Seasonal variation in leaf, wood and bark concentrations. *Environmental Pollution*, **131**, 485–94.
Leake, J. R. (2005) Plants parasitic on fungi: unearthing the fungi in myco-heterotrophs and debunking the 'saprophytic' plant myth. *Mycologist*, **19**, 113–22.
Lee, J. A. (1999) The calcicole-calcifuge problem revisited. *Advances in Botanical Research*, **29**, 1–30.
Le Houérou, H. N. (1997) Climate, flora and fauna changes in the Sahara over the past 500 million years. *Journal of Arid Environments*, **37**, 619–47.
Li, W. H. (2004) Degradation and restoration of forest ecosystems in China. *Forest Ecology and Management*, **201**, 33–41.

Liddle, M. J. (1997) *Recreation Ecology*. London: Chapman & Hall.
Likens, G. E. (2004) Some perspectives on long-term biogeochemical research from the Hubbard Brook Ecosystem Study. *Ecology*, **85**, 2355–62.
Likens, G. E., Bormann, F. H., Johnson, N. M., Fisher, D. W. and Pierce, R. S. (1970) Effects of forest cutting and herbicide treatment on nutrient budgets in the Hubbard Brook watershed-ecosystem. *Ecological Monographs*, **40**, 23–47.
Likens, G. E., Driscoll, C. T. and Buso, D. C. (1996) Long-term effects of acid rain: response and recovery of a forest ecosystem. *Science*, **272**, 244–6.
Likens, G. E., Driscoll, C. T., Buso, D. C. *et al.* (1998) The biogeochemistry of calcium at Hubbard Brook. *Biogeochemistry*, **41**, 89–173.
Likens, G. E., Driscoll, C. T., Buso, D. C. *et al.* (2002) The biogeochemistry of sulfur at Hubbard Brook. *Biogeochemistry*, **60**, 235–316.
Lincoln, R., Boxshall, G. and Clark, P. (1998) *A Dictionary of Ecology, Evolution and Systematics*. (Second Edition). Cambridge: Cambridge University Press.
Lindenmayer, D. B. and Franklin, J. F. (1997) Managing stand structure as part of ecologically sustainable forest management in Australian mountain ash forests. *Conservation Biology*, **11**, 1053–68.
Lindow, S. E. and Brandl, M. T. (2003) Microbiology of the phyllosphere. *Applied and Environmental Microbiology*, **69**, 1875–83.
Lindquist, B. (1931) Den Skandinaviska bogskogens biologi (The ecology of Scandinavian beechwoods). *Svenska Skogsvardsforeningens Tidskrift*, **29** (English digest), 486–520.
Lindström, J., Ranta, E., Kokko, H., Lundberg, P. and Kaitala, V. (2001) From arctic lemmings to adaptive dynamics: Charles Elton's legacy in population ecology. *Biological Review*, **76**, 129–58.
Lipson, D. and Näsholm, T. (2001) The unexpected versatility of plants: organic nitrogen use and availability in terrestrial ecosystems. *Oecologia*, **128**, 305–16.
Littlemore, J. and Barker, S. (2003) The ecological response of forest ground flora and soils to experimental trampling in British urban woodlands. *Urban Ecosystems*, **5**, 257–76.
Lodge, D. J. (1996) Microorganisms. In *The Food Web of a Tropical Forest*, Ed. Reagan, D. P. and Waide, R. B. Chicago: University of Chicago Press, pp. 54–108.
Lodge, D. J. (1997) Factors related to diversity of decomposer fungi in tropical forests. *Biodiversity and Conservation*, **6**, 681–8.
Lodge, D. J., McDowell, W. H. and McSwiney, C. P. (1994) The importance of nutrient pulses in tropical forests. *Trends in Ecology and Evolution*, **9**, 384–7.
Logan, J. A., Régnière, J. and Powell, J. A. (2003) Assessing the impacts of global warming on forest pest dynamics. *Frontiers in Ecology and the Evironment*, **1**, 130–7.
Lovett, G. M., Christenson L. M., Groffman, P. M. *et al.* (2002) Insect defoliation and nitrogen cycling in forests. *BioScience*, **52**, 335–41.
Lowry, W. P. (1966) Apparent meteorological requirements for abundant cone crop in Douglas-Fir. *Forest Science*, **12**, 185–92.
Lund, G. L. (2002) When is a forest not a forest? *Journal of Forestry*, **100**, 21–8.
Lundquist, J. E. and Hamelin, R. C. (2005) *Forest Pathology: From Genes to Landscapes*. St. Paul, MN: APS Press/The American Phytopathological Society.
Luxton, M. (1972) Studies of the oribatid mites of a Danish beech wood soil. I. Nutritional biology. *Pedobiologia*, **12**, 434–63.

Lyford, W. H. and Wilson, B. F. (1964) Development of the root system of *Acer rubrum* L. *Harvard Forest Paper*, **10**.

MacArthur, R. H. (1955) Fluctuations of animal populations and a measure of community stability. *Ecology*, **36**, 533–6.

MacArthur, R. H. (1972) *Geographical Ecology*. New York: Harper & Row.

MacArthur, R. H. and Wilson, E. D. (1967) *The Theory of Island Biogeography*. Princeton, NJ: Princeton University Press.

Macfayden, A. (1961) Metabolism of soil invertebrates in relation to soil fertility. *Annals of Applied Biology*, **49**, 216–19.

Mackensen, J., Bauhus, J. and Webber, E. (2003) Decomposition rates of coarse woody debris – a review with particular emphasis on Australian tree species. *Australian Journal of Botany*, **51**, 27–37.

Maclaren, J. P. (1996) *Environmental Effects of Planted Forests in New Zealand*. Bulletin No. 198. Rotorua: New Zealand Forest Research Institute.

Magill, A. H., Aber, D. A., Currie, W. S. *et al.* (2004) Ecosystem response to 15 years of chronic nitrogen additions at the Harvard Forest LTER, Massachusetts, USA. *Forest Ecology and Management*, **196**, 7–28.

Malmer, N., Lindgren, K. and Persson, S. (1978) Vegetational succession in a south-Swedish deciduous wood. *Vegetatio*, **36**, 17–29.

Maloof, J. (2005) Take a deep breath. *New Scientist*, **6 August**, 44–45.

Mann, C. C. (2002) The real dirt on rainforest fertility. *Science*, **297**, 920–3.

Mannan, R. W., Meslow, E. C. and Wight, H. M. (1980) Use of snags by birds in Douglas-fir forests. *Journal of Wildlife Management*, **44**, 787–97.

Manson, R. H., Ostfeld, R. S. and Canham, C. D. (2001) Long-term effects of rodent herbivores on tree invasion dynamics along forest-field edges. *Ecology*, **82**, 3320–9.

Martin, R. and Handasyde, K. (1999) *The Koala: Natural History, Conservation and Management*. (Second Edition). Malabar, FL: Krieger.

Marx, D. H. (1969) The influence of ectotrophic mycorrhizal fungi on the resistance of pine roots to pathogenic infections. I. Antagonism of mycorrhizal fungi to root pathogenic fungi and soil bacteria. *Phytopathology*, **59**, 153–63.

Mason, W. L., Kerr, G. and Simpson, J. (1999) *What is Continuous Cover Forestry?* Information Note 29. Edinburgh: Forestry Commission.

Matlack, G. R. (1994) Plant species migration in a mixed-history forest landscape in eastern North America. *Ecology*, **75**, 1491–502.

Matthews, E. (1997) Global litter production, pools, and turnover times: estimates from measurement data and regression models. *Journal of Geophysical Research: Atmospheres*, **102**, 18 771–800.

Matthews, J. D. (1989) *Silvicultural Systems*. Oxford: Clarendon Press.

Maurer, E. (1964) Buchen- und Eichensamenjahre in Unterfranken wahrend der letzen 100 Jahre. *Allgemeine Forstzeitschrift*, **31**, 469–70.

May, J. (2003) Soundwood. *Tree News* (spring/summer), 29–35.

May, R. M. (1972) Will a large complex system be stable? *Nature*, **238**, 413–14.

McCann, K. S. (2000) The diversity-stability debate. *Nature*, **405**, 228–33.

McGroddy, M. E., Daufresne, T. and Hedin, L. O. (2004) Scaling of C:N:P stoichiometry in forests worldwide: implications of terrestrial redfield-type ratios. *Ecology*, **85**, 2390–401.

McHale, P. J., Mitchell, M. J. and Bowles, F. P. (1998) Soil warming in a northern hardwood forest: trace gas fluxes and leaf litter decomposition. *Canadian Journal of Forest Research*, **28**, 1365–72.

McIntosh, P. and Laffan, M. (2005) Soil erodibility and erosion hazard: extending these cornerstone soil conservation concepts to headwater streams in the forestry estate in Tasmania. *Forest Ecology and Management*, **220**, 128–39.

McKane, R. B., Johnson, L. C., Shaver, G. R. *et al.* (2002) Resource-based niches provide a basis for plant species diversity and dominance in arctic tundra. *Nature*, **415**, 68–71.

Meggers, B. J. (1996) *Amazonia: Man and Culture in a Counterfeit Paradise*. Washington, DC: Smithsonian Institution Press.

Melillo, J. M., Aber, J. D. and Muratore, J. F. (1982) Nitrogen and lignin control of hardwood leaf litter decomposition dynamics. *Ecology*, **63**, 621–6.

Meng, J., Hu, Y., Wang, Y., Wang, X. and Li, C. (2006) A Mesozoic gliding mammal from northeastern China. *Nature*, **444**, 889–93.

Mesquita, R. C. G., Ickes, K., Ganade, G. and Williamson, G. B. (2001) Alternative successional pathways in the Amazon Basin. *Journal of Ecology*, **89**, 528–37.

Messina, M. G. and Conner, W. H. (1998) *Southern Forested Wetlands: Ecology and Management*. New York: Lewis Publishers.

Mielke, H. W. (1989) *Patterns of Life*. Boston: Unwin Hyman.

Mieth, A. and Bork, H. R. (2005) History, origin and extent of soil erosion on Easter Island (Rapa Nui). *Catena*, **63**, 244–60.

Miller, G. R. and Cummins, R. P. (2003) Soil seed banks of woodland, heathland, grassland, mire and montane communities, Cairngorm Mountains, Scotland. *Plant Ecology*, **168**, 255–66.

Minocha, R., Long, S., Magill, A. H., Aber, J. and McDowell, W. H. (2000) Foliar free polyamine and inorganic ion content in relation to soil and soil solution chemistry in two fertilized forest stands at the Harvard Forest, Massachusetts. *Plant and Soil*, **222**, 119–37.

Mitchell, A. (1974) *A Field Guide to the Trees of Britain and Northern Europe*. London: Collins.

Mitchell, E. J. G. (2005) How open were European primeval forests? Hypothesis testing using palaeoecological data. *Journal of Ecology*, **93**, 168–77.

Mitchell, P. L. and Woodward, F. L. (1988) Responses of three woodland herbs to reduced photosynthetically active radiation and low red to far-red ratio in shade. *Journal of Ecology*, **76**, 807–25.

Moberg, R. and Holmasen, I. (1990) *Lavar, en falthandbok*. (Third Edition). Stockholm: Interpublishing.

Moffat, A. and McNeill, J. (1994) *Reclaiming Disturbed Land for Forestry*. Bulletin 110, Forestry Commission. London: HMSO.

Molvar, E. M., Bowyer, R. T. and Vanballenberghe, V. (1993) Moose herbivory, browse quality, and nutrient cycling in an Alaskan treeline community. *Oecologia*, **94**, 472–9.

Moore, J. C., Walter, D. E. and Hunt, H. W. (1988) Arthropod regulation of micro- and mesobiota in below-ground detrital food webs. *Annual Review of Entomology*, **33**, 419–39.

Mori, S. A. and Prance, G. T. (1990) Taxonomy, ecology, and economic botany of the Brazil nut (*Bertholletia excelsa* Humb. and Bonpl.: Lecythidaceae). *Advances in Economic Botany*, **8**, 130–50.

Morris, P. (1997) *The Edible Dormouse (*Glis glis*)*. London: The Mammal Society.

Motzkin, G., Wilson, P., Foster, D. R. and Allen, A. (1999) Vegetation patterns in heterogeneous landscapes: the importance of history and environment. *Journal of Vegetation Science*, **10**, 903–20.

Mountford, E. P., Backmeroff, C. E. and Peterken, G. F. (2001) Long-term patterns of growth, mortality and natural disturbance in Wistman's Wood, a high altitude oakwood on Dartmoor. *Reports and Transactions of the Devonian Association for the Advancement of Science*, **133**, 227–62.

Mueller-Dombois, D. and Ellenberg, H. (1974) *Aims and Methods of Vegetation Ecology*. London: Wiley.

Muller, R. N. (2003) Landscape patterns of change in coarse woody debris accumulation in an old-growth deciduous forest on the Cumberland Plateau, southeastern Kentucky. *Canadian Journal of Forest Research*, **33**, 763–9.

Muller, R. N. and Bormann, F. H. (1976) Role of *Erythronium americanum* Ker. in energy flow and nutrient dynamics of a northern hardwood forest ecosystem. *Science*, **193**, 1126–8.

Muller, R. N. and Liu, Y. (1991) Coarse woody debris in an old-growth deciduous forest on the Cumberland Plateau, southeastern Kentucky. *Canadian Journal of Forest Research*, **21**, 1567–72.

Muñoz-Reinoso, J. C. (2004) Diversity of maritime juniper woodlands. *Forest Ecology and Management*, **192**, 267–76.

Nadelhoffer, K. J., Emmett, B. A., Gundersen, P. *et al.* (1999) Nitrogen deposition makes a minor contribution to carbon sequestration in temperate forests. *Nature*, **398**, 145–8.

Naeem, S., Thompson, L. J., Lawler, S. P., Lawton, J. H. and Woodfin, R. M. (1994) Declining biodiversity can alter the performance of ecosystems. *Nature*, **368**, 734–7.

Nara, K. and Hogetsu, T. (2004) Ectomycorrhizal fungi on established shrubs facilitate subsequent seedling establishment of successional plant species. *Ecology*, **85**, 1700–7.

Narukawa, Y. and Yamamoto, S.-I. (2001) Gap formation, microsite variation and the conifer seedling occurrence in a subalpine old-growth forest, central Japan. *Ecological Research*, **16**, 617–25.

Nee, S. (2005) The neutral theory of biodiversity: do the numbers add up? *Functional Ecology*, **19**, 173–6.

Neff, J. C., Chapin, III F. S. and Vitousek, P. M. (2003) Breaks in the cycle: dissolved organic nitrogen in terrestrial ecosystems. *Frontiers in Ecology and the Environment*, **1**, 205–11.

Nicholson, S. E. (2001) Climatic and environmental change in Africa during the last two centuries. *Climate Research*, **17**, 123–44.

Nielsen, B. O. (1978) Above ground food resources and herbivory in a beech forest ecosystem. *Oikos*, **31**, 273–9.

Nielsen, P. C. and de Muckadeli, M. S. (1954) Flower observations and controlled pollinations in *Fagus*. *Silvae Genetica*, **3**, 6–17.

Northup, R. R., Yu, Z., Dahlgren, R. A. and Vogt, K. A. (1995) Polyphenol control of nitrogen release from pine litter. *Nature*, **377**, 227–9.

Norton, D. A. and Kelly, D. (1988) Mast seeding over 33 years in *Dacrydium cupressinum* Lamb. (rimu) (Podocarpaceae) in New Zealand: the importance of economies of scale. *Functional Ecology*, **2**, 399–408.

Norton, J. M. and Firestone, M. K. (1996) N dynamics in the rhizosphere of *Pinus ponderosa* seedlings. *Soil Biology and Biochemistry*, **28**, 351–62.

Nosengo, N. (2003) Fertilized to death. *Nature*, **425**, 894–5.

O'Hanlon-Manners, D. L. and Kotanen, P. M. (2004) Evidence that fungal pathogens inhibit recruitment of a shade-intolerant tree, white birch (*Betula papyrifera*), in understory habitats. *Oecologia*, **140**, 650–3.

Olmsted, N. W. and Curtis, J. D. (1947) Seeds of the forest floor. *Ecology*, **28**, 49–52.

Olson, J. S. (1963) Energy storage and the balance of producers and decomposers in ecological systems. *Ecology*, **44**, 322–31.

Ortega, Y. K. and Capen, D. E. (2002) Roads as edges: effects on birds in forested landscapes. *Forest Science*, **48**, 381–90.

Orwig, D. A. (2002) Ecosystem response to regional impacts of introduced pests and pathogens: historical context, questions and issues. *Journal of Biogeography*, **29**, 1471–4.

O'Shea, M. and Halliday, T. (2001) *Reptiles and Amphibians*. London: Dorling Kindersley.

Ouis, D. (2003) Non-destructive techniques for detecting decay in standing trees. *Arboricultural Journal*, **27**, 159–77.

Ovington, J. D. (1958) Studies of the development of woodland conditions under different trees. VI. Soil calcium and magnesium. *Journal of Ecology*, **46**, 391–405.

Packham, J. R. (1979) Factors influencing the growth and distribution of the wood sorrel (*Oxalis acetosella* L.) on the Long Mynd, Shropshire. *Caradoc and Severn Valley Field Club*, Occasional paper No. 3, 1–14.

Packham, J. R. (1998) Entries on pp. 173, 662–3, 626, 673, 742–3, and 778. In *The Encyclopedia of Ecology and Environmental Management*, Ed. Calow, P. Oxford: Blackwell Science.

Packham, J. R. (2003) Regenerative ability in fungally infected ancient common beech (*Fagus sylvatica* L.). *Arboricultural Journal*, **27**, 155–7.

Packham, J. R. (2004) The Laurissylva of Madeira: contemporary status of a Tertiary Forest. *Arboricultural Journal*, **28**, 85–94.

Packham, J. R. and Cohn, E. V. J. (1990) Ecology of the woodland field layer. *Arboricultural Journal*, **14**, 357–71.

Packham, J. R. and Harding, D. J. L. (1982) *Ecology of Woodland Processes*. London: Edward Arnold.

Packham, J. R. and Harding, D. J. L. (1995) *Ecology, Management & History of the Wyre Forest*. University of Wolverhampton and the British Ecological Society.

Packham, J. R. and Hilton, G. M. (2002) Inter- and intra-site variation in the fruiting of common beech (*Fagus sylvatica* L.) over a twenty-two year period (1980–2001). *Arboricultural Journal*, **26**, 1–22.

Packham, J. R. and Willis, A. J. (1976) Aspects of the ecology of two woodland herbs *Oxalis acetosella* L. and *Galeobdolon luteum* Huds. *Journal of Ecology*, **64**, 485–510.

Packham, J. R. and Willis, A. J. (1977) The effects of shading on *Oxalis acetosella*. *Journal of Ecology*, **65**, 619–42.

Packham, J. R. and Willis, A. J. (1982) The influence of shading and of soil type on the growth of *Galeobdolon luteum*. *Journal of Ecology*, **70**, 491–512.

Packham, J. R., Harding, D. J. L., Hilton, G. M. and Stuttard, R. A. (1992) *Functional Ecology of Woodlands and Forests*. London: Chapman & Hall.

Packham, J. R., Cohn, E. V. J., Millett, P. and Trueman, I. C. (1995) Introduction of plants and manipulation of field layer vegetation. In *The Ecology of Woodland Creation*, Ed. Ferris-Kaan, R. Chichester: Wiley, pp. 129–148.

Paillet, F. L. (2002) Chestnut: history and ecology of a transformed species. *Journal of Biogeography*, **29**, 1517–30.

Paoli, G. D., Curran, L. M. and Zak, D. R. (2005) Phosphorus efficiency of Bornean rain forest productivity: evidence against the unimodal efficiency hypothesis. *Ecology*, **86**, 1548–61.

Pasek, J. E., Burdsall, H. H., Cavey, J. F. *et al.* (2000) *Pest Risk Assessment for Importation of Solid Wood Packing Materials into the United States*. US Department of Agriculture, Animal and Plant Health Inspection Service and US Department of Agriculture, Forest Service.

Peakall, R. (1998) Exceptionally low genetic diversity in an ancient relic, the Wollemi Pine: implications for conservation theory and practice. *Abstracts, Genetics Society of Australia, 45th Annual Conference*.

Penuelas, J. and Filella, I. (2003) Deuterium labelling of roots provides evidence of deep water access and hydraulic lift by *Pinus nigra* in a Mediterranean forest of NE Spain. *Environmental and Experimental Botany*, **49**, 201–8.

Perry, I. and Moore, P. D. (1987) Dutch elm disease as an analogue of Neolithic elm decline. *Nature*, **326**, 72–3.

Peterken, G. F. (1993) *Woodland Conservation and Management*. (Second Edition). London: Chapman & Hall.

Peterken, G. F. (1996) *Natural Woodland: Ecology and Conservation in Northern Temperate Regions*. Cambridge: Cambridge University Press.

Peterken, G. F. (1999) Applying natural forestry concepts in an intensively managed landscape. *Global Ecology and Biogeography*, **8**, 321–8.

Pfister, R. D., Kovalchik, B. L., Arno, S. F. and Presby, R. C. (1977) *Forest Habitat Types of Montana*. General Technical Report INT-34. Ogden, UT: US Department of Agriculture, Forest Service Intermountain Forest and Range Experimental Station.

Phillips, R. W., Hughes, J. H., Buford, M. A. *et al.* (1998) Atlantic White Cedar in North Carolina, USA: a brief history and current regeneration efforts. In *Coastally Restricted Forests*, Ed. Laderman, A. D. Oxford: Oxford University Press, pp. 156–170.

Pickett, S. T. A. and McDonnell, M. J. (1989) Changing perspectives in community dynamics – a theory of successional forces. *Trends in Ecology and Evolution*, **4**, 241–5.

Pickett, S. T. A. and White, P. S. (1985) *The Ecology of Natural Disturbance and Patch Dynamics*. New York: Academic Press.

Pigott, C. D. (1975) Natural regeneration of *Tilia cordata* in relation to forest-structure in the forest of Białowieża, Poland. *Philosophical Transactions of the Royal Society of London. Series B, Biological Sciences*, **270**, 151–79.

Piovesan, G., Di Filippo, A., Alessandrini, A., Biondi, F. & Schirone, B. (2005) Structure, dynamics and dendrochronology of an old-growth *Fagus* forest in the Appennines. *Journal of Vegetation Science*, **16**, 13–28.

Pitman, N. C. A., Terborgh, J., Silman, M. R. and Nuñez, V. P. (1999) Tree species distributions in an upper Amazonian forest. *Ecology*, **80**, 2651–61.

Ponge, J.-F. (2003) Humus forms in terrestrial ecosystems: a framework to biodiversity. *Soil Biology and Biochemistry*, **35**, 935–45.

Poorter, L., Bongers, F., Sterk, F. J. and Woll, H. (2005a) Beyond the regeneration phase: differentiation of height-light trajectories among tropical tree species. *Journal of Ecology*, **93**, 256–67.

Poorter, L., Zuidema, P. A., Pena-Claros, M. and Boot, R. G. A. (2005b) A monocarpic tree species in polycarpic world: how can *Tachigali vasqueszii* maintain itself in a tropical rain forest. *Journal of Ecology*, **93**, 268–78.

Poot, P. and Lambers, H. (2003) Are trade-offs in allocation patterns and root morphology related to species abundance? A congeneric comparison between rare and common species in the south-western Australian flora. *Journal of Ecology*, **91**, 58–67.

Popoola, T. O. S. and Fox, R. T. V. (2003) Effect of water stress on infection by species of honey fungus (*Armillaria mellea* and *A. gallica*). *Arboricultural Journal*, **27**, 139–54.

Potts, M. D. (2003) Drought in a Bornean everwet rainforest. *Journal of Ecology*, **91**, 467–74.

Potts, M. D., Ashton, P. S., Kaufman, L. S. and Plotkin, J. B. (2002) Habitat patterns in tropical rain forests: a comparison of 105 plots in northwest Borneo. *Ecology*, **83**, 2782–97.

Prance, G. T. (2003) From where do Brazil nuts come? *Trees*, **63**, 10–11.

Prescott, C. E. (1995) Does nitrogen availability control rates of litter decomposition in forests? *Plant and Soil*, **168–169**, 83–8.

Prescott, C. E., Maynard, D. G. and Laiho, R. (2000) Humus in northern forests: friend or foe? *Forest Ecology and Management*, **133**, 23–36.

Pridnya, M. V. and Cherpakov, V. V. (1996) Ecology and pathology of European chestnut (*Castanea sativa*) in the deciduous forests of the Caucasus Mountains in southern Russia. *Bulletin of the Torrey Botanical Club*, **123**, 223–39.

Punshon, T. and Dickinson, N. M. (1997) Acclimation of *Salix* to metal stress. *New Phytologist*, **137**, 303–14.

Purves, D. W., Caspersen, J. P., Moorcroft, P. R., Hurtt, G. C. and Pacala, S. W. (2004) Human-induced changes in US biogenic volatile organic compound emissions: evidence from long-term forest inventory data. *Global Change Biology*, **10**, 1737–55.

Rackham, O. (1975) *Hayley Wood: its History and Ecology*. Cambridge: Cambridgeshire and Isle of Ely Naturalists' Trust.

Rackham, O. (1990) *Trees and Woodlands in the British Landscape*. London: Dent.

Rackham, O. (2002) What is coppicing for? *Sylva/Tree News*, **1**, 1–3.

Rackham, O. (2003) *Ancient Woodland; its History, Vegetation and Uses in England*. (New Edition). Dalbeattie, Kircudbrightshire: Castlepoint Press.

Rajaniemi, T. K., Allison, V. J. and Goldberg, D. E. (2003) Root competition can cause a decline in diversity with increased productivity. *Journal of Ecology*, **91**, 407–16.

Rasanayagam, S. and Jeffries, P. (1992) Production of acid is responsible for antibiosis by some ectomycorrhizal fungi. *Mycological Research*, **11**, 971–6.

Raunkiaer, C. (1934) *The Life Forms of Plants and Statistical Plant Geography*. Oxford: Clarendon Press.

Rayner, A. D. M. and Boddy, L. (1988) Fungal communities in the decay of wood. *Advances in Microbial Ecology*, **10**, 115–66.

Redford, K. H. (1996) The empty forest. *BioScience*, **42**, 412–22.

Reed, R. A., Johnson-Barnard, J. and Baker, W. L. (1996) Contribution of roads to forest fragmentation in the Rocky Mountains. *Conservation Biology*, **10**, 1098–106.

Rees, M., Condit, R., Crawley, M., Pacala, S. and Tilman, D. (2001) Long-term studies of vegetation dynamics. *Science*, **293**, 650–5.

Reichle, D. E. (1981) *Dynamic Properties of Forest Ecosystems*. Cambridge: Cambridge University Press.

Rennenberg, V. H., Seiler, W., Matyssek, R., Gessler, A. and Kreuzwieser, J. (2004) European beech (*Fagus sylvatica* L.) – a forest tree without future in the south of Central Europe? *Allgemeine Forst Und Jagdzeitung*, **175**, 210–24.

Rhind, P. (2003) Comment: Britain's contribution to global conservation and our coastal temperate forest. *British Wildlife*, **15**, 97–102.

Rhoades, C. C., Brosi, S. L., Dattilo, A. J. and Vincelli, P. (2003) Effect of soil compaction and moisture on incidence of phytophthora root rot on American chestnut (*Castanea dentata*) seedlings. *Forest Ecology and Management*, **184**, 47–54.
Rice, S. K., Westerman, B. and Federici, R. (2004) Impacts of the exotic, nitrogen-fixing black locust (*Robinia pseudoacacia*) on nitrogen-cycling in a pine-oak ecosystem. *Plant Ecology*, **174**, 97–107.
Richards, P. W. (1996) *The Tropical Rain Forest*. (Second Edition). Cambridge: Cambridge University Press.
Richardson, D. M. and Rundel, P. W. (1998) Ecology and biogeography of *Pinus*: an introduction. In *Ecology and Biogeography of* Pinus, Ed. Richardson, D. M. Cambridge: Cambridge University Press, pp. 3–46.
Riitters, K. H. and Wickham, J. D. (2003) How far to the nearest road? *Frontiers in Ecology and the Environment*, **1**, 125–9.
Riitters, K. H., Wickham, J. D., O'Neill, R. V. *et al.* (2002) Fragmentation of continental United States forests. *Ecosystems*, **5**, 815–22.
Robertson, A. (1987) The centroid of tree crowns as an indicator of abiotic processes in a balsam fir wave forest. *Canadian Journal of Forest Research*, **17**, 746–55.
Roderick, M. L. (2006) The ever-flickering light. *Trends in Ecology and Evolution*, **21**, 3–5.
Rodin, L. E. and Bazilevich, N. I. (1967) *Production and Mineral Cycling in Terrestrial Vegetation*. Edinburgh: Oliver & Boyd.
Rodwell, J. S. (1991) *Woodlands and Scrub. British Plant Communities*. Vol. 1. Cambridge: Cambridge University Press.
Rollinson, T. J. D. (1985) *Thinning Control*. London: HMSO.
Rooney, T. P. and Waller, D. M. (2003) Direct and indirect effects of white-tailed deer in forest ecosystems. *Forest Ecology and Management*, **181**, 165–76.
Rose, F. (1988) Phytogeographical and ecological aspects of Lobarion communities in Europe. *Botanical Journal of the Linnean Society*, **96**, 69–79.
Rose, F. (1999) Indicators of ancient woodland. *British Wildlife*, **10**, 241–51.
Rothe, A. and Binkley, D. (2001) Nutritional interactions in mixed species forests: a synthesis. *Canadian Journal of Forest Research*, **31**, 1855–70.
Rothstein, D. E. (2000) Spring ephemeral herbs and nitrogen cycling in a northern hardwood forest: an experimental test of the vernal dam hypothesis. *Oecologia*, **124**, 446–53.
Rousseau, J. V. D., Sylvia, D. M. and Fox, A. J. (1994) Contribution of ectomycorrhiza to the potential nutrient-absorbing surface of pine. *New Phytologist*, **128**, 639–44.
Rydberg, D. and Falck, J. (2000) Urban forestry in Sweden from a silvicultural perspective: a review. *Landscape and Urban Planning*, **47**, 1–18.
Salisbury, E. J. (1916a) The oak-hornbeam woods of Hertfordshire. I and II. *Journal of Ecology*, **4**, 83–117.
Salisbury, E. J. (1916b) The emergence of aerial organs in woodland plants. *Journal of Ecology*, **4**, 121–8.
Salisbury, E. J. (1942) *The Reproductive Capacity of Plants*. London: Bell.
Salisbury, E. J. (1961) *Weeds and Aliens*. London: Collins.
Salmon, J. T. (1991) *Native New Zealand Flowering Plants*. Birkenhead, Auckland, New Zealand: Reed Books.
Satchell, J. E. (1962) Resistance in oak (*Quercus* spp.) to defoliation by *Tortrix viridana* L. in Roudsea Wood National Nature Reserve. *Annals of Applied Biology*, **50**, 431–42.
Sawidis, T., Marnasidis, A., Zachariadis, G. and Stratis, J. (1995) A study of air-pollution with heavy-metals in Thessaloniki City (Greece) using trees as

biological indicators. *Archives of Environmental Contamination and Toxicology*, **28**, 118–24.

Schaberg, P. G., DeHayes, D. H. and Hawley, G. J. (2001) Anthropogenic calcium depletion: a unique threat to forest ecosystem health? *Ecosystem Health*, **7**, 214–28.

Schemske, D. W. (2002) Ecological and evolutionary perspectives on the origins of tropical diversity. In *Foundations of Tropical Forest Biology*, Ed. Chazdon, R. L. and Whitmore, T. C. Chicago: Chicago University Press, pp. 163–73.

Schnurr, J. L., Ostfeld, R. S. and Canham, C. D. (2002) Direct and indirect effects of masting on rodent populations and tree seed survival. *Oikos*, **96**, 402–10.

Schroeder, D. (1969) *Bodenkunde in Stichworten*. Kiel: Hirt.

Schulze, E.-D. (1970) Der $CO_2$-Gaswechsel de Buche (*Fagus sylvatica* L.) in Abhangigkiet von Klimafaktoren im Freiland. *Flora, Jena*, **159**, 177–232.

Schulze, E.-D. and Mooney, H. A. (1993) Ecosystem function of biodiversity: a summary. In *Biodiversity and Ecosystem Function*, Ed. Schulze, E. D. and Mooney, H. A. Berlin: Springer, pp. 497–510.

Schulze, E.-D., Fuchs, M. I. and Fuchs, M. (1977a) Spatial distribution of photosynthetic capacity and performance in a mountain spruce forest of Northern Germany. I. Biomass distribution and daily $CO_2$ uptake in different crown layers. *Oecologia*, **29**, 43–61.

Schulze, E.-D., Fuchs, M. and Fuchs, M. I. (1977b) Spatial distribution of photosynthetic capacity and performance in a mountain spruce forest of Northern Germany. III. The significance of the evergreen habit. *Oecologia*, **30**, 239–48.

Schwaiger, H. and Schlamadinger, B. (1998) The potential of fuelwood to reduce greenhouse gas emissions in Europe. *Biomass and Bioenergy*, **15**, 369–77.

Schwarze, F. W. M. R. and Ferner, D. (2003) *Ganoderma* on trees – differentiation of species and studies of invasiveness. *Arboricultural Journal*, **27**, 59–77.

Sedjo, R. A. and Botkin, D. (1997) Using forest plantations to spare natural forests. *Environment*, **39**, 15–20, 30.

Sernander, R. (1936) The primitive forests of Granskar and Fiby: a study of the part played by storm-gaps and dwarf trees in the regeneration of the Swedish spruce forest. *Acta Phytogeographica Suecica*, **8**, 1–232 (English summary, pp. 220–7.)

Sfikas, G. (2002) *Wild Flowers of Crete*. Athens: Efstathiadis Group.

Sherrill, S. B. (2003) *Harvesting Urban Timber*. Fresno, CA: Linden Publishing.

Shugart, H. H., Leemans, R. and Bonan, G. B. (1992) *A Systems Analysis of the Global Boreal Forest*. Cambridge: Cambridge University Press.

Shultz. E. B., Bhatnagar, D., Jacobson, M. *et al.* (2002, reprinted from 1992 edition) *Neem – a Tree for Solving Global Problems*. New York: Books for Business.

Silver, W. L., Scatena, F. N., Johnson, A. H., Siccama, T. G. and Sánchez, M. J. (1994) Nutrient availability in a montane wet tropical forest in Puerto Rico: spatial patterns and methodological considerations. *Plant and Soil*, **164**, 129–45.

Silvertown, J. W. (1980) The evolutionary ecology of mast seeding in trees. *Biological Journal of the Linnean Society*, **14**, 567–87.

Simard, S. W., Perry, D. A., Jones, M. D. *et al.* (1997) Net transfer of carbon between ectomycorrhizal tree species in the field. *Nature*, **388**, 579–82.

Sirois, L. (1992) The transition between boreal forest and tundra. In *A Systems Analysis of the Global Boreal Forest*, Ed. Shugart, H. H., Leemans, R. and Bonan, G. B. Cambridge: Cambridge University Press, pp. 196–215.

Skarpe, C., Aarrestad, P. A., Andreassen, H. P. *et al.* (2004) The return of the giants: ecological effects of an increasing elephant population. *Ambio*, **33**, 276–82.

Skole, D. and Tucker, C. (1993) Tropical deforestation and habitat fragmentation in the Amazon: satellite data from 1978 to 1988. *Science*, **260**, 1905–10.

Slade, A. J. and Hutchings, M. J. (1987a) The effects of nutrient availability on foraging in the clonal herb *Glechoma hederacea*. *Journal of Ecology*, **75**, 95–112.

Slade, A. J. and Hutchings, M. J. (1987b) The effects of light intensity on foraging in the clonal herb *Glechoma hederacea*. *Journal of Ecology*, **75**, 639–50.

Slade, A. J. and Hutchings, M. J. (1987c) Clonal integration and plasticity in foraging behaviour in the clonal herb *Glechoma hederacea*. *Journal of Ecology*, **75**, 1023–36.

Smith, W. H. (1976) Character and significance of forest tree root exudates. *Ecology*, **57**, 324–31.

Sobey, D. G. (1995a, b) Analysis of the ground flora and other data collected during the 1991 Prince Edward Island Forest Inventory. II. Plant community analysis. III. A comparison of the vegetation and environmental factors of pre-1935 and post-1935 forested sites. *Report to the Prince Edward Island, Forestry Division*.

Sohlenius, B. (1980) Abundance, biomass and contribution to energy flow by soil nematodes in terrestrial ecosystems. *Oikos*, **34**, 186–94.

Solé, R. V., Bartumeus, F. and Gamarra, J. G. P. (2005) Gap percolation in rainforests. *Oikos*, **110**, 177–85.

Sork, V. L. (1993) Evolutionary ecology of mast-seeding in temperate and tropical oaks (*Quercus* spp.). *Vegetatio*, **107/108**, 133–47.

Sparks, T. H. (2000) The long-term phenology of woodland species in Britain. In *Long-term Studies in British Woodland*, Ed. Kirby, K. J. and Morecroft, M. D. English Nature Science No. 34. Peterborough: English Nature, pp. 98–105.

Sparks, T. H. and Smithers, R. J. (2002). Is spring getting earlier? *Weather*, **57**, 157–66.

Spears, J. D. H., Holub, S. M., Harmon, M. E. and Lajtha, K. (2003) The influence of decomposing logs on soil biology and nutrient cycling in an old-growth mixed coniferous forest in Oregon, U.S.A. *Canadian Journal of Forest Research*, **33**, 2193–201.

Speight, M. R. and Wylie, F. R. (2001) *Insect Pests in Tropical Forestry*. Wallingford: CABI Publishing.

Spellerberg, I. F. (1998) Ecological effects of roads and traffic: a literature review. *Global Change and Biogeography Letters*, **7**, 317–333.

Sprugel, D. G. (1976) Dynamic structure of wave-regenerated *Abies balsamea* forests in the north-eastern United States. *Journal of Ecology*, **64**, 889–911.

Spurr, S. H. and Barnes, B. V. (1980) *Forest Ecology*. New York: Wiley.

Stanhill, G. and Cohen, S. (2001) Global dimming: a review of the evidence for a widespread and significant reduction in global radiation with discussion of its probable causes and possible agricultural consequences. *Agricultural and Forest Meteorology*, **107**, 255–78.

Suresh, B. and Ravishankar, G. A. (2004) Phytoremediation – a novel and promising approach for environmental clean-up. *Critical Reviews in Biotechnology*, **24**, 97–124.

Suska, B., Muller, C. and Bonnet-Masimbert, M. (1996) *Seeds of Forest Broadleaves: from Harvest to Sowing*. Paris: INRA.

Sutton, W. R. J. (1999) Does the world need planted forests? *New Zealand Journal of Forestry*, **44**, 24–9.

Sutton, W. R. J. (2000) Wood in the third millennium. *Forest Products Journal*, **50**, 12–21.

Svenning, J.-C. (2002) A review of natural vegetation openness in north-western Europe. *Biological Conservation*, **104**, 133–48.

Swain, P. C. and Kearsley, J. B. (2000) *Classification of the Natural Communities of Massachusetts – draft*. Natural Heritage & Endangered Species Program, Massachusetts Division of Fisheries and Wildlife, Westborough, MA.

Swift, M. J., Heal, O. W. and Anderson, J. M. (1979) *Decomposition in Terrestrial Ecosystems*. Oxford: Blackwell Scientific.

Sziemer, P. (2000) *Madeira's Natural History in a Nutshell*. Funchal, Madeira: Francisco Ribeiro and Filhos Lda.

Tansley, A. G. (1935) The use and abuse of vegetational concepts and terms. *Ecology*, **16**, 284–307.

Tapias, R., Gil, L., Fuentes-Utrilla, P. and Pardos, J. A. (2001) Canopy seed banks in Mediterranean pines of south-eastern Spain: a comparison between *Pinus halepensis* Mill., *P. pinaster* Ait., *P. nigra* Arn. and *P. pinea L. Journal of Ecology*, **89**, 629–38.

Tattar, T. A., Berman, P. M., Gonzalez, M. S. and Dolloff, A. L. (1996) Biocontrol of the chestnut blight fungus *Cryphonectria parasitica*. *Arboricultural Journal*, **20**, 449–69.

Terborgh, J. (1992) *Diversity and the Tropical Rain Forest*. New York: Scientific American Library.

Tessier, J. T. and Raynal, D. J. (2003) Vernal nitrogen and phosphorus retention by forest understory vegetation and soil microbes. *Plant and Soil*, **256**, 443–53.

Thomas, P. A. (2000) *Trees: Their Natural History*. Cambridge: Cambridge University Press.

Thomas, P. A. and Polwart, A. (2003) Biological flora of the British Isles, *Taxus baccata* L. *Journal of Ecology*, **91**, 489–524.

Thomas, R. C., Kirby, K. J., and Reid, C. M. (1997) The conservation of a fragmented ecosystem within a cultural landscape – the case of ancient woodland in England. *Biological Conservation*, **82**, 243–352.

Thompson, K. and Grime, J. P. (1979) Seasonal variations in the seed banks of herbaceous species in ten contrasting habitats. *Journal of Ecology*, **67**, 893–921.

Thompson, K., Bakker, J. P. & Bekker, R. M. (1997) *The Soil Seed Banks of North West Europe: Methodology, Density and Longevity*. Cambridge: Cambridge University Press.

Tickell, O. (1994) Conifer forests are not the 'deserts' they seem. *New Scientist*, **17 September**, 16.

Tilgar, V., Mänd, R. and Mägi, M. (2002) Calcium shortage as a constraint on reproduction in great tits *Parus major*: a field experiment. *Journal of Avian Biology*, **33**, 407–13.

Tinker, D. B. and Knight, D. H. (2001) Temporal and spatial dynamics of coarse woody debris in harvested and unharvested lodgepole pine forests. *Ecological Modelling*, **141**, 125–49.

Tittensor, R. M. (1980) Ecological history of yew *Taxus baccata* L. in southern England. *Biological Conservation*, **17**, 243–65.

Toghill, P. (1990) *Geology in Shropshire*. Shrewsbury: Swan Hill Press.

Toghill, P. (2000) *The Geology of Britain: an Introduction*. Shrewsbury: Swan Hill Press.

Tokeshi, M. (1999) *Species Coexistence*. Oxford: Blackwell Science.

Tompkins, D. M., White, A. R. and Boots, M. (2003) Ecological replacement of native red squirrels by invasive greys driven by disease. *Ecology Letters*, **6**, 189–96.

Tranquillini, W. (1979) *Physiological Ecology of the Alpine Timberline*. Berlin: Springer.

Turner, A. P. and Dickinson, N. M. (1993) Survival of *Acer pseudoplatanus* L. (sycamore) seedlings on metalliferous soils. *New Phytologist*, **123**, 509–21.

Vallan, D. (2002) Effects of anthropogenic environmental changes on amphibian diversity in the rain forests of eastern Madagascar. *Journal of Tropical Ecology*, **18**, 725–42.

Vandenbeld, J. (1988) *Nature of Australia: a Portrait of the Island Continent*. New York: Facts on File and Autralian Broadcasting Corporation.

Vander Wall, S. B. (1997) Dispersal of single-leaf piñon pine (*Pinus monophylla*) by seed-caching rodents. *Journal of Mammalogy*, **78**, 181–91.

Vander Wall, S. B. (2001) The evolutionary ecology of nut dispersal. *Botanical Review*, **67**, 74–117.

Vander Wall, S. B. and Balda, R. P. (1977) Coadaptations of the Clark's nutcracker and piñon pine for efficient seed harvest and dispersal. *Ecological Monographs*, **47**, 89–111.

Vander Wall, S. B. and Longland, W. S. (2004) Diplochory: are two seed dispersers better than one? *Trends in Ecology and Evolution*, **19**, 155–61.

Varley, G. C. (1967) The estimation of secondary production in species with an annual life-cycle. In *Secondary Productivity of Terrestrial Ecosystems*, Ed. Petrusewicz, K. Warsaw: Polish Academy of Sciences, pp. 447–57.

Varley, G. C. (1970) The concept of energy flow applied to a woodland community. In *Animal Populations in Relation to their Food Resources*, Ed. Watson, A. Oxford: Blackwell Scientific, pp. 389–405.

Veblen, T. H. (1985) Stand dynamics in Chilean *Nothofagus* forests. In *The Ecology of Natural Disturbance and Patch Dynamics*, Ed. Pickett, S. T. A. and White, P. S. New York: Academic Press, pp. 33–51.

Vera, F. W. M. (2000) *Grazing Ecology and Forest History*. Wallingford: CABI Publishing.

Vera, F. W. M. (2002) The dynamic European forest. *Arboricultural Journal*, **26**, 179–211.

Vines, G. (2002) Gladrunners. *New Scientist*, **7 September**, 34–7.

Visser, M. E. and Holleman, L. J. (2001) Warmer springs disrupt the synchrony of oak and winter moth phenology. *Proceedings of the Royal Society of London B*, **268**, 289–94.

Visser, M. E., van Nordwijk, A. J., Tinbergen, J. M. and Lessells, C. M. (1998) Warmer springs lead to mistimed reproduction in great tits (*Parus major*). *Proceedings of the Royal Society of London B*, **265**, 1867–70.

Vitousek, P. M., Hättenschwiler, S., Olander, L. and Allison, S. (2002) Nitrogen and nature. *Ambio*, **31**, 97–101.

Vogt, K. A., Grier, C. C., Meier, C. E. and Keyes, M. R. (1983) Organic matter and nutrient dynamics in forest floors of young and mature *Abies amabilis* stands in western Washington, as affected by fine-root input. *Ecological Monographs*, **53**, 139–57.

Vogt, K. A., Grier, C. C. and Vogt, D. J. (1986) Production, turnover, and nutrient dynamics of aboveground and belowground detritus of world forests. *Advances in Ecological Research*, **15**, 303–77.

Wade, T. G., Riitters, K. H., Wickham, J. D. and Jones, K. B. (2003) Distribution and causes of global forest fragmentation. *Conservation Biology*, **7**(2), article 7.

Walker, L. R., Zasada, J. C. and Chapin, III F. S. (1986) The role of life history processes in primary succession on an Alaskan floodplain. *Ecology*, **67**, 1243–53.

Wallwork. J. A. (1970) *Ecology of Soil Animals*. London: McGraw-Hill.

Wallwork. J. A. (1983) Oribatids in forest ecosystems. *Annual Review of Entomology*, **28**, 109–30.

Walter, H. (1973) *Vegetation of the Earth in Relation to Climate and the Eco-physiological Conditions*. London: EUP-Springer.

Walther, G.-R. (2003) Plants in a warmer world. *Perspectives in Plant Ecology, Evolution and Systematics*, **6**, 169–85.

Wang, B. C. and Smith, T. B. (2002) Closing the seed dispersal loop. *Trends in Ecology and Evolution*, **17**, 379–85.

Wardle, D. A. (2002) *Communities and Ecosystems: Linking the Aboveground and Belowground Components*. Monographs in Population Biology, 34. Princeton, NJ: Princeton University Press.

Wardle, D. A., Barker, G. M., Yeates, G. W., Bonner, K. I. and Ghani, A. (2001) Introduced browsing mammals in New Zealand natural forests: aboveground and belowground consequences. *Ecological Monographs*, **71**, 587–614.

Watson, R. T., Noble, I. R., Bolin, B. *et al.* (2000) *Land Use, Land Use Changes and Forestry*. Cambridge: Cambridge University Press.

Watt, A. S. (1925) On the ecology of British beechwoods with special reference to their regeneration. Part II. Sections II and III. *Journal of Ecology*, **13**, 27–73.

Watt, A. S. (1947) Pattern and process in the plant community. *Journal of Ecology*, **35**, 1–22.

Welch, H. and Haddow, G. (1993) *The World Checklist of Conifers*. Landsman's Bookshop, Hertfordshire, UK: The World Conifer Data Pool.

Westman, W. E. (1978) Measuring inertia and resilience of ecosystems. *BioScience*, **28**, 705–10.

Westveld, M., Ashman, R. I., Baldwin, H. I. *et al.* (1956) Natural forest vegetation zones of New England. *Journal of Forestry*, **54**, 332–8.

Whild, S. (2003) Ancient woodland indicators in Shropshire. *Shropshire Botanical Society Newsletter* (Spring 2003), 18–19.

Whitmore, T. C. (1984) *Tropical Rain Forests of the Far East*. (Second Edition). Oxford: Clarendon Press/Oxford University Press.

Whitmore, T. C. (1998) *An Introduction to Tropical Rain Forests*. (Second Edition). Oxford: Oxford University Press.

Whittaker, R. H. (1975) *Communities and Ecosystems*. (Second Edition). London: Collier-Macmillan.

Whittaker, R. J., Willis, K. J. and Field, R. (2001) Scale and species richness: towards a general, hierarchical theory of species diversity. *Journal of Biogeography*, **28**, 453–70.

Willis, A. J. (1996) Obituary: Paul Westamacott Richards, C. B. E. (1908–95). *Journal of Ecology*, **84**, 795–8.

Willis, A. J. (1997) The ecosystem: an evolving concept viewed historically. *Functional Ecology*, **11**, 268–71.

Wingfield, M. J., Slippers, B., Roux, J. and Wingfield, B. D. (2001) Worldwide movement of exotic forest fungi, especially in the tropics and the southern hemisphere. *BioScience*, **51**, 134–40.

Wisniewski, L. and Dickinson, N. M. (2003) Toxicity of copper to *Quercus robur* (English oak) seedlings from a copper-rich soil. *Environmental and Experimental Botany*, **50**, 99–107.

Woldendorp, G. and Keenan, R. J. (2005) Coarse woody debris in Australian forest ecosystems: a review. *Austral Ecology*, **30**, 834–43.

Wood, M. (1989) *Soil Biology*. London: Blackie.

Woodford, J. (2002) *The Wollemi Pine: the Incredible Discovery of a Living Fossil from the Age of the Dinosaurs*. Melbourne, Australia: Text Publishing.

Woods, K. D. (2004) Intermediate disturbance in a late-successional hardwood forest. *Journal of Ecology*, **92**, 464–76.

Woodward, I. (1989) Plants in the greenhouse world. *New Scientist: Inside Science*, No. 21.
Woodward, S. L. (2003) *Biomes of Earth: Terrestrial, Aquatic and Human-Dominated*. Westport, CT: Greenwood Press.
Woollons, R. C. (2000) Comparison of growth of *Pinus radiata* over two rotations in central North Island of New Zealand. *The International Forestry Review*, **2**, 84–9.
Wright, S. J. (2002) Plant diversity in tropical forests: a review of mechanisms of species coexistence. *Oecologia*, **130**, 1–14.
Wyman, R. L. (1998) Experimental assessment of salamanders as predators of detrital food webs: effects on invertebrates, decomposition and the carbon cycle. *Biodiversity and Conservation*, **7**, 641–50.
Yalden, D. (2003) Letters: wolves and foxes. *British Wildlife*, **15**, 150.
Yamamoto, S.-I. (1993) Gap characteristics and gap regeneration in a subalpine coniferous forest on Mt Ontake, central Honshu, Japan. *Ecological Research*, **8**, 277–85.
Yamamoto, S.-I. (1998) Regeneration ecology of *Chamaecyparis obtusa* and *Chamaecyparis pisifera* (Hinoki and Sawara Cypress), Japan. In *Coastally Restricted Forests*, Ed. Laderman, A. D. Oxford: Oxford University Press, pp. 101–110.
Yanai, R. D. (1992) Phosphorus budget of a 70-year-old northern hardwood forest. *Biogeochemistry*, **17**, 1–22.
Yanai, R. D. (1998) The effect of whole-tree harvest on phosphorus cycling in a northern hardwood forest. *Forest Ecology and Management*, **104**, 281–95.
Yanai, R. D., Currie, W. S. and Goodale, C. L. (2003) Soil carbon dynamics after forest harvest: an ecosystem paradigm reconsidered. *Ecosystems*, **6**, 197–212.
Yang, L. H. (2004) Periodical cicada as resource pulses in North American forests. *Science*, **306**, 1565–7.
Young, J. E. (1975) Effects of spectral composition of light sources on the growth of a higher plant. In *Light as an Ecological Factor:* II, Ed. Evans, G. C. Bainbridge, R. and Rackham, O. Oxford: Blackwell Scientific, pp. 135–60.
Young, J. Z. (1950) *The Life of Vertebrates*. Oxford: Oxford University Press.
Zackrisson, O., Nilsson, M. C., Jaderlund, A. and Wardle, D. A. (1999) Nutritional effects of seed fall during mast years in boreal forest. *Oikos*, **84**, 17–26.
Zak, D. R., Groffman, P. M., Pregitzer, K. S., Christensen, S. and Tiedje, J. M. (1990) The vernal dam: plant-microbe competition for nitrogen in northern hardwood forests. *Ecology*, **71**, 651–6.
Zanne, A. E. and Chapman, C. A. (2005) Diversity of woody species in forest, treefall gaps, and edge in Kibale National Park, Uganda. *Plant Ecology*, **178**, 121–39.
Zohlen, A. and Tyler, G. (2004) Soluble inorganic tissue phosphorus and calcicole-calcifuge behaviour of plants. *Annals of Botany*, **94**, 427–32.

# Index

**Bold** numbers refer to pages giving a definition of key terms.
Note that members of several groups including the deer, ferns, firs, maples, oaks and sedges are listed together.